ISSI Scientific Report Series

Volume 16

The ISSI Scientific Report Series present the results of Working Groups (or Teams) that set out to assemble an expert overview of the latest research methods and observation techniques in a variety of fields in space science and astronomy. The Working Groups are organized by the International Space Science Institute (ISSI) in Bern, Switzerland. ISSI's main task is to contribute to the achievement of a deeper understanding of the results from space-research missions, adding value to those results through multi-disciplinary research in an atmosphere of international cooperation.

More information about this series at http://www.springer.com/series/10151

Idir Bouarar • Xuemei Wang • Guy P. Brasseur
Editors

Air Pollution in Eastern Asia: An Integrated Perspective

Springer

Editors
Idir Bouarar
Max Planck Institute for Meteorology
Hamburg, Germany

Guy P. Brasseur
National Center for Atmospheric Research
Boulder, CO, USA

Max Planck Institute for Meteorology
Hamburg, Germany

Xuemei Wang
Institute for Environment and Climate Research
Jinan University
Guangzhou, China

ISSI Scientific Report Series
ISBN 978-3-319-59488-0 ISBN 978-3-319-59489-7 (eBook)
DOI 10.1007/978-3-319-59489-7

Library of Congress Control Number: 2017946617

Cover figure: View of Shanghai city center, October 8, 2014. Credit: Hung Chung Chih / Shutterstock.com

Printed on acid-free paper

This Springer imprint is published by Springer Nature
The registered company is Springer International Publishing AG
The registered company address is: Gewerbestrasse 11, 6330 Cham, Switzerland

Preface

Air pollution has a considerable impact on society in different regions of the world. According to the World Health Organization (WHO), more than three million individuals die prematurely each year from diseases produced or exacerbated by outdoor air pollution: heart failures (40 %), strokes (40 %), pulmonary diseases (11 %), and lung cancers (6 %). These diseases are not limited to the elderlies. Other adverse effects include crop and forest damage, reduction in visibility, enhanced acidic precipitation, eutrophication of water resources, and damages to buildings including historical monuments. By interacting with solar and terrestrial radiation and affecting cloud microphysics, air pollutants and specifically aerosol particles also contribute directly or indirectly to climate change. In recent years, the importance of feedbacks between climate and atmospheric chemistry has been the subject of considerable attention by the scientific community.

The development of economic systems based on industrial and commercial activities as well as services has been at the source of rapid urban growth in the past decades. At the beginning of the nineteenth century, only 3 % of the population lived in urban areas, and in 1950, only 80 cities had a population exceeding one million. Since the end of the twentieth century, half of the world's population lives in urban areas and almost 500 cities host more than one million inhabitants. As of 2015, 35 cities were considered to be megacities (https://en.wikipedia.org/wiki/Megacity), which are large urban areas with a population exceeding ten million. In China, the largest cities are Shanghai, Guangzhou, Beijing, Shenzhen, Wuhan, and Chengdu. The fast development of the economy in Asia has not been without detrimental environmental consequences: the level of air pollution has increased dramatically in the last decades, specifically in the corridor extending from Beijing to Shanghai. Eastern China is indeed constituted of a multitude of closely located cities, which represent substantial sources of primary and secondary air pollutants. At the same time, the dense population in these urban areas is severely affected by high levels of fine particles and oxidants. Air pollution must be regarded as a vast regional problem rather than the addition of several localized urban-scale disturbances.

Until a few years ago, surface measurements of atmospheric pollutants in China were sparse and, in fact, not easily available. The adoption of ambient air quality standards and the development of a national air reporting system have considerably improved the situation. Today, a large number of measurements are available and are very useful not only to monitor the daily variations or long-term trends in air quality, but also to analyze the effectiveness of the measures taken to reduce the sources of pollutants. Further, such data have become the basis for the initialization and the evaluation of air quality forecast models that have been developed and are now used operationally. Figure 1, which reproduces the result of an analysis performed by Rohde and Muller (2015) in eastern China and is based on measurements made at more than 1,500 monitoring stations, provides a quantitative estimate of the magnitude of air pollution. If one considers the fine particles with a radius smaller than 2.5 μm ($PM_{2.5}$) that easily penetrate deep in the respiratory system, 38 % of the population living in this part of China is subject to unhealthy

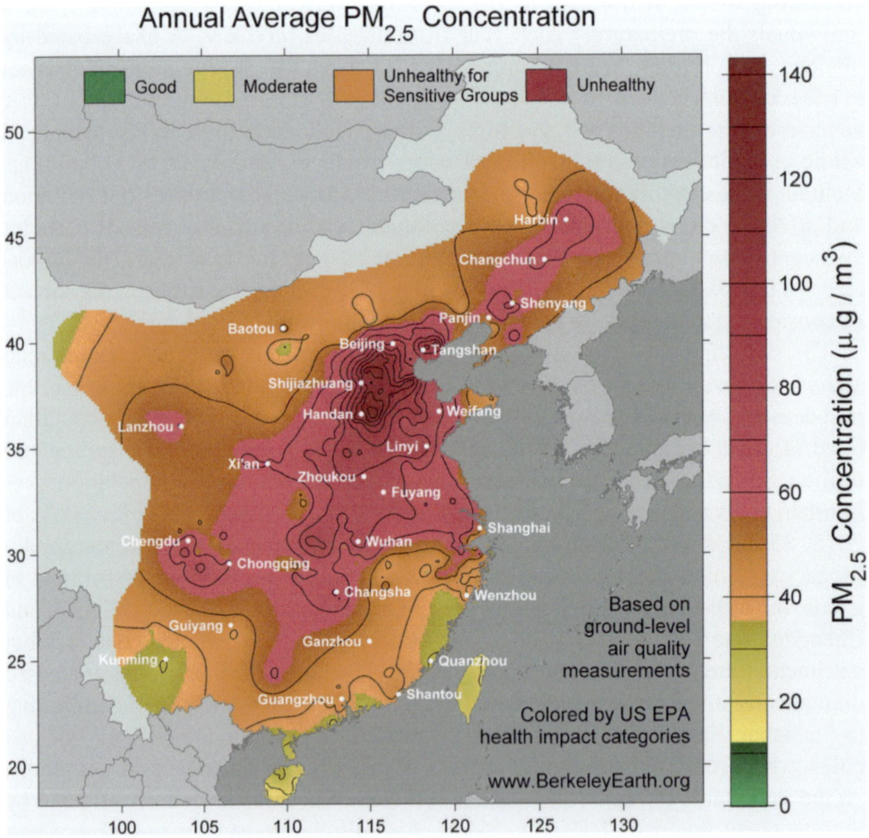

Fig. 1 Annual average pollutant concentration of $PM_{2.5}$ in eastern China in 2014 as determined by Berkeley Earth on the basis of surface measurements made at monitoring stations in China (Reproduced from http://berkeleyearth.org/)

conditions with average $PM_{2.5}$ concentrations exceeding 55 μg m^{-3} (red area on the figure). An additional 45 % of the population lives in areas labeled "unhealthy for sensitive groups" (orange area on the figure) with average $PM_{2.5}$ concentrations higher than 35 μg m^{-3}. The situation is less acute for PM_{10} particles (particles whose radius is less than 10 μm) and even less severe in the case of ozone, even though intermittent episodes of high ozone concentrations are observed during summertime. Concentrations of $PM_{2.5}$ higher than 500 μg m^{-3}, reaching in some rare occasions values of 600–1,000 μg m^{-3}, have been reported in urban areas during stable meteorological conditions in winter (Zheng et al. 2015). A challenge for the scientific community is to investigate the fundamental chemical, physical, and meteorological processes that produce and sustain elevated concentrations of secondary pollutants in eastern Asia. This question has motivated an interdisciplinary group of atmospheric scientists specialized in atmospheric chemistry and microphysics to contribute to two workshops supported by the International Space Science Institute (ISSI), one held in Beijing, China, and the second one in Bern, Switzerland. The presentations made during several intense days of discussions have been assembled in this volume.

The book is presenting an extended view on questions related to air pollution in Asia and specifically in China. It has been written by a group of experts from different continents for a broad audience involving scientists, educators and their students, environmental managers, policy-makers, as well as leaders in public administration and private corporations. It is organized around six distinct parts. The first five chapters in the first part of the volume offer a general perspective on issues related to air pollution including persistent haze events in eastern and southern Asia. The second part presents an overview of air pollution sources (i.e., anthropogenic and biomass burning sources). The third part analyzes in situ observations of chemical species in China, while the fourth part focuses on space observations of gas-phase and aerosol species. The modeling aspects are treated in the fifth part of the volume, which includes a presentation of several air quality forecast systems and an assessment of the role of urbanization on air pollution levels. Finally, the effects of air pollution on health and crop productivity in China are discussed in the last part of the book.

Even though the task is daunting, the problem of air quality can be solved. Experience acquired in other urban areas such as London and Los Angeles (Parrish et al. 2016) shows that decisive action can be taken to reduce the emissions of primary air pollutants and of the precursors of secondary pollutants. China has already taken important steps toward an effective mitigation of air pollution. The objective of this volume is to provide fundamental elements that will help decision-makers design effective science-informed policies that will lead to long-term improvements of air quality and to successfully manage short-term air pollution episodes that substantially affect the quality of life of the people and strongly impact the economy.

Acknowledgments The editors of this volume would like to thank the International Space Science Institute in Bern, Switzerland and in Beijing, China for their support of the two workshops that led to this book. They also thank the contributors to this volume and the reviewers of the different chapters. On behalf of several contributors, the editors gratefully acknowledge the support from the European Community's Seventh Framework Programme (FP7) under the PANDA project grant agreement n° 606719. Their gratitude goes also to Springer Publishing Company and specifically to Petra van Steenbergen who has greatly facilitated the preparation of this book.

Hamburg, Germany — Idir Bouarar
Guangzhou, China — Xuemei Wang
Hamburg, Germany — Guy P. Brasseur
March 2017

References

Parrish, D. D., Xu, J., Croes, B., Shao, M. (2016). Air quality improvement in Los Angeles – Perspectives for developing cities Front. *Environmental Science Engineering*, *10*(5), 1–13. doi:10.1007/s11783-016-0859-5.

Rohde, R. A., & Muller, R. A. (2015). Air pollution in China: Mapping concentrations and sources. *PLoS ONE*, *10*(8), e0135749. doi:10.1371/journal.pone.0135749.

Zheng, G. J., Duan, F. K., Su, H., Ma, Y. L., Cheng, Y., Zheng, B., Zhang, Q., Huang, T., Kimoto, T., Chang, D., Pöschl, U., Cheng, Y. F., He, K.B. (2015). Exploring the severe winter haze in Beijing: The impact of synoptic weather, regional transport and heterogeneous reactions. *Atmospheric Chemistry and Physics*, *15*(6), 2969–2983.

Contents

Part I
General Perspective

Chapter 1
Overview of Persistent Haze Events in China

Renyi Zhang, Pengfei Tian, Yemeng Ji, Yun Lin, Jianfei Peng, Bowen Pan, Yuan Wang, Gehui Wang, Guohui Li, Weigang Wang, Fang Zhang, Xidan Feng, Lian Duan, Jiaxi Hu, Wilmarie Marrero-Ortiz, Jeremiah Secrest, and Min Hu

Abstract As the world's second largest economy, China has experienced severe haze pollution, with fine particulate matter (PM) reaching unprecedentedly high levels across many cities. In addition to the profound impacts on human health and ecosystems, fine PM interacts directly and indirectly with the Earth's radiation budget, influencing weather and climate. An understanding of the PM formation

R. Zhang (✉) • J. Peng
Department of Atmospheric Sciences and Department of Chemistry, Texas A&M University, College Station, TX, USA

State Key Joint Laboratory of Environmental Simulation and Pollution Control, College of Environmental Sciences and Engineering, Peking University, Beijing, China
e-mail: Renyi-zhang@tamu.edu; pengjianfeipku@gmail.com

P. Tian
Department of Atmospheric Sciences and Department of Chemistry, Texas A&M University, College Station, TX, USA

Key Laboratory for Semi-Arid Climate Change of the Ministry of Education, Lanzhou, China
e-mail: tpftamu@gmail.com

Y. Ji
Department of Atmospheric Sciences and Department of Chemistry, Texas A&M University, College Station, TX, USA

Institute of Environmental Health and Pollution Control, School of Environmental Science and Engineering, Guangdong University of Technology, Guangdong, China
e-mail: jiym99@163.com

Y. Lin • B. Pan • J. Hu • W. Marrero-Ortiz • J. Secrest
Department of Atmospheric Sciences and Department of Chemistry, Texas A&M University, College Station, TX, USA
e-mail: yunlin121@tamu.edu; panb@email.tamu.edu; hujiaxi2012@gmail.com; wilmarie.ortiz@chem.tamu.edu; jeremiah.secrest@chem.tamu.edu

Y. Wang
Department of Atmospheric Sciences and Department of Chemistry, Texas A&M University, College Station, TX, USA

Jet Propulsion Laboratory, California Institute of Technology, Pasadena, CA, USA
e-mail: yuan.wang@caltech.edu

I. Bouarar et al. (eds.), *Air Pollution in Eastern Asia: An Integrated Perspective*, ISSI Scientific Report Series 16, DOI 10.1007/978-3-319-59489-7_1

mechanism is critical in the development of efficient mitigation policies to minimize the local, regional, and global impacts. The formation mechanisms leading to severe haze episodes with exceedingly high $PM_{2.5}$ levels in China remain uncertain, and the abundance and chemical constituents of $PM_{2.5}$ depend on the complex

G. Wang
Department of Atmospheric Sciences and Department of Chemistry, Texas A&M University, College Station, TX, USA

State Key Laboratory of Loess and Quaternary Geology, and Key Laboratory of Aerosol Chemistry and Physics, Institute of Earth Environment, Chinese Academy of Sciences, Xi'an 710075, China
e-mail: wanggh@ieecas.cn

G. Li
State Key Laboratory of Loess and Quaternary Geology, and Key Laboratory of Aerosol Chemistry and Physics, Institute of Earth Environment, Chinese Academy of Sciences, Xi'an 710075, China
e-mail: ligh@ieecas.cn

W. Wang
State Key Laboratory for Structural Chemistry of Unstable and Stable Species, Beijing National Laboratory for Molecular Sciences (BNLMS), Institute of Chemistry, Chinese Academy of Sciences, Beijing, China
e-mail: wangwg@iccas.ac.cn

F. Zhang
Department of Atmospheric Sciences and Department of Chemistry, Texas A&M University, College Station, TX, USA

State Key Laboratory of Earth Surface Processes and Resource Ecology, College of Global Change and Earth System Science, Beijing Normal University, Beijing, China
e-mail: fang.zhang@bnu.edu.cn

X. Feng
Department of Atmospheric Sciences and Department of Chemistry, Texas A&M University, College Station, TX, USA

College of Environmental Science and Engineering, Zhongkai University of Agriculture and Engineering, Haizhu District, Guangzhou 510225, China
e-mail: fengxidan@aliyun.com

L. Duan
Department of Atmospheric Sciences and Department of Chemistry, Texas A&M University, College Station, TX, USA

School of Resource and Environmental Engineering, East China University of Science and Technology, Shanghai, China
e-mail: duanlian1991@126.com

M. Hu
State Key Joint Laboratory of Environmental Simulation and Pollution Control, College of Environmental Sciences and Engineering, Peking University, Beijing 100871, China
e-mail: minhu@pku.edu.cn

interplay between meteorology, emission sources, and atmospheric chemical processes. This paper discusses the various aspects relevant to persistent haze in China, including the fundamental chemistry and meteorological conditions. The similarity and difference in PM formation between Beijing and other world regions are illustrated. Future research needs relevant to persistent haze formation in China are discussed.

Keywords Haze in China • Particulate matter • PM sources and formation • Aerosol precursors

1.1 Introduction

China has experienced persistent air pollution because of its fast-developing economy and urbanization (Zhang et al. 2015). The recent severe haze events with unprecedentedly high PM levels in China have attracted large public attention due to significantly reduced visibility and negative health effects. For example, as one of the most populous cities in the world with a population of over 21 million, Beijing is prone to develop stagnant conditions, because of surrounding mountains to the north of the city (Guo et al. 2014). In 2013 the annual level of $PM_{2.5}$ (particles with the aerodynamic diameter smaller than 2.5 μm) in Beijing was 89 μg m^{-3}, with the daily maximum $PM_{2.5}$ value reaching 568 μg m^{-3}; there were more half of the days when the $PM_{2.5}$ concentration exceeded 300 μg m^{-3}. Haze in China occurs in all seasons; the occurrence of haze episodes is more frequent and severe in wintertime, because of increased pollutant emissions from coal combustion for house heating and unfavorable meteorological conditions. The Chinese government has implemented major measures to improve air quality since the 1990s, by relocating power plants and industrial factories away from the city, replacing coal by clean energy such as electricity and natural gas, and promulgating stricter traffic emission standards (Xinhua 2012). Primary PM emissions from industry have been significantly reduced, but secondary fine PM has remained elevated, because of the presence of high levels of gaseous precursors from urban transportation and regional industrial facilities (Guo et al. 2014).

Fine PM profoundly impacts human health, visibility, the ecosystem, weather, and climate (IPCC 2013), and these PM effects are largely dependent of the aerosol properties, including the number concentration, size, and chemical composition. PM is emitted directly into (primary) or formed in the atmosphere through gas-to-particle conversion (secondary) (Fig. 1.1) (Zhang et al. 2012a; Seinfeld and Pandis 2006). Also, primary and secondary PM undergoes chemical and physical transformations and is subjected to cloud processing and transport in the atmosphere. The mechanisms leading to fine PM formation remain highly uncertain (Guo et al. 2014; Zhang et al. 2012a, 2013; Wang et al. 2010; Seinfeld and Pandis 2006; Khalisov et al. 2013), particularly for those processes related to the PM origin and growth. For example, the chemical constituents of organic carbon (OC),

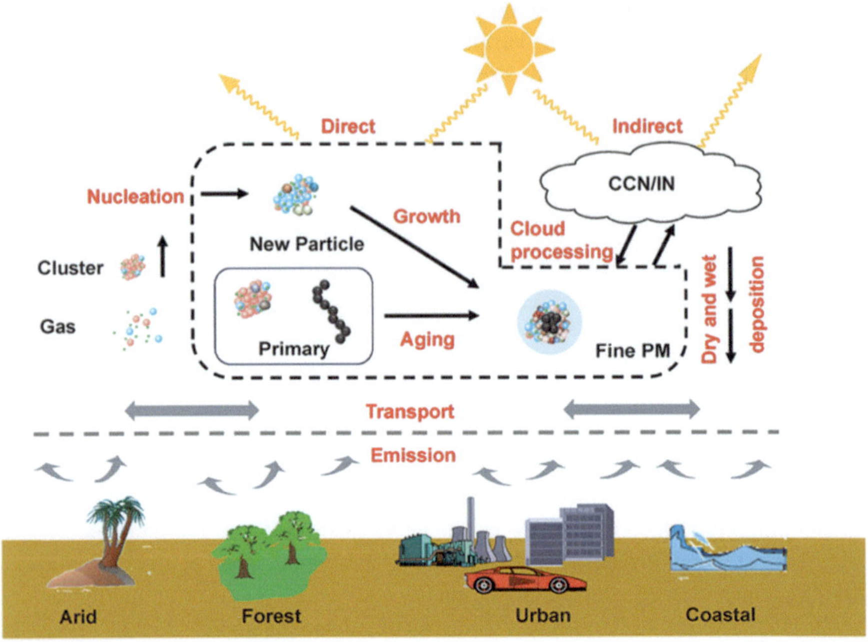

Fig. 1.1 Schematic representation of the formation, growth, and processing of atmospheric aerosols (After Zhang et al. 2015)

sulfate, nitrate, ammonium, trace metals, and elemental carbon (EC) have been commonly identified in fine PM, reflecting the diverse primary and secondary sources from the traffic, industry, biosphere, and other anthropogenic activities, along with regional components from biomass burning and agricultural activities (Guo et al. 2014; He et al. 2011; Zhang et al. 2008a, b). Since there exist a large number of natural and anthropogenic emission sources for PM, source apportionments are difficult and the currently available methodologies are rather inaccurate and often produce conflicting results (Zhang et al. 2008a, b, 2015; Guo et al. 2014; He et al. 2011).

The abundance and chemical constituents of $PM_{2.5}$ vary considerably in China, depending on complex interplay between meteorology, pollution sources, and atmospheric chemical processes (Zhang et al. 2013; Wang et al. 2011a, b, 2014; Wu et al. 2007; Suh et al. 2003; Guo et al. 2012, 2013). For example, on the basis of ambient measurements and receptor model analysis, the contribution to the annual mean $PM_{2.5}$ in Beijing has been suggested to be mainly from industrial pollution and secondary inorganic aerosol formation, but negligibly from traffic emissions (Zhang et al. 2013). In addition, meteorological conditions may govern regional and long-range transport of air pollutants (Dickerson et al. 2007; Guo et al. 2014).

Fine PM has been unequivocally associated with adverse human health impacts (NRC 2004). The human health effects of fine PM range from aggravating allergies, the development of serious chronic diseases, to premature death (Gauderman et al. 2004; Künzli et al. 2005; Pope and Dockery 2006; Correia et al. 2013; Matus et al. 2012). Both epidemiological and toxicological studies have indicated that smaller particles are more closely linked with adverse health outcomes than larger ones (Schlesinger et al. 2006). Although ultrafine particles (less than 100 nm) contribute negligibly to the total ambient PM mass, they are typically present in high number concentrations under urban environments and have a higher probability than larger particles to deposit in the pulmonary region after inhalation. Long-term exposure to fine PM has been associated with a 6% increase in the risk of premature death (Pope and Dockery 2006; COMEAP 2006). In China, more than 20 million cases of respiratory illnesses were reported in 2007 (Millman et al. 2008).

Aerosols also exhibit a broad range of impacts on the atmosphere, directly by interfering with the solar radiative transfer and indirectly by influencing cloud formation (IPCC 2013; NRC 2004). Specifically, those particles modify the lifetime and albedo of clouds, precipitation, and lightning (Fan et al. 2007a, 2008; Wang et al. 2011a, b; Orville et al. 2001; Williams et al. 1991), modulate photochemistry (Li et al. 2005; Tie et al. 2003, 2005), promote multiphase chemistry (Seinfeld and Pandis 2006; Fu et al. 2008; Zhao et al. 2005), degrade local, regional, and global air quality (Lin et al. 2014; Li 1998, 2010), and ultimately impact the Earth energy budget (IPCC 2013). By serving as cloud condensation nuclei (CCN) or ice nuclei (IN), aerosols influence the macro- and micro- properties of clouds (Li et al. 2008a, b, 2009) and regulate the cloud lifetimes and the precipitation efficiency (Rosenfeld et al. 2008; Tao et al. 2012; Wang et al. 2013a, b; Koren et al. 2005; Fan et al. 2007b; Yuan et al. 2008). Atmospheric measurements and modeling studies have revealed invigorated convective clouds resulting from elevated aerosol levels over urban regions (Rosenfeld et al. 2008; Koren et al. 2005). Also, the aerosol effects on urban precipitation have been shown to increase the rainfall rate under clean conditions but decrease the rainfall rate under polluted conditions, suggesting a plausible distinction of the aerosol effects on precipitation between developed and developing countries (Li et al. 2008a, b). Presently, the estimates of the cloud adjustment by aerosols range from 0.06 to 1.33 W m^{-2} in the global radiative forcing budget on the top of the atmosphere, representing the largest uncertainty in climate projections of anthropogenic activities (IPCC 2013). In addition, long-range transport of urban and regional PM from Asia has been implicated in climatically altered mid-latitude cyclones over the Pacific Ocean (Li et al. 2008a, b; Zhang et al. 2007; Wang et al. 2014). There has been a growing interest in the plausible interaction between regional and Asia Monsoon (Fig. 1.2) (Wu et al. 2016).

The aerosol-cloud interaction corresponds to one of the most poorly represented processes in atmospheric models, because of largely varying aerosol properties under diverse environmental conditions. For example, the CCN activation efficiency of aerosols to form cloud droplets is dependent of hygroscopicity, which is related to the particle size and chemical compositions (Petters and Kreidenweis 2007; Ma et al. 2013; Zhang et al. 2014). In addition, atmospheric aging of primary and secondary PM considerably modifies the hygroscopicity, because of increased

Fig. 1.2 Haze, weather, and climate interaction (After Wu et al. 2016)

coating by inorganic and organic species from the secondary formation processes (Zhang et al. 2008a, b). A change in the particle hygroscopicity also impacts the atmospheric lifetime of aerosols, since wet deposition represents one of the key removal processes for aerosols. On the other hand, cloud-processing of gaseous species has been suggested to account for an important fraction of the tropospheric aerosol loading (Seinfeld and Pandis 2006; Fu et al. 2008).

The optical effects of aerosols also impact visibility and air quality. In addition to scattering light, certain aerosol types, such as black carbon (BC), mineral particles, and brown carbon, are light absorbing (Wang et al. 2013a, b; Laskin et al. 2015). Enhanced light absorption and scattering by aerosols stabilize the atmosphere, resulting in a negative feedback on air quality and inhibition of cloud formation (Barbaro et al. 2013). Furthermore, the direct and indirect effects of aerosols on clouds often operate in the opposite directions (Fan et al. 2008).

The deficiencies in our understanding of fine PM formation hinder the developments of predictive atmospheric models to simulate pollution episodes and efficient mediation policies to minimize its local, regional, to global impacts. This paper reviews the various aspects relevant to persistent haze formation in China, including the fundamental chemistry and meteorological conditions. The similarities and differences in fine PM between China and other world regions are compared and discussed. Finally, the concluding remarks and future research needs relevant to persistent haze events in China are presented.

1.2 Emissions of Primary PM and PM Precursors

Most of primary PM in China is emitted from combustion sources, such as highway vehicles, industries, residential fuel combustion, and etc. Road dust, including paved road dust and construction dust, are also important in urban areas. In Beijing, residential fuel combustion dominates the local primary PM emissions (Zhao et al. 2009). Guo et al. (2013) summarized the source apportionment results from 2000 to 2008 in Beijing: secondary formation represents the major fine PM source, except in the winter heating period when the primary PM contribution is comparable to the secondary formation. Traffic emissions represent the most important primary PM sources in Beijing in the early part of the decade; however, several control measures of vehicle emissions have been implemented since the 2008 Olympic Games, resulting in a decreased total vehicle contribution. Coal combustion also significantly contributes to the ambient fine PM, particularly during the winter. Cooking emission contributes to up to 10% of fine PM (Wang et al. 2009). The contribution of biomass burning is highly seasonal and may be important during the harvest seasons (i.e., spring and fall).

Since VOCs, NO_x, SO_2, and NH_3 represent the major precursors for secondary fine PM formation, an accurate representation of their emission sources is critical for air quality modeling (Zhang et al. 2003). For VOC emissions, major anthropogenic sources include gasoline and diesel vehicles, solvent utilization, and industries, in addition to emissions from the biosphere and biomass mass burning. For NO_x emissions, diesel and gasoline vehicles and engines, industries, and coal combustion are the common sources in urban areas. In Beijing, NO_x emission from power plants, industries and motor vehicles is estimated to account for 40%, 20% and 20% of the particle-phase nitrate (Zhang et al. 2012b). Since there is little local emission of NO_x from power plants in Beijing, most of NO_x is likely due to local transportation and regional transport from upwind source areas (Ying et al. 2014). SO_2 in urban regions is mainly from fuel combustion. In Beijing, SO_2 from regional coal combustion dominates the emissions. In urban areas with heavy traffic, NH_3 emissions from catalyst-equipped motor vehicles can significantly contribute to NH_3 concentrations (Ying and Kleeman 2006; Zhang and Ying 2010).

For air quality modeling in China, the most widely used emission inventories are the Transport and Chemical Evolution over the Pacific (TRACE-P) (Streets et al. 2003) and the Intercontinental Chemical Transport Experiment (INTEX-B) inventories (Zhang et al. 2009). In addition, the regional emission inventory in Asia (REAS) (Ohara et al. 2007) and its successor REAS2 (Kurokawa et al. 2013) have also been used in air quality modeling studies. Recently, detailed emission inventories for the entire China have been complied and employed in a number of modeling studies (Mijling et al. 2013; Stavrakou et al. 2013).

1.3 Secondary PM Formation Processes

1.3.1 New Particle Formation

Aerosol nucleation has been measured under diverse environmental conditions, and globally new particle formation (NPF) accounts for about 50% of the aerosol number production in the troposphere (Zhang et al. 2012a; Zhang 2010). NPF events have been frequently observed in many urban areas of China (Peng et al. 2014). The efficient occurrence of NPF in urban regions may be explainable because of the presence of high levels of aerosol nucleation precursors, such as sulfur dioxide, ammonia, amines, and anthropogenic VOCs (Guo et al. 2014; Zhang et al. 2008a, b, 2012a, b; He et al. 2011).

NPF events occur frequently in Beijing, according to several field campaigns, such as the Campaign of Atmospheric Research in Beijing and Surrounding Areas (CAREBejing) in the summer of 2008 (Guo et al. 2014; Wu et al. 2007; Yue et al. 2010; Wang et al. 2011b, 2015). An annual statistics of NPF measurements from March 2004 to February 2005 shows that NPF events occur during all seasons in Beijing, with the frequency of 50%, 20%, 35%, and 45% for the spring, summer, fall, and winter, respectively (Wu et al. 2007). The nucleation rates at 1.5 nm ($J_{1.5}$) and 3 nm (J_3) in NPF events observed in Beijing range from several to 100 particles cm^{-3} s^{-1} (Wu et al. 2007; Yue et al. 2010; Wang et al. 2011b, 2015), comparable with the values observed in other cities (Kulmala et al. 2004). Gaseous sulfuric acid is shown to play an important role in NPF in Beijing (Yue et al. 2010; Wang et al. 2011a, b). For instance, Yue et al. (2010) showed that the average formation rates during the CAREBeijing-2008 campaign linearly correlate with the sulfuric acid concentration with a high correlation coefficient ($R^2 = 0.85$). In addition, Wang et al. (2015) suggested that organic compounds participate in the nucleation process in Beijing, since the nucleation rate exhibits a good correlation with the concentrations of sulfuric acid and organic vapors.

1.3.2 Evolutions in the Particle Properties During Haze Events

Typically, the measured $PM_{2.5}$ properties in China exhibit a periodic cycle of 4–7 days (Fig. 1.3) (Guo et al. 2014). For example, the particle mass concentration in Beijing is less than several tens of μg m^{-3} (clean) in the beginning of each cycle and reaches within 2–4 days several hundreds of μg m^{-3} (polluted). A higher number concentration of smaller particles exists during the clean period, and a slightly lower particle number concentration of larger particles exists during the polluted period. During a pollution episode, an average daily particle mass growth of 50–110 μg m^{-3} typically coincides with a daily increase of 40–65 nm in the mean diameter. The total particle number concentration is more than 200,000 cm^{-3}

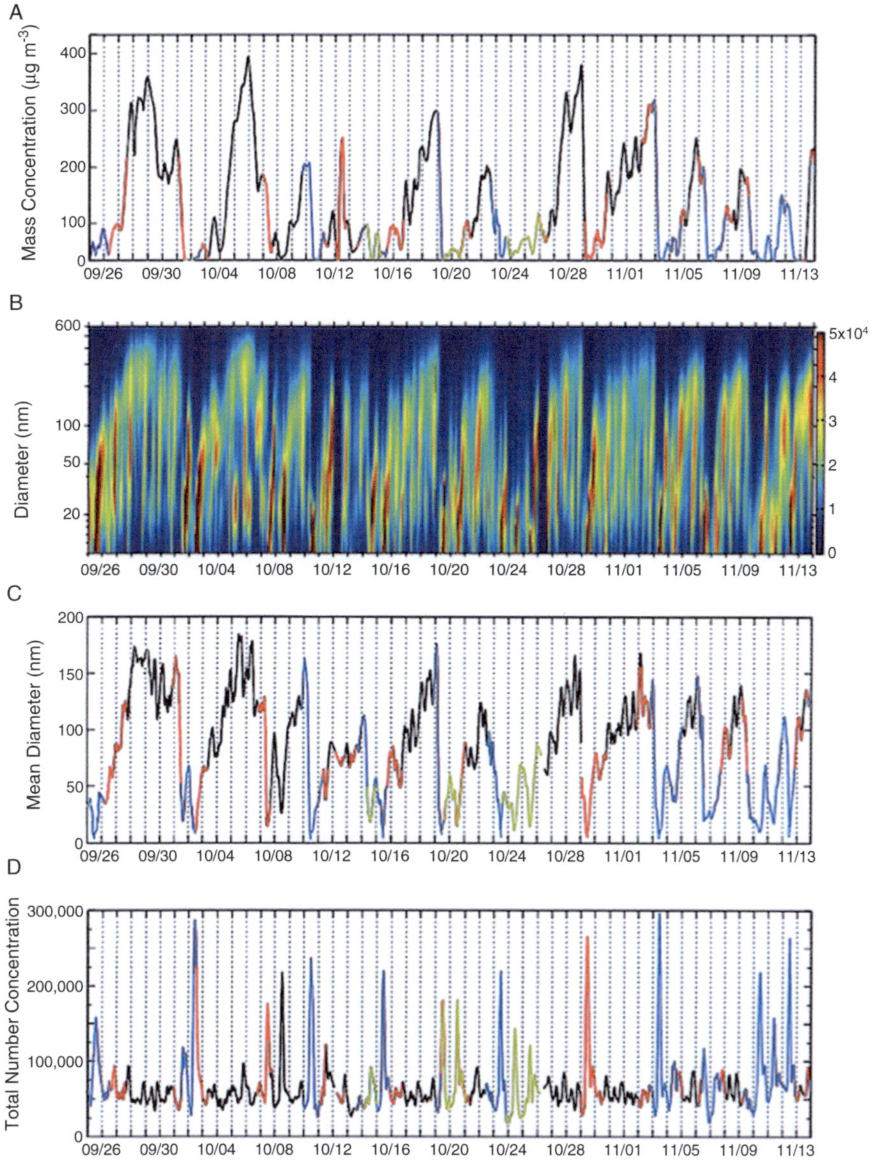

Fig. 1.3 Periodic $PM_{2.5}$ cycles in Beijing. Temporal evolutions of $PM_{2.5}$ mass concentration (**a**), number size distribution (**b**), mean diameter (**c**), and total number concentration (**d**) during the $PM_{2.5}$ events from 25 September through 14 November 2013 (After Guo et al. 2014)

during the clean period, and decreases slightly and remains at about 50,000 cm^{-3} throughout the polluted period. The periodic cycles of severe haze episodes in Beijing are largely driven by meteorological conditions; stagnation typically develops with weak southerly wind from polluted industrial source regions (Guo et al. 2014). The average particle mass concentrations of 35 and 114 μg m^{-3} during the clean and polluted periods correspond closely to the northerly and southerly wind conditions, respectively.

The development of PM episodes in Beijing is characterized by two distinct aerosol formation processes, i.e., nucleation and growth (Fig. 1.4) (Guo et al. 2014). Nucleation consistently occurs prior to a polluted period, producing a high number concentration of nanoparticles under clean conditions. The PM growth process is reflected by the particle mass concentration exceeding several hundreds of μg m^{-3} within 2–4 days, which is accompanied by a continuous size growth from the nucleation mode particles. Figure 1.4 shows very large particle mass increases of 270 μg m^{-3} (on Sept. 27, 2013) and 210 μg m^{-3} (on Oct. 4, 2013) during the daytime (6:00 a.m. to 6:00 p.m.), when the particle mean size increases by 60–70 nm. During the clean period, there exist high concentrations of ultrafine particles, but those particles contribute negligibly to the particle mass concentration; the severe pollution episodes in Beijing are attributable to the presence of numerous large particles.

The PM chemical composition in Beijing consists dominantly of organics (44%) and nitrate (22%), followed by sulfate (17%), ammonium (15%), and chloride (2%) (Fig. 1.5) (Guo et al. 2014). Measurements show continuously increasing mass concentrations of organics, sulfate, and nitrate during the transition and polluted periods, correlating closely with the evolutions of the $PM_{2.5}$ mass concentration and mean particle size (Fig. 1.5). The organic mass fraction dominates in the clean period and decreases slightly from the clean to polluted periods. In contrast, the contributions of sulfate and nitrate to the particle mass concentration increase slightly during the pollution episodes. The primary organic aerosol mass fraction decreases during the episodes, indicating small contributions of primary particle emissions to the pollution development (Guo et al. 2014).

The gaseous aerosol precursors (i.e., VOCs, NO_x, and SO_2) are high during the pollution episodes in Beijing (Fig. 1.6). For example, the SO_2 and NO_x peak concentrations are over 40 and 200 ppb, respectively, and the aromatic hydrocarbons (xylenes and toluene) represent the most abundant types of VOCs, with the xylene peak concentration of more than 10 ppb. The photochemical oxidation and/or multi-phase reactions of VOCs, SO_2, and NO_x lead to formation of less or non-volatile species (Zhao et al. 2006; He et al. 2014), contributing to the aerosol organic, sulfate, and nitrate constituents, respectively. Furthermore, the contributions from primary emissions and regional transport of particles to the formation of severe haze episodes in Beijing have been demonstrated to be small, suggesting that regulatory controls of gaseous emissions for VOCs and NO_x from local transportation and SO_2 from regional industrial sources represent the key steps to reduce the urban PM level in Beijing (Guo et al. 2014).

The density and hygroscopicity measurements also reveal an organic dominant composition in Beijing (Fig. 1.7). The peak effective density of 1.1 g cm^{-3} during the clean day is indicative of PM dominated by organics. The increased hygroscopicity and effective density from clean to polluted periods reveal the formation of an internal mixture of secondary organic and inorganic species, with increasing

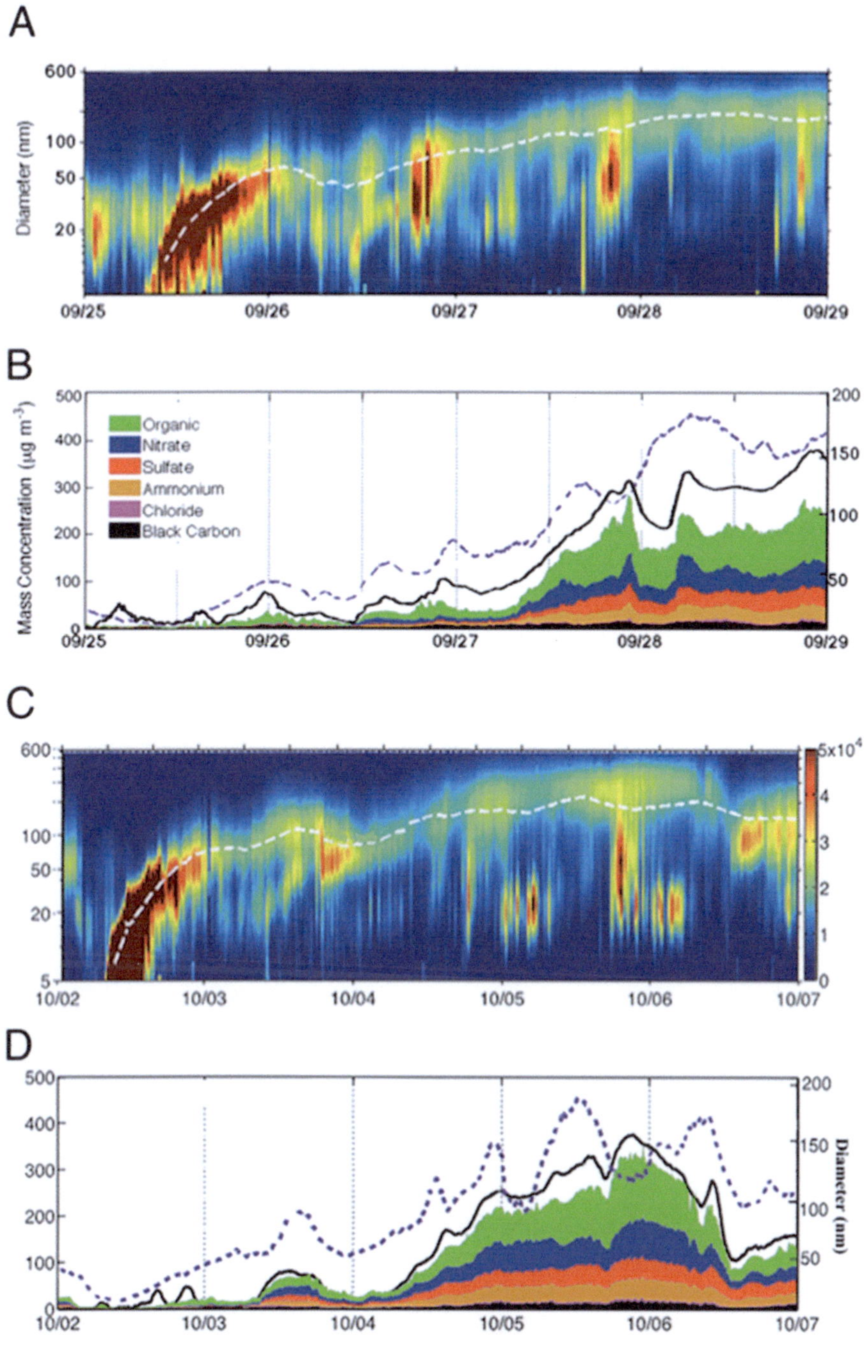

Fig. 1.4 Aerosol nucleation and growth during the PM2.5 episodes on 25–29 September and 2–7 October. (**a**, **c**) Temporal evolutions of particle number size distribution and mean diameter (*white dashed curve*) on 25–29 September (**a**) and 2–7 October (**c**). (**b**, **d**) PM2.5 mass concentration (*black solid line*), mean diameter (*purple dashed line*), and PM1 (particulate matter smaller than 1.0 μm) chemical composition on 25–29 September (**b**) and 2–7 October (**d**) (After Guo et al. 2014)

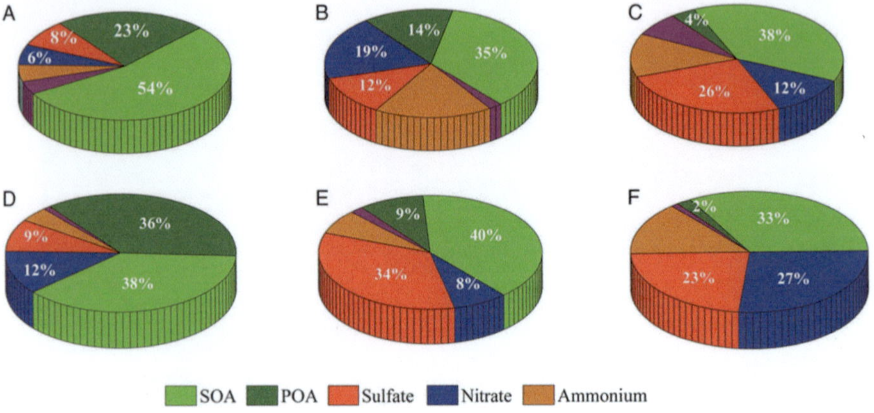

Fig. 1.5 Particle chemical compositions during the clean, transition, and polluted periods for the 25–29 September and 2–7 October episodes. (**a**–**c**) Chemical compositions for 80-nm (**a**), 100-nm (**b**), and 240-nm (**c**) particles measured by the AMS at 1500 h on 25 September, 1200 h on 27 September, and 1800 h on 28 September, respectively (After Guo et al. 2014)

contributions from sulfate and nitrate (Guo et al. 2014). The average SSA (at the wavelengths of 470 and 500 nm) in Beijing is near 0.87, indicating the presence of moderately strong absorbing aerosol and higher than those in the southern regions in China (Lee et al. 2007). The differences in SSA in urban regions may cause totally opposite aerosol radiative effects and lead to distinct feedback processes on PM formation and removal (Yang et al. 2013, Yang and Li 2014).

The efficient aerosol nucleation and growth in Beijing are clearly distinct from those in other urban cities. For example, the variations in the PM properties (i.e., particle number, size, and mass concentration) in Houston, LA, and Mexico City typically exhibit a clear diurnal characteristic. In contrast, the aerosol properties in Beijing exhibit continuous evolutions from the nucleation mode particles over an extended period of 2–4 days, yielding numerous large particles during the pollution episodes. The considerably more efficient aerosol nucleation and growth processes in Beijing are explained by much higher concentrations of aerosol precursor gases, i.e., anthropogenic VOCs (aromatics), NO_x, and SO_2 than those in the other three cities (i.e., Fig. 1.6). On the other hand, the particle compositions in Beijing exhibit a general similarity to those commonly measured in many global urban areas, consistent with the chemical constituents that are dominated by secondary aerosol formation (Fig. 1.8). For example, the PM mass fractions of organic, sulfate, and nitrate of 44%, 17%, and 22% in Beijing are nearly identical to those of 44%, 17%, and 24% in Los Angeles, respectively, likely reflecting the chemical constituents dominantly from traffic emissions (i.e., VOCs and NO_x). Table 1.1 summarizes the air pollutant characteristics between Beijing and other world cities.

Most recently, it has been shown that severe haze formation involves a coupling between atmospheric chemical processes and the height of the atmospheric planetary boundary (PBL): rapid aging of BC particles in China considerably enhances

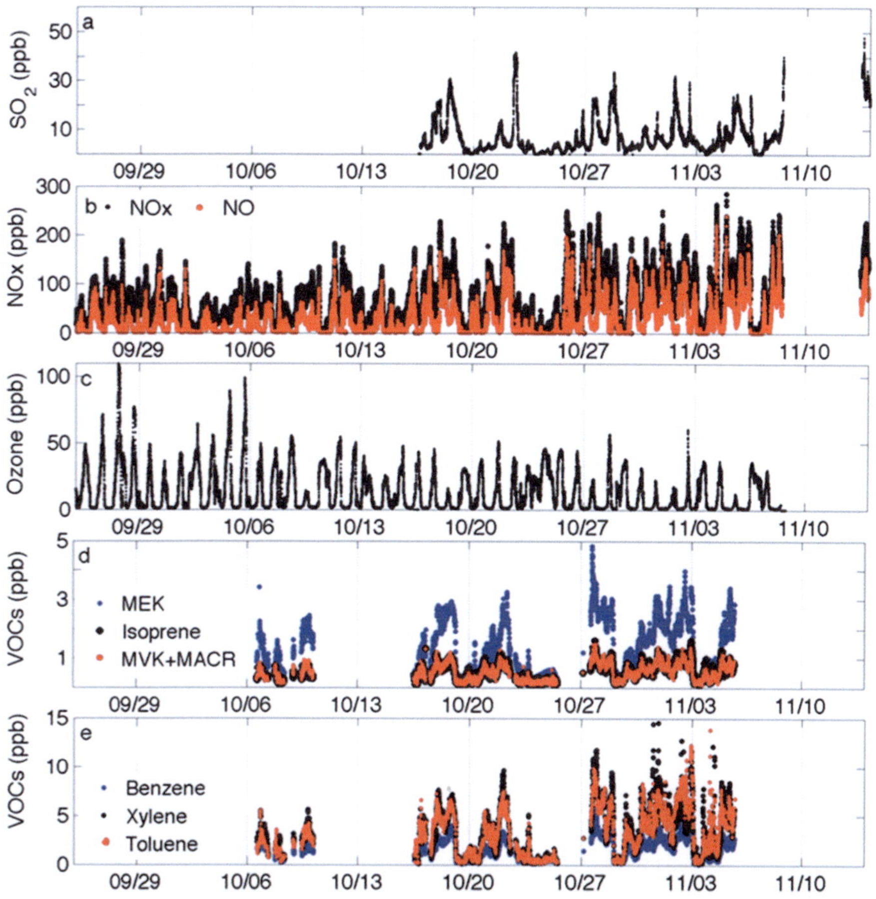

Fig. 1.6 Measurements of gaseous species. (**a–c**) Concentrations of SO2 (**a**), NOx (**b**), and O3 (**c**) measured between 25 September and 13 November. (**d**, **e**) Biogenic species, i.e., isoprene, methacrolein (MACR), methyl vinyl ketone (MVK), and methyl ethyl ketone (MEK) (**d**) and aromatic hydrocarbons, i.e., xylene, benzene, and toluene (**e**) (After Guo et al. 2014)

light absorption and contributes to stability and pollutant accumulation in China (Peng et al. 2016). Furthermore, there exists an anti-correlation between the photochemical activity and aqueous chemistry during the severe haze evolution (i.e., from the clean, transition, to polluted periods) in China, the sulfur to sulfate conversion plays a central role in facilitating aqueous production of the major secondary constituents (Wang et al. 2016). Clearly, the multi-sources and high-emissions of the PM precursors are the main reason for persistent haze occurrence in China (Zhang et al. 2015; Guo et al. 2014).

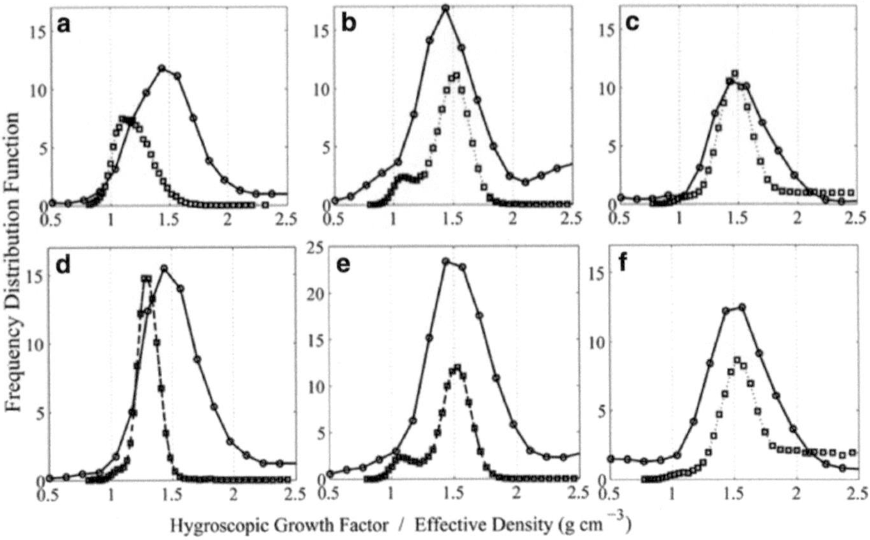

Fig. 1.7 Particle hygroscopicity and density during the clean, transition, and polluted periods for the 25–29 September and 2–7 October episodes. (**a–c**) Effective density (*solid line, circles*) and hygroscopicity (*dashed line, squares*) for 46-nm (**a**), 97-nm (**b**), and 240-nm (**c**) particles at 1500 h on 25 September, 1200 h on 27 September, and 1800 h on 28 September, respectively. **d**, **f** Effective density (*solid line, circles*) and hygroscopicity (*dashed line, squares*) for 46-nm (**d**), 97-nm (**e**), and 240-nm (**f**) particles measured at 1500 h on 2 October, 1200 h on 4 October, and 1200 h on 5 October, respectively (After Guo et al. 2014)

Table 1.1 Summary of air pollutant characteristics in Houston, Los Angeles, Mexico City, and Beijing (After Zhang et al. 2015)

	Houston, Texas	Los Angeles, California	Mexico City, Mexico	Beijing, China
Population	6.2 million	18 million	21.2 million	21.2 million
City (Metropolitan) area	1625 (10,062) km^2	1302 (87,940) km^2	2072 (3540) km^2	3820 (16,801) km^2
Number of vehicles	3.3 million	6.1 million	3.5 million	5.4 million
Primary pollutants	NO_x, VOCs (light olefins)	NO_x, VOCs (aromatics)	NO_x, VOCs (Alkanes, aromatics)	NO_x, SO_2, VOCs (aromatics)
Secondary pollutants	Ozone	Ozone, PM	Ozone, PM	PM
Average annual SO_2	0.35 ppb	0.37 ppb	2.50 ppb	7.8 ppb
Average annual $PM_{2.5}$ concentration	12 μg m^{-3}	18 μg m^{-3}	27 μg m^{-3}	102 μg m^{-3}
Average measured Particle concentration	10,700 cm^{-3}	20,000 cm^{-3}	21,000 cm^{-3}	72,900 cm^{-3}
Highest measured PM concentration	40,000 cm^{-3}	50,000 cm^{-3}	40,000 cm^{-3}	250,000 cm^{-3}
Dominant PM Composition	Organics (35%) and sulfate (30%)	Organics (44%) and nitrate (24%)	Organics (66%) and sulfate (14%)	Organics (44%) and nitrate (22%)

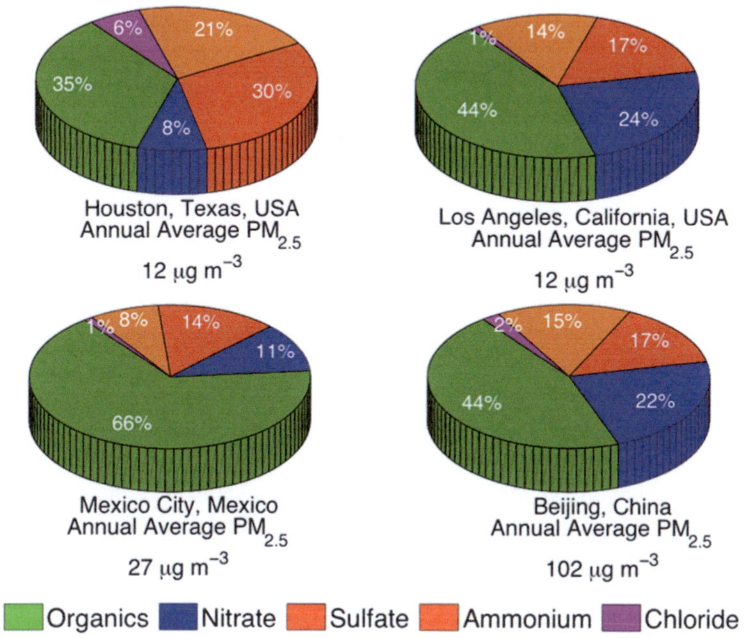

Fig. 1.8 Non-refractory compositions of particulate matter determined by an aerosol mass spectrometer in Houston, Los Angeles, Mexico City, and Beijing (After Zhang et al. 2015)

1.4 Challenges and Future Research Needs

The formation of persistent haze in China exhibits distinct characteristics in the particle properties, dependent of the emission sources, formation mechanisms, removal, and meteorological conditions. The secondary PM formation in China is most prominent, characterized by two aerosol formation mechanisms, i.e., nucleation and growth. Typically, there exist clear diurnal variations in the PM number, size, and mass concentration in other regions, reflecting the interplay between primary emissions, new particle formation, photochemical growth, removal, and the PBL variation. In contrast, haze events in China typically exhibit a periodic cycle of 4–7 days (Fig. 1.3). Aerosol nucleation consistently precedes a polluted period in Beijing, producing a high number concentration of nano-sized particles under clean conditions, and accumulation of the particle mass concentration exceeding several hundred micrograms per cubic meter is accompanied by a continuous size growth from the nucleation-mode particles over multiple days to yield numerous large particles (Fig. 1.3).

The efficient aerosol nucleation and growth in China are attributable to highly elevated concentrations of gaseous aerosol precursors, most noticeably anthropogenic VOCs, NO_x, and SO_2 emitted from local transportation and regional industrial activities (Fig. 1.6). For example, during pollution episodes the peak SO_2 and

NO_x concentrations exceed 40 and 200 ppb, respectively, and the peak xylene and toluene concentrations exceed 10 ppb in Beijing (Guo et al. 2014). These concentrations are significantly higher than those observed in the other urban regions worldwide.

The combination of the enormously efficient aerosol nucleation and growth over an extended period leading to the severe $PM_{2.5}$ episodes in Beijing is uniquely different from those typically observed in the urban regions of other countries and pristine environments worldwide. The efficient secondary formation in China is also reflected by the much higher annual $PM_{2.5}$ mass concentration (102 $\mu g\ m^{-3}$) than those in Houston (12 $\mu g\ m^{-3}$), Los Angeles (12 $\mu g\ m^{-3}$), and Mexico City (27 $\mu g\ m^{-3}$) (Fig. 1.8). On the other hand, the efficient secondary aerosol formation processes identified in China may also be characteristic of those in other urban cities of the developing world (such as India and other Asian developing countries), because of rapidly growing economy and fast urbanization leading to lower standards but higher rates for air pollutant emissions.

The currently available atmospheric chemical mechanisms in the gas and aqueous phases have yet to quantitatively account for the rapid accumulations of the PM chemical constituents under polluted conditions in China (i.e., Fig. 1.3), particularly for the formations of secondary organic matter, sulfate, and nitrate (Wang et al. 2013a, b). It is plausible that there exist synergetic effects among the various organic and inorganic compounds (i.e., organics, sulfate, nitrate, basic species, etc.) to enhance the particle growth (Guo et al. 2014). For example, the presence of basic species (i.e., ammonia and amines) may not only considerably enhance sulfate and nitrate formation, but also promote secondary organic matter production under polluted environments (Qiu and Zhang 2013).

The chemical mechanisms for the formation of secondary organic matter may also be distinct in China, leading to not only different PM production rates, but also different product types and aerosol properties. For example, while gas-particle partitioning of organic matter is linearly dependent of the gaseous reactant concentrations, oligomerization for hydration of small α-dicarbonyls and polymerization for aldol condensation of large aldehydes correspond to second- or higher-order reactions with respect to the organics to form high-molecular-weight species in the particle-phase. The latters are expected to occur more efficiently with elevated gaseous concentrations under polluted conditions. In addition, while the gas-phase oxidation leading to formation of semi- and low volatility and subsequent gas-particle partitioning increase particle hygroscopicity, oligomerization/polymerization generally leads to decreased hygroscopicity (Guo et al. 2014). Furthermore, organic matter formed from oligomerization/polymerization may exhibit enhanced light absorption in the ultraviolet and visible ranges (Wang et al. 2013a, b). Presently, very few of the multi-phase reactions have been incorporated into atmospheric models to assess their roles in the formation, growth, transformation, properties, and impacts of urban fine PM (Xu et al. 2014).

To improve the understanding on the formation of urban fine PM, future laboratory kinetic and mechanistic measurements of multi-phase chemistry are needed, particularly for those leading to secondary organic matter, sulfate, and nitrate formations under atmospherically relevant conditions, i.e., RH, temperature, and S concentrations. Laboratory experiments are also needed to quantify the different aerosol properties (i.e., hygroscopicity and optical properties) formed from the multi-phase reactions involving organic and inorganic species. Those experimental results on atmospheric multi-phase chemistry may assist in not only atmospheric modeling, but also interpretation and identification in field measurements for urban fine PM.

Further atmospheric measurements are needed to monitor simultaneously the gaseous aerosol precursors and a comprehensive set of aerosol properties (i.e., particle number, size, chemical composition, morphology, light scattering and absorption, and hygroscopicity, etc.). Atmospheric field measurements are crucial to providing the information on the temporal and spatial distributions of gaseous concentrations and PM properties under diverse urban environments. To achieve the highest level of chemical speciation, the development of more advanced analytical techniques is required, including identification and quantification of the diverse gaseous aerosol precursors (i.e., carbonyls, sulfuric acid, nitric acid, organic acids, basic species, etc.) present in ambient air at ppb or lower levels and detailed chemical composition characterization of aerosols from the molecular cluster (<1 nm), sub-nm, to sub-μm size ranges. Simultaneous measurements of the particle density, hygroscopicity, volatility, and optical properties can further assist in identification of the particle chemical compositions, since those particle properties are distinct for different PM types (Guo et al. 2014; Levy et al. 2013, 2014). Furthermore, improved physically based parameterizations of aerosol nucleation and growth developed on the basis of and validated against laboratory and field studies are required for incorporation into atmospheric models.

The human health effects of high concentrations of ultrafine particles produced from NPF under urban environments need to be carefully evaluated in China. For example, the PM episodes in Beijing typically consist of the clean and polluted periods, characterized by exceedingly high concentrations of ultrafine particles ($>10^5$ particles cm^{-3} for particles smaller than 100 nm) and numerous large particles (about 5×10^4 for particles larger than 150 nm), respectively. While many epidemiological studies have emphasized the correlations of the various health syndromes with the $PM_{2.5}$ levels (NRC 2004; Pope and Dockery 2006), little is known on the potential health outcomes of highly elevated concentrations of ultrafine particles produced from NPF under clean conditions.

The regional and climatic impacts of urban fine PM need to be further assessed to quantify their direct and indirect radiative forcing. Considering the profound societal implications of urban fine PM on human health, ecosystem, weather, and climate, it is imperative that sound science is employed to develop effective regulatory polices to mediate the local, regional, and global impacts of urban fine PM. The large urban centers/megacities may also represent the ideal locations to best achieve the co-benefits in simultaneously controlling air pollution and mitigating climate change (Zhu et al. 2012). Furthermore, it is essential that knowledge is transferred from well to less researched urban areas and the experience in mitigating urban fine PM for developed countries is considered in guiding development of regulatory policies for developing countries.

Acknowledgments This work was supported by the Robert A. Welch Foundation (Grant A-1417), the Ministry of Science and Technology of China (2013CB955800), and a collaborative research program by Texas A&M University.

References

Barbaro, E., de Arellano, J. V. G., Krol, M. C., & Holtslag, A. A. M. (2013). Impacts of aerosol shortwave radiation absorption on the dynamics of an idealized convective atmospheric boundary layer. *Boundary-Layer Meteorology, 148*(1), 31–49. doi:10.1007/s10546-013-9800-7.

Committee on the Medical Effects of Air Pollutants (COMEAP). (2006). Cardiovascular disease and air pollution, Department of Health, UK, https://www.gov.uk/government/uploads/system/uploads/attachment_data/file/304668/COMEAP_cardiovascular_disease_and_air_pollution.pdf.

Correia, A. W., Pope III, C. A., Dockery, D. W., Wang, Y., Ezzati, M., & Dominici, F. (2013). Effect of air pollution control on life expectancy in the United States an analysis of 545 US counties for the period from 2000 to 2007. *Epidemiology, 24*(1), 23–31. doi:10.1097/Ede.0b013e3182770237.

Dickerson, R. R., Li, C., Li, Z., Marufu, L. T., Stehr, J. W., McClure, B., Krotkov, N., Chen, H., Wang, P., Xia, X., & Ban, X. (2007). Aircraft observations of dust and pollutants over northeast China: Insight into the meteorological mechanisms of transport. *Journal of Geophysical Research, 112*(D24), doi:10.1029/2007jd008999.

Fan, J., Zhang, R., Li, G., & Tao, W.K. (2007a). Effects of aerosols and relative humidity on cumulus clouds. *Journal of Geophysical Research*, 112(D14), D14204. doi:10.1029/2006jd008136.

Fan, J., Zhang, R., Li, G., Tao, W. K., & Li, X. (2007b). Simulations of cumulus clouds using a spectral microphysics cloud-resolving model. *Journal of Geophysical Research, 112*(D4), D04201. doi:10.1029/2006jd007688.

Fan, J., Zhang, R., Tao, W. K., & Mohr, K. I. (2008). Effects of aerosol optical properties on deep convective clouds and radiative forcing. *Journal of Geophysical Research, 113*(D8), D08209. doi:10.1029/2007jd009257.

Fu, T. M., Jacob, D. J., Wittrock, F., Burrows, J. P., Vrekoussis, M., & Henze, D. K. (2008). Global budgets of atmospheric glyoxal and methylglyoxal, and implications for formation of secondary organic aerosols. *Journal of Geophysical Research, 113*(D15), doi:10.1029/2007jd009505.

Gauderman, W. J., Avol, E., Gilliland, F., Vora, H., Thomas, D., Berhane, K., McConnell, R., Kuenzli, N., Lurmann, F., Rappaport, E., & Margolis, H. (2004). The effect of air pollution on lung development from 10 to 18 years of age. *New England Journal Medicine, 351*(11), 1057–1067. doi:10.1056/Nejmoa040610.

Guo, S., Hu, M., Guo, Q., Zhang, X., Zheng, M., Zheng, J., Chang, C. C., Schauer, J. J., & Zhang, R. (2012). Primary sources and secondary formation of organic aerosols in Beijing, China. *Environmental Science and Technology, 46*(18), 9846–9853. doi:10.1021/es20425641.

Guo, S., Hu, M., Guo, Q., Zhang, X., Schauer, J. J., & Zhang, R. (2013). Quantitative evaluation of emission controls on primary and secondary organic aerosol sources during Beijing 2008 olympics. *Atmospheric Chemistry and Physics, 13*(16), 8303–8314. doi:10.5194/acp-13-8303-2013.

Guo, S., Hu, M., Zamora, M. L., Peng, J., Shang, D., Zheng, J., Du, Z., Wu, Z., Shao, M., Zeng, L., & Molina, M. J. (2014). Elucidating severe urban haze formation in China. *Proceedings of the*

National Academy of Sciences of the United States of America, 111(49), 17373–17378. doi:10.1073/pnas.1419604111.

He, L.-Y., Huang, X.-F., Xue, L., Hu, M., Lin, Y., Zheng, J., Zhang, R., & Zhang, Y.-H. (2011). Submicron aerosol analysis and organic source apportionment in an urban atmosphere in Pearl River Delta of China using high-resolution aerosol mass spectrometry. *Journal of Geophysical Research, 116*, D12304. doi:10.1029/2010JD014566.

He, H., Wang, Y., Ma, Q., Ma, J., Chu, B., Ji, D., Tang, G., Liu, C., Zhang, H., & Hao, J. (2014). Mineral dust and NOx promote the conversion of SO2 to sulfate in heavy pollution days. *Science Report, 4*, DOI: 10.1038/srep04172, doi:10.1038/srep04172.

IPCC. (2013). Climate change 2013: The physical science basis. *Contribution of Working Group I to the Fifth Assessment Report of the Intergovernmental Panel on Climate Change* (1535 pp). Cambridge, UK/New York: Cambridge University Press. http://www.ipcc.ch/report/ar5/wg1/. doi: 10.1017/CBO9781107415324.

Khalizov, A. F., Lin, Y., Qiu, C., Guo, S., Collins, D., & Zhang, R. (2013). The role of OH-initiated oxidation of isoprene in aging of combustion soot. *Environmental Science & Technology, 47*, 2254–2263. doi:10.1021/es3045339.

Koren, I., Kaufman, Y. J., Rosenfeld, D., Remer, L. A., & Rudich, Y. (2005). Aerosol invigoration and restructuring of Atlantic convective clouds. *Geophysical Research Letters, 32*(14). doi:10.1029/2005gl023187.

Kulmala, M., Vehkamäki, H., Petäjä, T., Dal Maso, M., Lauri, A., Kerminen, V. M., Birmili, W., & McMurry, P. H. (2004). Formation and growth rates of ultrafine atmospheric particles: A review of observations. *Journal of Aerosol Science, 35*(2), 143–176. doi:10.1016/j.jaerosci.2003.10.003.

Künzli, N., Jerrett, M., Mack, W. J., Beckerman, B., LaBree, L., Gilliland, F., Thomas, D., Peters, J., & Hodis, H. N. (2005). Ambient air pollution and atherosclerosis in Los Angeles. *Environmental Health Perspectives, 113*(2), 201–206. doi:10.1289/Ehp.7523.

Kurokawa, J., Ohara, T., Morikawa, T., Hanayama, S., Janssens-Maenhout, G., Fukui, T., Kawashima, K., & Akimoto, H. (2013). Emissions of air pollutants and greenhouse gases over Asian Regions during 2000–2008: Regional Emission Inventory in Asia (REAS) version 2. *Atmospheric Chemistry and Physics, 13*(21), 11019–11058. doi:10.5194/acp-13-11019-2013.

Laskin, A., Laskin, J., & Nizkorodov, S. A. (2015). Chemistry of atmospheric brown carbon. *Chemical Reviews*. doi:10.1021/cr5006167.

Lee, K. H., Li, Z., Wong, M. S., Xin, J., Wang, Y., Hao, W. M., & Zhao, F. (2007). Aerosol single scattering Albedo estimated across China from a combination of ground and satellite measurements. *Journal of Geophysical Research, 112*(D22), doi:10.1029/2007jd009077.

Levy, M. E., Zhang, R., Khalizov, A. F., Zheng, J., Collins, D. R., Glen, C. R., Wang, Y., Yu, X. Y., Luke, W., Jayne, J. T., & Olaguer, E. (2013). Measurements of submicron aerosols in Houston, Texas during the 2009 SHARP field campaign. *Journal of Geophysical Research, 118*(18), 10518–10534. doi:10.1002/Jgrd.50785.

Levy, M. E., Zhang, R., Zheng, J., Tan, H., Wang, Y., Molina, L. T., Takahama, S., Russell, L. M., & Li, G. (2014). Measurements of submicron aerosols at the California–Mexico border during the Cal–Mex 2010 field campaign. *Atmospheric Environment, 88*(0), 308–319. doi:http://dx.doi.org/10.1016/j.atmosenv.2013.08.062.

Li, Z. (1998). Influence of absorbing aerosols on the inference of solar surface radiation budget and cloud absorption. *Journal of Climate, 11*(1), 5–17. doi:10.1175/1520-0442(1998)011<0005:Ioaaot>2.0.Co;2.

Li, G., Zhang, R., Fan, J., & Tie, X. (2005). Impacts of black carbon aerosol on photolysis and ozone. *Journal of Geophysical Research, 110*(D23), D23206. doi:10.1029/2005jd005898.

Li, G., Wang, Y., Lee, K.H., Diao, Y., Zhang, R. (2008a), Increased winter precipitation over the North Pacific from 1984–1994 to 1995–2005 inferred from the global precipitation climatology project. *Geophysical Research Letters, 35*(13), L13821. doi:10.1029/2008gl034668.

Li, G., Wang, Y., & Zhang, R. (2008b), Implementation of a two-moment bulk microphysics scheme to the WRF model to investigate aerosol-cloud interaction. *Journal of Geophysical Research, 113*(D15), D15211. doi:10.1029/2007jd009361.

Li, G., Wang, Y., Lee, K. H., Diao, Y., & Zhang, R. (2009). Impacts of aerosols on the development and precipitation of a mesoscale squall line. *Journal of Geophysical Research,* 114, D17205. doi:10.1029/2008jd011581.

Li, Z., Lee, K.H., Wang, Y., Xin, J., Hao, W.M. (2010). First observation-based estimates of cloud-free aerosol radiative forcing across China. *Journal Geophysical Research, 115*(D7), doi:10.1029/2009JD013306.

Lin, J., Pan, D., Davis, S. J., Zhang, Q., He, K., Wang, C., Streets, D. G., Wuebbles, D. J., & Guan, D. (2014). China's international trade and air pollution in the United States. *Proceedings of the National Academy of Sciences of the United States of America, 111*(5), 1736–1741. doi:10.1073/pnas.1312860111.

Ma, Y., Brooks, S. D., Vidaurre, G., Khalizov, A. F., Wang, L., & Zhang, R. (2013). Rapid modification of cloud-nucleating ability of aerosols by biogenic emissions. *Geophysical Research Letters, 40*(23), 6293–6297. doi:10.1002/2013gl057895.

Matus, K., Nam, K. M., Selin, N. E., Lamsal, L. N., Reilly, J. M., & Paltsev, S. (2012). Health damages from air pollution in China. *Global Environmental Change, 22*(1), 55–66. doi:10.1016/j.gloenvcha.2011.08.006.

Mijling, B., van der A, R. J., & Zhang, Q. (2013). Regional nitrogen oxides emission trends in East Asia observed from space. *Atmospheric Chemistry and Physics, 13*(23), 12003–12012. doi:10.5194/acp-13-12003-2013.

Millman, A., Tang, D., & Perera, F. P. (2008). Air pollution threatens the health of children in China. *Pediatrics, 122*(3), 620–628. doi:10.1542/peds.2007-3143.

NRC. (2004). *Research priorities for airborne particulate matter, IV. Continuing research progress.* Washington, DC: National Academy Press.

Ohara, T. A. H. K., Akimoto, H., Kurokawa, J. I., Horii, N., Yamaji, K., Yan, X., & Hayasaka, T. (2007). An Asian emission inventory of anthropogenic emission sources for the period 1980–2020. *Atmospheric Chemistry and Physics Discussions, 7*(3), 6843–6902. doi:10.5194/acpd-7-6843-2007.

Orville, R. E., Huffines, G., Nielsen-Gammon, J., Zhang, R., Ely, B., Steiger, S., Phillips, S., Allen, S., & Read, W. (2001). Enhancement of cloud-to-ground lightning over Houston, Texas. *Geophysical Research Letters, 28*(13), 2597–2600. doi:10.1029/2001gl012990.

Peng, J. F., Hu, M., Wang, Z. B., Huang, X. F., Kumar, P., Wu, Z. J., Guo, S., Yue, D. L., Shang, D. J., Zheng, Z., & He, L. Y. (2014). Submicron aerosols at thirteen diversified sites in China: Size distribution, new particle formation and corresponding contribution to cloud condensation nuclei production. *Atmospheric Chemistry and Physics, 14*(18), 10249–10265. doi:10.5194/acp-14-10249-2014.

Peng, J., Hu, M., Guo, S., Du, Z., Zheng, J., Shang, D., Zamora, M. L., Zeng, L., Shao, M., Wu, Y., Zheng, J., Wang, Y., Glen, C. R., Collins, D. R., Molina, M. J., & Zhang, R. (2016). Markedly enhanced absorption and direct radiative forcing of black carbon under polluted urban environments. *Proceedings of the National Academy of Sciences of the United States of America, 113*, 4266–4271. doi:10.1073/pnas.1602310113.

Petters, M. D., & Kreidenweis, S. M. (2007). A single parameter representation of hygroscopic growth and cloud condensation nucleus activity. *Atmospheric Chemistry and Physics, 7*(8), 1961–1971.

Pope, C. A., & Dockery, D. W. (2006). Health effects of fine particulate air pollution: Lines that connect. *Journal of the Air & Waste Management Association (1995), 56*(6), 709–742. doi:10.1080/10473289.2006.10464485.

Qiu, C., & Zhang, R. (2013). Multiphase chemistry of atmospheric amines. *Physical Chemistry Chemical Physics, 15*(16), 5738–5752. doi:10.1039/C3cp43446j.

Rosenfeld, D., Lohmann, U., Raga, G. B., O'Dowd, C. D., Kulmala, M., Fuzzi, S., Reissell, A., & Andreae, M. O. (2008). Flood or drought: How do aerosols affect precipitation? *Science, 321* (5894), 1309–1313. doi:10.1126/science.1160606.

Schlesinger, R. B., Kunzli, N., Hidy, G. M., Gotschi, T., & Jerrett, M. (2006). The health relevance of ambient particulate matter characteristics: Coherence of toxicological and epidemiological inferences. *Inhalation Toxicology, 18*(2), 95–125. doi:10.1080/08958370500306016.

Seinfeld, J. H., & Pandis, S. N. (2006). *Atmospheric chemistry and physics – from air pollution to climate change* (2nd ed.). Hoboken: Wiley.

Stavrakou, T., Müller, J. F., Boersma, K. F., van der A, R. J., Kurokawa, J., Ohara, T., & Zhang, Q. (2013). Key chemical NOx sink uncertainties and how they influence top-down emissions of nitrogen oxides. *Atmospheric Chemistry and Physics, 13*(17), 9057–9082. doi:10.5194/acp-13-9057-2013.

Streets, D. G., Bond, T. C., Carmichael, G. R., Fernandes, S. D., Fu, Q., He, D., Klimont, Z., Nelson, S. M., Tsai, N. Y., Wang, M. Q., & Woo, J. H. (2003). An inventory of gaseous and primary aerosol emissions in Asia in the year 2000. *Journal of Geophysical Research, 108* (D21), doi:10.1029/2002jd003093.

Suh, I., Zhang, R., Molina, L. T., & Molina, M. J. (2003). Oxidation mechanism of aromatic peroxy and bicyclic radicals from OH-toluene reactions. *Journal of the American Chemical Society, 125*, 12655–12665.

Tao, W.K., Chen, J.P., Li, Z., Wang, C.. & Zhang, C. (2012). Impact of aerosols on convective clouds and precipitation. *Reviews of Geophysics, 50*(RG2001), doi:10.1029/2011rg000369.

Tie, X., Madronich, S., Walters, S., Zhang, R., Rasch, P., & Collins, W. (2003). Effect of clouds on photolysis and oxidants in the troposphere. *Journal of Geophysical Research, 108*(D20), 4642. doi:10.1029/2003jd003659.

Tie, X., Madronich, S., Walters, S., Edwards, D.P., Ginoux, P., Mahowald, N., Zhang, R., Lou, C., & Brasseur, G. (2005). Assessment of the global impact of aerosols on tropospheric oxidants. *Journal of Geophysical Research, 110*(D3), D03204. doi:10.1029/2004jd005359.

Wang, Q., Shao, M., Zhang, Y., Wei, Y., Hu, M., & Guo, S. (2009). Source apportionment of fine organic aerosols in Beijing. *Atmospheric Chemistry and Physics, 9*(21), 8573–8585.

Wang, L., Lal, V., Khalizov, A. F., & Zhang, R. (2010). Heterogeneous chemistry of alkylamines with sulfuric acid: Implications for atmospheric formation of alkylaminium sulfates. *Environmental Science & Technology, 44*, 2461–2465. doi:10.1021/es9036868.

Wang, Y., Wan, Q., Meng, W., Liao, F., Tan, H., & Zhang, R. (2011a). Long-term impacts of aerosols on precipitation and lightning over the Pearl River Delta Megacity Area in China. *Atmospheric Chemistry and Physics, 11*(23), 12421–12436. doi:10.5194/acp-11-12421-2011.

Wang, Z. B., Hu, M., Yue, D. L., Zheng, J., Zhang, R. Y., Wiedensohler, A., Wu, Z. J., Nieminen, T., & Boy, M. (2011b). Evaluation on the role of sulfuric acid in the mechanisms of new particle formation for Beijing case. *Atmospheric Chemistry and Physics, 11*(24), 12663–12671. doi:10.5194/acp-11-12663-2011.

Wang, Y., Fan, J., Zhang, R., Leung, L. R., & Franklin, C. (2013a). Improving bulk microphysics parameterizations in simulations of aerosol effects. *Journal of Geophysical Research, 118*(11), 5361–5379. doi:10.1002/Jgrd.50432.

Wang, Y., Khalizov, A., Levy, M., & Zhang, R. (2013b). New directions: Light absorbing aerosols and their atmospheric impacts. *Atmospheric Environment, 81*, 713–715. doi:10.1016/j.atmosenv.2013.09.034.

Wang, Y., Zhang, R., & Saravanan, R. (2014). Asian pollution climatically modulates mid-latitude cyclones following hierarchical modelling and observational analysis. *Nature Communications, 5*, 3098.

Wang, Z. B., Hu, M., Pei, X. Y., Zhang, R. Y., Paasonen, P., Zheng, J., Yue, D. L., Wu, Z. J., Boy, M., & Wiedensohler, A. (2015). Connection of organics to atmospheric new particle formation and growth at an Urban site of Beijing. *Atmospheric Environment, 103*, 7–17. doi:10.1016/j.atmosenv.2014.11.069.

Wang, G., Zhang, R., Gomez, M. E., Yang, L., Zamora, M. L., Hu, M., Lin, Y., Peng, J., Guo, S., Meng, J., Li, J., Cheng, C., Hu, T., Ren, Y., Wang, Y., Gao, J., Cao, J., An, Z., Zhou, W., Li, G., Wang, J., Tian, P., Marrero-Ortiz, W., Secrest, J., Du, Z., Zheng, J., Shang, D., Zeng, L., Shao, M., Wang, W., Huang, Y., Wang, Y., Zhu, Y., Li, Y., Hu, J., Pan, B., Cai, L., Cheng, Y., Ji, Y., Zhang, F., Rosenfeld, D., Liss, P. S., Duce, R. A., Kolb, C. E., & Molina, M. J. (2016). Persistent sulfate formation from London Fog to Chinese Haze. *Proceedings of the National Academy of Sciences of the United States of America, 113*, 13630–13635. doi:10.1073/pnas.1616540113.

Williams, E. R., Zhang, R., & Rydock, J. (1991). Mixed-phase microphysics and cloud electrification. *Journal of the Atmospheric Sciences, 48*(19), 2195–2203. doi:10.1175/1520-0469.

Wu, Z., Hu, M., Liu, S., Wehner, B., Bauer, S., Wiedensohler, A., Petäjä, T., Dal Maso, M., & Kulmala, M. (2007). New particle formation in Beijing, China: Statistical analysis of a 1-year data set. *Journal of Geophysical Research, 112*(D9), D09209. doi:10.1029/2006jd007406.

Wu, G., Li, Z., Fu, C., Zhang, X., Zhang, R., Zhang, R., Zhou, T., Li, J., Li, J., Zhou, D., & Wu, L. (2016). Advances in studying interactions between aerosols and monsoon in China. *Science China Earth Sciences, 59*(1), 1–16. doi:10.1007/s11430-015-5198-z.

Xinhua news (2012). Beijing car ownership exceeds 5 million, edited, http://news.xinhuanet.com/english/china/2012-02/16/c_122713279.htm

Xu, W., Gomez-Hernandez, M., Guo, S., Secrest, J., Marrero-Ortiz, W., Zhang, A. L., & Zhang, R. (2014). Acid-catalyzed reactions of epoxides for atmospheric nanoparticle growth. *Journal of the American Chemical Society, 136*(44), 15477–15480. doi:10.1021/Ja508989a.

Yang, X., & Li, Z. (2014). Increases in thunderstorm activity and relationships with air pollution in Southeast China. *Journal of Geophysical Research, 119*(4), 1835–1844. doi:10.1002/2013JD021224.

Yang, X., Yao, Z., Li, Z., & Fan, T. (2013). Heavy air pollution suppresses summer thunderstorms in Central China. *Journal of Atmospheric and Solar – Terrestrial Physics, 95-96*, 28–40. doi:10.1016/j.jastp.2012.12.023.

Ying, Q., & Kleeman, M. J. (2006). Source contributions to the regional distribution of secondary particulate matter in California. *Atmospheric Environment, 40*(4), 736–752. doi:10.1016/j.atmosenv.2005.10.007.

Ying, Q., Wu, L., & Zhang, H. (2014). Local and inter-regional contributions to PM2.5 nitrate and sulfate in China. *Atmospheric Environment*, *94*(0), 582–592. doi:http://dx.doi.org/10.1016/j.atmosenv.2014.05.078.

Yuan, T., Li, Z., Zhang, R., Fan, J. (2008). Increase of cloud droplet size with aerosol optical depth: An observation and modeling study. *Journal of Geophysical Research*, *113*(D4), 10.1029/2007jd008632.

Yue, D. L., Hu, M., Zhang, R. Y., Wang, Z. B., Zheng, J., Wu, Z. J., Wiedensohler, A., He, L. Y., Huang, X. F., & Zhu, T. (2010). The roles of sulfuric acid in new particle formation and growth in the Mega-City of Beijing. *Atmospheric Chemistry and Physics, 10*(10), 4953–4960. doi:10.5194/acp-10-4953-2010.

Zhang, R. (2010). Atmospheric science. Getting to the Critical Nucleus of Aerosol Formation. *Science, 328*(5984), 1366–1367. doi:10.1126/science.1189732.

Zhang, H., & Ying, Q. (2010). Source apportionment of airborne particulate matter in Southeast Texas using a source-oriented 3D air quality model. *Atmospheric Environment, 44*(29), 3547–3557.

Zhang, R., Tie, X., & Bond, D. W. (2003). Impacts of anthropogenic and natural NOx sources over the U.S. on tropospheric chemistry. *Proceedings of the National Academy of Sciences of the United States of America, 100*(4), 1505–1509. doi:10.1073/pnas.252763799.

Zhang, R., Li, G., Fan, J., Wu, D. L., & Molina, M. J. (2007). Intensification of pacific storm track linked to Asian pollution. *Proceedings of the National Academy of Sciences of the United States of America, 104*(13), 5295–5299. doi:10.1073/pnas.0700618104.

Zhang, R., Khalizov, A. F., Pagels, J., Zhang, D., Xue, H., & McMurry, P. H. (2008a). Variability in morphology, hygroscopicity, and optical properties of soot aerosols during atmospheric processing. *Proceedings of the National Academy of Sciences of the United States of America, 105*(30), 10291–10296. doi:10.1073/pnas.0804860105.

Zhang, H., Ye, X., Cheng, T., Chen, J., Yang, X., Wang, L., & Zhang, R. (2008b). A laboratory study of agricultural crop residue combustion in China: Emission factors and emission inventory. *Atmospheric Environment, 42*(36), 8432–8441. doi:10.1016/j.atmosenv.2008.08.015.

Zhang, Q., Streets, D. G., Carmichael, G. R., He, K. B., Huo, H., Kannari, A., Klimont, Z., Park, I. S., Reddy, S., Fu, J. S., & Chen, D. (2009). Asian emissions in 2006 for the NASA INTEX-B mission. *Atmospheric Chemistry and Physics, 9*(14), 5131–5153. doi:10.5194/acp-9-5131-2009.

Zhang, R., Khalizov, A., Wang, L., Hu, M., & Xu, W. (2012a). Nucleation and growth of nanoparticles in the atmosphere. *Chemical Reviews, 112*(3), 1957–2011. doi:10.1021/cr2001756.

Zhang, H., Li, J., Ying, Q., Yu, J.Z., Wu, D., Cheng, Y., He, K., & Jiang, J. (2012b), Source apportionment of PM2.5 nitrate and sulfate in China using a source-oriented chemical transport model. *Atmospheric Environment, 62*(0), 228–242. doi:http://dx.doi.org/10.1016/j.atmosenv.2012.08.014

Zhang, R., Jing, J., Tao, J., Hsu, S. C., Wang, G., Cao, J., Lee, C. S. L., Zhu, L., Chen, Z., Zhao, Y., & Shen, Z. (2013). Chemical characterization and source apportionment of $PM_{2.5}$ in Beijing: Seasonal perspective. *Atmospheric Chemistry and Physics, 13*(14), 7053–7074. doi:10.5194/acp-13-7053-2013.

Zhang, F., Li, Y., Li, Z., Sun, L., Li, R., Zhao, C., Wang, P., Sun, Y., Liu, X., Li, J., & Li, P. (2014). Aerosol hygroscopicity and cloud condensation nuclei activity during the AC3Exp campaign: Implications for cloud condensation nuclei parameterization. *Atmospheric Chemistry and Physics, 14*(24), 13423–13437. doi:10.5194/acp-14-13423-2014.

Zhang, R., Wang, G., Guo, S., Zamora, M. L., Ying, Q., Lin, Y., Wang, W., Hu, M., & Wang, Y. (2015). Formation of urban fine particulate matter. *Chemical Reviews, 115*(10), 3803–3855. doi:10.1021/acs.chemrev.5b00067.

Zhao, J., Levitt, N. P., & Zhang, R. (2005). Heterogeneous chemistry of octanal and 2,4-hexadienal with sulfuric acid. *Geophysical Research Letters, 32*(9), L09802. doi:10.1029/2004gl022200.

Zhao, J., Levitt, N. P., Zhang, R., & Chen, J. (2006). Heterogeneous reactions of methylglyoxal in acidic media: Implications for secondary organic aerosol formation. *Environmental Science & Technology, 40*(24), 7682–7687. doi:10.1021/Es060610k.

Zhao, J., Khalizov, A. F., Zhang, R., & McGraw, R. (2009). Hydrogen bonding interaction of molecular complexes and clusters of aerosol nucleation precursors. *The Journal of Physical Chemistry, 113*, 680–689.

Zhu, T., Melamed, M. L., Parrish, D., Gauss, M., Klenner, L. G., Lawrence, M. G., Konare, A., & Liousse, C. (2012). WMO/IGAC impacts of megacities on air pollution and climate (GAW Report No. 205), 1–296 pp., *WMO/IGAC*, http://www.wmo.int/pages/prog/arep/gaw/documents/Final_GAW_205.pdf.

Chapter 2
An Overview of Air Quality Modeling Activities in South Asia

Rajesh Kumar, Mary C. Barth, Luca Delle Monache, Sachin D. Ghude, Gabriele Pfister, Manish Naja, and Guy P. Brasseur

Abstract Air pollution is a growing major environmental concern in South Asia owing to rapid economic growth and expansion of urban, industrial, power generation and traffic sectors. This problem is widely recognized by the national and international community and many efforts have been made in the past 25 years to gain insight into the causes of this problem. These efforts include building ozone and aerosol monitoring networks, analysis of satellite datasets, development of emission inventories, setting-up and evaluating regional and global chemical transport models and using these models to address various scientific issues. Here, we provide a brief review of air quality related research activities in South Asia especially the evolution of air quality monitoring networks and air quality modeling activities. The chapter begins with an introduction to air quality problems in South Asia followed by a brief discussion of the evolution of aerosol and ozone monitoring networks and model evaluation efforts. This is followed by a summary of key findings from studies that focused on (i) understanding the seasonality of black carbon aerosols in India, (ii) on quantifying the sources of black carbon, carbon monoxide and ozone, (iii) the interaction of dust aerosols with tropospheric chemistry, (iv) the past and future changes in South Asian air quality, and (v) the effects of present-day pollution on human health and crop productivity in South Asia. We

R. Kumar (✉) • M.C. Barth • L.D. Monache • G. Pfister
National Center for Atmospheric Research, Boulder, CO, USA
e-mail: rkumar@ucar.edu; barthm@ucar.edu; lucadm@ucar.edu; pfister@ucar.edu

S.D. Ghude
Indian Institute of Tropical Meteorology, Pune, Maharashtra, India
e-mail: sachinghude@tropmet.res.in

M. Naja
Aryabhatta Research Institute of Observational Sciences, Nainital, Uttarakhand, India
e-mail: manish@aries.res.in

G.P. Brasseur
National Center for Atmospheric Research, Boulder, CO, USA

Max Planck Institute for Meteorology, Bundesstr. 53, Hamburg 20146, Germany
e-mail: guy.brasseur@mpimet.mpg.de

I. Bouarar et al. (eds.), *Air Pollution in Eastern Asia: An Integrated Perspective*,
ISSI Scientific Report Series 16, DOI 10.1007/978-3-319-59489-7_2

conclude with a list of modeling tasks that can be undertaken in the next 5 years to advance our understanding of air quality related issues in South Asia.

Keywords South Asia • Air quality • Chemical transport modeling • Ozone • Aerosols

2.1 Introduction

South Asia comprised of Afghanistan, Pakistan, India, Nepal, Bhutan, Bangladesh, Sri Lanka and Maldives is home to about one fifth of the world's population with about 32% of the population located in growing mega-cities. This large South Asian population is engaged in a wide range of industrial and residential activities (e.g. use of solid biofuels, kerosene and liquefied petroleum gas in cooking, fossil fuel burning in vehicles, coal burning in industries and power plants, and crop residue burning) that lead to large emissions of long-lived greenhouse gases (e.g. carbon dioxide (CO_2), nitrous oxide (N_2O), sulfur hexafluoride (SF_6), chlorofluorocarbons (CFCs) etc.), short-lived aerosol particles (e.g. black carbon, organic carbon and dust etc.) and trace gases (e.g. carbon monoxide (CO), methane (CH_4), non-methane volatile organic compounds (NMVOCs), nitrogen oxides (NO_x), sulfur dioxide (SO_2), ammonia (NH_3), etc.). Reactions among these short-lived primary pollutants in the presence of sunlight and water vapor also produce several secondary pollutants in the atmosphere such as ozone and secondary aerosols.

The atmospheric levels of air pollutants have been rising rapidly in South Asia especially since 1991 when economic liberalization in India initiated large-scale industrial and manufacturing operations which in turn led to a remarkable growth of the Indian Gross Domestic Product (GDP) from about 5.5 million Indian rupees (₹) in 1990 to about ₹1 billion in 2014. However, this economic growth led to environmental and air quality degradation. The World Health Organization (WHO) reports that 10 out of the 20 most polluted cities in the present-day world are located in South Asia (WHO 2016). The exposure to polluted air is estimated to reduce the lifespans of about 660 million people in South Asia by about 3 years (Greenstone et al. 2015) and damage crops worth $6 billion (Van Dingenen et al. 2009), which is enough to feed about 94 million people every year (Ghude et al. 2014). The rising air pollution in terms of aerosol particles is also suggested to affect the Himalayan glaciers (e.g. Yasunari et al. 2010; Menon et al. 2010) and the South Asian summer monsoon (e.g. Ramanathan et al. 2005; Lau et al. 2006; Ganguly et al. 2012), which are the main sources of water needed to sustain life in this region. Moreover, the South Asia pollutants have also been shown to affect other parts of the world especially the Indian Ocean (Lelieveld et al. 2001; Ramanathan et al. 2001), pristine Himalayas (e.g. Sagar et al. 2004; Kumar et al. 2010; Nair et al. 2013) and the Mediterranean Sea (e.g. Lawrence et al. 2003; Park et al. 2007).

To address the rapidly growing air pollution problem in South Asia, it is imperative to integrate the in situ and satellite based observations with chemical transport models and many efforts have been made in the past 25 years on both fronts. Here, we provide a brief overview of related progress. We refer the readers to Lawrence and Lelieveld (2010) for a detailed review on the emissions distributions and trends in South Asia and outflow of pollutants from South Asia. The majority of this review focuses on activities in India as India is the largest part of South Asia and has been the subject of most studies to date.

2.2 Brief Overview of Observational Studies

In this section, we briefly describe the evolution of surface ozone and aerosol in situ monitoring networks and use of satellite observations in air quality studies. All these in situ and satellite based observations have also been used in evaluating a variety of global and regional chemical transport models (CTM). First measurements of surface ozone over South Asia date back to pre-International Geophysical Year (IGY) when the Ehmerts instrument was used to measure surface ozone at Ahmedabad (Ramanathan 1956, Dave 1957). The Indian Meteorological Department (IMD) initiated surface ozone measurements at Trivandrum, Kodaikanal, Pune and Delhi using the Brewer "Bubbler" ozone sensor in the 1960s and 1970s. However, it was during the 1990s when systematic measurements of ozone and related gases were first initiated in different chemical environments namely at the urban site of Ahmedabad (Lal et al. 2000), the rural site of Gadanki (Naja and Lal 2002) and the high altitude site of Mt.-Abu (Naja et al. 2003). The number of sites measuring surface ozone and related gases in India has been growing rapidly since then under the Indian Space Research Organization's Geosphere-Biosphere Program (ISO-GBP) and ozone measurements at more than 20 stations have been reported so far (Fig. 2.1).

Similar to ozone and related gases, efforts to characterize aerosol optical properties in India date back to the 1950s and systematic measurements of aerosols began in the early 1980s under the Indian Middle Atmosphere Program (I-MAP). This program grew into a regional monitoring network called the "Aerosol Radiative Forcing over India" (ARFI) under the ISRO-GBP which measures aerosol properties at 35 stations in India (Moorthy et al. 2009, Fig. 2.1). Apart from ARFI, many sites have been established under the Aerosol Robotic Network (Holben et al. 1998), however, only the urban site of Kanpur has long-term (more than 5 years) data right now. In addition to establishing these in situ measurement networks, Indian scientists have also collaborated with international scientists and conducted a few national/international intensive field campaigns such as the Indian Ocean Experiment (INDOEX) (Ramanathan et al. 2001; Lelieveld et al. 2001), the Integrated Campaign for Aerosols, gases and Radiation Budget (ICARB) (Moorthy et al. 2008), TIGERZ (Giles et al. 2011) and Ganges Valley Aerosol Experiment (GVAX) (Moorthy et al. 2016). These missions have been conducted to study the export of

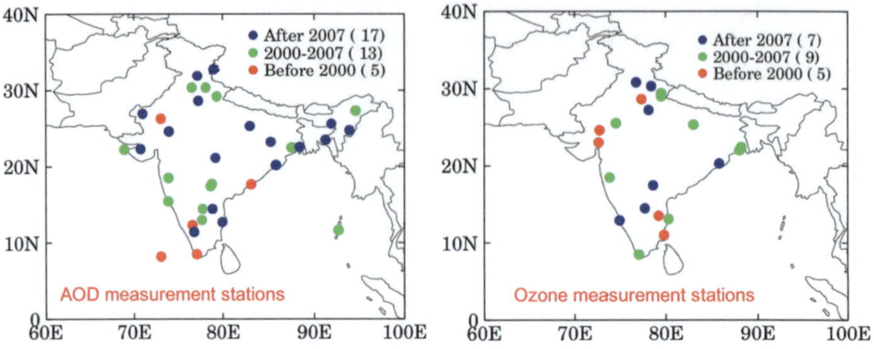

Fig. 2.1 Time evolution of surface ozone and aerosol optical depth measurements over India

pollutants from South Asia to the surrounding pristine oceanic and Himalayan environments and characterize the physical and chemical properties and radiative effects of trace gases and aerosols over the Indian subcontinent. All these measurements have significantly advanced our understanding of multi-scale spatial and temporal variability in trace species over the Indian region and a good summary of key findings from observation-based studies can be found in recent review papers (e.g. Lal 2007; Moorthy et al. 2009, 2016; Lawrence and Lelieveld 2010).

Satellite observations from a variety of space-borne sensors have been used to examine trends, spatial and temporal variability in trace species over the Indian region. Tropospheric column NO_2 retrievals from the Global Ozone Monitoring Experiment (GOME), the Scanning Imaging Absorption Spectrometer for Atmospheric Cartography (SCIAMACHY) and Ozone Monitoring Instrument (OMI) have shown increasing trends in tropospheric NO_2 abundances over India (Ghude et al. 2008, 2013) and South Asia (Ghude et al. 2009). Seasonal cycle of satellite-retrieved tropospheric column NO_2 is found to agree well with in situ observations (Ghude et al. 2008). OMI NO_2 retrievals have also been used to optimize NO_x emission over India (Ghude et al. 2012). OMI retrievals of formaldehyde and NO_2 have been used to show that ozone production over most of South Asia is NO_x limited (Mahajan et al. 2015). CO retrievals from the Measurement of Pollution in the Troposphere (MOPITT) have been used to demonstrate the transport of CO from South and Southeast Asia to the Bay of Bengal, Arabian Sea and the Indian Ocean during winter and to the upper troposphere during summer monsoon (Ghude et al. 2011). MOPITT CO has been reported to exhibit a decreasing tendency in the lower troposphere and increasing tendency in the upper troposphere during 2000–2014 (Girach and Nair 2014). Space-borne retrievals of AOD retrievals from different sensors (Moderate Resolution Imaging Spectroradiometer (MODIS), Multi-angle Imaging Spectroradiometer (MISR)) revealed elevated aerosol loadings over the Indo-Gangetic Plain (e.g. Di Girolamo et al. 2004; Jethva et al. 2005). MODIS and MISR AOD retrievals also show increasing trends in aerosol loadings over the urban centers and densely populated rural regions of the Indian subcontinent (Dey and Di Girolamo 2011; Ramachandran et al. 2012).

2.3 Model Evaluation

The efforts to evaluate the ability of chemistry transport models in simulating tropospheric ozone and precursor gases over South Asia started nearly two decades ago but only a few global and regional models have been tested so far. The evaluated global models include Model for Atmospheric Transport and Chemistry (MATCH), Max Planck Institute of Chemistry version of MATCH (MATCH-MPIC), Model for Ozone and Related Tracers (MOZART) and Frontier Research System for Global Change version of the University of California, Irvine global chemistry transport model (FRSGC/UCI), while the regional models include Weather Research and Forecasting model coupled with Chemistry (WRF-Chem), Regional model (REMO)-CTM, regional MATCH and HANK (an offline CTM using MOZARTv2 chemistry). All of these global and regional models qualitatively capture the observed seasonal cycle of surface ozone in South Asia except MATCH, which had problems in reproducing the winter maximum at Ahmedabad and monsoon minimum at Mt. Abu (Naja 1997). There is no clear overall bias in the modeled surface ozone as the models agree reasonably well with the observations at some sites but show high and low biases at other sites.

MOZART-2 reproduced the observed surface ozone within ± 10 ppbv over the remote Indian Ocean, Pune and Anantapur but is biased high over the oceanic region near continental India (Saraf et al. 2003; Beig et al. 2007; Ali et al. 2009), while MOZART-4 is biased high by about 1–30 ppbv at 13 stations in India (Surendran et al. 2015). The MATCH-MPIC is biased high by about 10–20 ppbv at a sub-urban site Pantnagar in the Indo-Gangetic Plain (Ojha et al. 2012). The FRSGC/UCI model underestimated surface ozone levels observed at a Himalayan site Nainital by 10–20 ppbv during spring and overestimated the observed levels by up to 30 ppbv during other times of the year (Kumar et al. 2010). The regional MATCH model is biased high by up to 10–30 ppbv at different urban, rural and coastal sites but agreed reasonably well with the observations at non-coastal rural sites and severely underestimated the ozone concentrations at Mt. Abu (Engardt 2008). The REMO-CTM generally underestimated the observed ozone concentrations during winter and spring months by about 5–20 ppbv while significantly overestimated the concentrations by up to 20 ppbv during the monsoon season at Pune, Anantapur and Darjeeling but the model is biased high throughout the year at Gadanki (Roy et al. 2008). The HANK model is biased high at Delhi and Ahmedabad and biased low over the Arabian Sea in ozone (Mittal et al. 2007). The WRF-Chem model agreed very well in ozone with the observations during the monsoon season with observed ozone levels at seven sites in India but was biased high outside the monsoon season (Kumar et al. 2012b). Both MOZART-4 and WRF-Chem reproduced the vertical distribution and seasonal variation in tropospheric ozone (surface-200 hPa) at Delhi and Trivandrum with average modeled values falling within one standard deviation of the observations (Kumar et al. 2012b; Surendran et al. 2015). The vertical distribution of ozone retrieved by Tropospheric Emission Spectrometer (TES) is also reproduced within ± 15 ppbv by the WRF-Chem model (Kumar et al. 2012b). The shortcoming in different models over South Asia can be attributed to use of different emission inventories, horizontal grid spacing used by different models and differences in their physical/chemical parameterizations. The WRF-Chem model captured the seasonal cycle and vertical distribution of MOPITT retrieved CO fairly well over different parts of South Asia but could not reproduce the seasonal cycle of NO_x very well (Kumar et al. 2012b). The global models (MOZART and EMAC) are also reported to capture the MOPITT observed variability in tropospheric CO over South Asia (Chandra et al. 2016).

Similar to trace gases, many studies have attempted to evaluate the spatial and temporal distribution of aerosols simulated by different global and regional models but most of these studies have noted large discrepancies between modeled and observed aerosol concentrations. BC and dust concentrations are underestimated by a factor of 2–4 by the Laboratoire de Météorologie Dynamique (LMDZT) model, however, the LMDZT model agreed better with observations for OC and sea-salt (Reddy et al. 2004). The atmospheric component of the Geophysical Fluid Dynamics Laboratory (GFDL-AM) underestimated the AOD over the IGP by a factor of 6 and BC mass concentrations by a factor of 10 (Ganguly et al. 2009). WRF-Chem and SPIRNTRAS models underestimated AOD and BC mass concentrations by a factor of 2 or more (Govardhan et al. 2015). Similar discrepancies have also been reported for Regional Climate Model system (RegCM4), CHIMERE, Goddard Chemistry Aerosol Radiation and Transport (GOCART) and seven other chemical transport models (Nair et al. 2012; Moorthy et al. 2013; Pan et al. 2015). The shortcomings in different models has been attributed mainly to uncertainties in emission estimates of different aerosols, errors in the simulations of key meteorological parameters such as temperature and relative humidity, and other model deficiencies such as heterogeneous/aqueous chemistry and secondary organic aerosol formation pathways. Efforts have been made to improve the emission estimates of black carbon aerosols by incorporating the missing emission sectors such as emissions from residential heating and cooking (Venkatraman et al. 2006) and develop emissions at high resolution of $0.1° \times 0.1°$. These latest emission inventories when combined with WRF-Chem have significantly reduced biases in the model simulation (Kumar et al. 2015a, b). Models also tend to under predict dust emissions (e.g. Cherian et al. 2012) and studies have also focused on tuning the dust emission schemes for South Asia (Kumar et al. 2014a).

2.4 Summary of Air Quality Modeling Studies in South Asia

The evaluation efforts clearly revealed discrepancies between various models and observations but different global/regional models also captured the spatial and temporal variability of trace species over India which in turn provides some confidence in using these models for scientific investigations. Several studies have focused on understanding different processes that control air quality in South Asia and these are summarized below.

2.4.1 Seasonality of Black Carbon Aerosols

The seasonal variation in black carbon (BC) aerosols over any region is determined by seasonal changes in emission sources and meteorological conditions. Anthropogenic emissions of BC over South Asia are highest during winter and lowest during summer but their seasonal change is estimated to be small (less than 1%). Biomass burning emissions, in contrast, show a large seasonal variability and a peak during spring. However, aside from a few regions in South Asia, their magnitude is about three times smaller than the anthropogenic emissions (Kumar et al. 2015b). Thus, seasonal variability in emission sources does not explain the large seasonal variability observed in BC aerosols over South Asia. To identify the meteorological processes that play important roles in controlling the BC variability in South Asia, Kumar et al. (2015b) examined the sensitivity of time rate change of black carbon mass mixing ratios (μg kg^{-1} day^{-1}) due to horizontal and vertical advection, convective transport and vertical mixing (Fig. 2.2).

At the surface, all meteorological processes act as a sink of BC except for horizontal advection during the summer monsoon season. Vertical mixing that includes contributions from both dry deposition and vertical diffusion is the main process removing BC from the surface layer. Dry deposition contributes $38 \pm 9\%$, $21 \pm 2\%$ and $24 \pm 6\%$ during the winter monsoon (WM), summer monsoon (SM) and monsoon transition (MT) seasons respectively, while the vertical diffusion contributes $62 \pm 9\%$, $79 \pm 2\%$, and $76 \pm 6\%$, respectively to the surface layer vertical mixing term.

BC lofted from the surface layer by vertical mixing enters into the lower troposphere throughout the year. A competition between horizontal and vertical advection determines whether BC particles from the lower troposphere are transported horizontally or vertically. Horizontal advection is much stronger than the vertical advection during the winter season and thus transports BC out of the Indian subcontinent, while vertical advection dominates during summer monsoon season and facilitates transport of BC to the middle and upper troposphere. A stronger convective transport during summer monsoon season further amplifies the vertical transport of BC. Horizontal and vertical advection has opposite effects on BC in the middle and upper troposphere. Horizontal advection dominates in the middle troposphere during winter and monsoon transition seasons while vertical advection dominates during the summer monsoon season. This analysis showed that the observed summertime minimum in near surface BC mass concentration over the Indian subcontinent is caused by reduced horizontal transport coupled with a stronger vertical transport.

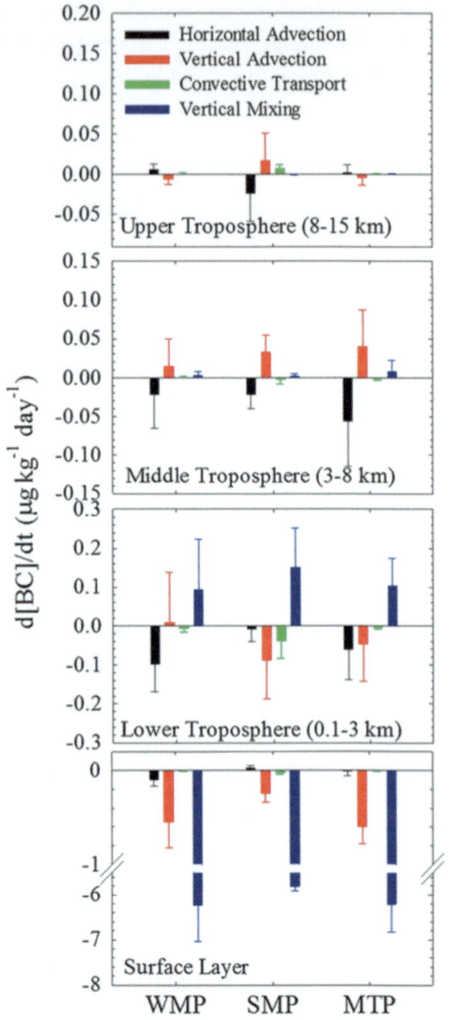

Fig. 2.2 Seasonal mean BC tendencies due to different meteorological processes at the surface and different tropospheric layers over the Indian region (68°–91°E, 7°–35°N). Vertical bars represent seasonal variability in mean tendency values. WMP, SMP and MTP stand for winter monsoon (November–March), summer monsoon (June–September) and monsoon transition (April–May and October) periods (Adapted from Kumar et al. 2015b)

2.4.2 Quantifying Sources of Air Pollution

Source contribution analysis has been conducted to identify the most important sources of BC throughout the year (Kumar et al. 2015b) and wintertime carbon monoxide (Kumar et al. 2013) by tracking BC and CO emissions by source type and region, domain boundaries and, in the case of CO, photochemical production. The ratios of chemical species have been used to understand whether surface ozone production is limited by volatile organic compounds (VOCs) or by nitrogen oxides (NO_x) (Kumar et al. 2012b). This section provides a summary of results from these studies. It is found that anthropogenic emissions are the most important sources of both BC and wintertime CO over South Asia as they contribute more than about

85% to total BC mass concentrations and more than 60% to surface CO mixing ratios. The contribution of biomass burning emissions is generally less than 15%. Long-range transport from sources located outside South Asia contribute little (2–3%) to surface BC but can contribute up to 35–40% to surface CO mixing ratios. The contribution of biogenic emissions and photochemical production from VOC emissions in South Asia to surface CO mixing ratios is estimated to be less than 10%.

The seasonal changes in regional meteorological circulation over South Asia play an important role in distributing the air pollutants among different parts of South Asia. The spatial distribution of BC emitted from anthropogenic sources in North, West, East and South India during winter, summer and monsoon transition seasons at the surface show that BC is transported from northern to southern parts of India during WM and MT seasons, and from southern to northern parts during the SM season (Fig. 2.3), which is in agreement with transport pathways described in Lawrence and Lelieveld (2010).

Because of the transport pathways, BC emitted from North India affects all other parts of India during winter season and South India plays a similar role during summer season. BC emitted from West India travels mostly to the Arabian Sea and some parts of South and East India during WM season and to North and East India during SM season. BC emitted from East India is transported to the Bay of Bengal, South India and Arabian Sea during WM season and mostly to North India during SM season. BC emitted from South India is transported mainly to the southern Arabian Sea during WM season and to the Bay of Bengal, East and North India during SM season. BC emitted from different regions of India during MT season follow transport pathways similar to winter season except that the spatial extent of their influence is smaller compared to the winter likely because of slower winds. However, BC emitted from South India shows a stronger impact in the Bay of Bengal during MT periods.

The contributions of local and regional anthropogenic emissions to total anthropogenic BC loadings in North, West, East and South India are estimated to quantify the contribution of regional transport to BC loadings in different regions of India. The amount of BC due to sources located in a given region itself (e.g. North India) is defined as the contribution from local sources, and BC coming from sources outside this region (e.g. sum of BC coming from West, East and South India and other regions for North India) is defined as the contribution from the regional sources. During winter and monsoon transition periods, local sources contribute about 94–96% to the anthropogenic BC loading in North India, while North India emissions contribute 68–81% to BC loading in West, East and South India indicating a significant contribution from regional transport (19–32%).

In contrast, during summer monsoon local sources dominate anthropogenic BC loadings in South India with a contribution of about 95%, while regional sources become important for North India with a contribution of about 28%. Regional sources remain important for West and East India during summer monsoon with a contribution of about 26–28%. The contribution of regional sources to wintertime surface CO mixing ratios are even higher than those estimated for BC (Kumar et al.

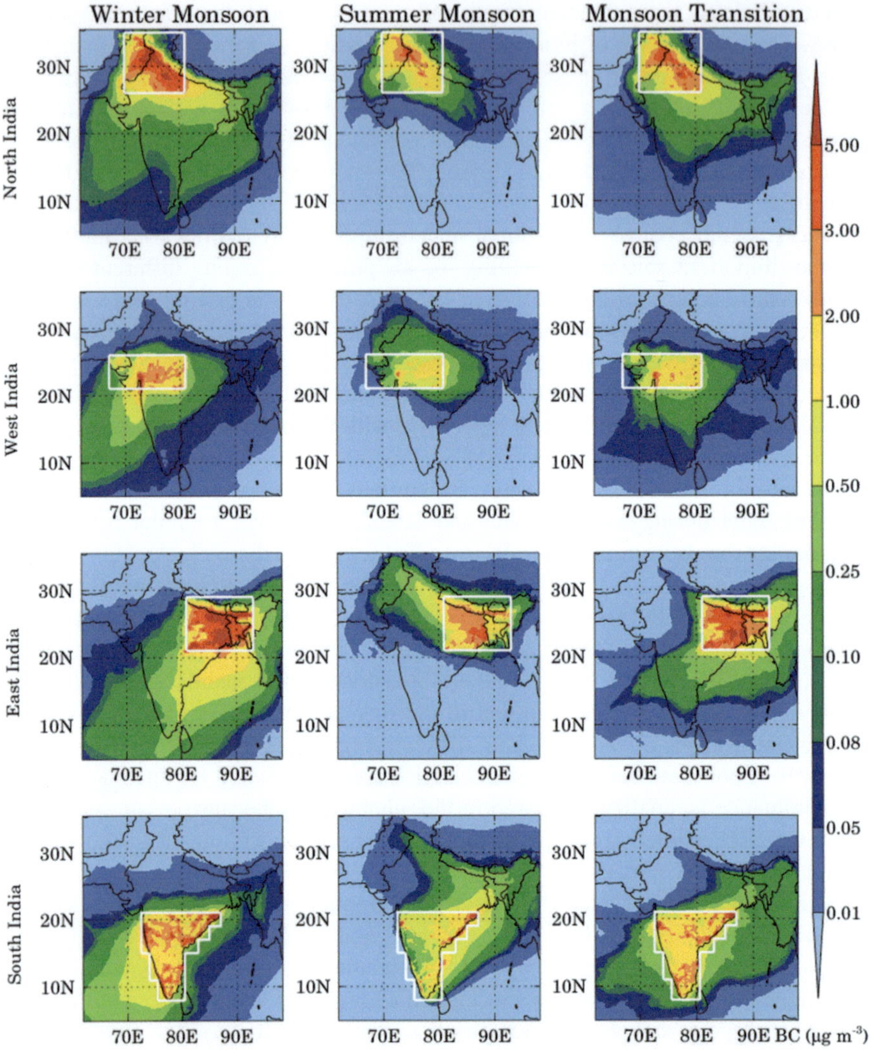

Fig. 2.3 Spatial distribution of anthropogenic BC concentrations (μg/m3) emitted from North, West, East and South India during the WM, SM and MT seasons at the surface. White solid lines mark the geographical boundaries of different regions (Adapted from Kumar et al. 2015b)

2013) because CO has a longer lifetime than BC as CO is not susceptible to dry and wet deposition and its chemical lifetime is typically 2–3 months. Local sources account for most of the wintertime surface CO in North (92%) and East (83%) India, but regional transport makes a significant contribution in South India (41%) and even contributes (58%) more than the local emissions in West India.

To understand whether surface ozone production in India is controlled by NO_x or VOCs, the modeled afternoon ratios of formaldehyde (CH_2O) to NO_y (Fig. 2.4), hydrogen peroxide (H_2O_2) to nitric acid (HNO_3) and ozone to (NO_y – NO_x) were analyzed (Kumar et al. 2012b). The critical values of the ratios CH_2O/NO_y, H_2O_2/HNO_3 and $O_3/(NO_y\text{-}NO_x)$ separating the two ozone production regimes are suggested to be 0.28, 0.4 and 7, respectively with lower values indicating a VOC-limited regime, while higher values correspond to a NO_x-sensitive regime (Sillman 1995).

The CH_2O/NO_y ratio is less than 0.28 over some parts of the IGP during winter, indicating VOC-limited ozone production regime over this region. The rest of the Indian region is NO_x-limited throughout the year. Interestingly, the ratio is seen to be lower over the shipping routes in the Arabian Sea and Indian Ocean, reflecting the critical role of shipping NO_x emissions in ozone production over the cleaner marine regions. The H_2O_2/HNO_3 ratio is estimated to be less than 0.4 only in a few grid cells over the IGP region during October, and the $O_3/(NO_y\text{-}NO_x)$ is estimated to be greater than 7 over the region for all seasons. These results confirm the dominance of a NO_x-limited ozone production regime over India. NO_x-limited

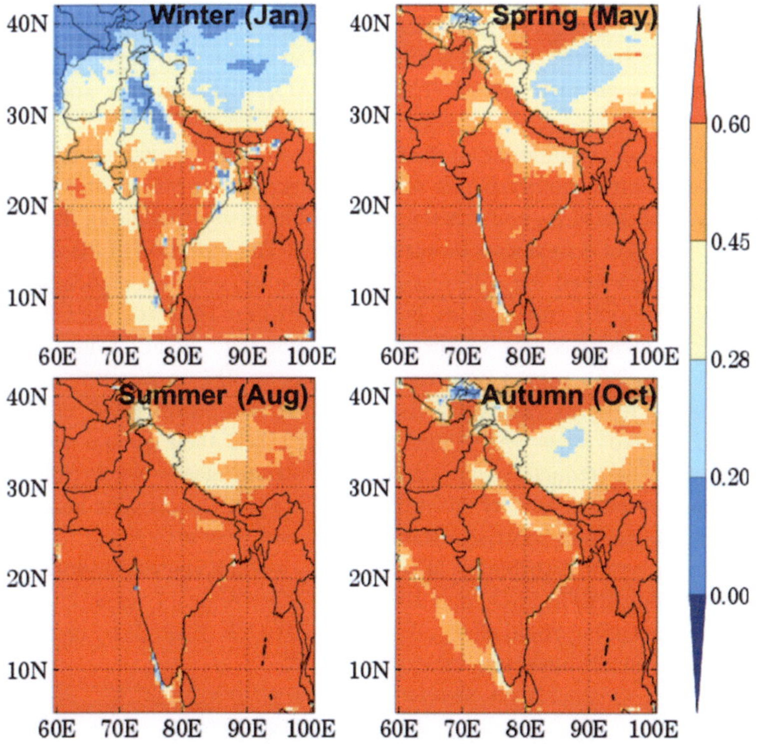

Fig. 2.4 WRF-Chem simulated afternoon (11:30–14:30 LT) CH_2O to NO_y ratios (dimensionless) during January, April, August and October of the year 2008

ozone production over South Asia might be associated with the fact that emissions in this region are influenced largely by incomplete combustion processes, particularly by biofuel burning, and thus have higher NMHC to NO_x emission ratio as compared to other regions of the Northern Hemisphere (Lawrence and Lelieveld 2010). This is also corroborated by previous studies that analyzed observed ozone-CO and ozone-NO_x correlation over some of the Indian sites to suggest that the emissions of ozone precursors and thus ozone levels are largely determined by incomplete combustion processes (Naja and Lal 2002; Naja et al. 2003).

2.4.3 Dust Aerosols and Tropospheric Chemistry

Dust storms occur frequently in South Asia during the pre-monsoon season (Prasad and Singh 2007) and significantly affect the regional aerosol loadings as well as tropospheric chemistry. Kumar et al. (2014a) updated the dust emission scheme in WRF-Chem to capture dust-storm induced changes on regional aerosol loadings over northern India for a typical dust storm that occurred during 17–22 April 2010. With those updates, WRF-Chem successfully simulated the temporal variations in Aerosol Robotic Network (AERONET) retrieved aerosol optical depth and Angstrom exponent at seven sites in South Asia. The model also captured the spatial distribution of satellite retrieved changes in AOD between low and high dust periods.

The WRF-Chem simulations of dust aerosols conducted by Kumar et al. (2014b) included an updated photolysis rate scheme to account for effects of dust aerosols on photolysis frequencies and a heterogeneous chemistry module in the model to assess the effects of dust storms on surface ozone. The inclusion of these processes in WRF-Chem reduced the difference between observed and modeled O_3 from 16 ± 9 to 2 ± 8 ppbv and that in NO_y from 2129 ± 1425 to 372 ± 1225 pptv compared to measurements at the high altitude site Nainital in the central Himalayas (Fig. 2.5). Biases were reduced by up to 30% in tropospheric column NO_2 compared to OMI retrievals. The simulated dust storm acted as a sink for ozone and related gases and significantly perturbed their spatial and vertical distributions. The reductions in ozone and related gases were estimated as 5–100% and more than 80% of this reduction was due to heterogeneous chemistry. The RH dependence of reactive uptake coefficients was also found to have substantial impact on the distribution of trace gases, with changes of up to 20–25% in ozone and HO_2, 50% in H_2O_2 and 100% in HNO_3. A set of sensitivity analyses showed that dust aging could change H_2O_2 and CH_3COOH levels by up to 50% but has a relatively small impact on other gases.

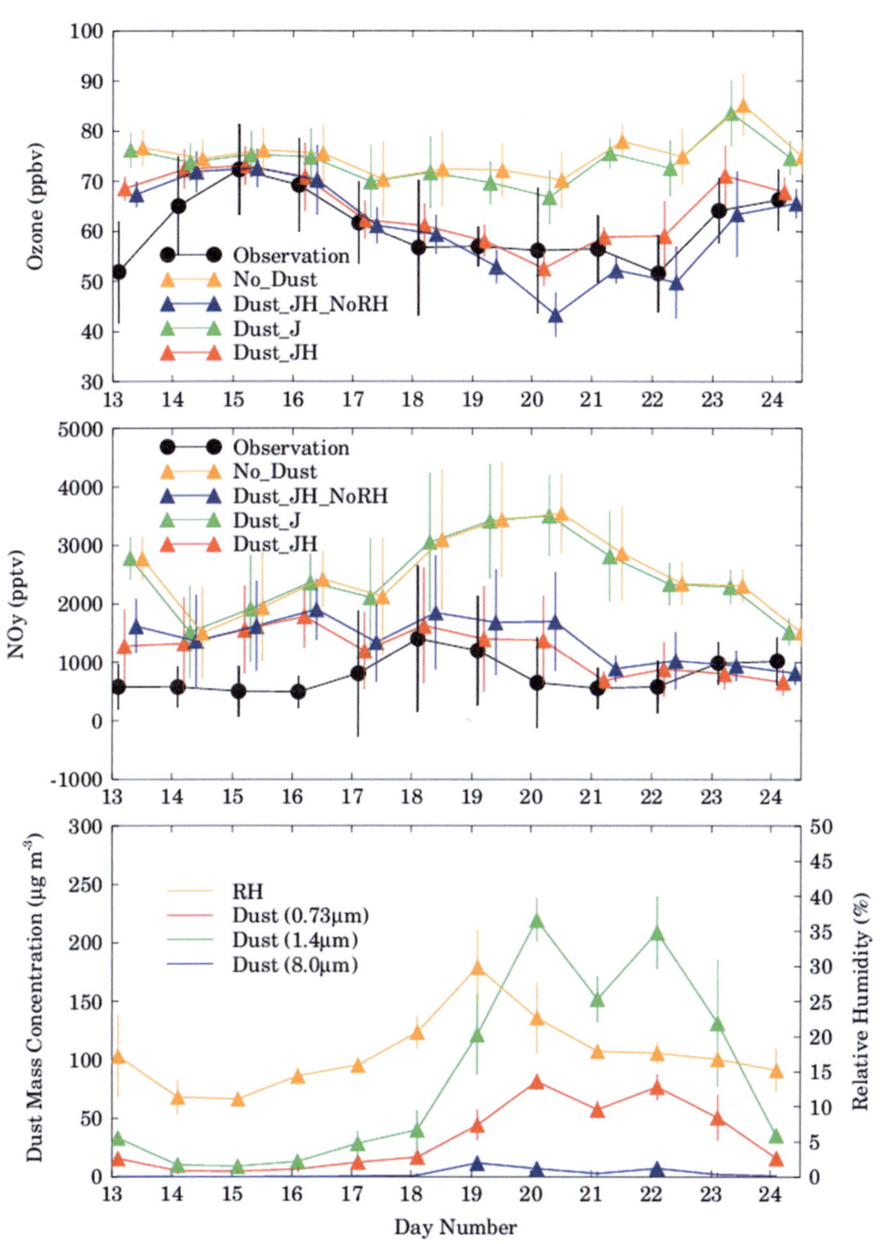

Fig. 2.5 Variations in observed and WRF-Chem simulated daily average O_3 (*top panel*) and NO_y (*middle panel*) at Nainital during 13–24 April 2010. WRF-Chem simulated daily average mass concentration of dust particles of 0.73, 1.4 and 8.0 μm effective radii and relative humidity at Nainital are also shown. The *vertical bars* represent standard deviation in the average values. Dust_JH represents WRF-Chem simulation with both photolysis and heterogeneous chemistry with RH dependence of uptake coefficients included. Dust_JH_NoRH is same as Dust_JH but without RH dependence of uptake coefficients and DUST-J includes only photolysis effects of dust (Adapted from Kumar et al. 2014b)

2.4.4 Past and Future Changes in Air Quality

In addition to process studies mentioned above, the chemistry transport models have also been used to understand how changes in anthropogenic emissions from past to present-day conditions and how future air quality in South Asia will respond to changes in climate and anthropogenic emissions. Beig and Brasseur (2006) examined changes in surface ozone due to changes in anthropogenic emissions between 1991 and 2001 and estimated that the decadal increase in CO (10–18%) and NO_x (20–50%) levels led to an increase of 2–9 ppbv (5–30%) in boundary layer ozone. Wild et al. (2012) developed a parameterization to estimate regionally averaged changes in surface ozone over different parts of the world and estimated that future surface ozone increases over South Asia under all Representative Concentration Pathways (RCPs) with increases of more than 5 ppbv under RCP8.5 and close to zero under RCP6.0. Chatani et al. (2014) estimated that surface ozone will increase in South Asia throughout the year by the year 2030 if the emissions follow a business as usual scenario.

2.4.5 Effect of Present-Day Air Pollution on Human Health and Food Agriculture

A few studies have also attempted to estimate the effect of present-day air pollution on human health and food security in South Asia. It is estimated that exposure to high levels of ozone adversely impact cotton, soybeans, rice and wheat crops especially in northern and central India with wheat being the most affected crop in terms of losses at 3.5 ± 0.8 million tons (Mt), followed by rice at 2.1 ± 0.8 Mt (Fig. 2.6; Ghude et al. (2014)).

They further estimated that surface ozone induced crop yield losses account for about 9% of the cereals required every year (61.2 Mt) under the Indian government's food security bill (2013) which is sufficient to feed about 94 million people living below the poverty line in India. Sinha et al. (2015) estimated surface ozone induced crop production loss to be 10–20 Mt. for wheat and 3–5 Mt. for rice in north Indian states of Haryana and Punjab for the years 2011–2013, which turns into total economic cost losses of $4–7 million. Van Dingenen et al. (2009) estimated that surface ozone induced losses in crop yields translate into an economic global loss of $14–$26 billion, 40% of which occurs in India and China.

Air pollution is ranked as the fifth largest killer in India according to the Global Burden of Diseases (GBD) estimates (Forouzanfar 2015). India is ranked second among the countries affected globally by $PM_{2.5}$ (Silva et al. 2013), and ozone (Lelieveld et al. 2015) induced respiratory mortalities. Ghude et al. (2016) combined the latest population estimates with ozone and $PM_{2.5}$ simulations from WRF-Chem to estimate exposure to air pollution on local and regional scales in India. They estimate that $PM_{2.5}$ exposure leads to about 570,000 premature

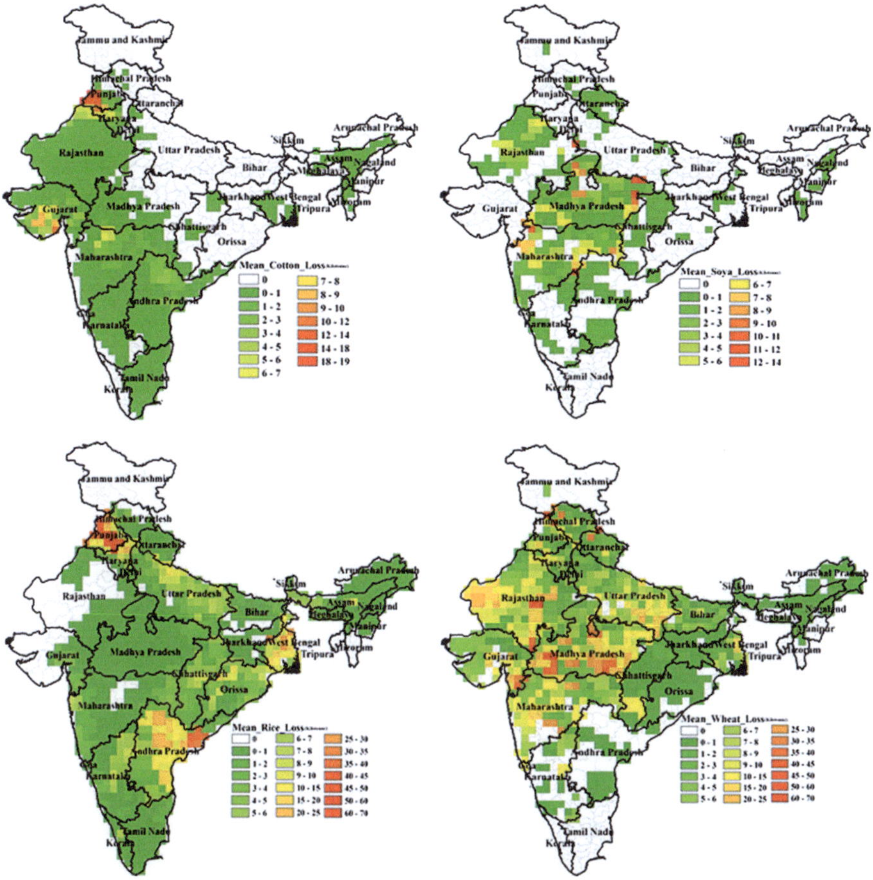

Fig. 2.6 Average O_3-induced crop production loss from AOT40 metrics for cotton, soybean, rice, and wheat during 2005. The production loss numbers are given in kilotons/grid box (Adapted from Ghude et al. 2014)

mortalities in 2011 and COPD mortalities due to ozone exposure is about 12,000 people.

2.5 Summary and Outlook

Rapid growth in population and economy over the past few decades has significantly elevated air pollution levels in South Asia. Recent studies have shown that air pollution is adversely affecting human health and food production in South Asia. In this chapter, we briefly discussed the evolution of surface ozone and aerosol optical depth monitoring networks over South Asia and how those measurements have been used to evaluate different global and regional climate models. Other parts of the chapter discussed a number of scientific questions addressed recently using global/regional models. Key results are listed here:

- The seasonal variations in near surface black carbon mass concentrations are driven mainly by seasonal changes in meteorology and not by seasonal variability in emissions.
- Long-range transport from sources located outside of South Asia contribute significantly to carbon monoxide levels in South Asia but not to black carbon aerosols.
- Regional transport among different regions of India contribute significantly to both CO and black carbon levels in other regions of India and should be considered in design of mitigation strategies.
- Surface ozone production in South Asia is mostly limited by NO_x emissions.
- Dust storms degrade the air quality by increasing the particulate matter loadings but act as a sink for key trace gases such as ozone.
- Surface ozone over South Asia has increased by about 5–30% since 1990 in South Asia and is expected to continue to increase in the future.
- Exposure to higher levels of ozone is significantly reducing wheat and rice production, and shortening the lifespan of people living in South Asia.

Chemical transport models have been very useful in advancing our understanding of air pollution in South Asia but further efforts are required to reduce uncertainties in model simulations. We compiled a list of tasks that can be undertaken in next 5 years.

- The dynamic economic growth of South Asia means rapid changes in the emission of air pollutants. Thus, bottom-top emission estimates usually lag behind a few years and are outdated by the time they become available. To address this limitation, satellite retrievals should be combined with bottom-up emission inventories in an inverse modeling framework for more timely inventories.
- The uncertainties in CTM simulated meteorology and NOx emissions are suggested to introduce errors in tropospheric ozone of about 10–25% and 5–40%, respectively (Kumar et al. 2012a; Jena et al. 2015). The contribution of other processes (uncertainties in VOC emissions and speciation, background ozone, model transport, chemistry and horizontal and vertical grid spacing) to the bias in tropospheric ozone over South Asia has not been explored in detail so far and should be focused in future studies. Detailed measurements of VOCs that have recently started in South Asia (e.g. Sinha et al. 2014; Sahu et al. 2016) will prove very useful in this regard. Aerosol chemical composition measurements should be used to evaluate the aerosol modules in state-of-the-art chemistry transport models. In addition, the coverage of trace gases and aerosol monitoring networks should also be expanded. The effects of air pollution on human health and crop productivity in South Asia are quantified using empirical relationships based on experiments conducted in North America and Europe except in Sinha

et al. (2015). We recommend developing these relationships for South Asia and reassessing the effects of air pollution on human health and crop productivity.
- Most of the studies focused on future air quality in South Asia either employed coarse resolution global models or offline regional models. We suggest that such future studies should use high resolution online chemistry-climate models in order to better resolve the growing urban sources in South Asia and simultaneously capture chemistry-climate interactions.
- Future studies should also explore how government policies will affect future air quality in South Asia. In particular, the effects of proposed energy policies, such as boosting power production by solar energy and increased usage of cleaner fuels (via subsidized Liquefied Petroleum Gas for people below poverty line), on emissions distribution and air quality should be quantified.

References

Ali, K., et al. (2009). Sink mechanism for significantly low level of ozone over the Arabian Sea. *Journal of Geophysical Research, 114*, D17306. (2009). doi:10.1029/2008JD011256.

Beig, G., & Brasseur, G. P. (2006). Influence of anthropogenic emissions on tropospheric ozone and its precursors over the Indian tropical region during a monsoon. *Geophysical Research Letters, 33*, L07808. doi:10.1029/2005GL024949.

Beig, G., Gunthe, S., & Jadhav, D. B. (2007). Simultaneous measurements of ozone and its precursors on a diurnal scale at a semi urban site in India. *Jounal of Atmospheric Chemistry, 57*: 239. doi:10.1007/s10874-007-9068-8.

Chandra, N., et al. (2016). Effects of convection and long-range transport on the distribution of carbon monoxide in the troposphere over India. *Atmospheric Pollution Research, 7*(5), 775–875.

Chatani, S., et al. (2014). Photochemical roles of rapid economic growth and potential abatement strategies on tropospheric ozone over South and East Asia in 2030. *Atmospheric Chemistry and Physics, 14*, 9259–9277. doi:10.5194/acp-14-9259-2014.

Cherian, R., Venkataraman, C., Ramachandran, S., Quaas, J., & Kedia, S. (2012). Examination of aerosol distributions and radiative effects over the Bay of Bengal and the Arabian Sea region during ICARB using satellite data and a general circulation model. *Atmospheric Chemistry and Physics, 12*, 1287–1305. doi:10.5194/acp-12-1287-2012.

Dave, J. V. (1957). Studies on twilight and atmospheric ozone PhD thesis. pp. III.1–III.29, *Gujrat University at Ahmedabad*, India.

Dey, S., & Di Girolamo, L. (2011). A decade of change in aerosol properties over the Indian subcontinent. *Geophysical Research Letters, 38*, L14811. doi:10.1029/2011GL048153.

Di Girolamo, L., et al. (2004). Analysis of Multi-angle Imaging SpectroRadiometer (MISR) aerosol optical depths over greater India during winter 2001–2004. *Geophysical Research Letters, 31*, L23115. doi:10.1029/2004GL021273.

Engardt, M. (2008). Modelling of near-surface ozone over South Asia. *Journal of Atmospheric Chemistry, 59*, 61–80. doi:10.1007/s10874-008-9096-z.

Forouzanfar, M. H., et al. (2015, December 5) Global, regional, and national comparative risk assessment of 79 behavioral, environmental and occupational, and metabolic risks or clusters of risks in 188 countries, 1990–2013: A systematic analysis for the Global Burden of Disease Study 2013. *The Lancet, 386*(10010):2287–2323.

Ganguly, D., et al. (2009). Inferring the composition and concentration of aerosols by combining AERONET and MPLNET data: Comparison with other measurements and utilization to evaluate GCM output. *Journal of Geophysical Research, 114*, D16203. doi:10.1029/2009JD011895.

Ganguly, D., Rasch, P. J., Wang, H., & Yoon, J. H. (2012). Climate response of the South Asian monsoon system to anthropogenic aerosols. *Journal of Geophysical Research, 117*, D13209. doi:10.1029/2012JD017508.

Ghude, S. D., Fadnavis, S., Beig, G., Polade, S. D., & van der A, R. J. (2008). Detection of surface emission hot spots, trends, and seasonal cycle from satellite-retrieved NO over India. *Journal of Geophysical Research, 113*, D20305. doi:10.1029/2007JD009615.

Ghude, S. D., et al. (2009). Satellite derived trends in NO_2 over the major global hotspot regions during the past decade and their inter-comparison. *Environmental Pollution, 157*(6), 1873–1878.

Ghude, S. D., et al. (2011). Regional CO pollution over the Indian-subcontinent and various transport pathways as observed by MOPITT. *International Journal of Remote Sensing, 32*(21), 6133–6148.

Ghude, S. D., et al. (2012). Satellite constraints of nitrogen oxide (NO_x) emissions from India based on OMI observations and WRF-Chem simulations. *Geophysical Research Letters, 40*, 423–428. doi:10.1029/2012GL053926.

Ghude, S. D., et al. (2013). Application of satellite observations for identifying regions of dominant sources of nitrogen oxides over the Indian Subcontinent. *Journal of Geophysical Research – Atmospheres, 118*, 1075–1089. doi:10.1029/2012JD017811.

Ghude, S. D., et al. (2014). Reductions in India's crop yield due to ozone. *Geophysical Research Letters, 41*, 5685–5691. doi:10.1002/2014GL060930.

Ghude, S. D., et al. (2016). Premature mortality in India due to $PM_{2.5}$ and ozone exposure. *Geophysical Research Letters, 43*, 4650–4658. doi:10.1002/2016GL068949.

Giles, D. M., et al. (2011). Aerosol properties over the Indo-Gangetic Plain: A mesoscale perspective from the TIGERZ experiment. *Journal of Geophysical Research, 116*, D18203. doi:10.1029/2011JD015809.

Girach, I. A., & Nair, P. R. (2014). Carbon monoxide over Indian region as observed by MOPITT. *Atmospheric Environment, 99*, 599–609.

Govardhan, G., Nanjundiah, R. S., Satheesh, S. K., Moorthy, K. K., & Kotamarthi, V. R. (2015). Performance of WRF-Chem over Indian region: Comparison with measurements. *Journal of Earth System Science, 124*(4), 875–896. doi:10.1007/s12040-015-0576-7.

Greenstone, M., et al. (2015). Lower pollution, longer lives. *Economic and Political Weekly, 8*, 40–46.

Holben, B. N., et al. (1998). AERONET-a federated instrument network and data archive for aerosol characterization. *Remote Sensing of Environment, 66*, 1–16.

Jena, C., et al. (2015). Inter-comparison of different NOx emission inventories and associated variations in simulated surface ozone in Indian region. *Atmospheric Environment, 117*, 61–73. doi:10.1016/j.atmosenv.2015.06.057.

Jethva, H., Satheesh, S. K., & Srinivasan, J. (2005). Seasonal variability of aerosols over the Indo-Gangetic basin. *Journal of Geophysical Research, 110*, D21204. doi:10.1029/2005JD005938.

Kumar, R., Naja, M., Venkataramani, V., & Wild, O. (2010). Variations in surface ozone at Nainital: A high-altitude site in the central Himalayas. *Journal of Geophysical Research, 115*, D16302. doi:10.1029/2009JD013715.

Kumar, R., Naja, M., Pfister, G. G., Barth, M. C., Wiedinmyer, C., & Brasseur, G. P. (2012a). Simulations over South Asia using the Weather Research and Forecasting Model with Chemistry (WRF-Chem): Chemistry evaluation and initial results. *Geoscientific Model Development, 5*, 619–648. doi:10.5194/gmd-5-619-2012.

Kumar, R., Naja, M., Pfister, G. G., Barth, M. C., & Brasseur, G. P. (2012b). Simulations over South Asia using the Weather Research and Forecasting model with Chemistry (WRF-Chem): Set-up and meteorological evaluation. *Geoscientific Model Development, 5*, 321–343. doi:10.5194/gmd-5-321-2012.

Kumar, R., et al. (2013). Source attribution of carbon monoxide in India and surrounding regions during wintertime. *Journal of Geophysical Research – Atmospheres, 118*, 1981–1995. doi:10. 1002/jgrd.50134.

Kumar, R., Barth, M. C., Pfister, G. G., Naja, M., & Brasseur, G. P. (2014a). WRF-Chem simulations of a typical pre-monsoon dust storm in northern India: Influences on aerosol optical properties and radiation budget. *Atmospheric Chemistry and Physics, 14*, 2431–2446. doi:10.5194/acp-14-2431-2014.

Kumar, R., et al. (2014b). Effects of dust aerosols on tropospheric chemistry during a typical pre-monsoon season dust storm in northern India. *Atmospheric Chemistry and Physics, 14*, 6813–6834. doi:10.5194/acp-14-6813-2014.

Kumar, R., et al. (2015a). Sources of black carbon aerosols in South Asia and surrounding regions during the Integrated Campaign for Aerosols, Gases and Radiation Budget (ICARB). *Atmospheric Chemistry and Physics, 15*, 5415–5428. doi:10.5194/acp-15-5415-2015.

Kumar, R., et al. (2015b). What controls the seasonal cycle of black carbon aerosols in India? *Journal of Geophysical Research – Atmospheres, 120*, 7788–7812. doi:10.1002/ 2015JD023298.

Lal, S. (2007). Trace gases over the Indian region. *Indian Journal of Radio & Space Physics, 36*, 556–570.

Lal, S., Naja, M., & Subbaraya, B. H. (2000). Seasonal variations in surface ozone and its precursors over an urban site in India. *Atmospheric Environment, 34*, 2713–2724.

Lau, K. M., Kim, M. K., & Kim, K. M. (2006). Asian summer monsoon anomalies induced by aerosol direct forcing: The role of the Tibetan Plateau. *Climate Dynamics, 26*, 855–864. doi:10. 1007/s00382- 006-0114-z.

Lawrence, M. G., & Lelieveld, J. (2010). Atmospheric pollutant outflow from southern Asia: A review. *Atmospheric Chemistry and Physics, 10*, 11017–11096. doi:10.5194/acp-10-11017-2010.

Lawrence, M. G., et al. (2003). Global chemical weather forecasts for field campaign planning: predictions and observations of large-scale features during MINOS, CONTRACE, and INDOEX. *Atmospheric Chemistry and Physics, 3*, 267–289. doi:10.5194/acp-3-267-2003.

Lelieveld, J., et al. (2001). The Indian Ocean experiment: Widespread air pollution from South and Southeast Asia. *Science, 291*, 1031–1036.

Lelieveld, J., Evans, J. S., Fnais, M., Giannadak, D., & Pozzer, A. (2015). The contribution of outdoor air pollution sources to premature mortality on a global scale. *Nature, 525*(7569), 367–371. doi:10.1038/nature15371.

Mahajan, A. S., et al. (2015). Inter-annual variations in satellite observations of nitrogen dioxide and formaldehyde over India. *Atmospheric Environment, 116*, 194–201. doi:10.1016/j. atmosenv.2015.06.004.

Menon, S., Koch, D., Beig, G., Sahu, S., Fasullo, J., & Orlikowski, D. (2010). Black carbon aerosols and the third polar ice cap. *Atmospheric Chemistry and Physics, 10*, 4559–4571. doi:10.5194/acp-10-4559-2010.

Mittal, M. L., et al. (2007). Surface ozone in the Indian region. *Atmospheric Environment, 41*, 6572–6584. doi:10.1016/j.atmosenv.2007.04.035.

Moorthy, K. K., Satheesh, S. K., & Babu, S. S. (2008). Integrated Campaign for Aerosols, Gases and Radiation Budget (ICARB): An overview. *Journal of Earth System Science, 117*, 243–262.

Moorthy, K. K., et al. (2009). Climate implications of atmospheric aerosols and trace gases: Indian Scenario. WCC-3 Climate Sense 3 Climate Sense. Ed GR Asrar. *World Meteorological Organization*, Tudor Rose, UK (pp. 157–160). ISBN 978-92-63-11403-5.

Moorthy, K. K., et al. (2013). Performance evaluation of chemistry transport models over India. *Atmospheric Environment, 71*, 210–225. doi:10.1016/j.atmosenv.2013.01.056.

Moorthy, K. K., Sathessh, S. K., & Kotamarthi, V. R. (2016). Evolution of aerosol research in India and the RAWEX-GVAX: An overview. *Current Science, 11*(1), 53–75.

Nair, V. S., Solmon, F., Giorgi, F., Mariotti, L., Babu, S. S., Moorthy, K., & K. (2012). Simulation of South Asian aerosols for regional climate studies. *Journal of Geophysical Research, 117*, D04209. doi:10.1029/2011JD016711.

Nair, V. S., Babu, S. S., Moorthy, K. K., Sharma, A. K., & Ajai, A. M. (2013). Black carbon aerosols over the Himalayas: Direct and surface albedo forcing. *Tellus B., 65*, 19738. doi:10.3402/tellusb.v65i0.19738.

Naja, M. (1997). Tropospheric chemistry in the Tropics. PhD Thesis, *Gujarat University*. pp. 1–186.

Naja, M. Lal, S. (2002). Surface ozone and precursor gases at Gadanki (13.5° N, 79.2° E), tropical rural site in India. *Journal of Geophysical Research 107*, 10.1029/2001JD000357.

Naja, M., Lal, S., & Chand, D. (2003). Diurnal and seasonal variabilities in surface ozone at a high altitude site Mt Abu (24.6° N, 72.7° E, 1680 m asl) in India. *Atmospheric Environment, 37*, 4205–4215.

Ojha, N., et al. (2012). Variabilities in ozone at a semi-urban site in the Indo-Gangetic Plain region: Association with the meteorology and regional processes. *Journal of Geophysical Research, 117*, D20301. doi:10.1029/2012JD017716.

Pan, X., et al. (2015). A multi-model evaluation of aerosols over South Asia: Common problems and possible causes. *Atmospheric Chemistry and Physics, 15*, 5903–5928. doi:10.5194/acp-15-5903-2015.

Park, M., Randel, W. J., Gettelman, A., Massie, S. T., & Jiang, J. H. (2007). Transport above the Asian summer monsoon anticyclone inferred from Aura Microwave Limb Sounder tracers. *Journal of Geophysical Research, 112*, D16309. doi:10.1029/2006JD008294.

Prasad, A. K., & Singh, R. P. (2007). Changes in aerosol parameters during major dust storm events (2001–2005) over the Indo-Gangetic Plains using AERONET and MODIS data. *Journal of Geophysical Research, 112*, D09208. doi:10.1029/2006JD007778.

Ramachandran, S., Kedia, S., & Srivastava, R. (2012). Aerosol optical depth trends over different regions of India. *Atmospheric Environment, 49*, 338–347.

Ramanathan, K. R. (1956). Atmospheric ozone and the general circulation of the atmosphere. *Science Proceedings International Association of Meteorology* (pp. 3–24). Rome: IUGG 10th General Assembly.

Ramanathan, V., et al. (2001). Indian Ocean Experiment: An integrated analysis of the climate forcing and effects of the great Indo-Asian haze. *Journal of Geophysical Research, 106*(D22), 28371–28398. doi:10.1029/2001JD900133.

Ramanathan, V., et al. (2005). Atmospheric brown clouds: Impacts on South Asian climate and hydrological cycle. *Proceedings National Academy Science USA, 102*, 5326–5333.

Reddy, M. S., et al. (2004). General circulation model estimates of aerosol transport and radiative forcing during the Indian Ocean Experiment. *Journal of Geophysical Research, 109*, D16205. doi:10.1029/2004JD004557.

Roy, S., Beig, G., & Jacob, D. J. (2008). Seasonal distribution of ozone and its precursors over the tropical Indian region using regional chemistry-transport model. *Journal of Geophysical Research, 113*, D21307. doi:10.1029/2007JD009712.

Sagar, R., Kumar, B., Dumka, U. C., Moorthy, K. K., & Pant, P. (2004). Characteristics of aerosol optical depths over Manora Peak: A high altitude station in the central Himalayas. *Journal of Geophysical Research, 109*, D06207. doi:10.1029/2003JD003954.

Sahu, L. K., Yadav, R., & Pal, D. (2016). Source identification of VOCs at an urban site of western India: Effect of marathon events and anthropogenic emissions. *Journal of Geophysical Research – Atmospheres, 121*, 2416–2433. doi:10.1002/2015JD024454.

Saraf, N., Beig, G., & Schultz, M. (2016). Tropospheric distribution of ozone and its precursors over the tropical Indian Ocean. *Journal of Geophysical Research* 108, D20, 4636 (2003), 10.1029/2003JD003521.

Sillman, S. (1995). The use of NOy, H2O2, and HNO3 as empirical indicators for ozone-NOx-hydrocarbon sensitivity in urban locations. *Journal of Geophysical Research, 100*, 14,175–14,188.

Silva, R. A., et al. (2013). Global premature mortality due to anthropogenic outdoor air pollution and the contribution of past climate change. *Environonment Research Letter, 8*, 034005. doi:10.1088/1748 9326/8/3/034005.

Sinha, V., Kumar, V., & Sarkar, C. (2014). Chemical composition of pre-monsoon air in the Indo-Gangetic plain measured using a new air quality facility and PTR-MS: High surface ozone and strong influence of biomass burning. *Atmospheric Chemistry and Physics, 14*, 5921–5941. doi:10.5194/acp-14-5921-2014.

Sinha, B., Singh Sangwan, K., Maurya, Y., Kumar, V., Sarkar, C., Chandra, B. P., & Sinha, V. (2015). Assessment of crop yield losses in Punjab and Haryana using 2 years of continuous in situ ozone measurements. *Atmospheric Chemistry and Physics, 15*, 9555–9576. doi:10.5194/acp-15-9555-2015.

Surendran, D. E., et al. (2015). Air quality simulation over South Asia using Hemispheric Transport of Air Pollution version-2 (HTAP-v2) emission inventory and Model for Ozone and Related chemical Tracers (MOZART-4). *Atmospheric Environment, 122*, 357–372. doi:10.1016/j.atmosenv.2015.08.023.

Van Dingenen, R., et al. (2009). The global impact of ozone on agricultural crop yields under current and future air quality legislation. *Atmospheric Environment, 43*(3), 604–618.

Venkataraman, C., et al. (2006). Emissions from open biomass burning in India: Integrating the inventory approach with high-resolution Moderate Resolution Imaging Spectroradiometer (MODIS) active-fire and land cover data. *Global Biogeochemical Cycles, 20*, GB2013. doi:10.1029/2005GB002547.

Wild, O., et al. (2012). Modelling future changes in surface ozone: A parameterized approach. *Atmospheric Chemistry and Physics, 12*, 2037–2054. doi:10.5194/acp-12-2037-2012.

World Health Organization (WHO). (2016). *WHO's urban ambient air pollution database – Update 2016*, pp. 1–7. Available online at http://www.who.int/phe/health_topics/outdoorair/databases/AAP_database_summary_results_2016_v02.pdf. Last accessed on 5 June 2017.

Yasunari, T. J., et al. (2010). Estimated impact of black carbon deposition during pre-monsoon season from Nepal climate observatory – pyramid data and snow albedo changes over Himalayan glaciers. *Atmospheric Chemistry and Physics, 10*, 6603–6615. doi:10.5194/acp-10-6603-2010.

Chapter 3
Sources and Chemical Composition of Particulate Matter During Haze Pollution Events in China

Ru-Jin Huang, Junji Cao, and Douglas R. Worsnop

Abstract Rapid industrialization and urbanization in China has led to an increase in severe haze pollution that is influencing air quality, regional and global climates, and human health. Understanding of the temporal and spatial profiles of chemical composition and sources of fine particulate matter (PM), the mechanisms and atmospheric processes of substantial wintertime secondary aerosol production and the constraint of sources and formation processes of secondary organic aerosol (SOA) during haze events are essential for the design of effective mitigation strategies. In this chapter, recent studies associated with the aforementioned aspects are summarized and discussed. Future studies directing to a better understanding of the chemical nature, formation mechanisms, atmospheric processes and sources of PM are also discussed.

Keywords Chemical composition • Source apportionment • Secondary aerosol • Particulate air pollution

3.1 Introduction

Particulate pollution in China is a serious environmental problem that is influencing air quality, regional and global climates and human health (Cao 2012; Wang et al. 2014). Especially the 2013-January large-scale and severe haze pollution that affected a total area of ~1.3 million km^2 and ~800 million people has brought China's particulate pollution to the forefront of world-wide media and evoked great scientific interest for air pollution studies. Measurements at a number of major

R.-J. Huang (✉) • J. Cao
Key Laboratory of Aerosol Chemistry and Physics, State Key Laboratory of Loess and Quaternary Geology, Institute of Earth Environment, Chinese Academy of Sciences, Xi'an 710061, China
e-mail: rujin.huang@ieecas.cn; cao@loess.llqg.ac.cn

D.R. Worsnop
Aerodyne Research, Inc., Billerica, MA, USA
e-mail: worsnop@aerodyne.com

I. Bouarar et al. (eds.), *Air Pollution in Eastern Asia: An Integrated Perspective*, ISSI Scientific Report Series 16, DOI 10.1007/978-3-319-59489-7_3

cities showed that the wintertime daily average mass concentrations of $PM_{2.5}$ (particulate matter with an aerodynamic diameter <2.5 μm) are approximately one to two orders of magnitude higher than those observed in urban areas in the US and European Countries (Cao 2012). Such acute particulate pollution is often accompanied by extremely poor visibility and air quality as well as a sharp increase in respiratory diseases. Long-term exposure to high levels of particulate pollution is estimated to result in 1.36 million premature deaths per year in China, ranked the top one in the world (Lelieveld et al. 2015). Moreover, long-range transport of air pollutants from China to e.g. the Pacific, North America and the Arctic can have serious consequences on climate and ecosystems on regional and global scales, such as changes in precipitation patterns, making Chinese air pollution a truly global problem (Wuebbles et al. 2007; Wang et al. 2014).

The urgent need for an improvement of the air quality in many cities in China has been recognized by the central and local governments as well as the public, and action has been taken by the authorities. For example, during the Asia-Pacific Economic Cooperation (APEC) summit in November 2014, various emission control measures were implemented in Beijing and surrounding areas to clean Beijing's sky. However, the so-called 'APEC blue' involved a variety of trade-offs which led to serious side effects on daily life and economic growth. For example, more than 11.7 million vehicles were ordered off the roads each day to reduce exhaust emissions, about 10,000 factories were shut and additional 39,000 ran on reduced schedules to minimize pollutants emissions, and hundreds of construction sites were also shut down to mitigate resuspension of dust particles. In addition to such short-term measures, the Chinese State Council released a new legislation 'Atmospheric Pollution Prevention and Control Action Plan' on 10 September 2013 which aims to reduce the $PM_{2.5}$ concentrations by up to 25% of the 2012 levels by 2017 and is backed by $277 billion investments from the central government. Some cities also released their local control regulation with additional investments, e.g., $123 billion from Beijing municipal government from 2014 to 2017. With the implementation of the new 'Air Pollution Prevention and Control Law' from 01 January 2016, effective air pollution control measures are imperative for cleaning air in many Chinese cities.

To clean China's air, targeted and optimized emission control strategies are required. However, the factors governing PM concentrations in China are still poorly constrained. PM consists of a highly complex mixture of inorganic and organic matter that is produced from a wide variety of natural and anthropogenic sources. Organic matter, water-soluble ions (e.g., NH_4^+, NO_3^-, SO_4^{2-}), and mineral dust have been found to be the major PM constituents (Yang et al. 2011). The organic matter is much less constrained than the inorganic fraction, and typically comprises 20–90% of the total mass of submicron aerosol (Zhang et al. 2007; Jimenez et al. 2009). Organic aerosol (OA) can be broadly classified as primary organic aerosol (POA), which is directly emitted into the atmosphere in the particle phase, or secondary organic aerosol (SOA), which is formed via gas-to-particle conversion processes such as nucleation, condensation and heterogeneous and multiphase chemical reactions. Recent studies have demonstrated the global dominance of SOA (Hallquist et al. 2009), which can contribute more than 50% of the OA mass even in heavily polluted urban environments such as Los Angeles and Mexico City (Jimenez et al. 2009). Huang et al. (2014) found that even in winter SOA can contribute 44–71% of organic aerosol (OA) mass in Chinese megacities during haze period. Such high SOA production was not expected because wintertime photochemical activities were thought to be low, highlighting the lack of understanding of the SOA formation mechanisms.

Several major emission sources including traffic, coal combustion, biomass burning, cooking, or dust have been quantified in China (e.g., Huang et al. 2014). Also high secondary aerosol is found to contribute to PM pollution in China. For example, our recent studies at Beijing, Shanghai, Guangzhou, and Xi'an show that the severe haze pollution was driven to a large extent by secondary aerosol formation which contributed to 30–77% of $PM_{2.5}$ mass (Huang et al. 2014). It should be noted that the precursors of secondary aerosol, including SO_2, NO_X, NH_3, and volatile organic compounds (VOCs), will continuously increase in a majority of China (see Fig. 3.1), if there are no improved energy efficiencies and strict environmental legislation (Xing et al. 2011). The sources of SO_2, NO_X, and NH_3 are in comparison to VOCs likely better constrained, i.e., SO_2 is emitted mainly from coal-fired power plants, NO_X from vehicle exhaust and power plants, and NH_3 from agricultural activities. The source of VOCs in China, however, can be multiple, including vehicle emission, coal and biomass burning, cooking and biogenic emission, which has led to the challenge in constraining the sources of SOA. Not only the emission sources are less constrained but also the knowledge on the yields and general understanding of the formation pathways of SOA is much lower. Another feature of $PM_{2.5}$ pollution in China is the large contribution from residential combustion emissions, which is often not the case in the US and most of the western European countries.

In general, our understanding in the chemical nature and sources of $PM_{2.5}$ in China is still very insufficient, hindering the efforts to mitigate $PM_{2.5}$ pollution and to diminish the environmental, economic and health costs of particulate pollution.

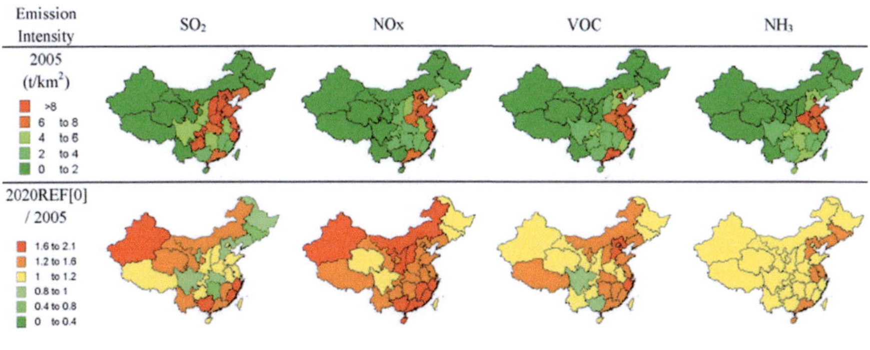

Fig. 3.1 Emission intensities of particular matter precursors in 2005 and changes in 2020 without improvement of energy efficiencies and environmental legislation (Modified from Xing et al. 2011)

This paper reviews the spatial and temporal profiles of chemical composition and sources of $PM_{2.5}$ in China. The importance of primary emissions versus secondary formation, the challenge in aerosol characterization and source apportionment, and future research needs related to the haze pollution in China, are discussed.

3.2 Chemical Composition

To quantify the PM sources measurements of inorganic and organic tracers and/or mass spectrometric fingerprints of ambient PM samples are required. This can be realized by either offline analysis of filter samples or online ambient measurements. Many PM relevant studies in China are still based on a time integrating sampling step (e.g., filters) and subsequent laboratory analysis. These often include the quantification of organic carbon (OC)/element carbon (EC), elements (e.g., Ni, Pb, Fe, As), cations and anions (e.g., NH_4^+, K^+, Ca^{2+}, NO_3^-, SO_4^{2-},), organics (e.g., polycyclic aromatic hydrocarbons (PAHs), oxygenated PAHs, carbonyls, organic acids, levoglucosan). With the development of online techniques, some state-of-the-art instruments have been used in China in recent years, providing near real-time concentration data of PM.

3.2.1 *Offline Analysis*

Numerous studies from China show that in general organic matter (OM) constitutes a major fraction of the total $PM_{2.5}$, followed by SO_4^{2-}, NO_3^-, NH_4^+, and EC. This profile is similar to studies from other regions in the world, but the concentrations are often much higher than those observed in urban areas in US and European countries. Measurements at 14 major cities show the following relative contribution to $PM_{2.5}$ mass: OM (25–45%), SO_4^{2-} (12–24%), NO_3^- (2–14%), NH_4^+ (6–11%) in wintertime; and OM (26–41%), SO_4^{2-} (14–22%), NO_3^- (3–8%), NH_4^+ (1–7%) in summertime (Fig. 3.2). Both absolute mass concentration and relative contribution of individual species/fractions show large difference at different cities, reflecting the difference in emissions and secondary formation in different regions. The seasonal variation is also significant, with enhanced secondary formation in summertime and increased primary emissions from residential heating in wintertime. The seasonal variation is also affected by temperature dependent gas-particle partitioning of species with relatively high vapour pressure. For example, the lower concentration of particulate NH_4NO_3 in summertime than in wintertime is a consequence of revaporization and decomposition of thermodynamically unstable NH_4NO_3.

Compared to inorganic compounds, OM is much more complex and only ~10 to 30% of the particulate OM has been identified as specific compounds despite years of effort and the use of the most sophisticated techniques available (Hoffmann et al.

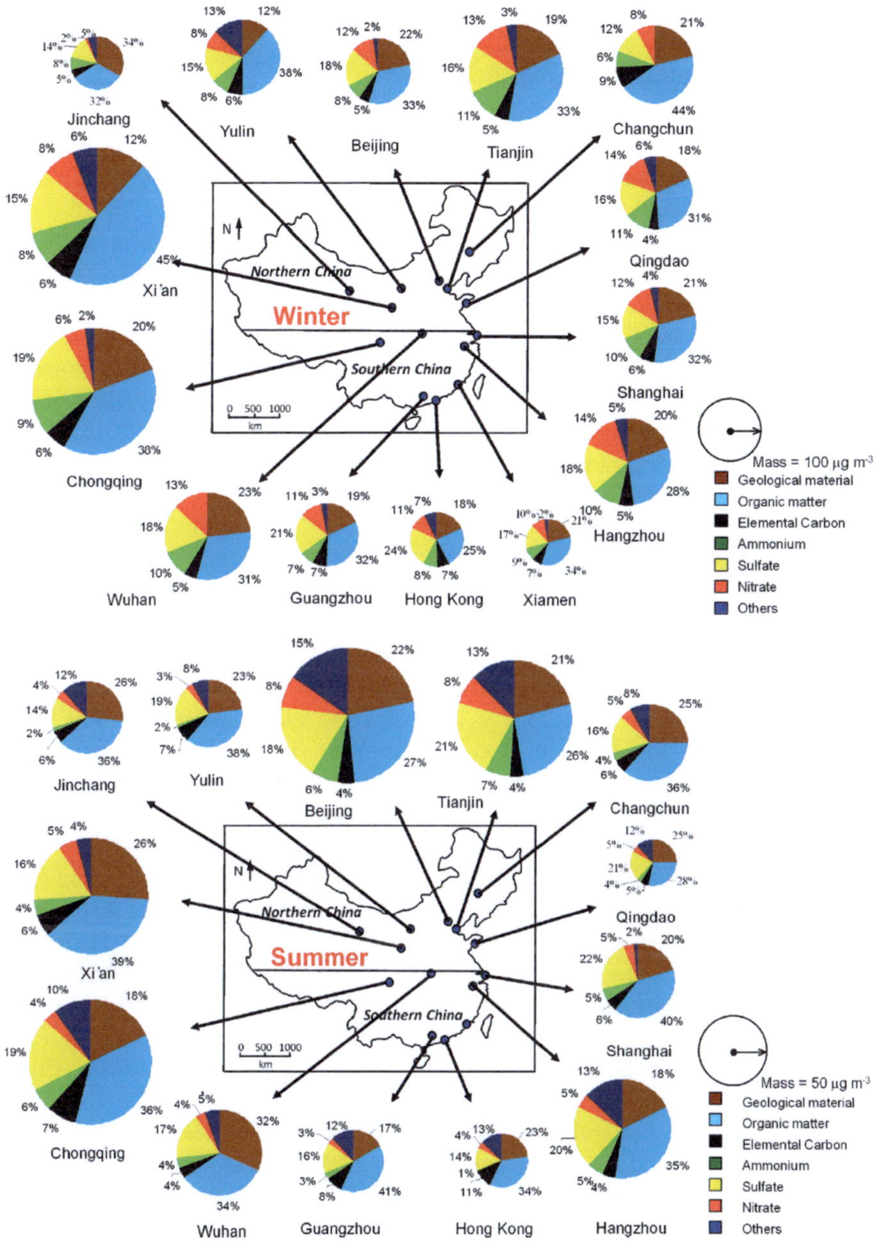

Fig. 3.2 Wintertime and summertime chemical composition of $PM_{2.5}$ in 14 Chinese cities (Cao et al. 2012)

2011). The OM relevant studies in China mainly include the quantification of organic acids, carbonyls, anhydrosugars, PAHs and their derivatives. One-year studies at Xi'an show that the concentrations of PAHs and their oxygenated and nitrogenated derivatives increased significantly during wintertime heating season, indicating enhanced emissions from domestic heating (Bandowe et al. 2014). This is consistent with enhanced biomass burning at Xi'an in winter, where the concentrations of levoglucosan, a specific marker of biomass burning, were 10–15 times higher in winter than in other seasons (Zhang et al. 2014a). However, not only the enhanced primary emissions but also the increased SOA contribute to the haze pollution. As shown in Fig. 3.3, the concentrations of water soluble organic carbon (WSOC), diacids and ketocarboxylic acids in PKU and Yufa are higher for heavier pollution episodes compared to the less polluted days, suggesting that high emission of carbonaceous aerosols and their precursor gases are one of the major sources responsible for the elevated particulate pollution in Beijing (Ho et al. 2015).

3.2.2 Online Analysis

It should be noted that many studies in China so far are realized by offline filter analysis. The time resolution of offline measurement is low, depending on the sampling time of filters, and thus hindering the capability for studying the haze formation processes. The Aerodyne aerosol mass spectrometer including high resolution time-of-flight aerosol mass spectrometer (HR-ToF-AMS) and aerosol chemical speciation monitor (ACSM) is an online instrument which has been widely used for field measurements in recent years (e.g., Jimenez et al. 2009; Canagaratna et al. 2015; Chen et al. 2015), providing quantitative mass spectra of non-refractory aerosol composition including OA, NH_4^+, NO_3^-, SO_4^{2-}, and Cl^-. Only very recently has Aerodyne HR-ToF-AMS or quadrupole ACSM (Q-ACSM) been deployed in few cities in China, e.g., Beijing, Shanghai, Xi'an, Shenzhen, Nanjing, Hong Kong (e.g., Huang et al. 2012; Sun et al. 2015; Zhang et al. 2015a, b; Elser et al. 2016), providing real time characterization of the particulate composition. Figure 3.4 shows the chemical evolution of different non-refractory PM_1 ($NR\text{-}PM_1$) species in January 2013 at Beijing, measured with a Q-ACSM. Compared to the clean period, the contribution of inorganic species increased and the organics decreased during the haze episodes. When organics decreased from 59% during the clean period to 43–50% during the four haze episodes, the total inorganic species (SO_4^{2-}+NO_3^-+NH_4^-+Cl^-) were correspondingly increased from 31% to 40–54%, of which SO_4^{2-} presented the largest increase from 9% to 16–22%. The concentration ratios of different species between the haze episodes and the clean period varied but generally decreased in an order of SO_4^{2-}>NO_3^->NH_4^->Cl^- >BC>Org. Like in other regions of the world, in general organics are dominant species in Beijing in all seasons, accounting for 40–51% of $NR\text{-}PM_1$ mass concentration (Sun et al. 2015). The dominance of organics in $NR\text{-}PM_1$ has been observed at other sites in China, e.g., 31–52% in the Yangtze River Delta (YRD) region

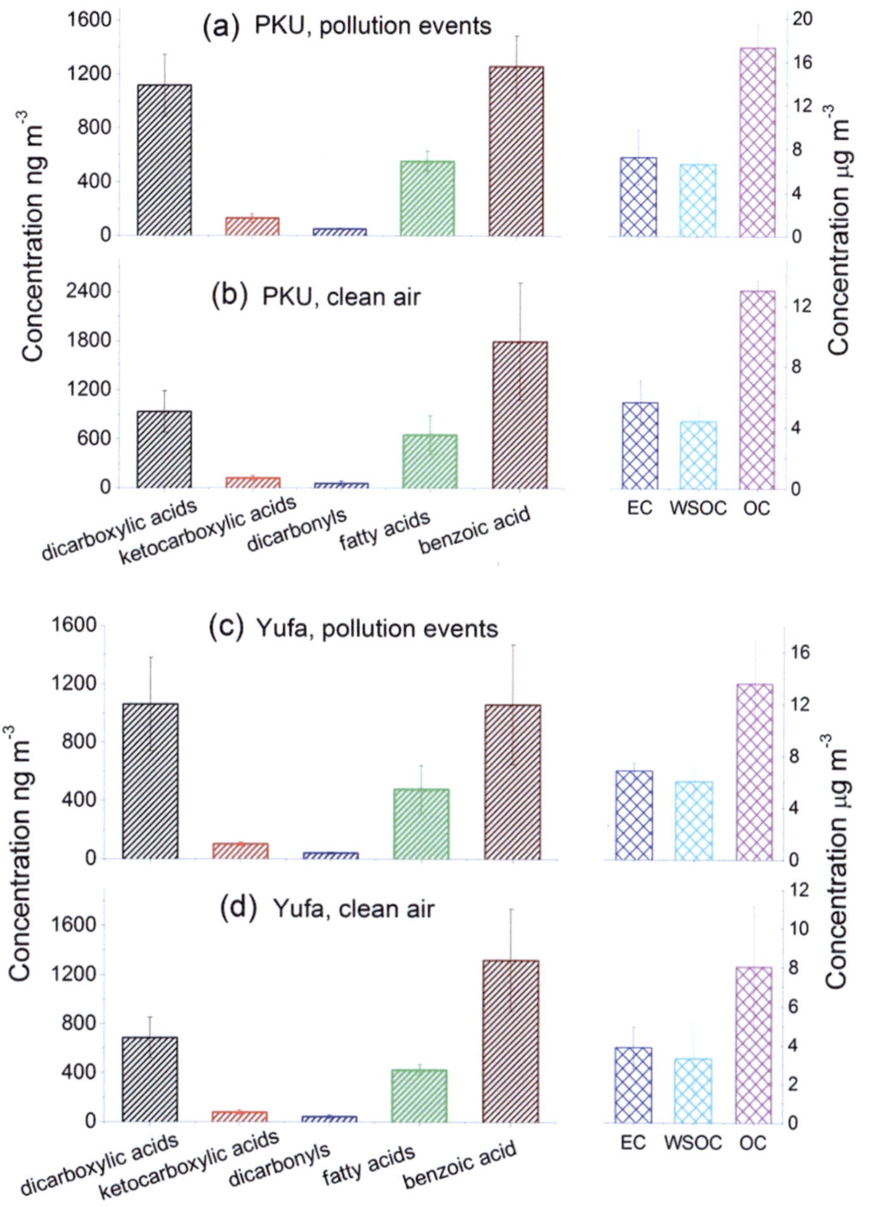

Fig. 3.3 The organic species, OC, WSOC, and EC during pollution events vs. less polluted events at PKU and Yufa (Ho et al. 2015). (**a**) PKU, pollution events. (**b**) PKU, clean air. (**c**) Yufa, pollution events. (**d**) Yufa, clean air

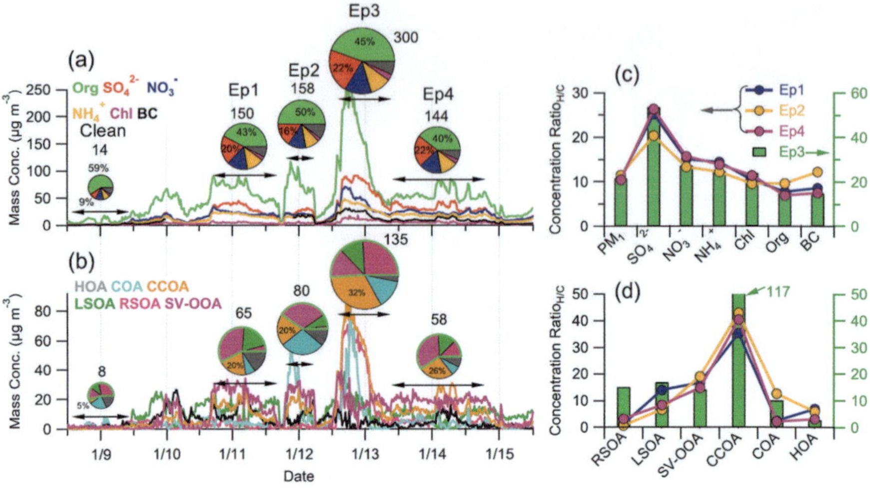

Fig. 3.4 Average chemical composition of PM_1 (**a**) and sources of organic aerosol (**b**) during the clean period and the four haze period in January 2013. The concentration ratios of submicron aerosol species and OA components between the four haze episodes and the clean period are shown in (**c**, **d**), respectively (Sun et al. 2014)

(Huang et al. 2012, 2013; Zhang et al. 2015a, b), 36–46% in the Pearl River Delta (PRD) region (He et al. 2011; Huang et al. 2011; Gong et al. 2012), and 47% in northwest China (Xu et al. 2014). However, it is noted that the relative contribution of each composition varies with the total PM concentration. As shown in Fig. 3.5, with the increase of PM_1 mass concentration in Shanghai, the relative contribution of sulfate showed a decreasing trend; the relative contribution of organics maintained at a stable level while the relative contribution of nitrate showed an increasing trend. The patterns of relative contribution from different studies show spatial and temporal difference, reflecting the differences in emissions, atmospheric processes, and meteorological conditions.

With the development of an aerodynamic lens which extends the measured particle size from PM_1 to $PM_{2.5}$, measurements of NR-$PM_{2.5}$ with HR-ToF-AMS are realized. Fig. 3.6 shows the temporal variations of the NR-$PM_{2.5}$ species and black carbon mass concentrations for measurement campaigns in Xi'an and Beijing from December 2013 to January 2014 (Elser et al. 2016). During the extreme haze periods a significant increase in the total particulate mass loading is observed, with total $PM_{2.5}$ mass reaching peak concentrations above 1,000 μg m^{-3} in Xi'an. The increase in the relative contribution of inorganic species is significant during the extreme haze periods, e.g., from 7.7% to 22.1% for sulfate, and from 10.1% to 13.1% for nitrate in Xi'an. The ratio of PM_1 to $PM_{2.5}$ mass, calculated from the integration of the size-resolved spectra (PToF mode), show a large decline from ~85% during clean period to ~60% during haze period, highlighting the importance of $PM_{2.5}$ lens for haze study in China where ~40% of the mass would have been lost in the PM_1 lens of AMS during haze events. This is associated with the growth of

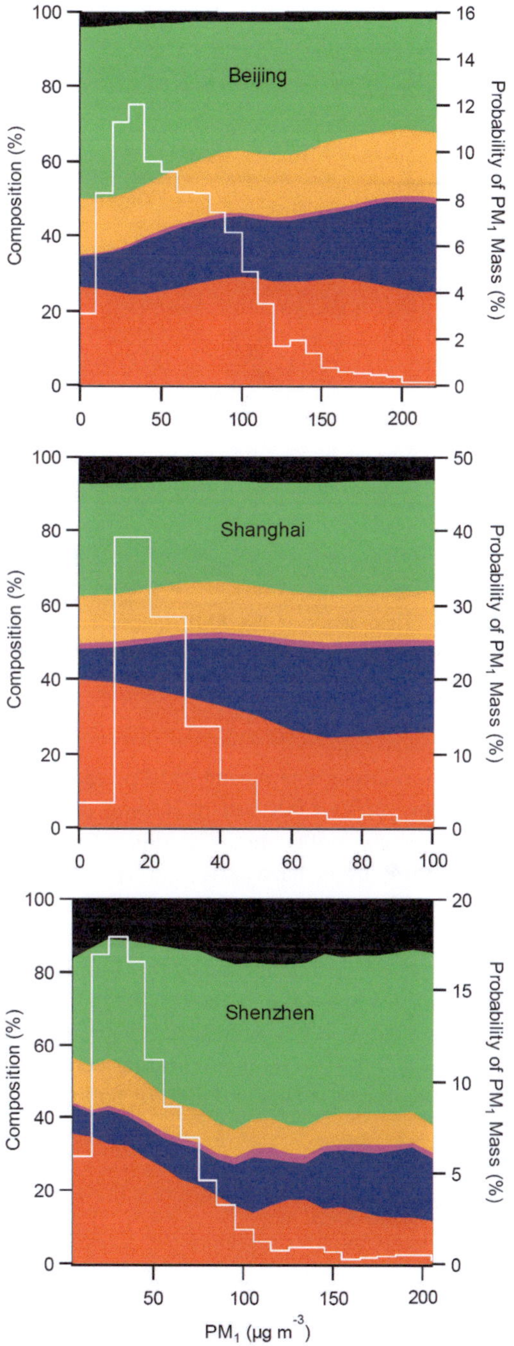

Fig. 3.5 The variation of percent composition with PM_1 mass concentration in Beijing, Shanghai, and Shenzhen (Huang et al. 2012)

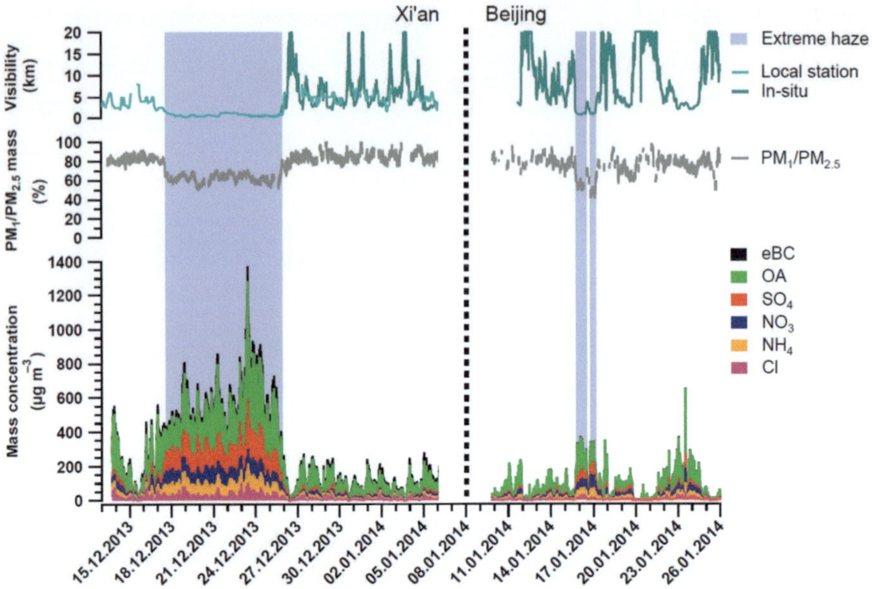

Fig. 3.6 Time series of the AMS species and eBC mass concentrations, ratio between the nonrefractory mass that would have been measured using a PM_1 lens and the total mass measured with the $PM_{2.5}$ lens, and the measured visibility during the measurement campaign period (Elser et al. 2016)

particle to larger size during haze periods. The particle size peaks at ~800 nm in Xi'an and ~800–1,000 nm in Beijing during haze periods, much larger than the clean periods (~400 nm). The diurnal trends of different species are rather flat during the extreme haze periods in Xi'an, and show more variation for the reference period with maximum concentrations at night. Such variation is strongly affected by the evolution of the planetary boundary layer height which governs the vertical dilution of the pollutants and by the diurnal cycle of the emissions. During extreme haze periods less solar radiation reaches the Earth's surface which depresses the development of the mixing layer during daytime and thus the dilution of particle concentrations. The stagnant weather conditions during haze periods, on the other hand, facilitate the accumulation of pollutants.

3.3 Sources

Source apportionment, performed with statistical techniques such as chemical mass balance (CMB) or positive matrix factorization (PMF), can provide identification and quantification of major PM sources and has been widely used in recent years (Zhang et al. 2011). Such factor analysis of time and compositionally-resolved dataset enables the extraction of "factors", of which each corresponds to many individual composition or mass spectrometric fingerprints and contains information about their sources, atmospheric processing histories, and/or chemical properties. The CMB model requires the input of *a priori* source profiles, and is unable to directly apportion secondary sources, but the fraction not attributed to primary sources considered in the model is considered as an upper limit estimate for secondary source. In contrast, in the PMF model both factor profiles and time series are determined within the model, and no *a priori* information is required to obtain a mathematical solution. The output factors are driven by internal correlations between the species contained within the analyzed dataset, providing great flexibility in the construction of the input data. The selection of an appropriate number of factors and adequate exploration of the multidimensional solution space are therefore important in PMF analysis to obtain environmentally meaningful solution. The PMF receptor model has been widely used for source apportionment in recent years. However, free PMF without *a priori* factor profiles could have yielded non-meaningful or mixed factors if some factors show temporal covariation, which is particularly true in China due to the complexity in multiple emission sources, atmospheric processes, and regional transport. The PMF model solved using the multi-linear engine (ME-2) has recently been demonstrated to be unique in directing the solution towards environmentally meaningful rotations, through the constraints of a subset of the elements of the factor profile matrix and the complete and efficient exploration of the solution space (e.g., Canonaco et al. 2013).

Many studies have used organic and inorganic tracers as input for the PMF model to apportion the sources of PAHs, organic aerosols or PM. Particular attention has been given to PAHs because of their carcinogenic or mutagenic properties. One-year measurements (from 2012 to 2013) of PAHs and their derivatives in $PM_{2.5}$ at Beijing show that the median concentration of 23 PAHs, 15 nitrated PAHs (NPAHs), 16 hydroxylated PAHs (OHPAHs), and 7 oxygenated PAHs (OPAHs) was 53.8, 1.14, 1.40, and 3.62 ng m^{-3}, respectively (Lin et al. 2015). PMF analysis shows that traffic emissions in the nonheating season (82.5% of PAHs, 59.6% of OHPAHs), and coal combustion and biomass burning in the heating season (78.7% of PAHs, 89.6% of OHPAHs, 96.6% of OPAHs), were the major primary sources of PAHs and their derivatives, while secondary formation contributed significantly to the nitrated (64.0%) and oxygenated (53.3%) derivatives of PAHs in the nonheating season in Beijing. In another 1-year study in Shanghai from 2011 to 2012 (Wang et al. 2015), it is found that the average concentration of 16 USEPA priority PAHs in $PM_{2.5}$ in the fall, winter, spring, and summer were 20.5, 27.2, 13.7, and 6.4 ng m^{-3}, respectively. The largest contributor for these PAHs in the fall and winter was gasoline engine emission (36.7%) and coal burning (41.9%), respectively; while in the spring and summer, it was diesel engine emission that contributed the most (52.1% and 43.5%, respectively). For apportioning the sources of $PM_{2.5}$, several studies have applied PMF to the dataset measured from $PM_{2.5}$ filter samples, including OC/EC, ions (Na^+, K^+, Mg^{2+}, Ca^{2+}, NH_4^+, SO_4^{2-}, NO_3^-, Cl^-), and elements (Al, Fe, Ba, Ti, Mn, Co, Ni, Cu, Zn, Mo, Cd, Sn, Sb, Pb, V, Cr, As, Se, Rb). These studies report the quantification of sources including coal combustion, biomass burning, traffic, dust, or secondary inorganic aerosol (Zhang et al. 2013;

Yang et al. 2016; Gao et al. 2016; Zikova et al. 2016). However, source-specific markers as the model input are very limited in these studies, which can lead to the biases of apportionment. In particular, the lack of organic markers in the model input explains the missing of secondary organic aerosol.

The addition of organic markers as model input is a way to obtain better constrained source profiles. These include levoglucosan as a specific marker for biomass burning, elemental carbon and hopanes as markers for vehicular emissions and coal burning, and PAHs as markers from incomplete combustion. Nevertheless, as many organic markers are not limited to one specific source, it is still a challenging work for source apportionment when two or more sources with common markers coexist, which is often the case in China. In a recent study, the mass spectrometric profiles of organic aerosol determined by offline AMS analysis of water extracts of $PM_{2.5}$ samples are included in the ME-2 model as additional input, which provides some chemical resolution across the entire bulk organic aerosol. These mass spectrometric profiles, in combination with source-specific markers, are demonstrated to provide a more robust apportionment for organic aerosol (Huang et al. 2014). By analyzing $PM_{2.5}$ samples collected simultaneously at Beijing, Shanghai, Guangzhou, and Xi'an in January 2013, this study provides a very comprehensive apportionment for $PM_{2.5}$ and organic aerosol. Two bilinear receptor models are used, including CMB and PMF with the latter solved using the ME-2, which are two complementary approaches and can be used to tackle their respective limitations. Five primary emission sources including traffic, coal burning, biomass burning, dust related emissions, and two secondary sources including secondary organic-rich and secondary inorganic-rich sources, are resolved and quantified in this study.

It is found (See Fig. 3.7) that during the severe haze period in January 2013 the relative contribution of primary sources is relatively low, i.e., ~6 to 9% from traffic, 5–7% from biomass burning, 1–2% from cooking, 3–26% from coal burning, and 3–10% from dust-related emission (with the exception of 46% in Xi'an). The relatively high contribution from coal burning in Beijing and Xi'an (interquartile range: 9–21% of $PM_{2.5}$ mass compared to 3–5% of $PM_{2.5}$ mass in Shanghai and Guangzhou) can be attributed to its extensive use in residential heating in North and West China. Fugitive dust is a very important source in West China like in Xi'an (46.3% of $PM_{2.5}$), in consist with the large unidentified fraction (35.5% of $PM_{2.5}$) related to crustal species (e.g., Al, Si). Compared to the primary emissions, the relative contribution from secondary formation is much larger, i.e., 51–77% of $PM_{2.5}$ mass in Beijing, Shanghai and Guangzhou, though this drops to ~30% in Xi'an due to higher dust levels in West China. The contribution of SOA and secondary inorganic aerosol (SIA) are found to be of similar importance, with higher values in North China (1.4 at Xi'an and 1.3 at Beijing) than in South China (0.6 at Shanghai and 0.7 at Guangzhou). Further, Fig. 3.8 clearly shows that high pollution events are characterized by an increasing secondary fraction, which accounts for up to 81% of $PM_{2.5}$ mass and up to 73% of OM mass. On average, compared to that in the lowest $PM_{2.5}$ bins, the secondary fraction in the highest $PM_{2.5}$ bins increases by a factor of 1.4 for $PM_{2.5}$ and 1.3 for OM,

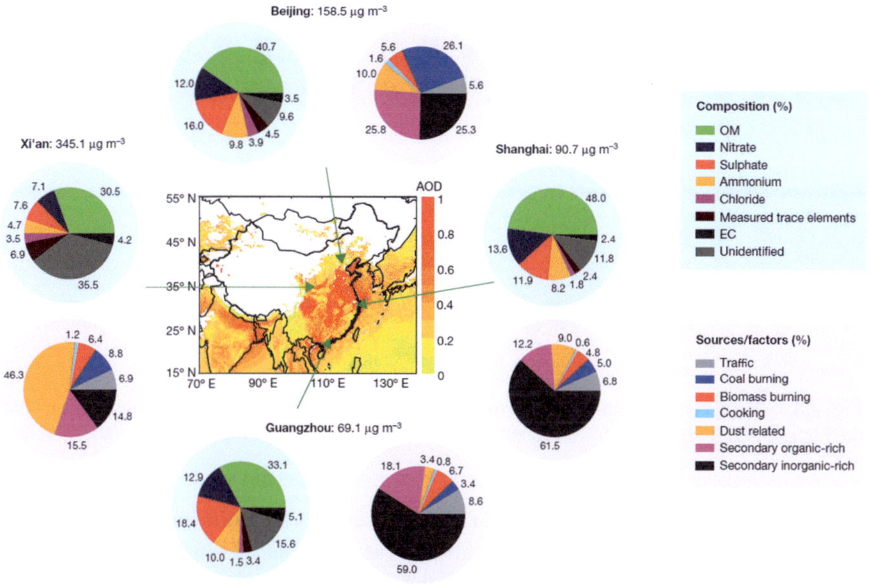

Fig. 3.7 Chemical composition and source apportionment of $PM_{2.5}$ collected during the high pollution events of 5–25 January 2013 at the urban sites of Beijing, Shanghai, Guangzhou, and Xi'an (Huang et al. 2014)

demonstrating the importance of secondary aerosol formation in driving PM pollution during haze events.

PMF analysis of mass spectra of AMS/ACSM has been proven effective in apportioning OA into different factors (Zhang et al. 2011), including traffic (HOA), biomass burning (BBOA), coal combustion (CCOA), cooking (COA), and oxygenated OA (OOA). Measurements at Beijing show that CCOA is a major primary source, contributing to on average 33% of the OA from November 2011 to January 2012 (Sun et al. 2013a). Lower contributions from CCOA were found in January 2013 (Sun et al. 2014; Zhang et al. 2014b); with coal combustion explaining 19% and 15% of the total OA, respectively. It is noted that all these three studies failed to resolve a factor related to biomass burning which is known to be a major source in winter. However, in another study performed in Beijing during the APEC period from 14 October to 12 November 2014, biomass burning OA was resolved, but coal combustion OA was not resolved (Xu et al. 2015). By applying ME-2 model to the AMS data measured at Beijing in January 2014, five primary sources including HOA, BBOA, CCOA, COA and one secondary factor OOA were successfully resolved. CCOA is found to contribute to 47% of OA during extreme haze period, which is much higher than previous studies (Elser et al. 2016). In all these studies carried out at Beijing, the average contribution of traffic to the OA varied from 9% to 18%, while cooking emissions varied from 6% to 32% of OA. Similar to the results in Huang et al. (2014), the importance of SOA was found during the extreme haze events in January 2013, i.e., 54% of OA in Zhang

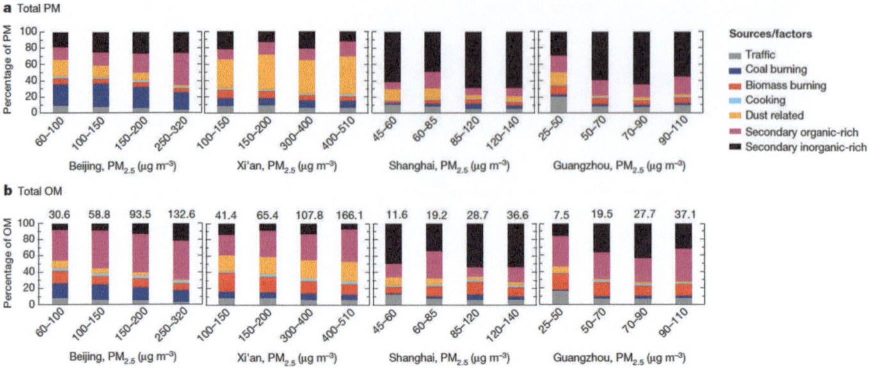

Fig. 3.8 Source contribution to total $PM_{2.5}$ and OM as a function of the $PM_{2.5}$ mass bins during the high pollution events of 5–25 January 2013 (Huang et al. 2014). (**a**) Total PM. (**b**) Total OM

et al. (2014b) and 55% in Sun et al. (2014). The reported contributions of SOA in other measurement periods are generally not as much as that in January 2013; but enhanced contribution is still found, e.g., the SOA contribution increased from 16% during clean period to 25% during pollution period in January 2014 at Beijing. Such large temporal variations in SOA contribution are likely due to the difference in atmospheric aging time, atmospheric oxidative capacity, meteorological conditions, and atmospheric transport. The reduction of emissions from 2013 due to the implementation of control measures could be another reason. For example, comparison measurements show large reduction in SOA of 37% (from 54% before APEC to 34% during APEC) and in SIA of 51–57% during the APEC period due to the implementation of strict emission measures (Sun et al. 2016). Similar studies with AMS/ACSM in other Chinese cities are still very scarce. Zhang et al. (2015a) measured the urban air with Q-ACSM during December 2013 in Nanjing in Yangtze River Delta of China and identified six OA factors including HOA, COA, BBOA and three OOA factors, i.e., a less-oxidized OOA (LO-OOA), a more-oxidized OOA (MO-OOA), and a low-volatility OOA (LV-OOA). OA is found to be dominated by secondary species during haze periods. In particular, MO-OOA and LV-OOA increased from 21% to 24% and from 18% to 23%, respectively, during haze periods compared to clean period. The importance of SOA for haze formation was also observed in Xi'an, with contribution of OOA to total OA increasing from 16% during clean period to 31% during haze period (Elser et al. 2016).

3.4 Conclusions and Outlook

Considerable advances have been made in the last few years in terms of better understanding of particulate air pollution in China. In terms of chemical nature of the fine PM, a number of inorganic and organic species have been measured, showing the dominance of organics in fine PM. However, most studies mainly measured OC/EC, ions, elements, with a substantial lack of measurement for organic compounds. This may lead to significant bias for apportioning sources of OA as discussed above. Therefore, organic compounds should be included in the chemical characterization of PM. This can improve the source apportionment as well as our understanding of the evolution of OA. Considering that many organic marker compounds are not solely limited to one specific source and that the emission sources in China are much more complex, future studies should pay more attention to the investigation of source-specific organic markers and their stability against atmospheric oxidation. In terms of the study regions, most of previous studies have focused on the Beijing-Tianjin-Hebei, Yangtze River Delta, Pearl River Delta and Guanzhong city clusters. However, there are many other cities which have also been experiencing heavy particulate air pollution but with limited studies so far. This is particularly true in the middle and eastern China where many cities suffer from severe particulate air pollution but the causes are nearly unexplored. Therefore, relevant studies should be extended to a larger scale, and consecutive studies should be carried out in each region in order to estimate the effect of emission control measures.

In terms of source apportionment, the key is better apportioning the sources of OA which are much more complex compared to the sources of inorganic aerosol. In particular, identification of SOA sources is a challenging task. SOA from different sources are usually reported as OOA, or classified into a relatively fresh semi-volatile OOA (SV-OOA) and a highly aged low-volatility OOA (LV-OOA), or LO-OOA and MO-OOA. The reaction products or AMS spectra of different types of SOA become more similar during aging and lose their original fingerprint when reaching the highly oxidized state (Jimenez et al. 2009), leading to the difficulty in identifying the contributions from different sources. The combination of factor analysis with ^{14}C analysis has been demonstrated to be a unique approach to quantify the fossil and non-fossil carbon contribution to SOA (see Fig. 3.9). This help to constrain the relative importance of specific sources to SOA production, for example, SOA precursors emitted by traffic/coal combustion would increase the fossil content while SOA precursors emitted by biomass burning/biogenic would increase the non-fossil content. Further, recent studies show that, for example, fresh SV-OOA still contains differences in ratios of individual ions that can be used for the differentiation of the different types of fresh SOA. More studies are therefore urgently needed to identify the source-specific oxidation products and/or mass spectrometric fingerprints of different types of SOA and therefore to apportion the sources of SOA for ambient measurements.

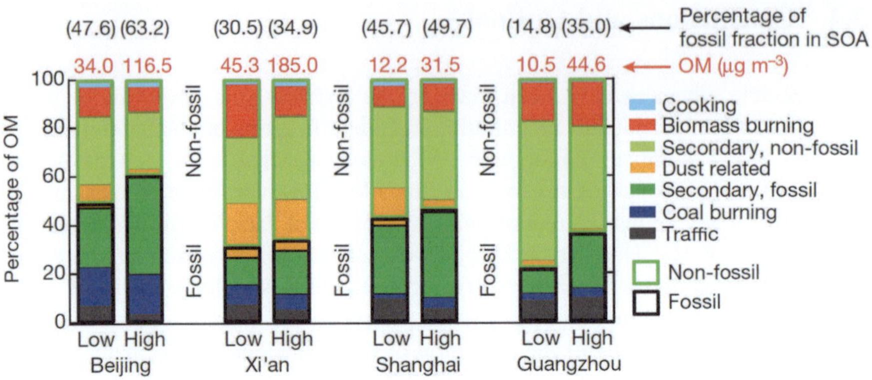

Fig. 3.9 Fossil and non-fossil fractional contributios of each source to total OM during low and high $PM_{2.5}$ levels obeserved in different cities in January 2013. The results are obtained by combining factor analysis with ^{14}C analysis (Huang et al. 2014)

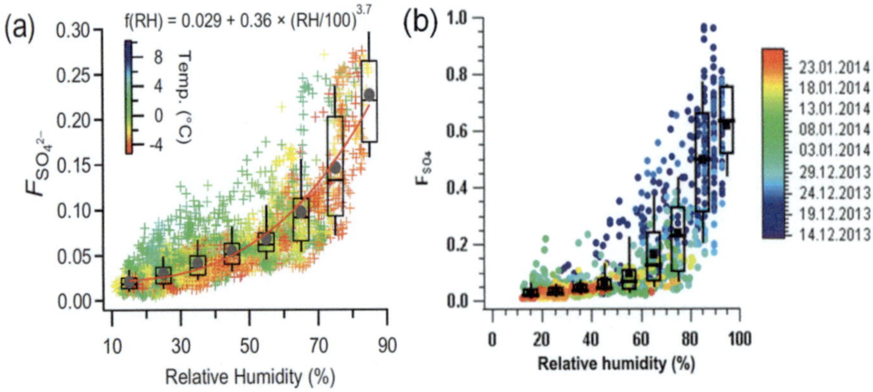

Fig. 3.10 Effect of RH on the formation of sulfate measured from November 2011 to January 2012 at Beijing (**a**) and from December 2013 to January 2014 at Xi'an and Beijing (**b**). $F_{SO4}{}^{2-}$ is the fraction of sulfate in total sulfur (Sun et al. 2013b; Elser et al. 2016)

The importance of secondary aerosol production in driving haze formation has been recognized in several recent studies. The haze pollution is often characterized with high relative humidity (RH). The strong increase of the normalized sulfate at high RH suggest that aqueous-phase oxidation of SO_2 could be an important process during haze events (Fig. 3.10). However, it is still not clear if and to what extent aqueous-phase chemistry is an important process for SOA production during haze events, although Ervens et al. (2011) provide evidence in their review paper that aqueous-phase reaction does apply to organic species. Turpin and co-workers (Lim et al. 2010) report that water-soluble organic products of gas-phase photochemistry dissolve into the aqueous-phase where they can react further to form low volatility products that are largely retained in the particle phase. Organic acids, oligomers and other products can form via radical and non-radical reactions, including hemiacetal formation during droplet evaporation, acid/base catalysis, and reaction of organics with other constituents (e.g., NH_4^+). The nighttime NO_3 chemistry is another process that may contribute to SOA production. Especially, during wintertime heating period, many SOA precursors emitted at nighttime which likely contribute substantial SOA production. Further, the OH radical chemistry is believed to be the main atmospheric oxidation process responsible for SOA production. The wintertime OH radical concentrations are traditionally thought to be low because of low O_3 levels in China during haze events. The OH production from HONO photolysis is likely another channel. However, the wintertime OH concentrations in China are not clear so far. Therefore, OH relevant field measurements are urgently needed to explore the contribution of OH chemistry to SOA production during wintertime haze events.

Finally, previous studies typically take a few months or even years to get the source apportionment results of a measurement campaign, which is much delay for policy makers. Online chemical characterization and online source apportionment (with high time resolution) are required for environment protection agencies to take efficient measures when haze events occur. In terms of receptor model, future studies should improve the algorithm to realize automatic model calculation and online validation of the results. The prerequisite for online source apportionment is the real-time online quantification of mass concentration of main PM composition. The Aerodyne aerosol mass spectrometer has been proven to be a very useful instrument providing quantitative information of chemical composition and sources. However, the HR-ToF-AMS measurements are often limited to few weeks at specific sites because simultaneous and long-term deployment of AMS at many sites is impractical due to the high cost and complex operation/maintenance required by the instrument. Instead, Q-ACSM and ToF-ACSM, which are simplified versions of AMS, can provide long-term measurements with much less maintenance. The ACSM has been successfully used in many sites of the US and Europe, e.g., the European ACTRIS network, which has greatly helped to understand the PM chemical nature and sources and therefore improve the air quality. It can be expected that an ACSM China network will help to solve the haze pollution in China.

In general, the temporal and spatial profiles of chemical composition and sources of PM, a better constraint of the SOA sources, the mechanisms for substantial secondary aerosol production during wintertime haze events, the online chemical characterization and online source apportionment of PM, all these aspects require more detailed studies in terms of better understanding of the causes of particulate air pollution in China and its effects on air quality, visibility and regional climate. This will help the Chinese central and local governments to design effective mitigation strategies to meet the goal of improving urban air quality to levels typical of the United States and Europe. The severe particulate air pollution in China, on the other hand, provides a unique platform for scientists worldwide to obtain new insights into many aspects of atmospheric chemistry and physics. This calls for international collaboration.

Acknowledgment This work was supported by the National Natural Science Foundation of China (NSFC) under Grant No. 91644219, No. 41403110 and No. 41673134.

References

Bandowe, B. A. M., Meusel, H., Huang, R. J., Ho, K. F., Cao, J. J., Hoffmann, T., & Wilcke, W. (2014). PM2.5-bound oxygenated PAHs, nitro-PAHs and parent-PAHs from the atmosphere of a Chinese megacity: Seasonal variation, sources and cancer risk assessment. *Science of the Total Environment, 473–474*, 77–87.

Canagaratna, M. R., Jimenez, J. L., Kroll, J. H., Chen, Q., Kessler, S. H., Massoli, P., Hildebrandt Ruiz, L., Fortner, E., Williams, L. R., Wilson, K. R., et al. (2015). Elemental ratio measurements of organic compounds using aerosol mass spectrometry: Characterization, improved calibration, and implications. *Atmospheric Chemistry and Physics, 15*, 253–272.

Canonaco, F., Crippa, M., Slowik, J. G., Baltensperger, U., & Prevot, A. S. H. (2013). SoFi, an IGOR-based interface for the efficient use of the generalized multilinear engine (ME-2) for the source apportionment: ME-2 application to aerosol mass spectrometer data. *Atmospheric Measurement Techniques, 6*, 3649–3661.

Cao, J. J. (2012). Pollution status and control strategies of PM2.5 in China. *Journal of Earth Environment, 3*, 1030–1036.

Cao, J. J., Shen, Z. X., Chow, J. C., Watson, J. G., Lee, S. C., Tie, X. X., Ho, K. F., Wang, G. H., & Han, Y. M. (2012). Winter and summer $PM_{2.5}$ chemical compositions in fourteen Chinese cities. *Journal of the Air & Waste Management Association, 62*, 1214–1226.

Chen, Q., Heald, C. L., Jimenez, J. L., Canagaratna, M. R., Zhang, Q., He, L.-Y., Huang, X.-F., Campuzano-Jost, P., Palm, B. B., Poulain, L., et al. (2015). Elemental composition of organic aerosol: The gap between ambient and laboratory measurements. *Geophysical Research Letters, 42*, 4182–4189. doi:10.1002/2015GL063693.

Elser, M., Huang, R. J., Wolf, R., Slowik, J. G., Wang, Q. Y., Canonaco, F., Li, G. H., Bozzetti, C., Daellenbach, K. R., Huang, Y., Zhang, R. J., Li, Z. Q., Cao, J. J., Baltensperger, U., & Prevot, A. S. H. (2016). New insights into PM2.5 chemical composition and sources in two major cities in China during extreme haze events using aerosol mass spectrometry. *Atmospheric Chemistry and Physics, 16*, 3207–3225.

Ervens, B., Turpin, B. J., & Weber, R. J. (2011). Secondary organic aerosol formation in cloud droplets and aqueous particles (aqSOA): A review of laboratory, field and model studies. *Atmospheric Chemistry and Physics, 11*, 11069–11102.

Gao, J., Peng, X., Chen, G., Xu, J., Shi, G. L., Zhang, Y. C., & Feng, Y. C. (2016). Insights into the chemical characterization and sources of $PM_{2.5}$ in Beijing at a 1-h time resolution. *Science of the Total Environment, 542*, 162–171.

Gong, Z., Lan, Z., Xue, L., Zeng, L., He, L., & Huang, X. F. (2012). Characterization of submicron aerosols in the urban outflow of the central Pearl River Delta region of China. *Frontiers of Environmental Science & Engineering, 6*, 725–733.

Hallquist, M., Wenger, J. C., Baltensperger, U., Rudich, Y., Simpson, D., Claeys, M., Dommen, J., Donahue, N. M., George, C., Goldstein, A. H., et al. (2009). The formation, properties and impact of secondary organic aerosol: Current and emerging issues. *Atmospheric Chemistry and Physics, 9*, 5155–5236.

He, L. Y., Huang, X. F., Xue, L., Hu, M., Lin, Y., Zheng, J., Zhang, R. Y., & Zhuang, Y. H. (2011). Submicron aerosol analysis and organic source apportionment in an urban atmosphere in Pearl River Delta of China using high resolution aerosol mass spectrometry. *Journal of Geophysical Research-Atmospheres, 116*, D12304. doi:10.1029/2010JD014566.

Ho, K. F., Huang, R. J., Kawamura, K., Tachibana, E., Lee, S. C., Ho, S. S. H., Zhu, T., & Tian, L. (2015). Dicarboxylic acids, ketocarboxylic acids, α-dicarbonyls, fatty acids and benzoic acid in PM2.5 aerosol collected during CAREBeijing-2007: An effect of traffic restriction on air quality. *Atmospheric Chemistry and Physics, 15*, 3111–3123.

Hoffmann, T., Huang, R. J., & Kalberer, M. (2011). Atmospheric analytical chemistry. *Analytical Chemistry, 83*, 4649–4664.

Huang, X. F., He, L. Y., Hu, M., Canagaratna, M. R., Kroll, J. H., Ng, N. L., Zhang, Y. H., Lin, Y., Xue, L., Sun, T. L., Liu, X. G., Shao, M., Jayne, J. T., & Worsnop, D. R. (2011). Characterization of submicron aerosols at a rural site in Pearl River Delta of China using an Aerodyne high resolution aerosol mass spectrometer. *Atmospheric Chemistry and Physics, 11*, 1865–1877.

Huang, X. F., He, L. Y., Xue, L., Sun, T. L., Zeng, L. W., Gong, Z. H., Hu, M., & Zhu, T. (2012). Hignly time-resolved chemical characterization of atmospheric fine particles during 2010 Shanghai World Expo. *Atmospheric Chemistry and Physics, 12*, 4897–4907.

Huang, X. F., Xue, L., Tian, X. D., Shao, W. W., Sun, T. L., Gong, Z. H., Ju, W. W., Jiang, B., Hu, M., & He, L. Y. (2013). Highly time-resolved carbonaceous aerosol characterization in Yangtze River Delta of China: Composition, mixing state and secondary formation. *Atmospheric Environment, 64*, 200–207.

Huang, R. J., Zhang, Y. L., Bozzetti, C., Ho, K. F., Cao, J. J., Han, Y. M., Dällenbach, K. R., Slowik, J. G., Platt, S. M., Canonaco, F., Zotter, P., Wolf, R., Pieber, S. M., Bruns, E. A., Crippa, M., Ciarelli, G., Piazzalunga, A., Schwikowski, M., Abbaszade, G., Schnelle-Kreis, J., Zimmermann, R., An, Z. S., Szidat, S., Baltensperger, U., El Haddad, I., & Prévôt, A. S. H. (2014). High secondary aerosol contribution to particulate pollution during haze events in China. *Nature, 514*, 218–222.

Jimenez, J. L., Canagaratna, M. R., Donahue, N. M., Prévôt, A. S. H., Zhang, Q., Kroll, J. H., DeCarlo, P. F., Allan, J. D., Coe, H., Ng, N. L., et al. (2009). Evolution of organic aerosols in the atmosphere. *Science, 11*, 1525–1529.

Lelieveld, J., Evans, J. S., Fnais, M., Giannadaki, D., & Pozzer, A. (2015). The contribution of outdoor air pollution sources to premature mortality on a global scale. *Nature, 525*, 367–271.

Lim, Y. B., Tan, Y., Perri, M. J., Seitzinger, S. P., & Turpin, B. J. (2010). Aqueous chemistry and its role in secondary organic aerosol (SOA) formation. *Atmospheric Chemistry and Physics, 10*, 10521–10539.

Lin, Y., Ma, Y., Qiu, X., Li, R., Fang, Y., Wang, J., Zhu, Y., & Hu, D. (2015). Sources, transformation, and health implications of PAHs and their nitrated, hydroxylated, and oxygenated derivatives in $PM_{2.5}$ in Beijing. *Journal of Geophysical Research – Atmospheres, 120*, 7219–7228.

Sun, Y. L., Wang, Z. F., Fu, P. Q., Yang, T., Jiang, Q., Dong, H. B., Li, J., & Jia, J. J. (2013a). Aerosol composition, sources and processes during wintertime in Beijing, China. *Atmospheric Chemistry and Physics, 13*, 4577–4592.

Sun, Y. L., Wang, Z. F., Fu, P. Q., Jiang, Q., Yang, T., Li, J., & Ge, X. L. (2013b). The impact of relative humidity on aerosol composition and evolution processes during wintertime in Beijing, China. *Atmospheric Environment, 77*, 927–934.

Sun, Y., Jiang, Q., Wang, Z., Fu, P., Li, J., Yang, T., & Yin, Y. (2014). Investigation of the sources and evolution processes of severe haze pollution in Beijing in January 2013. *Journal of Geophysical Research-Atmospheres, 119*, 4380–4398.

Sun, Y. L., Wang, Z. F., Zhang, Q., Wang, Q. Q., Fu, P. Q., Pan, X. L., Li, J., Jayne, J., & Worsnop, D. R. (2015). Long-term real-time measurements of aerosol particle composition in Beijing, China: Seasonal variations, meteorological effects, and source analysis. *Atmospheric Chemistry and Physics, 15*, 10149–10165.

Sun, Y. L., Wang, Z. F., Wild, O., Xu, W. Q., Chen, C., Fu, P. Q., Du, W., Zhou, L. B., Zhang, Q., Han, T. T., Wang, Q. Q., Pan, X. L., Zheng, H. T., Li, J., Guo, X. F., Liu, J. G., & Worsnop, D. R. (2016). "APEC blue": Secondary aerosol reductions from emission controls in Beijing. *Scientific Reports, 6*, 20668. doi:10.1038/srep20668.

Wang, Y., Zhang, R. Y., & Saravanan, R. (2014). Asian pollution climatically modulates midlatitude cyclones following hierarchical modeling and observational analysis. *Nature Communications, 5*, 3098. http://dx.doi.org/10.1038/ncomms4098.

Wang, F. W., Lin, T., Feng, J. L., Fu, H. Y., & Guo, Z. G. (2015). Source apportionment of polycyclic aromatic hydrocarbons in $PM_{2.5}$ using positive matrix factorization modeling in Shanghai, China. *Environmental Science: Processes & Impacts, 17*, 197–205.

Wuebbles, D. J., Lei, H., & Lin, J. T. (2007). Intercontinental transport of aerosols and photochemical oxidants from Asia and its consequences. *Environmental Pollution, 150*, 65–84.

Xing, J., Wang, S. X., Chatani, S., Zhang, C. Y., Wei, W., Hao, J. M., Klimont, Z., Cofala, J., & Amann, M. (2011). Projections of air pollutant emissions and its impacts on regional air quality in China in 2020. *Atmospheric Chemistry and Physics, 11*, 3119–3136.

Xu, J., Zhang, Q., Chen, M., Ge, X., Ren, J., & Qin, D. (2014). Chemical composition, sources, and processes of urban aerosols during summertime in northwest China: Insights from high-resolution aerosol mass spectrometry. *Atmospheric Chemistry and Physics, 14*, 12593–12611.

Xu, W. Q., Sun, Y. L., Chen, C., Du, W., Han, T. T., Wang, Q. Q., Fu, P. Q., Wang, Z. F., Zhao, X. J., Zhou, L. B., Ji, D. S., Wang, P. C., & Worsnop, D. R. (2015). Aerosol composition, oxidation properties, and sources in Beijing: Results from the 2014 Asia-Pacific Economic Cooperation summit study. *Atmospheric Chemistry and Physics, 15*, 13681–13698.

Yang, F., Tan, J., Zhao, Q., Du, Z., He, K., Ma, Y., Duan, F., Chen, G., & Zhao, Q. (2011). Characteristics of PM2.5 speciation in representative megacities and across China. *Atmospheric Chemistry and Physics, 11*, 5207–5219.

Yang, H. N., Chen, J., Wen, J. J., Tian, H. Z., & Liu, X. G. (2016). Composition and sources of $PM_{2.5}$ around the heating periods of 2013 and 2014 in Beijing: Implications for efficient mitigation measures. *Atmospheric Environment, 124*, 378–386.

Zhang, Q., Jimenez, J. L., Canagaratna, M. R., Allan, J. D., Coe, H., Ulbrich, I., Alfarra, M. R., Takami, A., Middlebrook, A. M., Sun, Y. L., et al. (2007). Ubiquity and dominance of oxygenated species in organic aerosols in anthropogenically-influenced Northern Hemisphere midlatitudes. *Geophysical Research Letters, 34*, L13801. doi:10.1029/2007GL029979.

Zhang, Q., Jimenez, J. L., Canagaratna, M. R., Ulbrich, I. M., Ng, N. L., Worsnop, D. R., & Sun, Y. L. (2011). Understanding atmospheric organic aerosols via factor analysis of aerosol mass spectrometry: A review. *Analytical and Bioanalytical Chemistry, 401*, 3045–3067.

Zhang, R., Jing, J., Tao, J., Hsu, S. C., Wang, G., Cao, J., Lee, C. S. L., Zhu, L., Chen, Z., Zhao, Y., & Shen, Z. (2013). Chemical characterization and source apportionment of $PM_{2.5}$ in Beijing: Seasonal perspective. *Atmospheric Chemistry and Physics, 13*, 7053–7074.

Zhang, T., Cao, J. J., Chow, J. C., Shen, Z. X., Ho, K. F., Ho, S. S. H., Liu, S. X., Han, Y. M., Watson, J. G., Wang, G. H., & Huang, R. J. (2014a). Characterization and seasonal variations of PM2.5 levoglucosan in Xi'an, China. *Journal of the Air & Waste Management Association, 64*, 1317–1327.

Zhang, J. K., Sun, Y., Liu, Z. R., Ji, D. S., Hu, B., Liu, Q., & Wang, Y. S. (2014b). Characterization of submicron aerosols during a month of serious pollution in Beijing, 2013. *Atmospheric Chemistry and Physics, 14*, 2887–2903.

Zhang, Y. J., Tang, L. L., Yu, H. X., Wang, Z., Sun, Y. L., Qin, W., Chen, W. T., Chen, C. H., Ding, A. J., Wu, J., Ge, S., Chen, C., & Zhou, H. C. (2015a). Chemical composition, sources and evolution processes of aerosol at an urban site in Yangtze River Delta, China during wintertime. *Atmospheric Environment, 123*, 339–349.

Zhang, Y. J., Tang, L. L., Wang, Z., Yu, H. X., Sun, Y. L., Liu, D., Qin, W., Canonaco, F., Prévôt, A. S. H., Zhang, H. L., & Zhou, H. C. (2015b). Insights into characteristics, sources, and evolution of submicron aerosols during harvest seasons in the Yangtze River delta region, China. *Atmospheric Chemistry and Physics, 15*, 1331–1349.

Zikova, N., Wang, Y. G., Yang, F. M., Li, X. H., Tian, M., & Hopke, P. K. (2016). On the source contribution to Beijing $PM_{2.5}$ concentrations. *Atmospheric Environment, 134*, 84–95.

Chapter 4
Photochemical Smog in Southern China: A Synthesis of Observations and Model Investigations of the Sources and Effects of Nitrous Acid

Tao Wang, Yutong Liang, Qiaozhi Zha, Li Zhang, Zhe Wang, Weihao Wang, and Steven Poon

Abstract Recent studies have revealed potentially important effects of additional source(s) of hydroxyl radicals on the atmosphere's oxidative capacity and, in turn, the production of secondary air pollutants. In this paper, we give an overview of our recent efforts in investigating the sources and effects of nitrous acid (HONO) on ozone and some secondary aerosols in southern China by combining field measurements and model simulations. Beginning in 2011, a series of field measurements of HONO were conducted at five sites, with diverse land use and different effects of emission sources. We observed the seasonal characteristics, emission ratios, heterogeneous production, and made simulations with a chemical transport model for the photochemical effects of HONO. The key findings are as follows. The derived emission ratios from vehicles exhibited wide variability and were mostly higher than the more uniform value of 0.8% reported in the literature. Larger nocturnal heterogeneous conversion rates of NO_2 to HONO were observed when air masses were passing over sea surfaces, compared with land surfaces. Widely reported daytime sources of HONO also exist in Hong Kong. Moreover, the revised WRF-Chem model with comprehensive HONO sources significantly improved the simulations of the observed HONO, which enhanced regional hydroxyl radicals, O_3, and $PM_{2.5}$ by 10–20, 8–15, and 10–15% over urban areas in the Pearl River

T. Wang (✉) • Y. Liang • L. Zhang • Z. Wang • W. Wang • S. Poon
Department of Civil and Environmental Engineering, Hong Kong Polytechnic University, Kowloon, Hong Kong, China
e-mail: cetwang@polyu.edu.hk; liang.yutong@connect.polyu.hk; alex.zhang@connect.polyu.hk; z.wang@polyu.edu.hk; weihao.wang91@gmail.com; steven.poon@polyu.edu.hk

Q. Zha
Department of Civil and Environmental Engineering, Hong Kong Polytechnic University, Kowloon, Hong Kong, China

Department of Physics, University of Helsinki, Helsinki, Finland
e-mail: qiaozhi.zha@helsinki.fi

I. Bouarar et al. (eds.), *Air Pollution in Eastern Asia: An Integrated Perspective*, ISSI Scientific Report Series 16, DOI 10.1007/978-3-319-59489-7_4

Delta region, respectively. Our studies highlight the importance of considering HONO sources when simulating secondary pollutants in polluted atmospheres.

Keywords Nitrous acid (HONO) • Field measurement • WRF-Chem • Secondary pollutants

4.1 Introduction

Air pollution is a serious problem in many Chinese urban areas. The frequent outbreaks of extremely heavy haze in Beijing attract major attention from the media, citizens, and regulatory agencies. This then prompts the central government to revise air-quality standards and launch stringent control measures in China's three most developed urban clusters. Previous research and control efforts have focused on particulate matter, which is most severe in the winter, but processes affecting photochemical pollutants such as ozone and other secondary pollutants have not been comprehensively examined. One emerging issue is the effects of new radical(s) such as chlorine (e.g., Osthoff et al. 2008), or additional source(s) of hydroxyl radicals (Kleffmann 2007) on the atmosphere's oxidative capacity and, subsequently, the production of secondary air pollutants such as ozone and some aerosol constituents. Studies conducted worldwide in the last 10 years have confirmed that HONO can play an important role in the chemistry of a polluted atmosphere, not only in the morning, but also throughout the day. However, the sources of HONO in different environments and its effect on secondary pollutants in China are not well understood.

HONO in the atmosphere comes from a wide range of sources. Direct emissions include the release of HONO from fossil fuel combustion (e.g., Kurtenbach et al. 2001) and the microbial activities in soil (Su et al. 2011). The chemical formation of HONO includes gas-phase reactions of NO + OH and heterogeneous processes on surfaces, such as the redox reaction of nitrogen dioxide (NO_2) on Black Carbon (BC) and semi-volatile species (e.g., Ammann et al. 1998), and the conversion of NO_2 to HONO on wet surfaces—the latter of which is often the dominant source of nocturnal HONO (Kleffmann 2007).

One topic under intense research is the sources/processes contributing to elevated levels of daytime HONO, which cannot be explained by the homogenous reaction between NO and OH and the sources just mentioned (Acker and Möller 2007; Czader et al. 2012, 2013; Li et al. 2012; Qin et al. 2009; Sörgel et al. 2011; Su et al. 2008; Vandenboer et al. 2014). Several possible photo-enhanced sources have been proposed, including the light-dependent heterogeneous reactions of NO_2 with aerosol and ground surfaces, the photolysis of HNO_3 absorbed on surfaces, the reaction of excited NO_2 with water vapors, and the reaction of NO_2 with $HO_2{*}H_2O$ (Kleffmann 2007; Monge et al. 2010; Li et al. 2008, 2014; Wong et al. 2011; Zhou et al. 2011).

Current state-of-the-art photochemical models treat various HONO sources with potentially large uncertainties. The HONO/NO_x emission ratios have only been measured in limited locations. The exact mechanisms of non-gaseous HONO pathways remain unclear, and parameterizations of these sources in the models are highly simplified, with large variations in the key parameters (e.g., the uptake coefficient of NO_2 on surfaces (γ_{NO_2}) varying from 10^{-6} to 10^{-4}) in different modeling studies. This has led to different conclusions on the importance of atmospheric aerosols in HONO formation (An et al. 2013; Li et al. 2010, 2011; Sarwar et al. 2008; Aumont et al. 2003). In addition, emissions from soil microbial processes and the conversion of NO_2 on the ocean surface have not been considered in most of the previous model studies.

In this paper, we provide an overview of our recent efforts in investigating the sources and effects of HONO on ozone and some secondary aerosols in southern China by combining field measurements and modeling simulations. Due to their sub-tropic location, Hong Kong and the part of the Pearl River Delta in Guangdong Province on China's mainland have long suffered year-round photochemical pollution (e.g., Wang et al. 2001; Xue et al. 2014). Since 2011, a series of field studies on HONO have been conducted in five Hong Kong locations: a tunnel, a roadside site, a suburban area, a coastal area, and a mountain top (974 m a.s.l.). The data represent a wide range of conditions, from the source to the background, and from the surface to the top of the Planetary Boundary Layer (PBL). This paper provides an overview of the main characteristics of this dataset and synthesizes the key findings from the analysis, including seasonal characteristics, emission ratios, heterogeneous production, and model simulations of photochemical effects.

4.2 Measurement Sites and Instrumentation

4.2.1 Measurement Sites

The locations of the five measurement sites are shown in Fig. 4.1. The times the measurements were taken and the mean, maximum, and minimum concentrations of HONO are listed in Table 4.1. The Shing Mun Tunnel (SMT) site, located deep inside the north bore of the SMT (1.6 km in length), hosts 25,910 vehicles (over 40% diesel) per day, on average. The Mong Kok (MK) site, with a sampling point of less than 5 m from traffic and 2 m a.g.l., is situated at the junction of two major roads in Kowloon. The Tung Chung (TC) site (16 m a.g.l.) is in a newly developed residential area 3 km south of Hong Kong International Airport. This site is 80 m south of the North Lantau Highway, which is the only road connecting the airport and Tung Chung New Town with the city center. The Hok Tsui (HT) site is located in a relatively remote coastal area 10 km southeast of downtown Hong Kong Island,

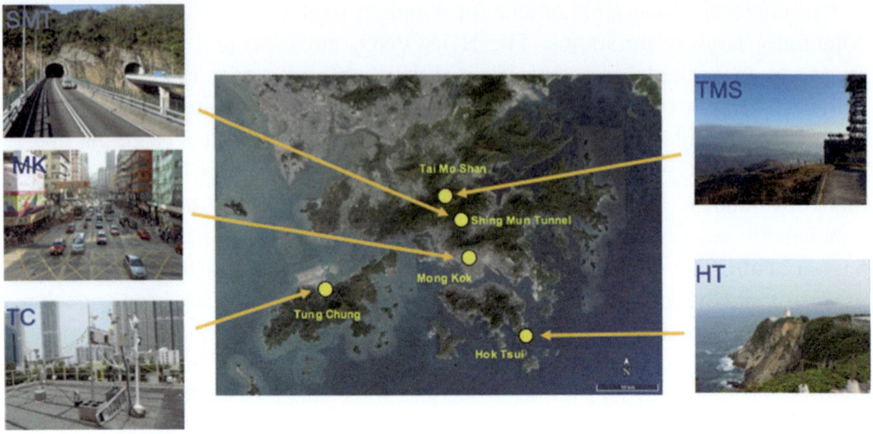

Fig. 4.1 Locations and views of the five Hong Kong measurement sites used in this study

Table 4.1 Site information with mean and range of concentrations

Site reference	Time	Description	Mean NO_x (ppb)	Mean HONO (ppb)	Range of HONO (ppb)
Shing Mun Tunnel (SMT)	Mar 11–21, 2015	Tunnel with dense traffic	1117.3	15.79	7.78–30.67
Mong Kok (MK)	Mar 28–May 4, 2015	Roadside	129.0	3.26	0.12–15.44
Tung Chung (TC)	Aug and Nov 2011 Feb and May 2012	Suburban	23.2	0.71	BDL-5.00
Hok Tsui (HT)	Sep 1–Dec 19, 2012	Coastal	5.48	0.16	BDL-1.15
Tai Mo Shan (TMS)	Nov 15–Dec 6, 2013	Mountain-top	3.24	0.14	BDL-0.58

BDL Below Detection Limit

and does not feature any strong emission sources close by. The mountain-top Tai Mao Shan (TMS) site sits on Tai Mo Shan, the highest point in Hong Kong. The altitude (974 m a.s.l.) is slightly beneath the average height of the PBL, which varies between 1.2 km (autumn) and 1.0 km (winter) (Yang et al. 2013).

4.2.2 *Instrumentation*

The HONO was measured with a commercial LOng Path Absorption Photometer (QUMA, Model LOPAP-03) (Heland et al. 2001). The ambient air was sampled using two temperature-controlled stripping coils in series with a mixture reagent of

100 g sulfanilamide and 1 L HCl (37% v/v) in 9 L of pure water (R1). In the first stripping coil, most of the HONO and a fraction of interfering substances were absorbed in solution R1 of Channel 1. In the second stripping coil, the remaining HONO and most of the interfering species were absorbed in solution R1 of Channel 2. After adding a reagent of 1.6 g N-naphtylethylendiamine-dihydrochloride in 9 L of pure water (R2) to both coils, colored azo dye was formed in both channels, which were then separately detected via long path absorption in special Teflon tubing. The HONO signal was the difference between the signals in the two channels. Compressed air was injected into the instrument to correct for the small drifts in baseline, and a span check was conducted to check the sensitivity of the instrument. Before each campaign, an HONO-source generator (QUMA, Model QS-03) was used to determine the sampling efficiency of the HONO in the sampling unit, which was found to be 99.95%. For more details, the reader is referred to Xu et al. (2015).

4.3 Results and Discussion

4.3.1 Concentrations at the Four Ambient Sites

The concentrations of HONO revealed a cascade trend from the tunnel site (mean = 15.79 ppb), to the roadside site (3.26 ppb), to the suburban site (0.71 ppb), to the coastal site (0.16 ppb), and to mountain-top background site (0.14 ppb) (Table 4.1). This result indicates that road traffic is an important source of HONO (either from direct emissions or from reaction(s) involving NO_2 emitted from vehicles). Diurnal patterns of HONO, NO_x, and O_3 at the four ambient sites are shown in Fig. 4.2. The influence of vehicles was clearly seen in the roadside and suburban sites, as evidenced by the morning peaks (and afternoon rush hour at the roadside site). In contrast, the two remote sites showed HONO peaking at noon or in the early afternoon. At the TC site, HONO began to decline immediately after sunrise—a typical pattern in urban (Pusede et al. 2015; Lee et al. 2016; Wang et al. 2015), suburban (Tong et al. 2016; Michoud et al. 2014), and remote areas (Ren et al. 2010; Wojtal et al. 2011). However, this trend was not observed at the other three ambient sites. The bimodal curve of HONO at MK can be attributed to the high traffic volume during rush hours, whereas the daytime peaks in HT and TMS can be explained by the heterogeneous conversion of NO_2 to HONO and variations in the height of the PBL, respectively.

At the TC site, HONO data were collected for 1 month in each of the four seasons. The highest concentration of HONO was found in late autumn (November), followed by late winter (February), late summer (August), and late spring (May). The mean HONO mixing ratios for these four seasons were 0.93 ± 0.78, 0.91 ± 0.74, 0.66 ± 0.53, and 0.35 ± 0.30 ppb, respectively. The seasonal profiles of the HONO were similar to those of the other gases, such as CO and NO_x (Fig. 4.3).

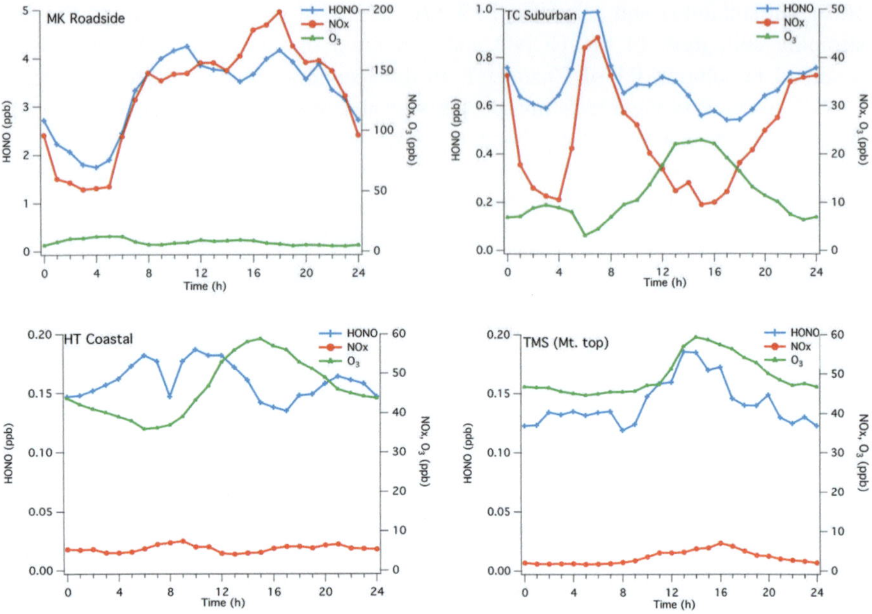

Fig. 4.2 Average mixing ratios of *HONO*, NO_x, and O_3 as a function of time of day at four ambient sites: MK (roadside), TC (suburban, August), HT (coastal background), and TMS (mountain top)

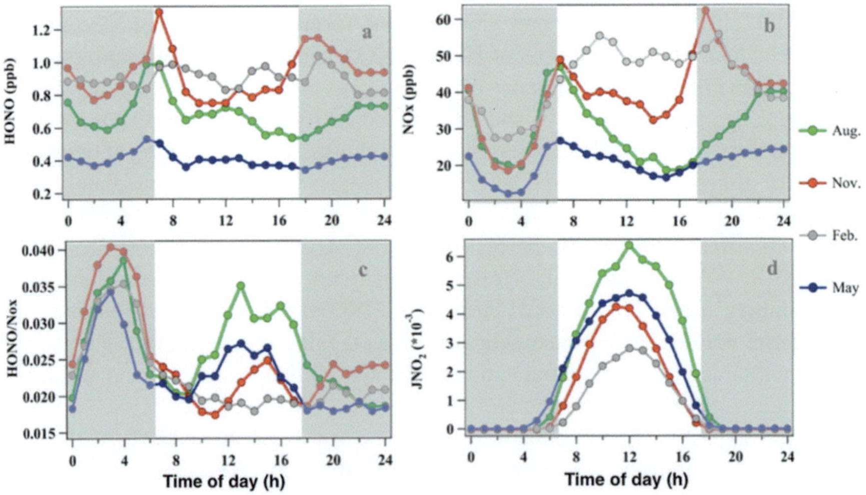

Fig. 4.3 Diurnal variations of (**a**) *HONO*, (**b**) NO_x, (**c**) $HONO/NO_x$, and (**d**) J_{NO2} in February, May, August, and November at the TC sites. The *gray* shading refers to the nighttime period (18:00–06:00) (Figure adapted from Xu et al. 2015)

The mean concentration of HONO at the MK roadside site was more than four times larger than that measured at a roadside site in Houston (Rappenglück et al. 2013), while the averaged HONO/NO_x ratio was only 12% larger. The mixing ratios of HONO and NO_x at our urban-influenced coastal background site (HT) were much higher, compared with those observed within the Marine Boundary Layer (MBL) of the Atlantic Ocean near North Carolina (Ye et al. 2016). HONO and NO_2 at our TMS site exhibited similar levels and trends to those measured on a rural mountain in Germany (Acker et al. 2006), although the noon peak of HONO in our measurement was much sharper. This may have been related to an upward transport of polluted air mass from the urban area to the hilltop once the PBL was broken up.

4.3.2 *Emission Ratio from Road Traffic*

The emission ratios of HONO/NO_x can be derived from measurements taken in the tunnel and in fresh plumes at a receptor site. Figure 4.4 shows a moderately positive correlation between HONO and NO_x (10-min data) during the high-traffic period (7:00–22:00) inside the SMT. The emission ratio (HONO/NO_x) was highly variable, but the majority of the data points fell between the [HONO] = 0.6% [NO_x] and [HONO] = 2.3% [NO_x] lines, which is in line with the measurements taken in gasoline vehicle-dominated tunnels (Kirchstetter et al. 1996; Kurtenbach et al. 2001), and the upper limit of HONO/NO_x obtained in a lab study with diesel exhausts (Gutzwiller et al. 2002).

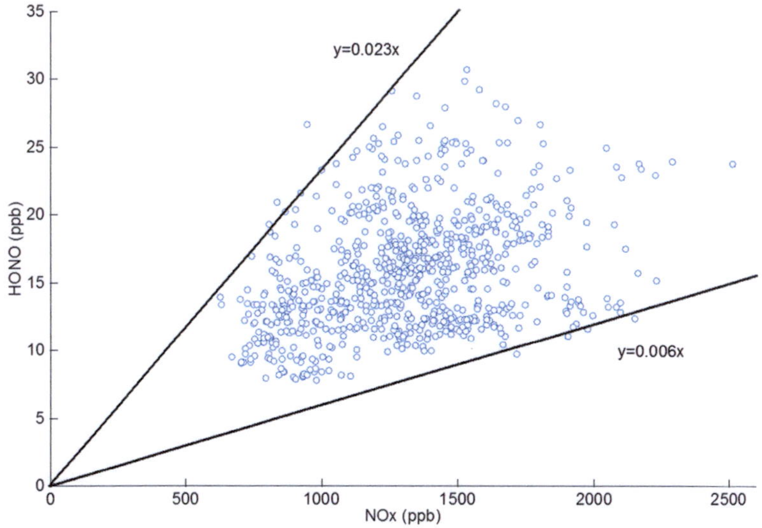

Fig. 4.4 Scatter plot of 10-min averaged *HONO* vs. NO_x in the Shing Mun Tunnel

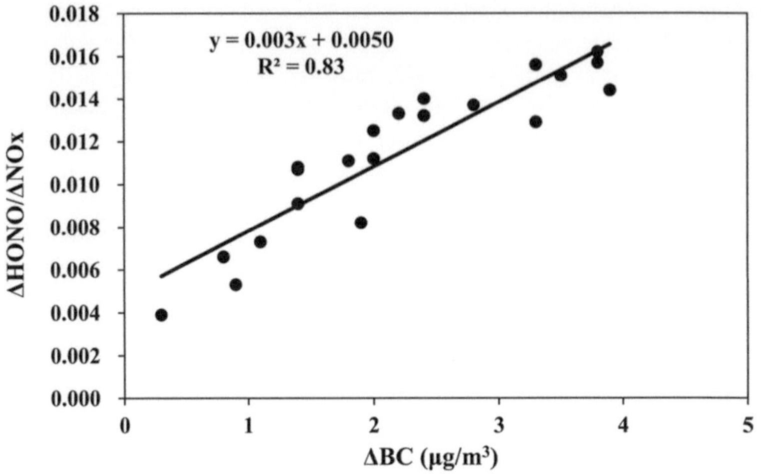

Fig. 4.5 Correlation of $\Delta HONO/\Delta NO_x$ with freshly emitted BC in the 21 fresh plumes (Figure adapted from Xu et al. 2015)

At the TC suburban site, 21 freshly emitted plumes in the dark were selected based on the sharp increase in NO compared with NO_2, and the good correlation between HONO and NO_x (Xu et al. 2015). The derived emission ratios Δ[HONO]/Δ[NO_x] from vehicular plumes were mostly higher than the commonly adopted value of 0.8% (Kurtenbach et al. 2001), and were found to be positively and almost linearly related to the emission of BC (Fig. 4.5). This result suggests that BC seems to enhance the formation of HONO in fresh emissions, and this process (i.e., the heterogeneous reduction of NO_2 on fresh BC) must be considered in modeling studies, especially in locations that are close to road traffic. An empirical formula—$\frac{\Delta HONO}{\Delta NOx} = 0.0050 + 0.003\Delta BC$ (where ΔBC is in μg/m^3)—was proposed to account for this effect (Xu et al. 2015).

4.3.3 *Derived Heterogeneous Production Rates of HONO*

The heterogeneous production rates of HONO can be derived from ambient measurements of HONO and NO_2. At the coastal HT site, we selected six nighttime air masses in which the HONO concentration showed a steady increase while other gases such as ozone, CO, and NO_x were less variable, to rule out the photolytic reactions and the change in air mass (Fig. 4.6). The conversion rates of NO_2 to HONO were calculated in these air masses using a slightly improved linear regression method based on the formula below (see Zha et al. 2014 for details).

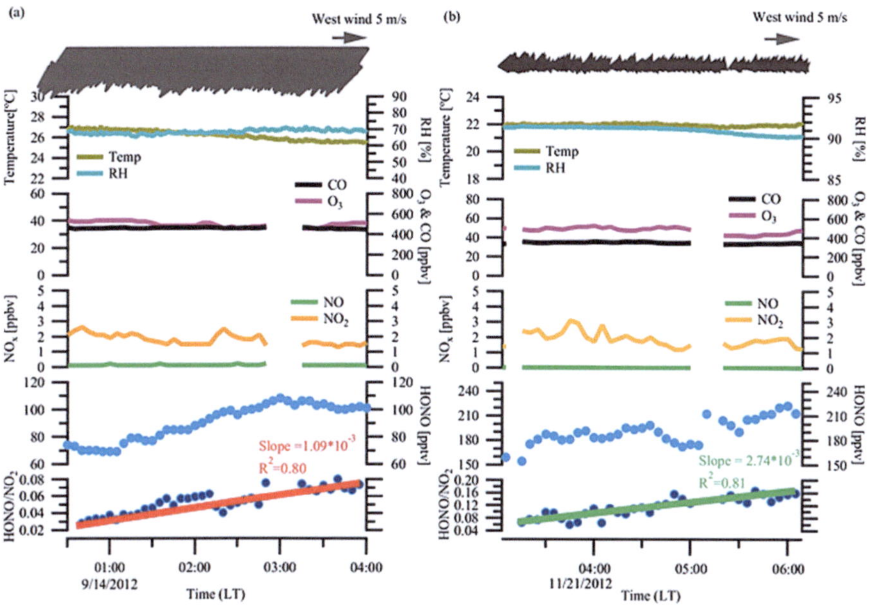

Fig. 4.6 (**a**) Land case from the night of September 13–14, 2012; (**b**) sea case from the night of November 20–21, 2012 (Figure adapted from Zha et al. 2014)

$$C_{HONO} = \frac{[HONO](t_2) - [HONO](t_1)}{(t_2 - t_1) \times \overline{[NO_2]}}$$

where t_1 and t_2 are the starting and ending times of the case and $\overline{[NO_2]}$ is the average concentration of NO_2 during the period between t_1 and t_2.

Six-hourly backward trajectories using the Hybrid Single-Particle Lagrangian Integrated Trajectory (HYSPLIT) model identified whether the air mass came over land or sea. Larger nocturnal heterogeneous conversion rates of NO_2 to HONO were observed when air masses passed over sea surfaces, than those passed over land surfaces (~3 times), suggesting that air–sea interactions may be a significant source of atmospheric HONO.

Figure 4.7 compares the NO_2 to HONO conversion rate in different atmospheric environments (Xu et al. 2015). The air masses that came over the sea at the HT site had the highest conversion rate, whereas the “land” case at the same site resembled those measured in other remote areas. The low conversion rate at the TC site could be due to nocturnal traffic emissions. The large variability in conversion rates suggests that air quality models should consider inhomogeneity of surface type when simulating NO_2 formation on ground/sea surfaces.

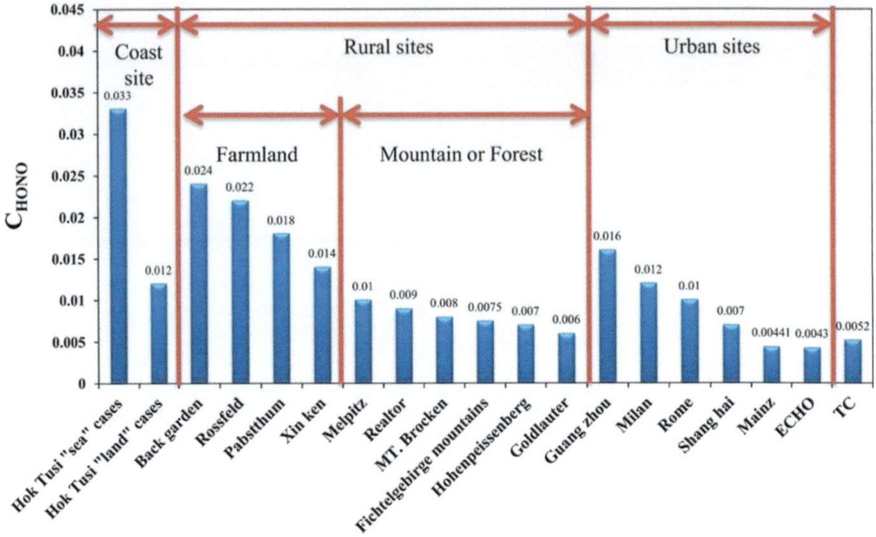

Fig. 4.7 Observed NO_2 to HONO conversion rates at HT, TC, and other sites (Figure adapted from Xu et al. 2015)

4.3.4 WRF-Chem Simulations on HONO Sources and Their Effects on Ozone and $PM_{2.5}$

We parameterized the up-to-date HONO sources into a widely used regional chemistry transport model (WRF-Chem). These sources included (1) heterogeneous reactions on ground surfaces, (2) photo-enhanced reactions on aerosol surfaces, (3) direct vehicle and vessel emissions, (4) potential conversion of NO_2 at the ocean surface, and (5) emissions from soil bacteria. Detailed parameterizations of the sources in the model can be found in Zhang et al. (2016). Seven simulation cases considering different HONO sources were designed in that study, as listed in Table 4.2.

Figure 4.8 illustrates the mean observed and simulated HONO at the TC site in each case during August 20–31, 2011 when a multi-day photochemical episode occurred in the region. The heterogeneous conversion of NO_2 on the land surface was the dominant source (~42%) of the HONO observed at the TC site, followed by emissions from soil bacteria (23%), the oceanic source (9%), the gaseous formation via photochemical reaction consuming OH and NO (6%), and aerosol surfaces (3%). The results suggest that HONO sources in suburban areas could be more complex and diverse than those in urban or rural areas, indicating the need to consider the bacterial and ocean processes in HONO production in forested or coastal areas.

The inclusion of HONO sources has improved the WRF-Chem simulations of both daily and peak O_3 concentrations during noontime at multiple stations across

Table 4.2 WRF-Chem simulation cases considering different sources

Case	Additional HONO sources considered
BASE	Without additional HONO processes (i.e., with NO + OH only)
L	Heterogeneous sources from **L**and surfaces
LO	Heterogeneous sources from **L**and and **O**cean surfaces
LOA	Heterogeneous sources from **L**and, **O**cean, and **A**erosol surfaces
LOAE	Heterogeneous sources from **L**and, **O**cean, and **A**erosol surfaces and traffic **E**missions
LOAES	Heterogeneous sources from **L**and, **O**cean, and **A**erosol surfaces, traffic **E**missions, and **S**oil emissions
LOAESG	Heterogeneous sources from **L**and, **O**cean, and **A**erosol surfaces, traffic **E**missions, **S**oil emissions, and additional **G**as-phase formations

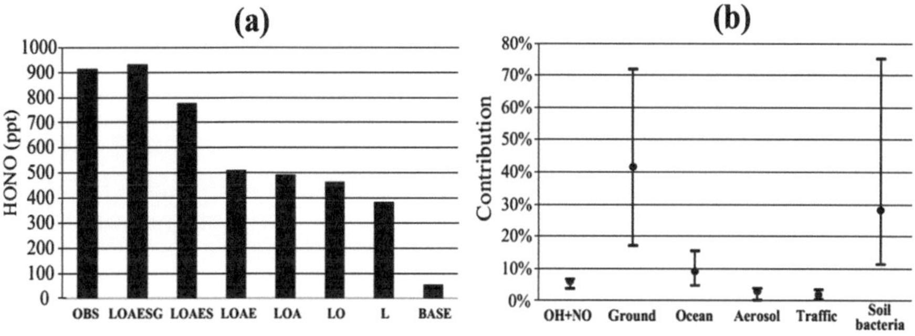

Fig. 4.8 (**a**) Averages of observed and simulated HONO and (**b**) contributions from homogeneous source (*OH + NO*), heterogeneous production on ground surface (*Ground*), oceanic source (*Ocean*), heterogeneous production on aerosol surface (*Aerosol*), traffic emissions (*Traffic*), and soil bacteria emissions (*Soil bacteria*) at the TC site (Figure adapted from Zhang et al. 2016)

Hong Kong (see Table 4.3). The mean biases between the simulated and observed values of 8- and 1-h maxima O_3 in the LOAES case have appreciable improvements, decreasing from −10.03 ppb in the BASE case to −0.53 ppb, and from −21.77 ppb in the BASE case to −9.17 ppb, respectively. The daily average O_3 concentration at the measurement stations also increased from 30.34 ppb in the BASE case to 31.99 ppb in the LOAES case, much closer to the actual observations.

We further evaluated the effects on simulated O_3 and $PM_{2.5}$ due to the considered heterogeneous sources of HONO. Figures 4.9 and 4.10 present the averaged distributions of modeled O_3 14:00 LTC and daily $PM_{2.5}$ in the BASE and LOAES cases over the PRD-HK region during August 25–31, 2011 (see Zhang et al. 2016 for details). As shown in Fig. 4.8, high levels of O_3 of up to 80–100 ppb occurred over the northern parts of the PRD during the episode. Higher O_3 concentrations were shown over the downwind areas of the PRD in the LOAES case, especially over Hong Kong (Fig. 4.10c), with an enhancement that reached up to 5–10 ppb (8–15%) over the urban areas in this region. The simulated $PM_{2.5}$ was up to

Table 4.3 Statistics of model performance in BASE and LOAES cases for hourly O_3, 8-h maximum O_3, and 1-h maximum O_3 at 12 air-quality monitoring stations in Hong Kong (unit: ppb)

	Metrics	OBS	BASE	LOAES
Hourly	Mean	30.97	30.34	31.99
	COR	/	0.75	0.78
	MB	/	−0.63	1.02
8-h maximum	Mean	64.25	54.22	63.71
	COR	/	0.51	0.56
	MB	/	−10.03	−0.53
1-h maximum	Mean	80.58	58.81	71.41
	COR	/	0.42	0.45
	MB	/	−21.77	−9.17

OBS observation, *Mean* averaged value, *COR* correlation, *MB* mean bias

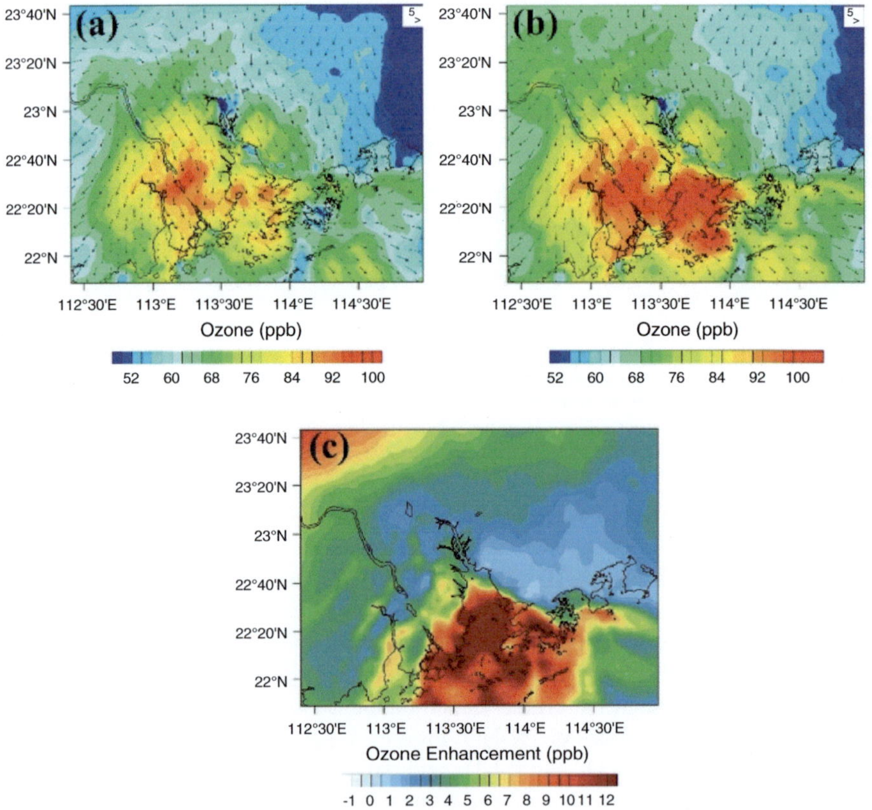

Fig. 4.9 Spatial distributions of simulated O_3 at 14:00 LTC over the PRD-HK in (**a**) BASE case and (**b**) LOAES case during the polluted period (August 25–31, 2011). Differences between BASE and LOAES cases are shown in (**c**)

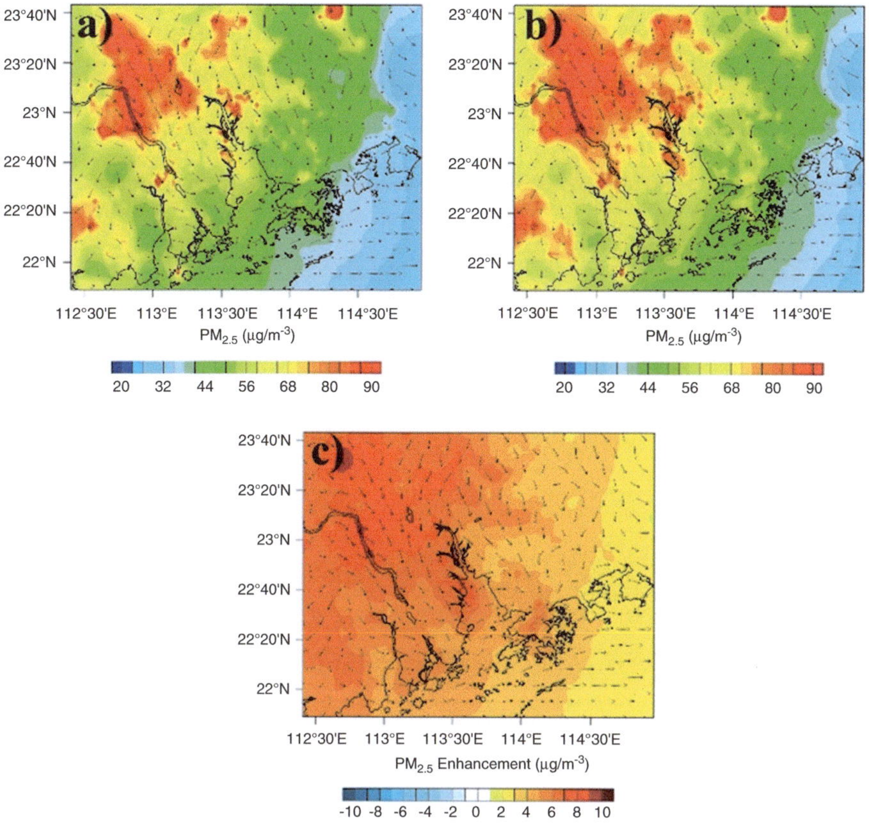

Fig. 4.10 Same as Fig. 4.9, but for daily $PM_{2.5}$

80–90 μg/m^3 in urban Guangzhou and Foshan and 50–60 μg/m^3 in Shenzhen and Hong Kong in the BASE case. With inclusion of the additional HONO sources, the average $PM_{2.5}$ increased to 60–70 μg/m^3 in Shenzhen and Hong Kong. As shown in Fig. 4.11c, the enhancements in total $PM_{2.5}$ were around 8–10 μg/m^3 (10–15%) in Guangzhou and Foshan, and 4–7 μg/m^3 (5–15%) in Shenzhen, Dongguan, and Hong Kong. The increase in $PM_{2.5}$ was mainly due to additional production of aerosol nitrate (see Fig. 4.11). It is worth noting that the simulated effect of HONO on PM2.5 should be a lower limit of the actual effect because it is known that the current WRF-Chem model (and most other chemistry transport models) tends to under-simulate secondary organic aerosols. Thus, it is probable that the simulated increase in organic aerosols due to additional HONO sources may have been underestimated.

Overall, the incorporation of the aforementioned HONO sources into the model appreciably improved the ozone predictions at multiple monitoring stations in Hong Kong, and led to an 8–15% enhancement in averaged ozone and 10–15% in daily

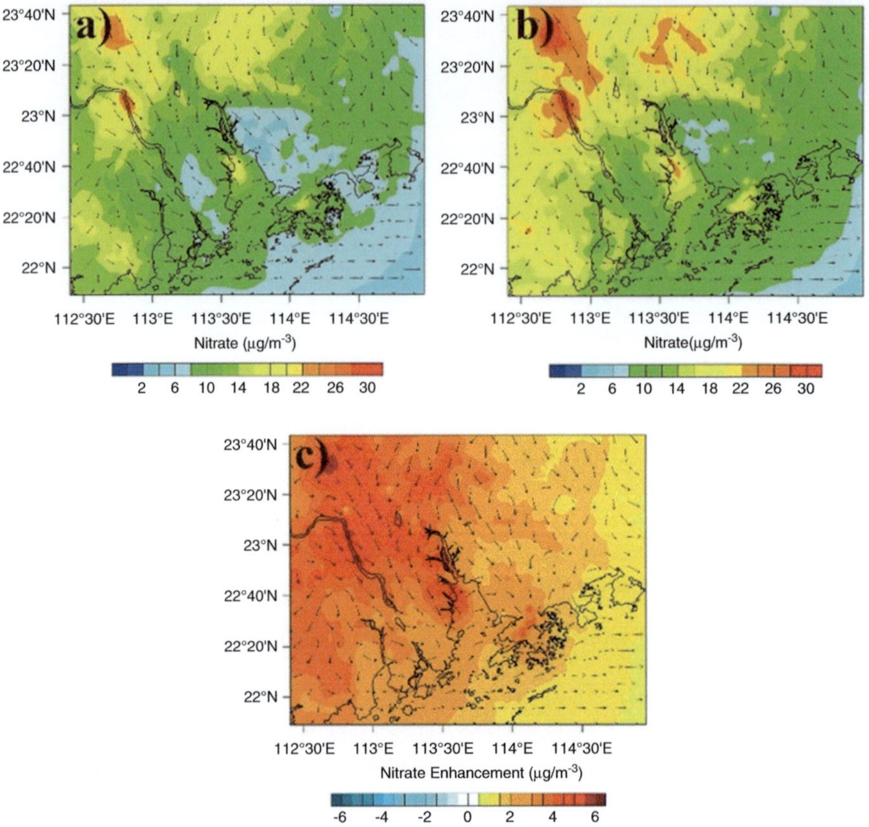

Fig. 4.11 Same as Fig. 4.9, but for daily PM2.5 nitrate

$PM_{2.5}$ over the Pearl River Delta region and Hong Kong. Our results highlight the importance of accurately representing HONO sources in simulations of secondary pollutants over polluted regions.

4.4 Concluding Remarks

As illustrated in our research in the Hong Kong region and others' findings, it is clear that HONO can be a very important source of hydroxyl radicals in polluted regions, which in turn play a critical role in atmospheric photochemistry and air pollution problems such as high ground-level ozone and haze. However, there are still significant uncertainties in quantifying HONO sources. Emissions from fuel combustion for various sources/under different conditions must be more fully understood, and soil emissions should be studied for a wide range of soil types and then properly represented in current models. The uptake processes of NO_2 on various surfaces (aerosol, terrestrial, and oceanic) should be better quantified, and our understanding of photo-related (daytime) sources could be improved. Emission-based air-quality models should consider the additional daytime sources of HONO (apart from reaction of OH and NO and vehicle emissions) in predicting secondary pollutants.

Acknowledgements The contributions to field measurements from Zheng Xu, Likun Xue, Shun Cheng Lee, Kin Fai Ho, Long Cui, Peter Louie, and Connie Luk are gratefully acknowledged. This work was supported by the Hong Kong Environmental Protection Department, Hong Kong Research Grants Council (C5022-14G), and the Hong Kong Polytechnic University.

References

Acker, K., & Möller, D. (2007). Atmospheric variation of nitrous acid at different sites in Europe. *Environmental Chemistry, 4*, 242–255.

Acker, K., Moller, D., Wieprecht, W., Meixner, F. X., Bohn, B., Gilge, S., Plass-Dulmer, C., & Berresheim, H. (2006). Strong daytime production of OH from HNO_2 at a rural mountain site. *Geophysical Research Letters, 33*.

Ammann, M., Kalberer, M., Jost, D. T., Tobler, L., Rossler, E., Piguet, D., Gaggeler, H. W., & Baltensperger, U. (1998). Heterogeneous production of nitrous acid on soot in polluted air masses. *Nature, 395*, 157–160.

An, J., Li, Y., Chen, Y., Li, J., Qu, Y., & Tang, Y. (2013). Enhancements of major aerosol components due to additional HONO sources in the North China Plain and implications for visibility and haze. *Advances in Atmospheric Sciences, 30*, 57–66.

Aumont, B., Chervier, F., & Laval, S. (2003). Contribution of HONO sources to the $NO_x/HO_x/O_3$ chemistry in the polluted boundary layer. *Atmospheric Environment, 37*, 487–498.

Czader, B. H., Rappenglück, B., Percell, P., Byun, D. W., Ngan, F., & Kim, S. (2012). Modeling nitrous acid and its impact on ozone and hydroxyl radical during the Texas Air Quality Study 2006. *Atmospheric Chemistry and Physics, 12*, 6939–6951.

Czader, B. H., Li, X., & Rappenglueck, B. (2013). CMAQ modeling and analysis of radicals, radical precursors, and chemical transformations. *Journal of Geophysical Research-Atmospheres, 118*, 11376–11387.

Gutzwiller, L., Arens, F., Baltensperger, U., Gäggeler, H. W., & Ammann, M. (2002). Significance of semivolatile diesel exhaust organics for secondary HONO formation. *Environmental Science & Technology, 36*, 677–682.

Heland, J., Kleffmann, J., Kurtenbach, R., & Wiesen, P. (2001). A new instrument to measure gaseous nitrous acid (HONO) in the atmosphere. *Environmental Science & Technology, 35*, 3207–3212.

Kirchstetter, T. W., Harley, R. A., & Littlejohn, D. (1996). Measurement of nitrous acid in motor vehicle exhaust. *Environmental Science & Technology, 30*, 2843–2849.

Kleffmann, J. (2007). Daytime sources of nitrous acid (HONO) in the atmospheric boundary layer. *Chem Phys Chem, 8*, 1137–1144.

Kurtenbach, R., Becker, K. H., Gomes, J. A. G., Kleffmann, J., Lörzer, J. C., Spittler, M., Wiesen, P., Ackermann, R., Geyer, A., & Platt, U. (2001). Investigations of emissions and heterogeneous formation of HONO in a road traffic tunnel. *Atmospheric Environment, 35*, 3385–3394.

Lee, J. D., Whalley, L. K., Heard, D. E., Stone, D., Dunmore, R. E., Hamilton, J. F., Young, D. E., Allan, J. D., Laufs, S., & Kleffmann, J. (2016). Detailed budget analysis of HONO in central London reveals a missing daytime source. *Atmospheric Chemistry and Physics, 16*, 2747–2764.

Li, S., Matthews, J., & Sinha, A. (2008). Atmospheric hydroxyl radical production from electronically excited NO_2 and H_2O. *Science, 319*, 1657–1660.

Li, G., Lei, W., Zavala, M., Volkamer, R., Dusanter, S., Stevens, P., & Molina, L. T. (2010). Impacts of HONO sources on the photochemistry in Mexico City during the MCMA-2006/ MILAGO Campaign. *Atmospheric Chemistry and Physics, 10*, 6551–6567.

Li, Y., An, J., Min, M., Zhang, W., Wang, F., & Xie, P. (2011). Impacts of HONO sources on the air quality in Beijing, Tianjin and Hebei Province of China. *Atmospheric Environment, 45*, 4735–4744.

Li, X., Brauers, T., Häseler, R., Bohn, B., Fuchs, H., Hofzumahaus, A., Holland, F., Lou, S., Lu, K. D., Rohrer, F., Hu, M., Zeng, L. M., Zhang, Y. H., Garland, R. M., Su, H., Nowak, A., Wiedensohler, A., Takegawa, N., Shao, M., & Wahner, A. (2012). Exploring the atmospheric chemistry of nitrous acid (HONO) at a rural site in Southern China. *Atmospheric Chemistry and Physics, 12*, 1497–1513.

Li, X., Rohrer, F., Hofzumahaus, A., Brauers, T., Häseler, R., Bohn, B., Broch, S., Fuchs, H., Gomm, S., Holland, F., Jäger, J., Kaiser, J., Keutsch, F. N., Lohse, I., Lu, K., Tillmann, R., Wegener, R., Wolfe, G. M., Mentel, T. F., Kiendler-Scharr, A., & Wahner, A. (2014). Missing gas-phase source of HONO inferred from zeppelin measurements in the troposphere. *Science, 344*, 292–296.

Michoud, V., Colomb, A., Borbon, A., Miet, K., Beekmann, M., Camredon, M., Aumont, B., Perrier, S., Zapf, P., Siour, G., Ait-Helal, W., Afif, C., Kukui, A., Furger, M., Dupont, J. C., Haeffelin, M., & Doussin, J. F. (2014). Study of the unknown HONO daytime source at a European suburban site during the MEGAPOLI summer and winter field campaigns. *Atmospheric Chemistry and Physics, 14*, 2805–2822.

Monge, M. E., D'anna, B., Mazri, L., Giroir-Fendler, A., Ammann, M., Donaldson, D. J., & George, C. (2010). Light changes the atmospheric reactivity of soot. *Proceedings of the National Academy of Sciences, 107*, 6605–6609.

Osthoff, H. D., Roberts, J. M., Ravishankara, A. R., Williams, E. J., Lerner, B. M., Sommariva, R., Bates, T. S., Coffman, D., Quinn, P. K., Dibb, J. E., Stark, H., Burkholder, J. B., Talukdar, R. K., Meagher, J., Fehsenfeld, F. C., & Brown, S. S. (2008). High levels of nitryl chloride in the polluted subtropical marine boundary layer. *Nature Geoscience, 1*, 324–328.

Pusede, S. E., Vandenboer, T. C., Murphy, J. G., Markovic, M. Z., Young, C. J., Veres, P. R., Roberts, J. M., Washenfelder, R. A., Brown, S. S., Ren, X., Tsai, C., Stutz, J., Brune, W. H., Browne, E. C., Wooldridge, P. J., Graham, A. R., Weber, R., Goldstein, A. H., Dusanter, S., Griffith, S. M., Stevens, P. S., Lefer, B. L., & Cohen, R. C. (2015). An atmospheric constraint on the NO_2 dependence of daytime near-surface nitrous acid (HONO). *Environmental Science & Technology, 49*, 12774–12781.

Qin, M., Xie, P., Su, H., Gu, J., Peng, F., Li, S., Zeng, L., Liu, J., Liu, W., & Zhang, Y. (2009). An observational study of the HONO–NO_2 coupling at an urban site in Guangzhou City, South China. *Atmospheric Environment, 43*, 5731–5742.

Rappenglück, B., Lubertino, G., Alvarez, S., Golovko, J., Czader, B., & Ackermann, L. (2013). Radical precursors and related species from traffic as observed and modeled at an urban highway junction. *Journal of the Air & Waste Management Association, 63*, 1270–1286.

Ren, X., Gao, H., Zhou, X., Crounse, J. D., Wennberg, P. O., Browne, E. C., Lafranchi, B. W., Cohen, R. C., Mckay, M., Goldstein, A. H., & Mao, J. (2010). Measurement of atmospheric nitrous acid at Bodgett Forest during BEARPEX2007. *Atmospheric Chemistry and Physics, 10*, 6283–6294.

Sarwar, G., Roselle, S. J., Mathur, R., Appel, W., Dennis, R. L., & Vogel, B. (2008). A comparison of CMAQ HONO predictions with observations from the Northeast Oxidant and Particle Study. *Atmospheric Environment, 42*, 5760–5770.

Sörgel, M., Regelin, E., Bozem, H., Diesch, J. M., Drewnick, F., Fischer, H., Harder, H., Held, A., Hosaynali-Beygi, Z., Martinez, M., & Zetzsch, C. (2011). Quantification of the unknown HONO daytime source and its relation to NO_2. *Atmospheric Chemistry and Physics, 11*, 10433–10447.

Su, H., Cheng, Y. F., Shao, M., Gao, D. F., Yu, Z. Y., Zeng, L. M., Slanina, J., Zhang, Y. H., & Wiedensohler, A. (2008). Nitrous acid (HONO) and its daytime sources at a rural site during the 2004 PRIDE-PRD experiment in China. *Journal of Geophysical Research-Atmospheres, 113*, D14132.

Su, H., Cheng, Y., Oswald, R., Behrendt, T., Trebs, I., Meixner, F. X., Andreae, M. O., Cheng, P., Zhang, Y., & Pöschl, U. (2011). Soil nitrite as a source of atmospheric HONO and OH radicals. *Science, 333*, 1616–1618.

Tong, S., Hou, S., Zhang, Y., Chu, B., Liu, Y., He, H., Zhao, P., & Ge, M. (2016). Exploring the nitrous acid (HONO) formation mechanism in winter Beijing: Direct emissions and heterogeneous production in urban and suburban areas. *Faraday Discussions.*

Vandenboer, T. C., Markovic, M. Z., Sanders, J. E., Ren, X., Pusede, S. E., Browne, E. C., Cohen, R. C., Zhang, L., Thomas, J., Brune, W. H., & Murphy, J. G. (2014). Evidence for a nitrous acid (HONO) reservoir at the ground surface in Bakersfield, CA, during CalNex 2010/. *Journal of Geophysical Research-Atmospheres, 119*, 9093–9106.

Wang, T., Cheung, V. T. F., Anson, M., & Li, Y. S. (2001). Ozone and related gaseous pollutants in the boundary layer of eastern China: Overview of the recent measurements at a rural site. *Geophysical Research Letters, 28*, 2373–2376.

Wang, L., Wen, L., Xu, C., Chen, J., Wang, X., Yang, L., Wang, W., Yang, X., Sui, X., Yao, L., & Zhang, Q. (2015). HONO and its potential source particulate nitrite at an urban site in North China during the cold season. *Science of the Total Environment, 538*, 93–101.

Wojtal, P., Halla, J. D., & Mclaren, R. (2011). Pseudo steady states of HONO measured in the nocturnal marine boundary layer: A conceptual model for HONO formation on aqueous surfaces. *Atmospheric Chemistry and Physics, 11*, 3243–3261.

Wong, K. W., Oh, H. J., Lefer, B. L., Rappenglück, B., & Stutz, J. (2011). Vertical profiles of nitrous acid in the nocturnal urban atmosphere of Houston, TX. *Atmospheric Chemistry and Physics, 11*, 3595–3609.

Xu, Z., Wang, T., Wu, J., Xue, L., Chan, J., Zha, Q., Zhou, S., Louie, P. K. K., & Luk, C. W. Y. (2015). Nitrous acid (HONO) in a polluted subtropical atmosphere: Seasonal variability, direct vehicle emissions and heterogeneous production at ground surface. *Atmospheric Environment, 106*, 100–109.

Xue, L., Wang, T., Louie, P. K. K., Luk, C. W. Y., Blake, D. R., & Xu, Z. (2014). Increasing external effects negate local efforts to control ozone air pollution: A case study of Hong Kong and implications for other Chinese cities. *Environmental Science & Technology, 48*, 10769–10775.

Yang, D., Li, C., Lau, A. K., & Li, Y. (2013). Long-term measurement of daytime atmospheric mixing layer height over Hong Kong. *Journal of Geophysical Research-Atmospheres, 118*(5), 2422–2433. doi:10.1002/jgrd.50251.

Ye, C., Zhou, X., Pu, D., Stutz, J., Festa, J., Spolaor, M., Tsai, C., Cantrell, C., Mauldin, R. L., Campos, T., Weinheimer, A., Hornbrook, R. S., Apel, E. C., Guenther, A., Kaser, L., Yuan, B., Karl, T., Haggerty, J., Hall, S., Ullmann, K., Smith, J. N., Ortega, J., & Knote, C. (2016). Rapid cycling of reactive nitrogen in the marine boundary layer. *Nature, 532*, 489–491.

Zha, Q., Xue, L., Wang, T., Xu, Z., Yeung, C., Louie, P. K. K., & Luk, C. W. Y. (2014). Large conversion rates of NO_2 to HNO_2 observed in air masses from the South China Sea: Evidence of strong production at sea surface? *Geophysical Research Letters, 41*, 7710–7715.

Zhang, L., Wang, T., Zhang, Q., Zheng, J., Xu, Z., & Lv, M. (2016). Potential sources of nitrous acid (HONO) and their impacts on ozone: A WRF-Chem study in a polluted subtropical region. *Journal of Geophysical Research-Atmospheres, 21*, 3645–3662.

Zhou, X., Zhang, N., Teravest, M., Tang, D., Hou, J., Bertman, S., Alaghmand, M., Shepson, P. B., Carroll, M. A., Griffith, S., Dusanter, S., & Stevens, P. S. (2011). Nitric acid photolysis on forest canopy surface as a source for tropospheric nitrous acid. *Nature Geoscience, 4*, 440–443.

Chapter 5
Connection Between East Asian Air Pollution and Monsoon System

Mian Chin, Huisheng Bian, Tom Kucsera, Thomas Diehl, Zhining Tao, Dongchul Kim, and Xiaohua Pan

Abstract We present in this chapter a study on connections between the wintertime East Asian air pollution phenomenon and the monsoon strength. East Asia has been experiencing a fast worsening of air quality in recent years, particularly in winter, a problem commonly attributed to the increase of pollutant emissions associated with the rapid economic development. Meanwhile, previous studies have shown that the decadal-scale weakening of the Asian monsoon also contributed to the increase of $PM_{2.5}$ (particulate matter with diameter less than 2.5 μm), a major pollutant that determines the air quality. Using a global modeling system, we investigate the emission and meteorological effects on the wintertime surface $PM_{2.5}$ concentrations in East Asia in the past 30 years and find their relationship to the monsoon strength. We also examine the feedbacks between aerosols and meteorological fields via aerosol-radiation interaction to estimate the effects of such interaction on air quality.

M. Chin (✉)
NASA Goddard Space Flight Center, 8800 Greenbelt Rd, Greenbelt, MD, USA
e-mail: mian.chin@nasa.gov

H. Bian
NASA Goddard Space Flight Center, 8800 Greenbelt Rd, Greenbelt, MD, USA

University of Maryland Baltimore County, 100 Hilltop Cir., Baltimore, MD 21250, USA
e-mail: huisheng.bian@nasa.gov

T. Kucsera • Z. Tao • D. Kim
NASA Goddard Space Flight Center, 8800 Greenbelt Rd, Greenbelt, MD, USA

Universities Space Research Association, 7178 Columbia Gateway Dr., Columbia, MD 21046, USA
e-mail: tom.l.kucsera@nasa.gov; zhining.tao@nasa.gov; dongchul.kim@nasa.gov

T. Diehl
Joint Research Center, Via Enrico Fermi 2749, I - 21027 Ispra, VA, Italy
e-mail: thomas.diehl@jrc.ec.europa.eu

X. Pan
NASA Goddard Space Flight Center, 8800 Greenbelt Rd, Greenbelt, MD, USA

University of Maryland College Park, College Park, MD 20742, USA
e-mail: xiaohua.pan@nasa.gov

I. Bouarar et al. (eds.), *Air Pollution in Eastern Asia: An Integrated Perspective*,
ISSI Scientific Report Series 16, DOI 10.1007/978-3-319-59489-7_5

Keywords East Asian air pollution • Winter monsoon • Connection

5.1 Introduction

East Asia's rapid economic growth over the past several decades has brought a remarkable increase of air pollution levels in the region. It is now one of the most polluted regions in the world with pollutant "hot spots" frequently observed by the satellite remote sensing from space (e.g., Veefkind et al. 2011; Streets et al. 2013; Duncan et al. 2016). The major weather system in East Asia, namely the monsoon, has been widely reported to exhibit a weakening trend since the latter half of the last century (e.g., Xu et al. 2006; Dash et al. 2009; Zhu et al. 2012a). Although the build-up of long-lived greenhouse gases and the resulting global warming are deemed the major culprit for the regional climate change in Asia, the increase of pollution levels, especially aerosol particles, has caused considerable concerns over their impacts on air quality, water cycle, and regional climate (Ramanathan et al. 2008). Furthermore, the outflow of Asian pollution through long-range transport and monsoon convection is evident from the satellite data (e.g., Park et al. 2007; Yu et al. 2008; Vernier et al. 2011), which affect the atmospheric composition, global climate, and downwind regions' air quality.

Air pollution in East Asia has become a serious problem that threatens the dense population living in the region. Aerosol and precursor emissions from fossil fuel and biofuel combustion in Asia have increased significantly in recent decades, with high emissions concentrated in areas of high population density (Fig. 5.1). China is now among the world's largest emitters of aerosols and precursor species such as SO_2, black carbon (BC), and organic matter (OM) from anthropogenic sources according to several leading estimates of emission datasets (e.g., Streets et al. 2009; Granier et al. 2011; Smith et al. 2011). It has been recorded that pollution levels in several big cities in China were ten times above the threshold level of "good air quality". For example, the concentrations of $PM_{2.5}$ – a form of aerosol or particulate matter with a diameter less than 2.5 μm produced mainly by burning fuels – in Beijing's air hit a few hundreds of $\mu g\ m^{-3}$ in several episodes in recent years with an annual average value close to 100 $\mu g\ m^{-3}$ (compared with the US EPA's air quality standard of annual average less than 15 $\mu g\ m^{-3}$).

The rising pollution levels in East Asia are commonly attributed to the fast increase in emissions from the rapid economic development in the region. On the other hand, the observed multi-decadal declining trend of the East Asian summer monsoon could have also contributed to the increase in aerosol concentrations in summer (Zhu et al. 2012b; Chin 2012). So far, there is only very limited studies that link the East Asian monsoon circulation to air pollution, which have mostly focused in the summer, largely due to its connection to the South Asian summer monsoon strength and the active convective transport to the upper troposphere and lower stratosphere. However, the East Asian air pollution is usually worse in winter than in summer (e.g., Fig. 5.2), mostly because of more stagnation of the air, weaker vertical mixing, and more residential energy use during the cold months especially

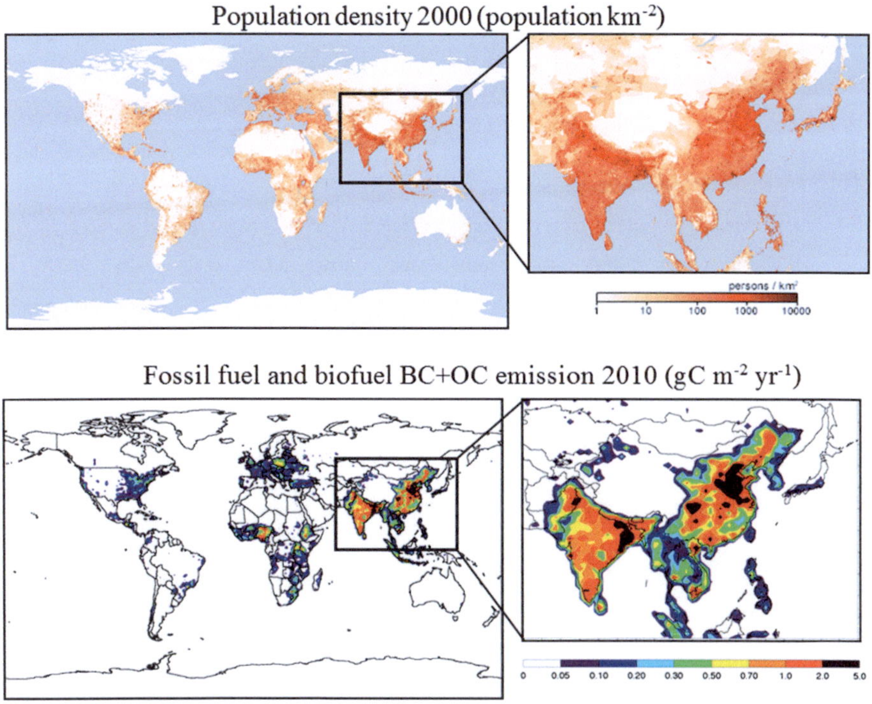

Fig. 5.1 *Top*: Population density in 2000 (Source: http://neo.sci.gsfc.nasa.gov/view.php?datasetId=SEDAC_POP). *Bottom*: Fossil fuel and biofuel BC and OC emissions in 2010 in gC m^{-2} $year.^{-1}$ (Source: Hemispheric Transport of Air Pollution, http://htap.org)

in the northern part of East Asia. Furthermore, the absorbing aerosol particles from pollution sources, such as BC, interact with radiation that cool the surface but heat the atmosphere, leading to an increase of boundary layer stability (Ramanathan et al. 2005; Xu et al. 2006) to effectively trap the pollution within the boundary layer, thus worsening the air quality. A recent study analyzed the phenomenon of increased winter haze over eastern China with a global chemistry transport model GEOS-Chem to conclude that anthropogenic emissions dominated the increase in winter haze days whereas the change of meteorological parameters dominated by the weakening of wind speed explained 17% of such increase (Yang et al. 2016). Yet, that study was not able to tackle any aspects on the effects of aerosol-meteorology interactions on wintertime air quality.

We present here a modeling study focusing on the wintertime pollution phenomenon in China using a NASA modeling system. First, similar to the study of Yang et al. (2016), we simulate the 30-year variations of aerosols with an off-line model, the Goddard Chemistry Aerosol Radiation and Transport (GOCART) model, driven by the meteorological data from the NASA Modern Era Reanalysis for Research and Applications (MERRA) (Rienecker et al. 2011) to examine the role of anthropogenic emission and meteorological conditions affecting the

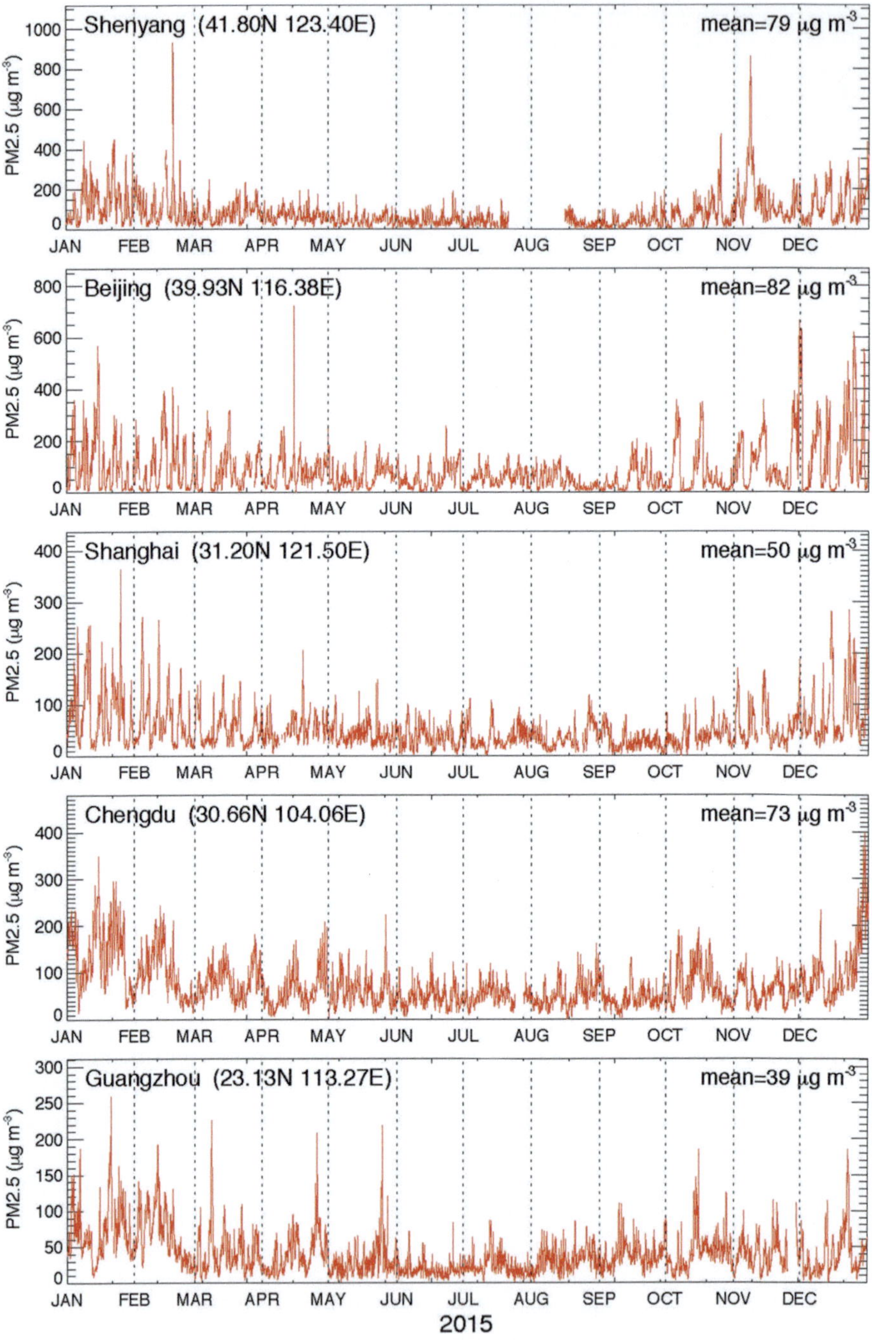

Fig. 5.2 Hourly surface $PM_{2.5}$ concentrations in five big cities in China in 2015 (Source: US State Department Mission China program (http://stateair.net/web/mission/1/))

wintertime particle pollution levels over China. We then examine the variation of anthropogenic emission and meteorology that causes the change of the particle pollution levels in five megacities in China in the past three decades. We further analyze the relationship between particle concentrations, key meteorological variables, and the East Asian winter monsoon index (EAWMI) using the GOCART simulated pollution aerosols and the MERRA meteorological fields. Lastly, the aerosol effects on meteorology and pollution levels due to aerosol-radiation interactions are estimated through sensitivity simulations with the NASA GEOS-5 Atmospheric General Circulation Model (AGCM) coupled with GOCART aerosol simulations.

In the rest of this paper, we first describe the method used in this study in Sect. 5.2, including model simulations, the meteorological reanalysis fields, and the calculation of EAWMI. We then present in Sect. 5.3 the GOCART simulated 30-year (1980–2009) variability of wintertime pollution aerosol levels at five megacities in China and attribute it to the change of anthropogenic emission or the variations of meteorology; we also show the relationships between pollution aerosol, key meteorological variables, and EAWMI over China. In Sect. 5.4 we examine the aerosol effects on wintertime meteorology with the GEOS-5 AGCM simulation that include the feedbacks between pollution and aerosol-radiation interaction. Summary and remarks are given in Sect. 5.5.

5.2 Methods

5.2.1 Description of 30-Year Global Aerosol Simulation and Model Experiments

Aerosols are simulated with the GOCART model from 1980 to 2009. Details of the GOCART model multi-decadal simulations have been presented in Chin et al. (2014) and references therein. Briefly, GOCART simulates atmospheric aerosols and their precursors globally with meteorological fields from MERRA. There are 72 vertical levels extending from surface to 0.01 hPa. In this study, the horizontal resolution of GOCART is at 2.5 longitude by 2 latitude. The model accounts for emissions from fossil fuel and biofuel combustions, biomass burning, biogenic, volcanic eruptions, wind-blown dust, and sea-salt. We consider eight particle size bins for dust and four size bins for sea salt, and assume bulk aerosol for sulfate, organic matter (OM), and black carbon (BC) with lognormal size distributions. A major limitation in the current version of the GOCART model is that it does not include the nitrate and anthropogenic secondary organic aerosols, which are important $PM_{2.5}$ components over Asia. Therefore, in this study we do not try to match the magnitude of measured surface $PM_{2.5}$ over China, but rather to use the sum of sulfate, OM, and BC from anthropogenic sources (SU+BC+OM_an) as proxy of pollution $PM_{2.5}$. We consider biomass burning aerosols "anthropogenic" since most

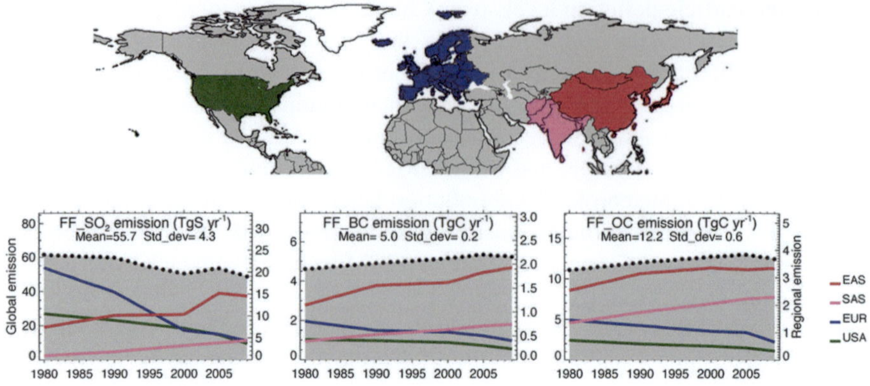

Fig. 5.3 *Top*: Region domains of East Asia (EAS, *red*), South Asia (SAS, *magenta*), Europe (EUR, *blue*), and USA (*green*). *Bottom*: Emissions from fossil fuel/biofuel (FF) sources shown as global total (*black circles* on *top* of the *grey* shaded area, *left* y axis scale) and from major source regions (*colored lines*, *right* y axis scale) of SO_2 (*left*), BC (*middle*), and OC (*right*) (Figure is adapted from Chin et al. 2014)

open fire in mid-latitudes and tropics is related to human activities (agriculture burning, forest cleaning, etc.).

We simulate the aerosol concentrations over three decades from 1980 to 2009 (details are given in Chin et al. 2014). Here, emissions of SO_2, BC, and OM from fossil fuel and biofuel combustion and biomass burning are taken from the emission data set A2-ACCMIP (Diehl et al. 2012). Natural sources from volcanic emissions, wind-blown dust and sea salt, and biogenic OM are also included in the model, but will not be a part of this study focusing on pollution aerosol. Figure 5.3 shows the annual global emissions from fossil fuel/biofuel (FF) sources as well as regional emissions from East Asia, South Asia, Europe, and the USA used in this study. Clearly, anthropogenic emissions in East Asia and South Asia have been steadily increasing in the 30-year period, in contrast with the continuously decreasing trends in Europe and the U.S. Such trends are consistent among all available global anthropogenic emission datasets for the past three decades, and the A2-ACCMIP estimated magnitudes shown in Fig. 5.3 generally within the range of those datasets (see Figs. 5.2, 5.3, 5.4, and 5.5 in Granier et al. 2011 for comparisons).

We perform three sets of 30-year model simulations: the "Base" simulation that includes time-varying emissions from fuel combustion, biomass burning, and natural sources, the "FixEmi" simulation that uses fixed emission from anthropogenic at year 2000 level but allows meteorology and natural emissions to vary, and "Natural" simulation that includes only emissions from natural sources. Difference between "Base" and "Natural" simulations, referred as "P1" hereafter, is the anthropogenic aerosol component (SU+BC+OM_an) with temporal variations driven by the changes from both anthropogenic emission and meteorology, in contrast with the difference between the "Base" and "FixEmi" simulations, referred as "P2" hereafter, that is regarded as the pollution aerosol whose variability is

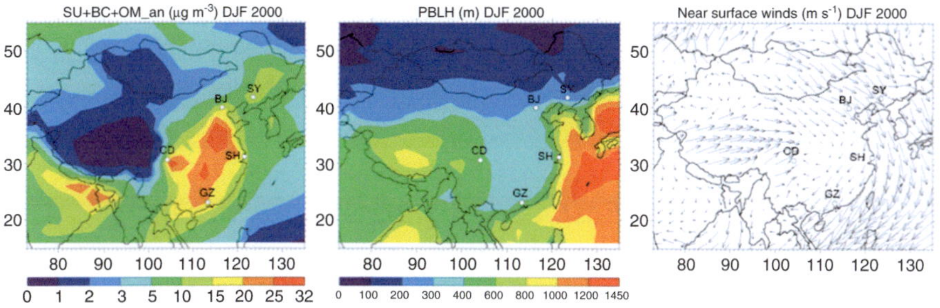

Fig. 5.4 Seasonal averaged (*left*) SU+BC+OM_an concentration, (*middle*) PBLH, and (*right*) near surface wind speed and direction in DJF 2000. Five megacity locations are marked in *circle*, where *SY* Shenyang, *BJ* Beijing, *SH* Shanghai, *CD* Chengdu, and *GZ* Guangzhou

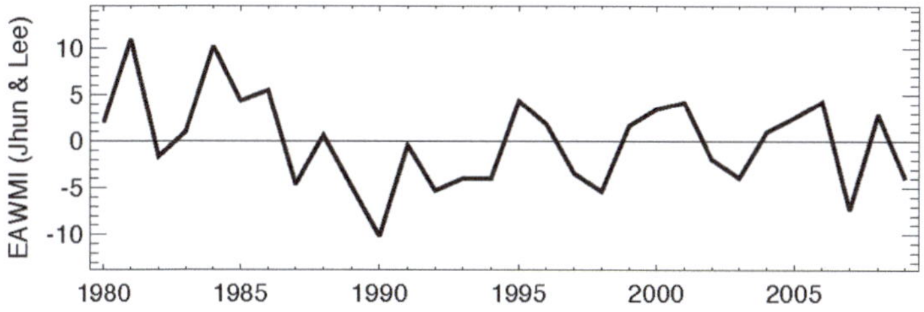

Fig. 5.5 East Asian winter monsoon index (*EAWMI*) calculated with MERRA zonal winds at 300 hPa, following the definition in Jhun and Lee (2004).

Table. 5.1 Model experiments 1980–2009 and methods of determination on the changes of anthropogenic aerosol concentrations controlled by the change of emission or meteorology

Simulation	Emission	Product	Information
Base	All emission from anthropogenic and natural sources	P1 = Base – Natural	Change of anthropogenic aerosol due to the changes of both anthropogenic emission and meteorology
FixEmi	Fossil fuel/biofuel/biomass burning emission fixed at 2,000 level	P2 = Base – FixEmi	Change of anthropogenic aerosol due to the change of anthropogenic emission
Natural	Only include emissions from natural sources	P3 = FixEmi – Natural	Change of anthropogenic aerosol due to the change of meteorology

controlled only by the change of emissions since the influences of meteorology has been removed in P2. Further, the change of pollution aerosol that is solely because of the change of meteorology can be derived from the difference between “FixEmi” and “Natural”, referred as “P3”. The model experiments and derived products are summarized in Table 5.1.

In this study, our analysis is focused on the East Asia in winter, which is defined as December-January-February (DJF) with December from the previous calendar year (e.g., winter 2000 consists of December 1999 and January and February 2000).

Model simulation of winter 2000 concentration of SU+BC+OM_an over the study region is shown in Fig. 5.4 together with the planetary boundary layer height (PBLH) and near surface winds from MERRA for the same time period.

5.2.2 Calculation of East Asian Winter Monsoon Index (EAWMI)

East Asian winter monsoon (EAWM) is the most important circulation system in the boreal winter in East Asia. The change of local weather and regional climate is strongly influenced by the anomalous monsoon circulation. The monsoon strength is usually indicated by a monsoon index that reflects the general characteristics of the monsoon circulation. There are several EAWMIs calculated with selected one or several members of EAWM circulations, such as sea level pressure (Shi 1996), geopotential height (Sun and Sun 1995), zonal or meridional winds at different altitudes (Chen et al. 2000; Li and Yang 2010), etc. A study on intercomparisons of four typical EAWMIs has concluded that all of them show similar interannual and multi-decadal variations, and also noted the weakening trends since the 1980s (Gao 2007). It should be pointed out that any EAWM can only well represent the monsoon characteristics over part of the East Asia, because the large domain of East Asia that contains different and complex climate zones. Therefore, no one EAWMI can capture the anomalies of monsoon circulation at all local scales. Nonetheless, the EAWMI serves as a general indicator of the winter meteorology in large part of the region, as we will discuss later.

In this study, we use the EAWMI introduced by Jhun and Lee 2004), which is defined as the differences of the averaged zonal winds at 300 hPa between two domains at 27.5–37.5N, 110–170E and 50–60N, 80–140E to describe the variability of the winter monsoon strength in mid-latitude East Asia. This EAWMI is associated with the East Asian jet stream such that a strong (weak) winter monsoon is characterized by an enhanced (reduced) upper-level jet stream south of Japan (Jhun and Lee 2004). The Jhun-Lee EAWMI from 1980 to 2009 calculated with the MERRA winds at 300 hPa is shown in Fig. 5.5.

5.3 Multi-decadal Variations of Pollution PM Levels Over China

5.3.1 Change of Pollutant Level in Five Megacities

We plot in Fig. 5.6 the differences of wintertime average SU+BC+OM_an between a given year in 1980 and 2009 and the reference year 2000 at five major megacities in China: Shenyang, Beijing, Shanghai, Chengdu, and Guangzhou (location marked

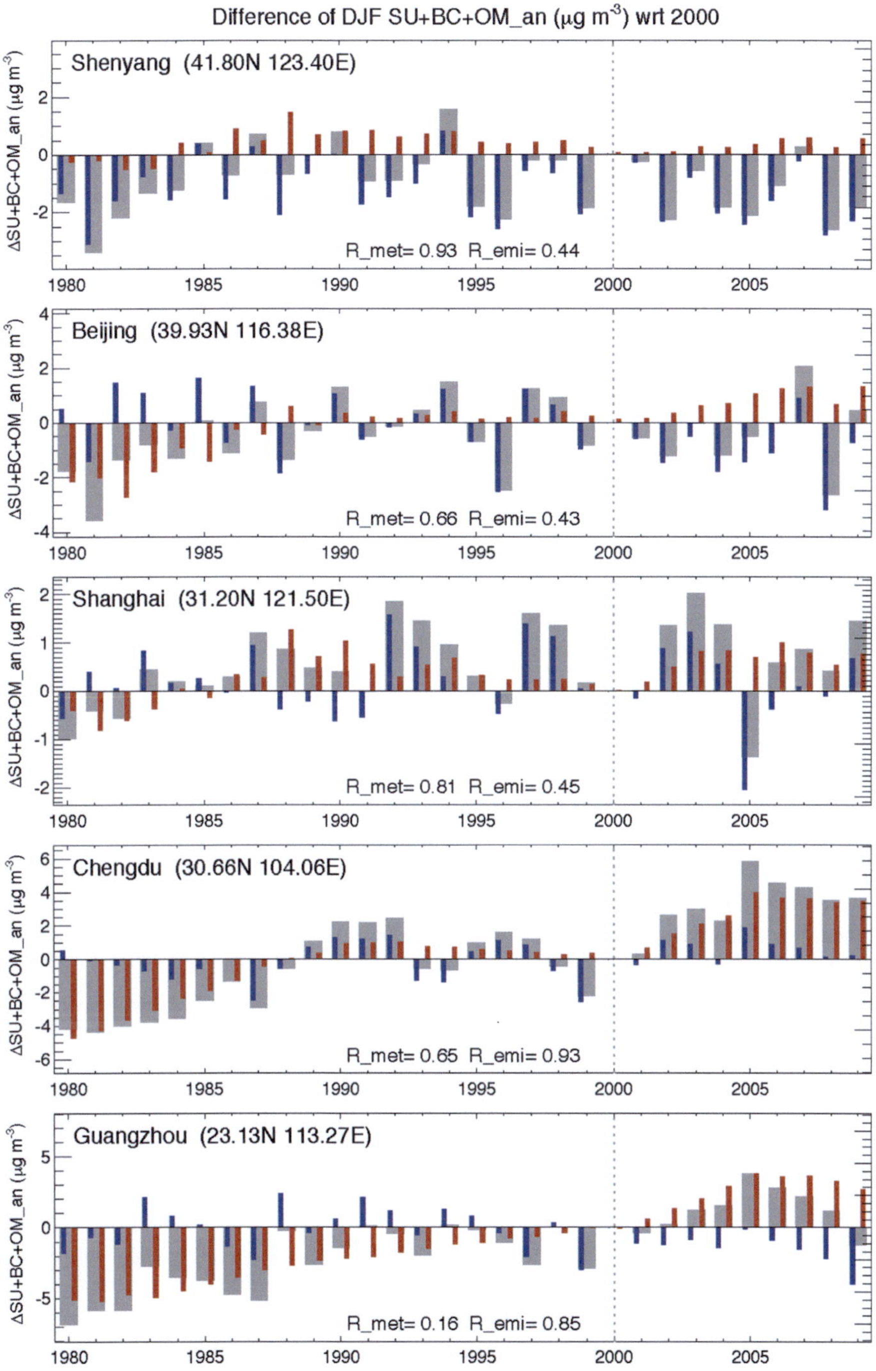

Fig. 5.6 Difference of wintertime SU+BC+OM_an surface concentrations between a given year in 1980–2009 and the year 2000 from product P1 (*thick grey bar*, change due to both meteorology and emission), P2 (*thin red bar*, change due to emission), and P3 (*thin blue bar*, change due to meteorology)

in Fig. 5.4). These cities are chosen to demonstrate the variation of pollution levels in most populated cites located in different parts of China, where the PM measurements in recent years are readily available (e.g., Fig. 5.2). The population of these megacities is more than or close to ten million in mid-2010s. The thick grey bars in Fig. 5.4 represent the level of SU+BC+OM_an in each winter during 1980–2009 that is different from the winter of 2000 (from product P1, see Table 5.1), whereas the thin red bars indicate such difference that is caused by the change of anthropogenic emission (product P2) and the thin blue bar represents such difference that is caused by the change of meteorology (product P3). Concentrations of SU+BC+OM_an in these cities are not necessarily only determined by the regional anthropogenic emission, but they can also be affected by extra regional sources via long-range transport.

It is interesting to see the differences of the relative importance of anthropogenic emission and meteorology affecting SU+BC+OM_an from north to south. Over Shenyang, located in northeast China (41.8N, 123.4E), the interannual variation of wintertime SU+BC+OM_an is predominantly controlled by the internannual difference of meteorology. The situation in Shenyang is in sharp contrast with that in Guangzhou, which is a coastal city located in southeast China (23.1N, 113.3E) where the increasing trend of pollution concentrations over the 30-year period is predominantly controlled by the increase of anthropogenic emission. The relative importance of anthropogenic emission and meteorology can be seen on the relative magnitudes of red and blue bars in Fig. 5.6 as well as by the correlation coefficients between the concentration of SU+BC+OM_an and meteorology (R_met) or pollutant emissions (R_emi) listed on each panel. Anthropogenic emission is also the most important factor determining the pollutant levels and increasing trend in Chengdu, an inland city in south-central China (30.7N, 104.1E) within the Sichuan Basin. Although the meteorology also moderately correlated with the pollutant levels, its absolute influence is much smaller compared to the emission. Over China's two most populated cities, Shanghai (31.2N, 121.5E, population>22 million) and Beijing (39.9N, 116.4E, population near 20 million), the change of pollutant levels are controlled by both anthropogenic emission and meteorology with similar magnitudes, but the correlation is stronger with the change of meteorology.

In Chengdu, the changes of anthropogenic emission and meteorology usually work in the same direction to influence the change of SU+BC+OM_an concentration relative to 2000. However, in other cities, these two factors act against each other for at least half of the winters during the 30-year time period. For example, in Beijing in 2006, the increase of anthropogenic emission would make the SU+BC+OM_an concentration 1.3 μg m^{-3} higher than that in 2000; however, due to the counteract of meteorology, such would-be-increase is avoided to make the concentration level in 2006 the same as that in 2000.

We have examined a few key meteorological variables from MERRA that affect the pollution levels in winter, including PBLH, total precipitation, and near surface temperature and wind speed. We use the SU+BC+OM_an from product P3 in order to eliminate the effects from the emission change. Among those meteorological variables, PBLH is most closely anti-correlated with the wintertime SU+BC+OM_an concentrations in five megacities with correlation coefficients between

−0.4 (Guangzhou) and −0.8 (Shenyang) (see Fig. 5.7a). The near-surface wind speed at 10-m (W10m) are also generally anti-correlated with the pollution PM concentrations (from 0 in Guangzhou to −0.6 in Beijing) (Fig. 5.7b), although the correlation coefficients are not as negative as those between PBLH and concentrations. The negative correlation means that a shallower PBL and a weaker wind speed lead to higher pollutant concentrations near the surface. Precipitation and temperature, on the other hand, do not show clear correlation patterns with the wintertime surface concentrations. We will further discuss the relationship between the surface concentrations and meteorological variables in the next section.

We have calculated the correlation coefficients between EAWMI and SU+BC+OM_an to examine the relationship between them over the five cities. The strength of the anti-correlation decreases from north to south with the highest degree of anti-correlation ($R = -0.44$) in Shenyang and no correlation in Guangzhou ($R = 0.05$) (Fig. 5.7c). Such relationships indicate that the change of meteorological condition described by the Jhon and Lee EAWMI have different influences on pollutant concentrations at different locations in China. Therefore it is not always applicable to estimate the change of pollutant levels with EAWMI. The regional correlation patterns will be discussed in next section (Sect. 5.3.2).

5.3.2 Relationships Between Pollutant Concentration, Key Meteorological Variables, and EAWM Strength Over China in Winter

Figure 5.7 provides a broader view of the relationships between pollutant concentrations and wintertime meteorology over China. The pollutant concentrations are from product P3 (see Table 5.1) in order to highlight the role of meteorology. Without any change of anthropogenic emission, the pollutant concentration is negatively correlated with PBLH or W10m over eastern China (east of 100E, Fig. 5.7a, b). For PBLH, the strongest negative correlation is found in northeast China. Similar negative correlation patterns are found between pollutant concentrations and W10m for the same region. The correlation between concentration and EAWMI also displays a negative relationship in eastern China (Fig. 5.7c), although the correlation is weaker than that between the concentration and PBLH or W10m. The negative correlations over eastern China shown in Fig. 5.7a–c mean that a shallower (deeper) PBL layer or lower (higher) surface winds, both are associated with a weaker (stronger) EAWM, will lead to higher (lower) pollutant concentrations over eastern China. Interestingly, over western China/Tibet Plateau where local anthropogenic emissions are very low (see Fig. 5.1), these correlations either become positive or none, in contrast with the negative pattern in eastern China. We will further discuss it in the next paragraph. The last two panels in Fig. 5.7d, e show the correlation coefficients between EAWMI and two key meteorological variables, PBLH and W10m, respectively. Both meteorological variables are positively correlated

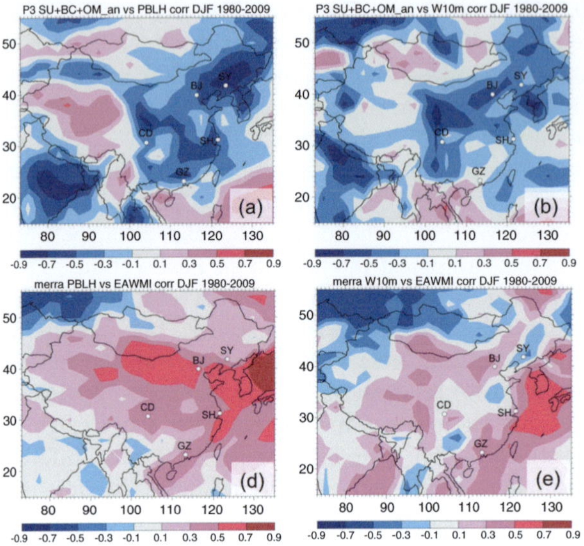

Fig. 5.7 *Top row*: correlation coefficients between 1980 and 2009 wintertime SU+BC+OM_an from P3 and (**a**) PBLH, (**b**) W10m, or **c** EAWMI. *Bottom row*: correlation coefficients between (**d**) PBLH or (**e**) W10m and EAWMI (See Fig. 5.4 for city names marked in *white circles*)

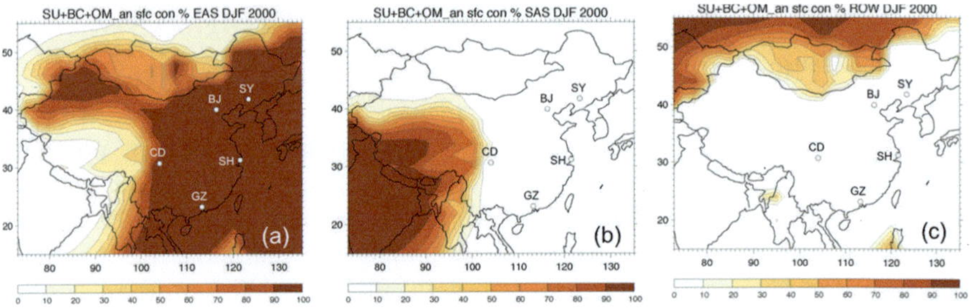

Fig. 5.8 Percentage of wintertime (DJF) surface SU+BC+OM_an concentrations originated from anthropogenic emissions in (**a**) East Asia (EAS), (**b**) South Asia (SAS), and (**c**) rest of the world (ROW) in 2000. Domains of tagged EAS and SAS emissions are shown in Fig. 5.3

with EAWMI over most area in China and adjacent ocean area. These correlations just confirm the general understanding: A weaker EAWM usually indicates shallower PBL and weaker winds, causing accumulation of more pollutants at the surface in the heavily polluted regions to worsen the air quality even without the increase of anthropogenic emission.

To understand the different relationships of surface pollutant concentrations and meteorological variables between eastern and western China, we have done GOCART simulations that tagged anthropogenic emissions in different source regions to attribute the origin of pollutant concentrations in China. Figure 5.8

displays the results showing the percentage of wintertime surface SU+BC +OM_an that are from anthropogenic emissions in East Asia (EAS), South Asia (SAS), and the rest of the world (ROW) in 2000. In eastern China (east of 100E), concentration of SU+BC+OM_an is overwhelmingly (90–100%) from the regional anthropogenic emissions; whereas in the western China over the Tibet plateau it is mostly (more than 50%) from long-range transport of pollutants originated in SAS (Fig. 5.8b). The extra regional pollutants are less likely to response to the local meteorology, such as PBLH and winds, than the pollutants directly produced within the region such as the situation in eastern China. It is also clear that, from Figs. 5.7c and 5.8, the EAWMI is not a reliable indicator for pollutant level change over the locations where significant fraction of pollutant is from extra regional sources.

5.4 Interaction Between Pollution Aerosols and Meteorology

In previous section we have discussed the results of the effects of meteorology on wintertime pollutant aerosol concentrations over China. To examine the aerosol effects on meteorology, we have performed a GEOS-5/GOCART AGCM simulation (with prescribed sea surface temperature) that allows meteorology to change as a result of aerosol-radiation interaction (ARI). The simulation is done for winter 2010 with two experiments: one with no ARI and one with ARI. Note that the aerosol-cloud interactions capability is under development in GEOS-5 and is not included in this study.

Figure 5.9 shows the surface concentration of SU+BC+OM, PBLH, and W10m in winter 2010 in the first row (Fig. 5.9a–c) from the GEOS-5 AGCM simulation without including ARI, and the changes of these quantities due to the inclusion of ARI (calculated as the simulation with ARI minus that without ARI) are shown in the second row (Fig. 5.9d–f). Absorbing aerosols (such as BC) absorb sunlight to cool the surface but heat the atmosphere, causing an increase of stability in the lower atmosphere. The reduction of PBLH over the Northern China Plain and other areas (Fig. 5.9e) and the weakening of W10m over the Eastern China Plain and Hexi Corridor (in north-central China) (Fig. 5.9f) are results of the ARI interaction, which lead to more PM (1–20 $\mu g\ m^{-3}$) confined at the surface over most of the polluted areas. Therefore, interaction between air pollution and radiation will likely weaken the EAWM to intensify the air quality problem. These results are consistent with previous findings, for example, several studies have shown that through the aerosol-radiation interaction, absorbing aerosols stabilize the lower atmosphere to significantly decrease the PBLH (e.g., Wendisch et al. 2008; Barbaro et al. 2013) and lengthen the duration of air stagnation.

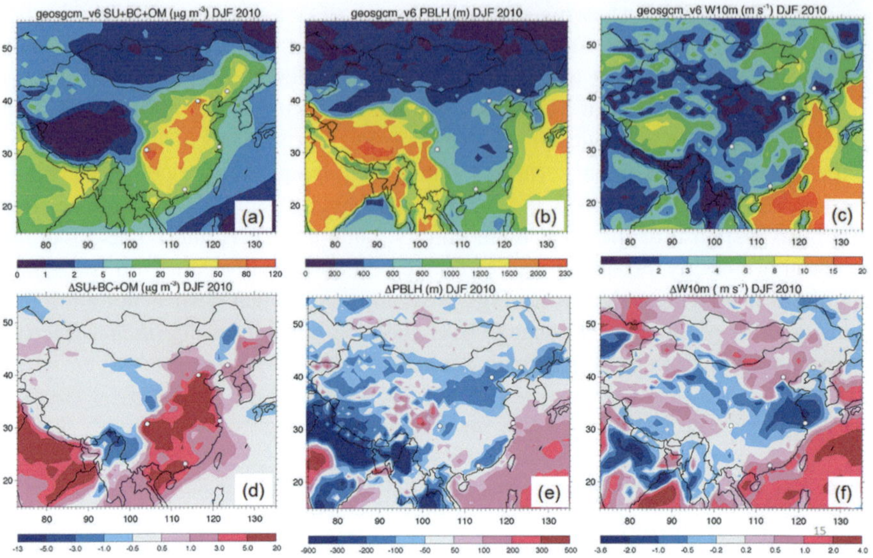

Fig. 5.9 *Top row*: (**a**) surface concentration of SU+BC+OM, (**b**) PBLH, and (**c**) W10m for winter 2010 from the GEOS-5 GCM simulation without considering aerosol-radiation interaction (ARI). *Bottom row*: the difference between the simulation with and without ARI for (**d**) surface concentration of SU+BC+OM, (**e**) PBLH, and (**f**) W10m

5.5 Summary and Concluding Remarks

We use the NASA GEOS-5/GOCART modeling system and MERRA reanalysis of meteorological fields to analyze the impact of anthropogenic emissions and meteorology on the multi-decadal variations of wintertime surface pollutant concentrations in China, to examine the link between air pollution level and EAWM, and to estimate the effects of air pollution on meteorology through the aerosol-radiation interactions.

Using the model experiments targeted to distinct the effects of emission and meteorology, we attribute the causes of wintertime SU+BC+OM_an concentrations changes in five major megacities in the 30-year time period (1980–2009) relative to that in the reference year 2000. These cities include (from north to south) Shenyang, Beijing, Shanghai, Chengdu, and Guangzhou. We find that the meteorology-induced differences of SU+BC+OM_an are within ± 3 μg m^{-2} averaged over DJF, a magnitude that is similar for all five cites. However, the importance of meteorology is remarkably different among the cities, generally decreases from north to south. The magnitude of SU+BC+OM_an change induced by the variation of meteorology is a factor five of that caused by the change of anthropogenic emission in the northeast city of Shenyang, whereas the ratio of the magnitudes is reversed in the southern city of Guangzhou.

By examining the correlations among SU+BC+OM_an, EAWMI, and several key meteorological variables in the 30 winters in China without the change of anthropogenic emission, we find that the PBLH and near surface winds are the most important meteorological variables affecting SU+BC+OM_an, since a shallower PBL tends to trap more pollutants near the surface and weaker winds makes the air more stagnant. The SU+BC+OM_an is negatively correlated with PBLH or W10m for the eastern half of China (east of 100E) where pollution level is high and most pollutants are from regional emissions. Such correlation falls apart in the western China, especially over the Tibet Plateau, where the pollution level is very low and most of them are originated from outside of EAS that is being transported to western China. The extra regional pollutants do not respond to the meteorology the same way as the pollutants generated locally (e.g., eastern China). Similar spatial correlation pattern is found between SU+BC+OM_an and EAWMI, suggesting that the EAWMI is not an effective indicator for air quality change over the Tibet Plateau where the pollutants are mostly from extra regional sources.

Over eastern Chin, PBLH or W10m and EAWMI are positively correlated over most area in China, although the strength varies with location. Even though a single EAWMI does not represent the monsoon strength everywhere in eastern China, it indicates the direction of the change of key meteorological variables to provide information for estimating the pollutant concentration change for the most polluted part of China. To summarize in a general sense: the weaker the EAWMI is, the shallower the PBL and the weaker the near surface winds become, and the higher the wintertime pollutant concentration develops.

Furthermore, in a broader view, the pollution aerosols in East Asia can modify monsoon strength and its onset, alter the precipitation patterns, change the wind direction and velocity, and modify the PBLH; all these will in turn affect the air quality. Our model sensitivity simulation with the GEOS-5 AGCM confirms these results: aerosol-radiation interaction leads to pollutants more concentrated at the surface, because such interaction reduces PBLH and weakens the winds over major polluted areas. Such positive feedback between aerosol and meteorology would make the bad air quality even worse. Studies have also suggested that the build-up of pollution aerosols in East Asia may have contributed to the weakening of the monsoon system (e.g., Ramanathan et al. 2005), although these studies mainly focused on the summer monsoon.

Acknowledgments This work is funded by NASA Atmospheric Composition Modeling and Analysis Program (ACMAP).

References

Barbaro, E., Vilà-Guerau de Arellano, J., Krol, M. C., & Holtslag, A. A. M. (2013). Impacts of aerosol shortwave radiation absorption on the dynamics of an idealized convective atmospheric boundary layer. *Boundary-Layer Meteorology, 148*, 31–49.

Chen, W., Graf, H. F., & Huang, R. H. (2000). The interannual variability of East Asian winter monsoon and its relation to the summer monsoon. *Advances in Atmospheric Sciences, 17*(1), 48–60.

Chin, M. (2012). Dirtier air from a weaker monsoon. *Nature Geoscience, 5*, 449–450.
Chin, M., Diehl, T., Tan, Q., Prospero, J. M., Kahn, R. A., Remer, L. A., Yu, H., Sayer, A. M., Bian, H., Geogdzhayev, I. V., Holben, B. N., Howell, S. G., Huebert, B. J., Hsu, N. C., Kim, D., Kucsera, T. L., Levy, R. C., Mishchenko, M. I., Pan, X., Quinn, P. K., Schuster, G. L., Streets, D. G., Strode, S. A., Torres, O., & Zhao, X.-P. (2014). Multi-decadal variations of atmospheric aerosols from 1980 to 2009: A perspective from observations and a global model. *Atmospheric Chemistry and Physics, 14*, 3657–3690.
Dash, S. K., Kulkarni, M. A., Mohanty, U. C., & Prasad, K. (2009). Changes in the characteristics of rain events in India. *Journal of Geophysical Research, 114*, D10109. doi:10.1029/2008JD010572.
Diehl, T., Heil, A., Chin, M., Pan, X., Streets, D., Schultz, M., & Kinne, S. (2012). Anthropogenic, biomass burning, and volcanic emissions of black carbon, organic carbon, and SO2 from 1980 to 2010 for hindcast model experiments. *Atmospheric Chemistry and Physics Discussions, 12*, 24895–24954.
Duncan, B., Lamsal, L. N., Thompson, A. M., Yoshida, Y., Lu, Z., Streets, D. G., Hurwitz, M. M., & Pickering, K. E. (2016). A space-based, high-resolution view of notable changes in urban NOx pollution around the world (2005–2014). *Journal of Geophysical Research, 121*. doi:10.1002/2015JD024121.
Gao, H. (2007). Comparison of East Asian winter monsoon indices. *Advances in Geosciences, 10*, 31–37.
Granier, C., et al. (2011). Evolution of anthropogenic and biomass burning emissions of air pollutants at global and regional scales during the 1980–2010 period. *Climatic Change, 109*. doi:10.1007/s10584-011-0154-1.
Jhun, J. G., & Lee, E. J. (2004). A new East Asian winter monsoon index and associated characteristics of the winter monsoon. *Journal of Climate, 17*(4), 711–726.
Li, Y., & Yang, S. (2010). A dynamical index for the East Asian winter monsoon. *Journal of Climate, 23*, 4255–4262.
Park, M., Randel, W. J., Gettelman, A., Massie, S. T., & Jiang, J. H. (2007). Transport above the Asian summer monsoon anticyclone inferred from Aura Microwave Limb Sounder tracers. *Journal of Geophysical Research, 112*, D16309. doi:10.1029/2006JD008294.
Ramanathan, V., Chung, C., Kim, D., Bettge, T., Buja, L., Kiehl, J. T., Washington, W. M., Fu, Q., Sikka, D. R., & Wild, M. (2005). Atmospheric brown clouds: Impacts on South Asian climate and hydrological cycle. *Proceedings National Academy Sciences, 102*, 5326–5333.
Ramanathan, V. M., et al. (2008). *Atmospheric brown clouds: Regional assessment report with focus on Asia*. Nairobi: United Nations Environment Programme.
Rienecker, M. M., Suarez, M. J., Gelaro, R., Todling, R., Bacmeister, J., Liu, E., Bosilovich, M. G., Schubert, S. D., Takacs, L., Kim, G.-K., Bloom, S., Chen, J., Collins, D., Conaty, A., da Silva, A., Gu, W., Joiner, J., Koster, R. D., Lucchesi, R., Molod, A., Owens, T., Pawson, S., Pegion, P., Redder, C. R., Reichle, R., Robertson, F. R., Ruddick, A. G., Sienkiewicz, M., & Woollen, J. (2011). MERRA – NASA's Modern-Era Retrospective Analysis for Research and Applications. *Journal of Climate, 24*, 3624–3648.
Shi, N. (1996). Features of the East Asian winter monsoon intensity on multiple time scale in recent 40 years and their relation to climate. *Journal of Applied Meteorological Sciences, 7*(2), 175–182.
Smith, S. J., van Aardenne, J., Klimont, Z., Andres, R. J., Volke, A., & Delgado Arias, S. (2011). Anthropogenic sulfur dioxide emissions: 1850–2005. *Atmospheric Chemistry and Physics, 11*, 1101–1116.
Streets, D. G., Yan, F., Chin, M., Diehl, T., Mahowald, N., Schultz, M., Wild, M., Wu, Y., & Yu, C. (2009). Anthropogenic and natural contributions to regional trends in aerosol optical depth, 1980–2006. *Journal of Geophysics Research, 114*, D00D18. doi:10.1029/2008JD011624.
Streets, D. G., Canty, T., Carmichael, G. R., de Foy, B., et al. (2013). Emissions estimation from satellite retrievals: A review of current capability. *Atmospheric Environment, 77*, 1011–1042.
Sun, S. Q., & Sun, B. M. (1995). The relationship between the anomalous winter monsoon circulation over East Asia and summer drought/flooding in the Yangtze and Huaihe River valley. *Acta Meteorologica Sinica, 57*(4), 513–522.

Veefkind, J. P., Boersma, K. F., Wang, J., Kurosu, T. P., Krotkov, N., Chance, K., & Levelt, P. F. (2011). Global satellite analysis of the relation between aerosols and short-lived trace gases. *Atmospheric Chemistry and Physics, 11*, 1255–1267.

Vernier, J. P., Thomason, L. W., & Kar, J. (2011). CALIPSO detection of an Asian tropopause aerosol layer. *Geophysical Research Letters, 38*, L07804. doi:10.1029/2010GL046614.

Wendisch, M., Hellmuth, O., Ansmann, A., Heintzenberg, J., Engelmann, R., Althausen, D., Eichler, H., Müller, D., Hu, M., Zhang, Y., & Mao, J. (2008). Radiative and dynamic effects of absorbing aerosol particles over the Pearl River Delta, China. *Atmospheric Environment, 42*, 6405–6416.

Xu, M., Chang, C.-P., Fu, C., Qi, Y., Robock, A., Robinson, D., & Zhang, H. (2006). Steady decline of east Asian monsoon winds, 1969–2000: Evidence from direct ground measurements of wind speed. *Journal of Geophysical Research, 111*, D24111. doi:10.1029/2006JD007337.

Yang, Y., Liao, H., & Lou, S. (2016). Increase in winter haze over eastern China in recent decades: Roles of variations in meteorological parameters and anthropogenic emissions. *Journal of Geophysical Research, 121*. Accepted. doi:10.1002/2016JD025136.

Yu, H., Remer, L. A., Chin, M., Bian, H., Kleidman, R. G., & Diehl, T. (2008). A satellite-based assessment of trans-Pacific transport of pollution aerosol. *Journal of Geophysical Research, 113*, D14S12. doi:10.1029/2007JD009349.

Zhu, C., Wang, B., Qian, W., & Zhang, B. (2012a). Recent weakening of northern East Asian summer monsoon: A possible response to global warming. *Geophysical Research Letters, 39*, L09701. doi:10.1029/2012GL051155.

Zhu, J., Liao, H., & Li, J. (2012b). Increases in aerosol concentrations over eastern China due to the decadal-scale weakening of the East Asian summer monsoon. *Geophysical Research Letters, 39*, L09809. doi:10.1029/2012GL051428.

Part II
Sources of Air Pollution

Chapter 6
Anthropogenic Emissions in Asia

Claire Granier, Thierno Doumbia, Louise Granier, Katerina Sindelarova, Gregory J. Frost, Idir Bouarar, Catherine Liousse, Sabine Darras, and Jenny Stavrakou

Abstract Accurate information on surface emissions is critical for understanding the current chemical composition of the atmosphere and for forecast simulations. Many research teams worked during the past years to better quantify emissions in

---

C. Granier (✉)
Laboratoire d'Aérologie, Toulouse, France

NOAA Earth System Research Laboratory, Boulder, CO, USA

LATMOS/IPSL, UPMC University Paris 06 Sorbonne Universities, Paris, France

Cooperative Institute for Research in Environmental Sciences, University of Colorado, Boulder, CO, USA
e-mail: claire.granier@latmos.ipsl.fr

T. Doumbia • L. Granier
LATMOS/IPSL, UPMC University Paris 06 Sorbonne Universities, Paris, France
e-mail: thiernodoumbia@yahoo.fr; louise.granier@latmos.ipsl.fr

K. Sindelarova
LATMOS/IPSL, UPMC University Paris 06 Sorbonne Universities, Paris, France

Charles University of Prague, Prague, Czech Republic
e-mail: katerina.sindelarova@latmos.ipsl.fr

G.J. Frost
NOAA Earth System Research Laboratory, Boulder, CO, USA
e-mail: Gregory.J.Frost@noaa.gov

I. Bouarar
Max Planck Institute for Meteorology, Bundesstr. 53, Hamburg 20146, Germany
e-mail: idir.bouarar@mpimet.mpg.de

C. Liousse
Laboratoire d'Aérologie, Toulouse, France
e-mail: catherine.liousse@aero.obs-mip.fr

S. Darras
Observatoire Midi-Pyrénées, Toulouse, France
e-mail: sabine.darras@obs-mip.fr

J. Stavrakou
Belgian Institute for Space Aeronomy, Brussels, Belgium
e-mail: jenny@aeronomy.be

I. Bouarar et al. (eds.), *Air Pollution in Eastern Asia: An Integrated Perspective*,
ISSI Scientific Report Series 16, DOI 10.1007/978-3-319-59489-7_6

different parts of the world, and to develop inventories of anthropogenic emissions for different periods. An evaluation of the most recent datasets providing emissions for Asia for the 1960–2014 period is discussed in this chapter for different gaseous and particulate compounds, i.e. carbon monoxide, nitrogen oxides, non-methane volatile organic compounds, sulfur dioxide, ammonia, black carbon, organic carbon, and particulate matter PM_{10} and $PM_{2.5}$. The comparison between the inventories is used to quantify the range of the emissions for several regions and different periods. Since very little information is available concerning the uncertainties on emissions in the different regions of the world, we use the comparison between the different datasets to provide some information on these uncertainties.

Keywords Anthropogenic emissions • Air pollution in Asia • Surface emission inventories

6.1 Introduction

Accurate knowledge of surface emissions of air pollutants is necessary to understand the origin and evolution of the severe air pollution episodes that are more and more frequent in China and other regions in Asia. In order to understand the origin and evolution of these events, it is necessary to have an accurate knowledge of the surface emissions of the compounds involved in air pollution. The spatial and temporal distributions of the emissions need to be determined, as well as their evolution over the past decades.

The chemical species involved in pollution episodes do not all originate from local surface emissions: they can be produced either by chemical processes or be transported far away from their sources. The formation of these chemical compounds involves species that are directly emitted at the surface. For example, ozone formation results from chemical reactions involving several surface emitted compounds such as carbon monoxide, methane, nitrogen oxides and volatile organic compounds. Particulate matter (PM) represents a mixture of different compounds from different origins. It can result from the direct emissions of black carbon and organic carbon, as well as of dust or sea salt. A large part of the PM concentrations also results from the oxidation of sulfur dioxide, ammonia, black carbon, organic carbon, as well as volatile organic compounds as detailed in other chapters of the book.

The understanding of the formation of air pollution in Asia and in other regions relies therefore on an accurate characterization of the direct emissions of air pollutants and their precursors at the local, regional and global scales. In this chapter, we present a review of the datasets providing emissions at the global scale and at the regional scale in Asia. The chapter focuses on anthropogenic emissions, which are the main drivers of changes in the atmospheric composition in Asia (He et al. 2002; Kan et al. 2012).

6.2 Anthropogenic Emissions at the Global Scale

Several datasets have been developed during the past years, which provide emissions of many atmospheric chemical compounds at the global scale. Up to now, all inventories do not provide emissions after 2008 or 2010. The only global emissions data available after 2010 are provided by future emissions scenarios, and we have included in the analysis of the emissions the so-called RCPs (Representative Concentration Pathways) scenarios (van Vuuren et al. 2011) developed for the IPCC (Intergovernmental Panel on Climate Change) Assessment Report 5.

In addition to global emissions inventories that will be reviewed in this chapter, we have also used the anthropogenic emissions of nitrogen oxides and volatile organic compounds optimized by inverse modeling based on the OMI and GOME-2 satellite observations (Mijling et al. 2013; Stavrakou et al. 2015).

A short description of the global datasets included in this study is provided below. The list of these global inventories together with their reference or websites is given in Table 6.1.

6.2.1 ACCMIP and the RCPs

The ACCMIP emissions inventory (Lamarque et al. 2010) is developed as part of a community effort to develop consistent gridded emissions of reactive gases and aerosol for use in model simulations performed in the framework of the Assessment Report 5 (AR5) of the Intergovernmental Panel on Climate Change (IPCC). ACCMIP provides monthly, sectoral gridded emissions for the 1850–2000 period

Table 6.1 Inventories providing global anthropogenic emissions

Acronym	Period	Reference and/or website
MACCity	1980–2010	Granier et al. (2011) (http://pole-ether.fr/eccad)
ACCMIP	1980–2010	Lamarque et al. (2010) (http://pole-ether.fr/eccad)
RCPs	2000–2010	Van Vuuren et al. (2011) http://www.iiasa.ac.at/web-apps/tnt/RcpDb
EDGAR v4.2	1970–2008	Janssens-Maenhout et al. (2013) (http://edgar.jrc.europa.eu/)
EDGAR v4.3	1970 and 2010	Crippa et al. (2016) (http://edgar.jrc.europa.eu/pegasos)
HTAPv2	2008 and 2010	Janssens-Maenhout et al. (2015) http://edgar.jrc.europa.eu/htap_v2
RETRO	1980–2000	Schultz et al. (2007) (http://pole-ether.fr/eccad)
ECLIPSE v4a	2005–2050	Stohl et al. (2015) (http://eclipse.nilu.no)
ECLIPSE v5	1990–2020	Klimont et al. (in preparation, 2016) http://eclipse.nilu.no
Bond	1850–2000	Bond et al. (2007) (http://hiwater.org)
Junker&Liousse	1860–1997	Junker and Liousse (2008)
PKU	2002–2013	Huang et al. (2014) (http://inventory.pku.edu.cn)

on a decadal basis, at a spatial resolution of $0.5 \times 0.5°$ in latitude and longitude. Anthropogenic emissions are given for nine sectors: energy, industries, transportation, shipping, residential, solvents, waste, agriculture and agriculture waste.

The RCPs (Representative Concentration Pathways) scenarios provide emissions for four different scenarios, based on assessment models from four different groups. The RCPs are named according to the levels of the 2100 radiative forcing, i.e. RCP2.6 corresponds to a 3.6 W/m^2 forcing in 2100. In order to ensure a good continuity between the model simulations for past decades and for the future, the future emissions were forced to agree with the year 2000 values.

6.2.2 MACCity

MACCity (Granier et al. 2011) represents an extension of the ACCMIP historical emissions to the year 2015. When the MACCity dataset was developed, no inventory existed, which provided emissions of the main tropospheric compounds for each year during the 1960–2015 period. The emissions for each year between 1960 and 2000 are obtained through a linear interpolation of the decadal ACCMIP emissions. The emissions for the 2000–2015 period were obtained through a linear regression of the emissions provided by the RCP8.5 scenario for the years 2005, 2010 and 2020. The RCP8.5 scenario was chosen as it includes information on regional emissions for the year 2005.

6.2.3 EDGARv4.2 and v4.3

The Emissions Database for Global Atmospheric Research (EDGAR) provides global past and present day anthropogenic emissions of greenhouse gases and air pollutants (Janssens-Maenhout et al. 2013). EDGARv4.2 includes country- and sector-specific emissions of greenhouse gases and air pollutants. EDGAR uses a technology-based approach and combines scientific information and data from international statistics on energy production and consumption, industrial manufacturing, agricultural production, waste treatment and disposal and burning of biomass to provide a complete historical emission trend from 1970 to 2008 for all countries in the world in a comparable and consistent manner. The emissions are gridded for each species, at a 0.1×0.1 degree resolution.

The EDGARv4.3 inventory provides emissions of gaseous and particulate air pollutants for 1970 and 2010 (Crippa et al. 2016). This dataset has been developed to quantify the effectiveness of emissions reduction measures, change in fuel consumptions and technological developments on air quality emissions, and their impact on health, crops, and climate. Three retrospective scenarios are available: the first simulating the complete stagnation of technology (STAG_TECH: lack of abatement measures and no improvement in emission standards), the second assuming constant fuel mixture and consumption as they were in 1970 (STAG_FUEL: no change in human activities), and the third considering unchanged energy consumption since 1970, but assuming the technological development, end-of-pipe reductions, fuel mix and energy efficiency of 2010 (STAG_ENERGY).

6.2.4 *HTAPv2*

The HTAPv2 dataset consists of 0.1 × 0.1 degree gridmaps of several gaseous and particulate compounds for the years 2008 and 2010, in order to provide consistent information to global and regional scale modeling activities (Janssens-Maenhout et al. 2015). HTAPv2 uses nationally reported emissions combined with regional scientific inventories and is given as sector-specific gridmaps. This compilation of different regional gridded inventories includes the data from the Environmental Protection Agency (EPA) for the USA, the emissions from Environment Canada for Canada, the European Monitoring and Evaluation Programme (EMEP) and Netherlands Organisation for Applied Scientific Research (TNO) emissions for Europe, and the Model Intercomparison Study for Asia (MICS-Asia III) emissions for China, India and other Asian countries. Emission grid maps from the Emissions Database for Global Atmospheric Research (EDGARv4.3) are used for the rest of the world (mainly South America, Africa, Russia and Oceania).

6.2.5 *ECLIPSE v4 and ECLIPSE v5*

The ECLIPSE emissions have been developed as part of the Evaluating the CLimate and Air Quality ImPacts of Short-livEd Pollutants (ECLIPSE, Stohl et al. 2015) project. The anthropogenic emissions are developed with the GAINS (Greenhouse gas – Air pollution Interactions and Synergies) framework (Amann et al. 2011). This model calculates emissions for about 170 regions by all major economic sectors, including energy and industrial production, transport, residential combustion, agriculture, and waste. Several detailed subsectors, fuels, and emission control options are distinguished.

The ECLIPSEv5 dataset is an update and extension of the version 4 of the dataset: it extends its time horizon and emissions are available in 5-year intervals from 1990 to 2030. In ECLIPSEv5, the historical data used in version 4 for the period 1990–2010 were revised using the latest statistics extending to 2010 and updated country reporting where available.

6.2.6 RETRO

Global gridded datasets were generated by the RETRO (Reanalysis of the TROpospheric chemical composition) project for the 1960–2000 period, on a monthly basis. Anthropogenic and vegetation fire emissions are provided on a monthly basis at a 0.5 $\times$0.5 degree resolution by RETRO for several gaseous compounds. RETRO provides the emissions for several individual volatile organic compounds. The anthropogenic emissions in the RETRO inventory cover combustion sources only. Detailed information on the RETRO dataset are found in Schultz et al. (2007).

6.2.7 Bond

The Bond inventory for black and primary organic carbon (Bond et al. 2007) provides historical emissions from 1850 to 2000. Emissions are calculated as a function of technology choices and are based on a sectoral fuel-specific reconstruction of fossil fuel consumption and historical estimates of bioenergy consumption. The methodology includes a detailed analysis of the evolution of emission factors for different sectors.

6.2.8 Junker & Liousse (J&L)

An historical inventory of carbonaceous and organic aerosol for the period 1860–2003 has been developed by Junker and Liousse (2008). Historical fuel production, trade historical datasets and changes in emission factors during the past decades have been used to develop emissions for traffic, domestic and industrial emissions. The changes over time are based on the evolution of the levels of economic and technological development in each country during the past decades. The inventory is gridded at a 1 $\times$ 1 degree resolution, based on a gridded map of the population during the past century.

6.2.9 Huang Y. (PKU)

A global inventory at a 0.1 $\times$0.1 degree for the 1960–2009 period has been developed by Huang et al. (2014) for $PM_{2.5}$ and PM_{10} for 77 different sectors. These emissions are based on a global fuel data product which includes 64 fuel types, based on the PKU-FUEL-2007 dataset, which uses subnational fuel consumption for different countries. The uncertainties on the emissions were characterized.

We have also used in this work older determinations of surface emissions from the HYDE inventory (van Aardenne et al. 2001) for the 1890–1990 period and the POET emissions (Granier et al. 2005) for 1990–2000.

6.3 Regional Anthropogenic Emissions in Asia

Several inventories providing emissions for Asia have been developed during the past decade (Table 6.2). Some of these inventories provide gridded data for several years or for just a specific year for China or the whole of Asia. Several other datasets provide data only for a single species or for a short period: this type of dataset concerns mostly the emissions for China. As part of this work, we have compiled information on these inventories, and the following paragraphs provide a short summary of the characteristics of each dataset.

6.3.1 *REASv1 and REASv2*

The REAS (Regional Emission Inventory in Asia) inventory is providing emissions of gases and particles for Asia. REASv1 (Ohara et al. 2007) considers historical emissions for the 1980–2003 period, and projected emissions in 2010 and 2020. REASv2 (Kurokawa et al. 2013) gives an update of the previous version, and includes most major air pollutants and greenhouse gases from each year during 2000 to 2008. The regions considered in REAS include East, Southeast, South and Central Asia and the Asian part of Russia. Emissions are estimated for each country and region using updated activity data and parameters. Emissions are provided on a monthly basis, at a 0.25×0.25 degree spatial resolution. As an example of the spatial distribution of emissions in Asia, Fig. 6.1 shows the emissions of nitrogen oxides from the REASv2.1 inventory for June 2008.

6.3.2 *TRACE-P and INTEX-B*

As part of the NASA TRACE-P (Transport and Chemical Evolution over the Pacific) campaign, an inventory providing emissions for Asia in 2000 was developed by Streets et al. (2003), based on technology information for this region. In order to take into account the strong growth of the economy, and therefore of the emissions in this region, a new inventory was developed in support of the NASA INTEX-B campaign, which provides emissions in Asia for 2006 (Zhang et al. 2009). In the plots comparing the emissions, both inventories are shown under the acronym INTEX-B.

Table 6.2 Inventories providing Asian anthropogenic emissions

Acronym	Period	Reference and/or website
REAS v1 (Asia)	1980–2010	Ohara et al. (2007)
		http://www.jamstec.go.jp/frcgc/research/d4/emission.htm
REAS v2 (Asia)	2000–2008	Kurokawa et al. (2013)
		http://web.nies.go.jp/REAS/
MEIC 1.0 (China)	2008, 2010	http://meicmodel.org
MEIC 1.2 (China)	2008, 2010 and 2012	
TRACE-P	2000	Streets et al. (2003) and Zhang et al. (2009)
INTEX-B (Asia)	2006	http://cgrer.uiowa.edu/projects/emmison-data
MIX (Asia)	2008 and 2010	Li et al. (2015)
ZhaoB (China)	1995–2010	Zhao et al. (2013a) (NO_x)
ZhaoY (China)	2005 and 2010	Zhao et al. (2011, 2013b)
Cao (China)	2000	Cao et al. (2006) (BC and OC)
HuangX (China)	2006	Huang et al. (2012) (NH_3)
Kang (China)	1980–2012	Kang et al. (2015) (NH_3)
Bo (China)	1980–2005	Bo et al. (2008) (NMVOC)
Lei (China)	1990–2005	Lei et al. (2011) (BC)
Wei (China)	2005–2020	Wei et al. (2011, 2008) (NMVOC)
Su (China)	1990–2007	Su et al. (2011)
Lu (China/India)	1996–2010	Lu et al. (2011) (SO_2 and BC)
WangR (China)	1949–2007	Wang et al. (2012) (BC)
WangS (Asia)	2005–2010	Wang et al. 2014
MPolo (China)	2005–2013	Mijling et al. 2013 (NO_x) [NO_x Inverse modeling]
		http://www.marcopolo-panda.eu/products/toolbox/emission-data/
Stavrakou (China)	2005–2014	Stavrakou et al. (2009, 2015) [VOCs Inverse modeling]
		http://www.marcopolo-panda.eu/products/toolbox/emission-data/

6.3.3 MEIC

The MEIC emissions (Multi-resolution Emission Inventory for China) have been developed to support chemical transport and climate model simulations at different resolution and time scales. The first version of MEIC, v1.0 gives emissions for 2008 and 2010, while version v1.2 gives emissions for 2008, 2010 and 2012. Monthly emissions can be downloaded from the MEIC website (meicmodel.org), for the three different spatial resolutions 0.25×0.25, 0.5×0.5 and 1×1 degree

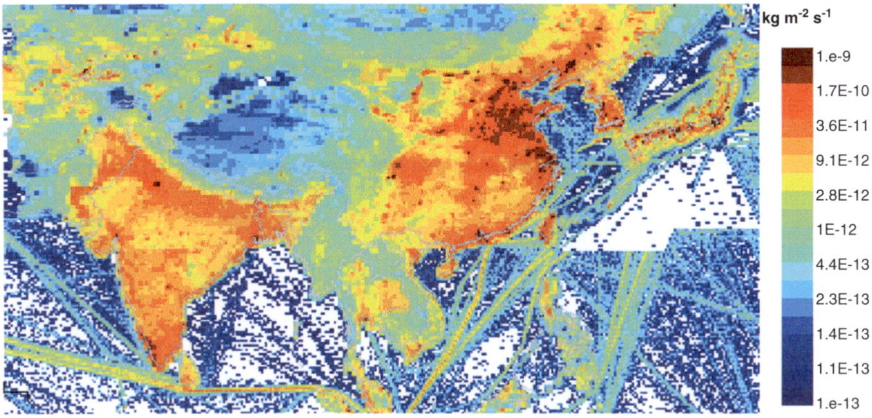

Fig. 6.1 NO_x emissions (in kg/m2/s) in Asia in 2008 in the REASv2 inventory

resolution. Speciations of VOCs emissions are available for five different chemical schemes: CBIV, CB05, SAPRC99, SAPRC07 and RADM2.

6.3.4 MIX

The MIX anthropogenic emissions inventory for Asia has been developed for the years 2008 and 2010 to support the Model Inter-Comparison Study for Asia (MICS-Asia) and the Task Force on Hemispheric Transport of Air Pollution (TF HTAP) projects by a mosaic of up-to-date regional emission inventories (Li et al. 2015). Emissions are estimated for all major anthropogenic sources in 30 countries and regions in Asia. Emissions of speciated NMVOCs for two chemical mechanisms, SAPRC-99 and CB05, are available. The MIX emissions are provided on a monthly basis, at a 0.25×0.25 degree resolution.

6.3.5 Zhao B. (ZhaoB)

NO_x emissions in China have increased rapidly during the past few years in Asia and Zhao et al. (2013a) have estimated the past distribution of NO_x emissions in China for 1995–2010 period, and future emissions up to 2030 for different scenarios. The emissions are based on historical energy consumption data and on assumptions on the future driving forces in China. The emissions are not gridded and are given as total emissions for China for different sectors.

6.3.6 *Zhao Y. (ZhaoY)*

Uncertainties on emissions in Asia are very large, and they have been discussed by Zhao et al. (2011, 2013b). The authors have first determined the emissions for 2005 and 2010, and then determined the uncertainties as probability distributions. They have also examined the effects of China's policies on reducing emissions during the past years.

6.3.7 *Cao*

Cao et al. (2006) have developed an inventory of BC and OC emissions for China for the year 2000. The inventory is developed on the basis of fuel consumption data and socio-economic statistics from government agencies. The data are gridded, at a 0.2×0.2 degree resolution.

6.3.8 *Huang X. (HuangX)*

Huang et al. (2012) have compiled a comprehensive inventory of ammonia emissions for China for 2006 at a 1×1 km^2 resolution. The inventory takes into consideration different observations, as well as parameterizations of emission factors as a function of ambient temperature, soil acidity and other factors. A seasonal variation of the emissions is also provided.

6.3.9 *Kang*

In order to characterize the impact of the large increase in agriculture activities in China since the early 1980s, an inventory of NH_3 emissions for the 1980–2012 period was developed by Kang et al. (2015). The inventory shows increase in NH_3 emissions until 1996, followed by stabilization and a small decrease in the following years. The decrease is attributed to a reduction of livestock and fertilizer emissions in the recent years. The resolution of the inventory is 1×1 km^2.

6.3.10 *Bo*

Bo et al. (2008) have compiled a multiyear emission inventory of anthropogenic emissions of NMVOCs for China, for the 1980–2005 period. The quantification of emissions is based on statistical data, review of published information in the literature, and model calculated emission factors. The emissions are gridded at a spatial resolution of 40 × 40 km^2.

6.3.11 Lei

Using a technology-based approach, Lei et al. (2011) developed an inventory providing anthropogenic emissions of primary aerosols (PM_{10}, $PM_{2.5}$, BC and OC) for China. The inventory takes into account changes in the penetration of technology in industrial activities and transportation.

6.3.12 Wei

The emissions of NMVOCs for China for 2005 were compiled by Wei et al. (2008). Several sectors are considered including industrial and domestic solvent use, road transportation and biofuel combustion. Emissions have been developed at a resolution of 36 × 36 km^2. Uncertainties on the emissions are also determined. Wei et al. (2011) also proposed projections of the emissions of NMVOCs for the 2010–2020 for different scenarios.

6.3.13 Su

Levels of SO_2 emissions for different Chinese provinces were estimated by Su et al. (2011) for the 1990–2007 period. The sulfur content of the coal used in different provinces was used as a basis for the calculation of the emissions. A decrease of SO_2 emissions after 2006 is identified, which is the result of the application of new technologies in power plants and the phase out of small high polluting units.

6.3.14 Lu

An inventory of the emissions of SO_2, as well as of black and organic carbon for China and India was developed for the 1996–2010 period, through a technology-based methodology (Lu et al. 2011). Trends in activity data and in emission factors are taken into account in the calculations.

6.3.15 Wang R. (WangR)

Based on the PKU dataset mentioned in the previous section, an inventory of BC emissions for China was developed by Wang et al. (2012) for the 1949–2007 period, together with scenarios for the 2008–2050 period. The emission factors database was updated and the emissions are provided at a 0.1×0.1 degree resolution. The uncertainties on the emissions were also quantified.

6.3.16 Wang S. (WangS)

The emissions of SO_2, NO_x, PMs and NMVOCs for the period 2005–2010 were evaluated by Wang et al. (2014), together with projections for 2020 and 2030. The region considered is East Asia, i.e. China, Taiwan, Mongolia, the Koreas and Japan. The inventory takes into account the different air pollution control measures taken in East Asia since 2005.

6.3.17 MarcoPolo Inverse Modeling (MPolo)

In this work, we have also used the emissions of nitrogen oxides optimized by inverse modeling, as part of the European MarcoPolo project, using satellite observations from the OMI and GOME2 satellites. The optimization is done using the DECSO algorithm (Daily Emission Estimation Constrained by Satellite Observations) described in Mijling et al. (2013). Different versions of the MarcoPolo inverse emissions were used, i.e. DECSO v3a (OMI and GOME-2), DECSO v4 and DECSO v5 (OMI only). In the figures presented in the next section, the NO_x emissions optimized using observations from the GOME-2 observations are indicated by *MPoloG*, and the optimized emissions using OMI are indicated by *MPoloOv3, MPoloOv4, MPoloOv5*, for each DECSO version, respectively.

Space observations from the OMI satellite instrument of the vertical formaldehyde (HCHO) columns have been used to infer the emissions of volatile organic compounds (VOCs) using inverse modeling techniques (Stavrakou et al. 2009, 2015). Since anthropogenic VOC emissions have a rather minor contribution to the HCHO concentrations, they are only weakly constrained by an inversion system at the global scale, except in strongly polluted regions, like China, where the optimized VOCs emissions are used in the analysis. The VOCs emissions in China optimized by inverse modeling were performed as part of the MarcoPolo project, and are reported as MPolo on the figures in the next section.

6.4 Trends in Anthropogenic Emissions in Asia

Using all the inventories described in the previous section, we have analyzed the trends in the emissions for different species in Asia from 1960 to 2014. We present in this section the results, first for the emissions in China, for which many datasets are available, and a few selected results for other countries in Asia.

6.4.1 Emissions Trends in China

Figures 6.2, 6.3, 6.4, and 6.5 show the emissions of NO_x, CO, NMVOCs, SO_2, PM_{10}, BC and NH_3 from 1960 to 2010 in China. The years for which emissions are represented are 1960, 1970, 1980, 1990, 1995 and every year from 2000 to 2014.

In all the figures, the emissions from the RCPs, i.e. the future emissions scenarios, are indicated in dashed lines. The names and acronyms of each inventory displayed on the figures are the ones indicated in Table 6.2.

During the past 40 years, emissions have increased significantly for most species. The comparisons show very large differences for all species among the inventories, for all species, and the trends in emissions given by the different inventories differ also significantly and more particularly for the past 10 years.

NO_x is the compound for which the best agreement between different inventories is found. All emissions show a constant increase during the full selected time period, with a difference of about 30 % between the highest and lowest values in 2008. The emissions optimized using inverse modeling techniques are in close agreement with the other inventories. They show however a decrease after 2012 and such a feature will have to be confirmed in the coming years, when more recent data on emissions and more satellite observations will be available. It should be noted that the emissions optimized by inverse modeling techniques have large differences which are as large as the differences between all the different inventories used Such differences are related to the differences between the satellite data used to constrain the emissions.

CO emissions show large differences among the datasets, with a factor of 2 difference between the highest and lowest emissions in 2008. All the datasets show a constant increase in emissions, except the MEIC and HTAPv2 emissions, which show a decrease starting in 2008. HTAPv2 is based on MEIC, which is itself based on local information on emissions. It should also be noted that the EDGAR4.2 inventory, which is used in many modeling studies at the global scale, provides the lowest emissions in China. The most recent version of EDGAR, EDGAR4.3, provides emissions which are very close to MACCity.

Total NMVOCs emissions show a constant increase in all datasets, except for the emissions optimized with inverse modeling techniques, which show rather constant values since 2005: the decrease obtained in 2011–2012 is not present in the 2013

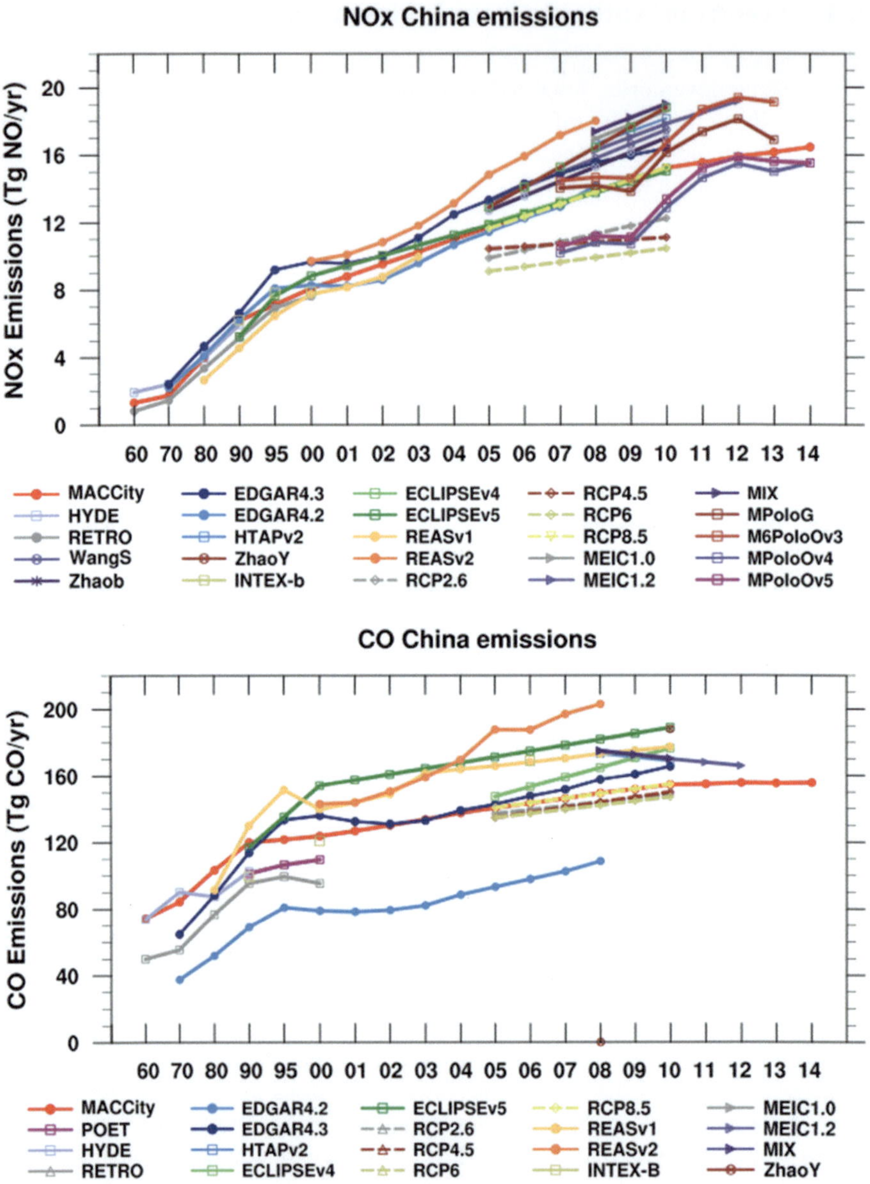

Fig. 6.2 Emissions of NO_x (*top*) and CO (*bottom*) in China

emissions. It should also be noted that the MACCity inventory provides emissions significantly higher than all the other datasets.

There is a large number of inventories providing emissions of SO_2. All of them show an increase of emissions until 2005, followed by a decrease shown in most datasets. EDGARv4.2 shows a significant increase in the recent years, while

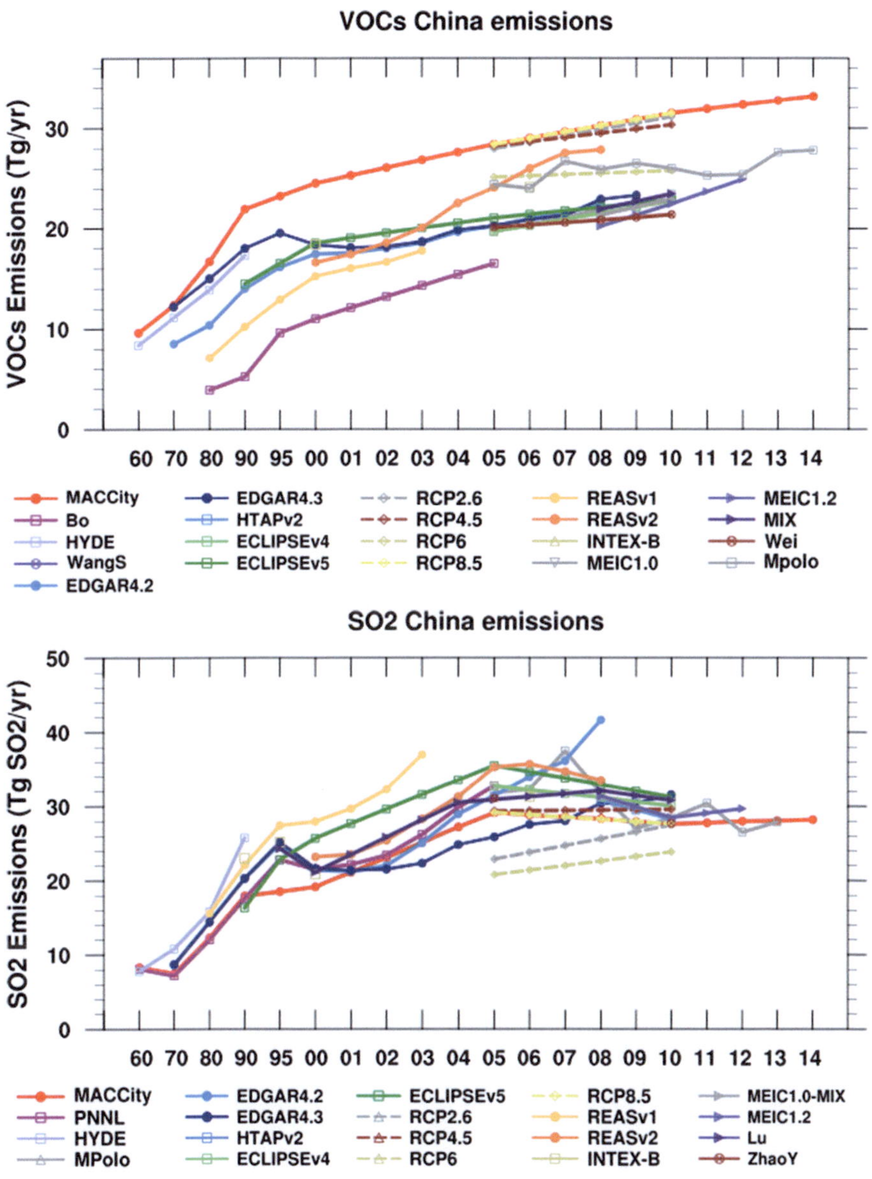

Fig. 6.3 Emissions of VOCs (*top*) and SO_2 (*bottom*) in China

EDGAR4.3 provides lower values than EDGAR4.2, increasing for the full period. Lu et al. (2011) show almost constant emissions, with a slight decrease after 2008. The emissions optimized by inverse modeling show also a slight decrease in the emissions, and provide very similar values to the MACCity estimates after 2008.

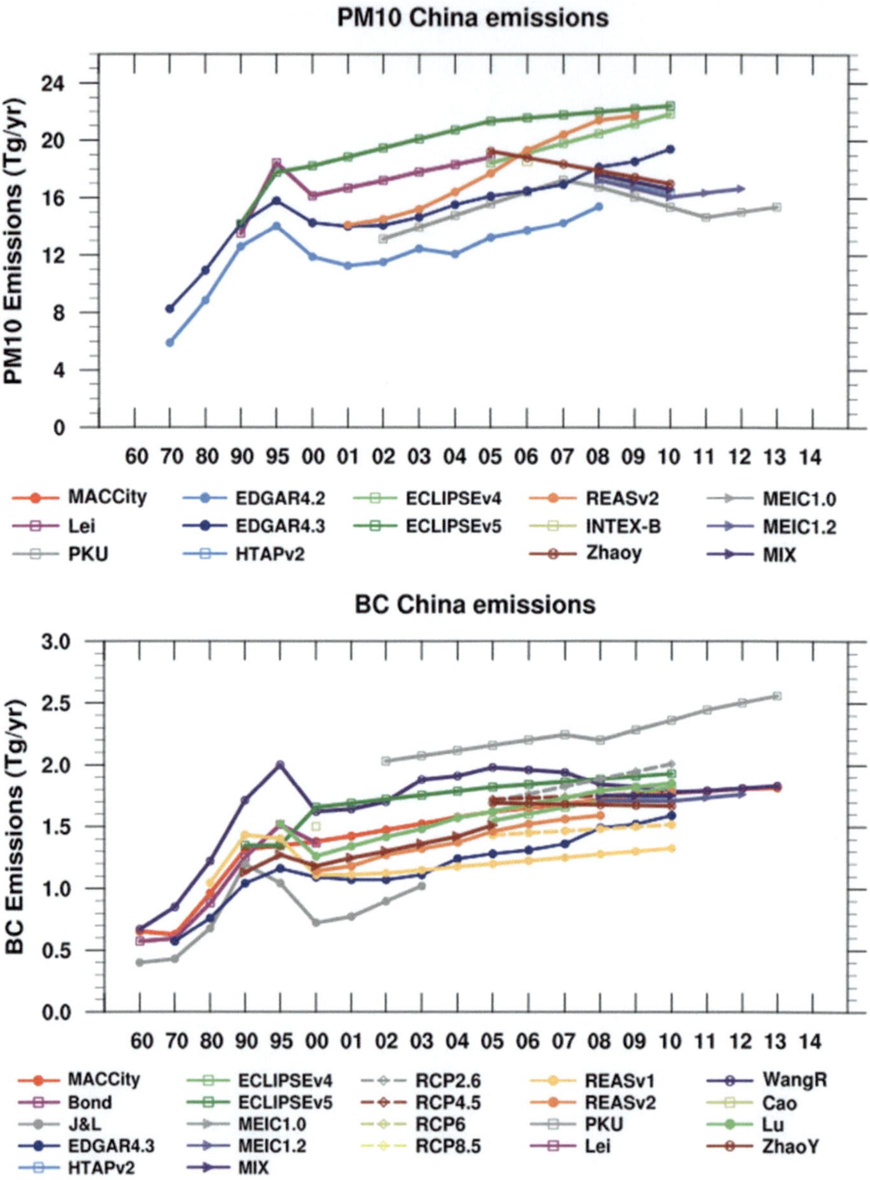

Fig. 6.4 Emissions of PM_{10} (*top*) and BC (*bottom*) in China

The emissions of PM_{10} (and also $PM_{2.5}$, not shown) show large differences among the datasets in the amounts emitted as well as in the trends. There is a difference of about 35–50 % between the highest and lowest emissions in 2008. For PM_{10} (as well as for $PM_{2.5}$), the highest emissions are given by the ECLIPSEv5 global inventory, and the lowest values by the PKU and EDGARv4.2 datasets. The

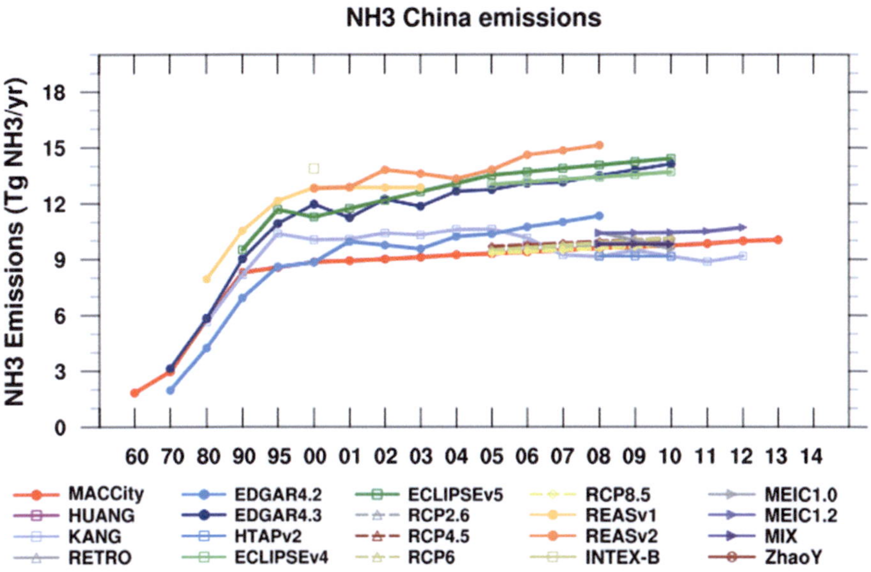

Fig. 6.5 Emissions of NH_3 in China from 1960 to 2010

trend in emissions is not consistent among the datasets, with constant, increasing or decreasing emissions, depending on the inventory.

BC emissions strongly increase in all datasets until 1995. After that, almost all inventories show a constant increase in the emissions of BC for the past 20 years, except for the WangR dataset, which show emissions in 2010 close to their 2000 values. The emissions provided by the PKU dataset are significantly higher than the other emissions, and display a larger increase after 2008.

For NH_3, emissions have increased significantly until about 2000, and are rather stable afterwards. A group of inventories, including ECLIPSEv4 and v5, EDGAR 4.2 and 4.3 and REAS, provide almost similar trend and amount of emissions, which are about 40 % higher than the other inventories, i.e. MACCity, MEIC, MIX and Kang. It should be noted that NH_3 emissions are mostly related to agriculture practices and depend strongly on meteorology. All the inventories available up to now are static inventories, which provide only monthly- or yearly-averaged emissions.

6.4.2 Trends in Emissions in Other Regions in Asia

Emissions in different regions in Asia have also been evaluated. The emissions have been grouped for different regions, as shown in Fig. 6.6. Emissions for the regions called India + (in yellow in Fig. 6.6), South-East Asia (light blue), Indonesia (orange), Japan and Korea have been compared.

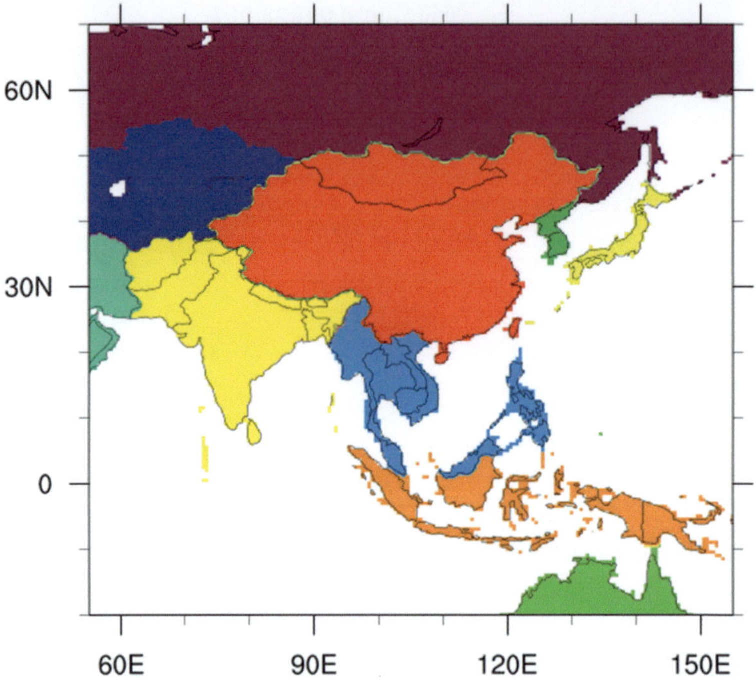

Fig. 6.6 Regions used in the comparisons of emissions

Figures 6.7 and 6.8 show examples of the comparisons of the emissions inventories in the selected regions, and highlight the differences between the inventories for the different species considered. Contrarily to China, there are very few or no regional datasets providing emissions for these regions, and most of the emissions shown in these plots are given by global inventories.

In India, all inventories agree about an increase of the NO_x emissions during the period considered. There are however large differences on the amount emitted by about a factor of 2 between some inventories. The emissions developed by local groups provide either the lowest emissions (Garg), or emissions very close to MACCity and ECLIPSE (Sahu).

CO emissions in Korea provided by MACCity and REASv1 are higher compared to inventories developed more recently such as REASv2, ECLIPSEv5m and EDGAR 4.3. The emissions provided by these recently developed datasets show a decrease in the 1990s, and rather constant emissions in the 2000s–2010s, with values about a factor of 2 lower than MACCity and REASv1.

VOCs emissions in India (Fig. 6.8) are shown to be either rather constant since 2000 (MACCity, ECLIPSEv4 and v5, EDGAR4.2 and 4.3, REASv1), or increasing significantly, as given by REASv2, HTAPv2 or MIX. No regional inventory developed by local groups is available for VOCs emissions in this region.

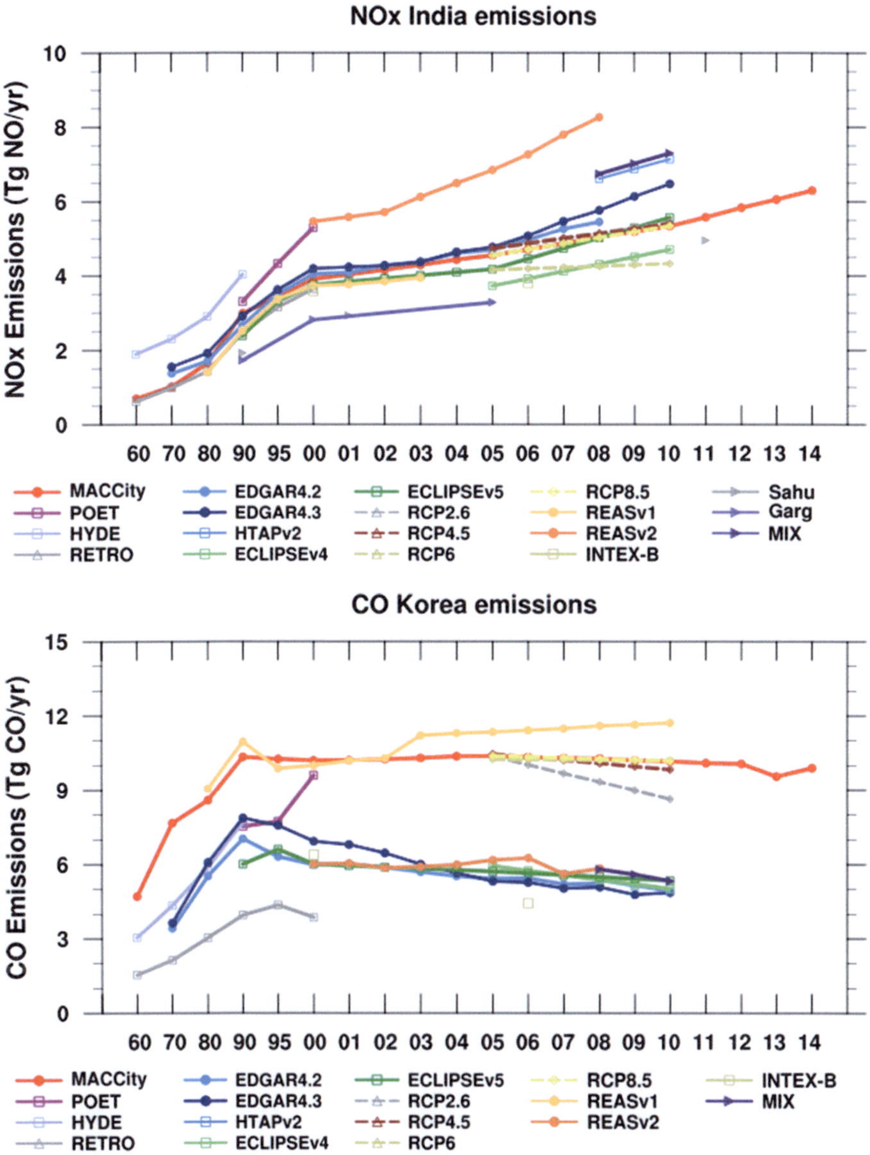

Fig. 6.7 Emissions of NO_x in India (*top*) and of CO in Korea (*bottom*)

All inventories providing emissions of PM_{10} in Indonesia (Fig. 6.8) provide very different values, with a factor of about 3 between the highest and lowest emissions. Trends in the emissions are not very large in all datasets. All these figures, which give only a few examples for a few species and a few regions, show large

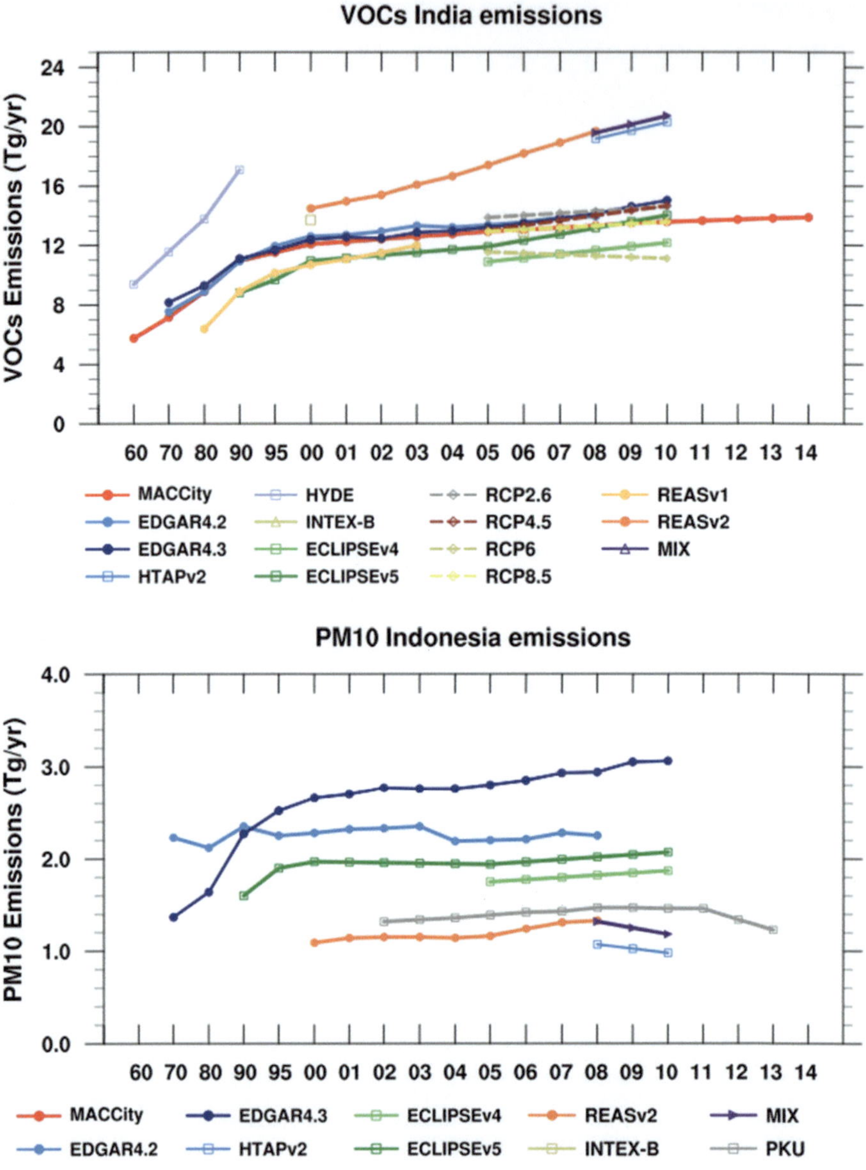

Fig. 6.8 Emissions of VOCs in India (*top*) and PM_{10} in Indonesia (*bottom*)

differences among the emissions inventories. No systematic differences, such as an inventory always being higher than the others in a specific region could be found.

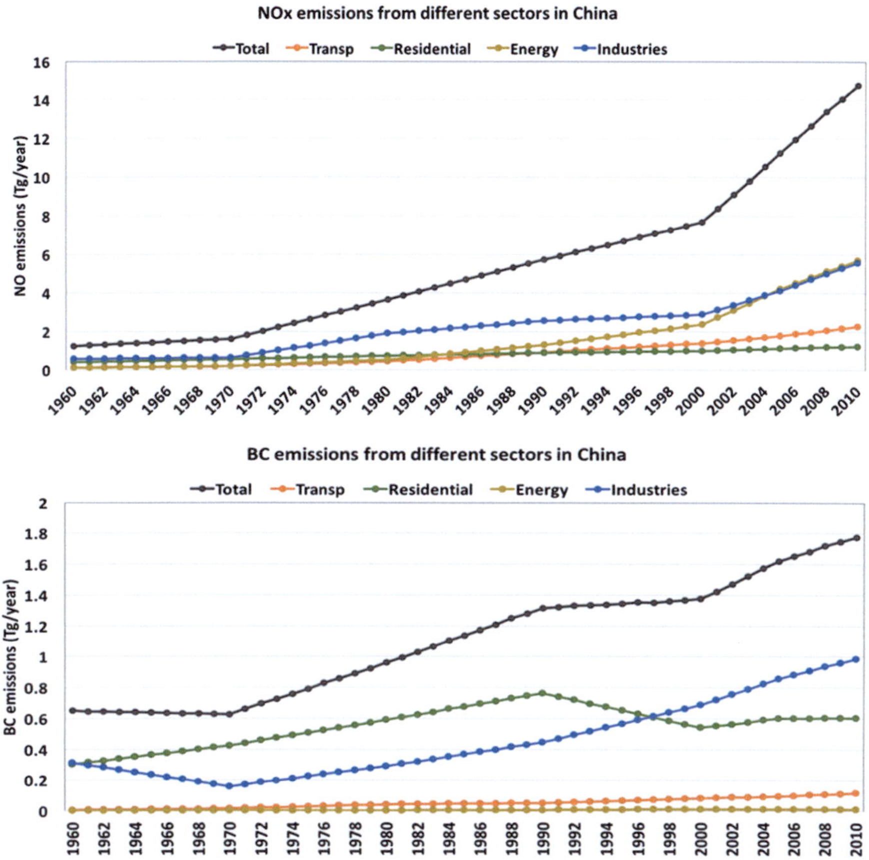

Fig. 6.9 Changes in the emissions for different sectors for NO_x (*top*) and BC (*bottom*) in China from 1960 to 2010, from the MACCity inventory

6.5 Contributions of Different Sectors to the Changes in Emissions in China

In order to better understand the origin of the evolution of the emissions in Asia, a preliminary analysis of the changes in the emissions for different sectors was done for the MACCity inventory. The results for NO_x and BC are shown in Fig. 6.9: this figure shows the changes in the emissions from the energy, industry, transportation and residential sectors from 1960 to 2010.

The large increase in NO_x emissions during the past decades is mainly due to the rapid growth in energy demand and industrial activities in this part of the world. For BC, a constant increase in the emissions from the industrial sector is mainly responsible for the increase of the BC total emissions. A large increase in the emissions from the residential sector is seen until the 1990s, followed by a stabilization of the emissions from this sector.

More studies of the evolution of the emissions for different sectors from different inventories will be performed in the future. It should however be noted that the definition of the sectors can be different from inventory to inventory, which makes the results and the differences between the datasets difficult to analyze.

6.6 Summary

Almost all inventories are provided without any information on the data used to calculate the emissions (such as data on energy production and use in each country, on industrial activities, or on emissions factors and their evolution). Information on the uncertainties in the estimated emissions is also not always provided. It is therefore very difficult to understand the differences between the inventories highlighted in the previous sections.

The results of the comparisons discussed in the previous sections can however be used to provide some information on the uncertainty of the emissions, using the minimum and maximum value of the emissions provided by the different inventories in each region.

Figures 6.10 and 6.11 show the minimum and maximum values of the emissions in China (top) and India (bottom) for six different periods, 1970, 1980, 1990, 2000, 2005 and 2008. In order to be able to accommodate all the results on the same figure, several adjustments have been made, as indicated on the legends on the horizontal axis: on Fig. 6.9, which shows the emissions in China, CO emissions have been divided by 10, SO_2 emissions have been divided by 2 and BC emissions have multiplied by 10. On Fig. 6.10 for India, CO emissions have been divided by 10, VOCs emissions have been divided by 2, and BC emissions have been multiplied by 10.

On both figures, the range of the emissions in 1970 and 1980 is quite small: this is due to the fact that very few inventories are available for these years. Furthermore, the few datasets available for these decades are not independent of each other, which makes the data rather similar due to the fact that few information are available and all datasets use the same information.

These figures show that the emissions of NO_x and SO_2 have the lowest vertical extent, which means that these emissions seem to have the best characterization. On the contrary, VOCs, BC and CO show a rather large range for all periods, with values reaching a factor 2 or 3.

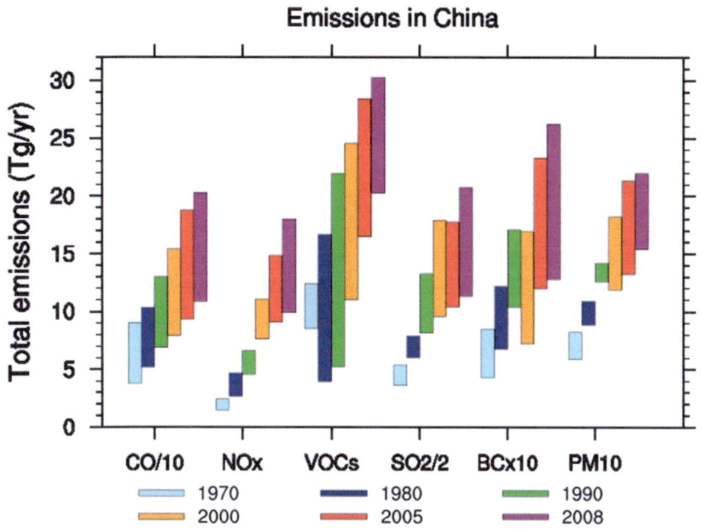

Fig. 6.10 Range of the emissions for several compounds and different years in China

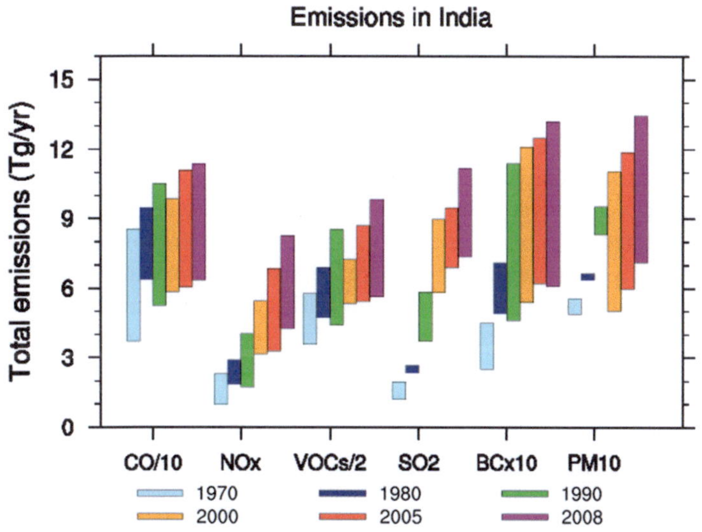

Fig. 6.11 Range of the emissions for several compounds and different years in India

6.7 Conclusions and Perspectives

We have compiled in this chapter the most recent information on anthropogenic emissions in Asia, and have shown large differences among some of the inventories for specific compounds and/or periods. These differences are very difficult to explain, since most of the inventories are not provided with the data that have been used to calculate the emissions for the different regions, sectors, and periods. Information on uncertainties is not provided either and we have used the comparisons between the different datasets to give some information on these uncertainties.

In the future, several recommendations and indications for future work can be proposed as a result of the evaluation performed in this analysis:

- The analysis has focused on yearly-averaged emissions, which correspond to the data generally provided by many inventories. When used in atmospheric models, seasonal, weekly and diurnal changes in the emissions would be very useful, for example for forecasting the atmospheric composition.
- We have compared the emissions of total VOCs. However, this species which corresponds to a mixture of different hydrocarbons is not always well defined. When used in models, the emissions of VOCs have to be speciated to much the chemistry schemes used in the models. Not much information of speciation is yet available however. There has been recent work providing speciation of VOCs in China (MIX and Wei inventories) for a few years, which should be extended to several decades and to other regions.
- Similarly, emissions of particulate matter could be given with more details on the composition of the particles. Many inventories, which are developed for climate or atmospheric composition studies, provide only the emissions of aerosol precursors. Inventories developed for air quality studies provide generally only the emissions of PMs. Therefore, the consistency of global and regional inventories providing emissions of particles is very difficult to assess. Since climate and air quality studies will have more and more links, it would be very useful to develop inventories providing PM emissions and their speciation.
- Several global inventories and most regional inventories included in this study consider only anthropogenic continental emissions and do not provide information on ship emissions. However, ship traffic has increased significantly in the past decades, especially in Asia. Therefore, emissions from ships should be quantified in more details, since they have the potential to affect strongly the atmospheric composition in coastal areas and be transported over long distances.

Acknowledgments This work has received research funding from the European Community's Seventh Framework Programme (FP7) under the PANDA project grant agreement n° 606719.

References

Amann, M., Bertok, I., Borken-Kleefeld, J., Cofala, J., Heyes, C., Höglund-Isaksson, L., Klimont, Z., Nguyen, B., Posch, M., Rafaj, P., Sander, R., Schöpp, W., Wagner, F., & Winiwarter, W. (2011). Cost-effective control of air quality and greenhouse gases in Europe: Modeling and policy applications. *Environmental Modelling and Software, 26*, 1489–1501. doi:10.1016/j.envsoft.2011.07.012.

Bo, Y., Cai, H., & Xie, S. D. (2008). Spatial and temporal variation of historical anthropogenic NMVOCs emission inventories in China. *Atmospheric Chemistry and Physics, 8*, 7297–7316. doi:10.5194/acp-8-7297-2008.

Bond, T. C., Bhardwaj, E., Dong, R., Jogani, R., Jung, S., Roden, C., Streets, D. G., Fernandes, S., & Trautmann, N. (2007). Historical emissions of black and organic carbon aerosol from energy-related combustion, 1850–2000. *Global Biogeochemical Cycles, 21*, GB2018. doi:10.1029/2006GB002840.

Cao, G., Zhang, X., & Zheng, F. (2006). Inventory of black carbon and organic carbon emissions from China. *Atmospheric Environment, 40*, 6516–6527.

Crippa, M., Janssens-Maenhout, G., Dentener, F., Guizzardi, D., Sindelarova, K., Muntean, M., Van Dingenen, R., & Granier, C. (2016). Forty years of improvements in European air quality: Regional policy-industry interactions with global impacts. *Atmospheric Chemistry and Physics, 16*, 3825–3841. doi:10.5194/acp-16-3825-2016.

Granier, C., Lamarque, J. F., Mieville, A., Muller, J. F., Olivier, J., Orlando, J., Peters, J., Petron, G., Tyndall, G., & Wallens, S. (2005). POET, a database of surface emissions of ozone precursors. Available at http://www.aero.jussieu.fr/projet/ACCENT/POET.php.

Granier, C., Bessagnet, B., Bond, T., D'Angiola, A., Denier van der Gon, H., Frost, G., Heil, A., Kaiser, J., Kinne, S., Klimont, Z., Lamarque, J.-F., Liousse, C., Masui, T., Meleux, F., Mieville, A., Ohara, T., Riahi, K., Schultz, M., Smith, S., Thomson, A. M., van Aardenne, J., & van der Werf, G. (2011). Evolution of anthropogenic and biomass burning emissions at global and regional scales during the 1980–2010 period. *Climatic Change*. doi:10.1007/s10584-011-0154-1.

He, K., Huo, H., & Zhang, Q. (2002). Urban air pollution in China: Current status, characteristics and progress. *Annual Review of Energy and the Environment, 27*, 397–431. doi:10.1146/annurev.energy.27.122001. 083421.

Huang, X., Song, Y., Li, M., Li, J., Huo, Q., Cai, X., Zhu, T., Hu, M., & Zhang, H. (2012). A high-resolution ammonia emission inventory in China. *Global BiogeochemCycles, 26*, GB1030. doi:10.1029/2011GB004161.

Huang, Shen, Y. H., Chen, H., Wang, R., Zhang, Y., Su, S., Chen, Y., Lin, N., Zhuo, S., Zhong, Q., Wang, X., Liu, J., Li, B., Liu, W., & Tao, S. (2014). Quantification of global primary emissions of $PM_{2.5}$, PM_{10}, and TSP from combustion and industrial process sources. *Environmental Science & Technology, 48*(23), 13834–13843.

Janssens-Maenhout, G., Pagliari, V., Guizzardi, D., & Muntean, M. (2013). Global emission inventories in the Emission Database for Global Atmospheric Research (EDGAR) – Manual (I): Gridding: EDGAR emissions distribution on global gridmaps, JRC Report, EUR 25785 EN, ISBN 978-92-79-28283-6, doi:10.2788/81454.

Janssens-Maenhout, G., Crippa, M., Guizzardi, D., Dentener, F., Muntean, M., Pouliot, G., Keating, T., Zhang, Q., Kurokawa, J., Wankmüller, R., Denier van der Gon, H., Kuenen, J. J. P., Klimont, Z., Frost, G., Darras, S., Koffi, B., & Li, M. (2015). HTAP_v2.2: A mosaic of regional and global emission grid maps for 2008 and 2010 to study hemispheric transport of air pollution. *Atmospheric Chemistry and Physics, 15*, 11411–11432. doi:10.5194/acp-15-11411-2015.

Junker, C., & Liousse, C. (2008). A global emission inventory of carbonaceous aerosol from historic records of fossil fuel and biofuel consumption for the period 1860–1997. *Atmospheric Chemistry and Physics, 8*, 1195–1207. doi:10.5194/acp-8-1195-2008.

Kan, H., Chen R., & Tong S. (2012), Ambient air pollution, climate change, and population health in China, Environment International, 42, 10–19, ISSN 0160-4120, http://dx.doi.org/10.1016/j.envint.2011.03.003.

Kang, Y., Liu, M., Song, Y., Huang, X., Yao, H., Cai, X., Zhang, H., Kang, L., Liu, X., Yan, X., He, H., Shao, M., & Zhu, T. (2015). High-resolution ammonia emissions inventories in China from 1980–2012. *Atmospheric Chemistry and Physics Discussions, 15*, 26959–26995. doi:10.5194/acpd-15-26959-2015.

Kurokawa, J., Ohara, T., Morikawa, T., Hanayama, S., Janssens-Maenhout, G., Fukui, T., Kawashima, K., & Akimoto, H. (2013). Emissions of air pollutants and greenhouse gases over Asian regions during 2000–2008: Regional Emission inventory in ASia (REAS) version 2. *Atmospheric Chemistry and Physics, 13*, 11019–11058. doi:10.5194/acp-13-11019-2013.

Lamarque, J. F., Bond, T. C., Eyring, V., Granier, C., Heil, A., Klimont, Z., Lee, D., Liousse, C., Mieville, A., Owen, B., Schultz, M. G., Shindell, D., Smith, S. J., Stehfest, E., Van Aardenne, J., Cooper, O. R., Kainuma, M., Mahowald, N., McConnell, J. R., Naik, V., Riahi, K., & van Vuuren, D. P. (2010). Historical (1850–2000) gridded anthropogenic and biomass burning emissions of reactive gases and aerosols: Methodology and application. *Atmospheric Chemistry and Physics, 10*, 7017–7039. doi:10.5194/acp-10-7017-2010.

Lei, Y., Zhang, Q., He, K. B., & Streets, D. G. (2011). Primary anthropogenic aerosol emission trends for China, 1990–2005. *Atmospheric Chemistry and Physics, 11*, 931–954. doi:10.5194/acp-11-931-2011.

Li, M., Zhang, Q., Kurokawa, J., Woo, J.-H., He, K. B., Lu, Z., Ohara, T., Song, Y., Streets, D. G., Carmichael, G. R., Cheng, Y. F., Hong, C. P., Huo, H., Jiang, X. J., Kang, S. C., Liu, F., Su, H., & Zheng, B. (2015). MIX: A mosaic Asian anthropogenic emission inventory for the MICS-Asia and the HTAP projects. *Atmospheric Chemistry and Physics Discussions, 15*, 34813–34869. doi:10.5194/acpd-15-34813-2015.

Lu, Z., Zhang, Q., & Streets, D. G. (2011). Sulfur dioxide and primary carbonaceous aerosol emissions in China and India, 1996–2010. *Atmospheric Chemistry and Physics, 11*, 9839–9864. doi:10.5194/acp-11-9839-2011.

Mijling, B., van der, A. R. J., & Zhang, Q. (2013). Regional nitrogen oxides emission trends in East Asia observed from space. *Atmospheric Chemistry and Physics, 13*, 12003–12012. doi:10.5194/acp-13-12003-2013.

Ohara, T., Akimoto, H., Kurokawa, J., Horii, N., Yamaji, K., Yan, X., & Hayasaka, T. (2007). An Asian emission inventory of anthropogenic emission sources for the period 1980–2020. *Atmospheric Chemistry and Physics, 7*, 4419–4444. doi:10.5194/acp-7-4419-2007.

Schultz, M., Rast, S., van het Bolscher, M., Pulles, T., Brand, R., Pereira, J., Mota, B., Spessa, A., Dalsøren, S., van Nojie, T., & Szopa, S. (2007). Emission data sets and methodologies for estimating emissions. Report of the RETRO European project, EU-Contract No. EVK2-CT-2002-00170, available at: http://retro-archive.iek.fz-juelich.de/data/documents/reports/.

Stavrakou, T., Müller, J.-F., De Smedt, I., Van Roozendael, M., van der Werf, G. R., Giglio, L., & Guenther, A. (2009). Global emissions of non-methane hydrocarbons deduced from SCIAMACHY formaldehyde columns through 2003–2006. *Atmospheric Chemistry and Physics, 9*, 3663–3679. doi:10.5194/acp-9-3663-2009.

Stavrakou, T., Müller, J.-F., Bauwens, M., De Smedt, I., Van Roozendael, M., De Mazière, M., Vigouroux, C., Hendrick, F., George, M., Clerbaux, C., Coheur, P.-F., & Guenther, A. (2015). How consistent are top-down hydrocarbon emissions based on formaldehyde observations from GOME-2 and OMI? *Atmospheric Chemistry and Physics, 15*, 11861–11884. doi:10.5194/acp-15-11861-2015.

Stohl, A., Aamaas, B., Amann, M., Baker, L. H., Bellouin, N., Berntsen, T. K., Boucher, O., Cherian, R., Collins, W., Daskalakis, N., Dusinska, M., Eckhardt, S., Fuglestvedt, J. S., Harju, M., Heyes, C., Hodnebrog, Ø., Hao, J., Im, U., Kanakidou, M., Klimont, Z., Kupiainen, K., Law, K. S., Lund, M. T., Maas, R., MacIntosh, C. R., Myhre, G., Myriokefalitakis, S., Olivié, D., Quaas, J., Quennehen, B., Raut, J.-C., Rumbold, S. T., Samset, B. H., Schulz, M., Seland, Ø., Shine, K. P., Skeie, R. B., Wang, S., Yttri, K. E., & Zhu, T. (2015). Evaluating the climate and air quality impacts of short-lived pollutants. *Atmospheric Chemistry and Physics, 15*, 10529–10566. doi:10.5194/acp-15-10529-2015.

Streets, D. G., Bond, T. C., Carmichael, G. R., Fernandes, S. D., Fu, Q., Klimont, Z., Nelson, S. M., Tsai, N. Y., Wang, M. Q., Woo, J.-H., & Yarber, K. F. (2003). An inventory of gaseous and primary aerosol emissions in Asia in the year 2000. *Journal of Geophysical Research, 108*, D21. doi:10.1029/2002JD003093.

Su, S., Bengang, L., Cui, S., & Tao, S. (2011). Sulfur dioxide emissions from combustion in China: From 1990 to 2007. *Environmental Science & Technology, 45*(19), 8403–8410. doi:10.1021/es201656f.

van Aardenne, J. A., Dentener, F. J., Olivier, J. G. J., Klein Goldewijk, C. G. M., & Lelieveld, J. (2001). A 1x1 degree resolution dataset of historical anthropogenic trace gas emissions for the period 1890–1990. *Global Biogeochemical Cycles, 15*(4), 909–928.

van Vuuren, D. P., Edmonds, J., Kainuma, M., Riahi, K., Thomson, A., Hibbard, K., Hurtt, G. C., Kram, T., Krey, V., Lamarque, J.-F., Masui, T., Meinshausen, M., Nakicenovic, N., Smith, S. J., & Rose, S. K. (2011). The representative concentration pathways: An overview. *Climatic Change, 109*, 5–31. doi:10.1007/s10584-011-0148-z.

Wang, R., Tao, S., Wang, W., Liu, J., Shen, H., Shen, G., Wang, B., Liu, X., Li, W., Huang, Y., Zhang, Y., Lu, Y., Chen, H., Chen, Y., Wang, C., Zhu, D., Wang, X., Li, B., Liu, W., & Ma, J. (2012). Black carbon emissions in China from 1949 to 2050. *Environmental Science & Technology, 46*(14), 7595–7603.

Wang, S. X., Zhao, B., Cai, S. Y., Klimont, Z., Nielsen, C. P., Morikawa, T., Woo, J. H., Kim, Y., Fu, X., Xu, J. Y., Hao, J. M., & He, K. B. (2014). Emission trends and mitigation options for air pollutants in East Asia. *Atmospheric Chemistry and Physics, 14*, 6571–6603. doi:10.5194/acp-14-6571-2014.

Wei, W., Wang, S., Chatani, S., Klimont, Z., Cofala, J., & Hao, J. (2008). Emission and speciation of non-methane volatile organic compounds from anthropogenic sources in China. *Atmospheric Environment, 42*, 4976–4988.

Wei, W., Wang, S. J., & Hao, S. (2011). Projection of anthropogenic volatile organic compounds (VOCs) emissions in China for the period 2010–2020. *Atmospheric Environment.*, 2011*, 45*, 6863–6871. doi:10.1016/j.atmosenv.2011.01.013.

Zhang, Q., Streets, D. G., Carmichael, G. R., He, K. B., Huo, H., Kannari, A., Klimont, Z., Park, I. S., Reddy, S., Fu, J. S., Chen, D., Duan, L., Lei, Y., Wang, L. T., & Yao, Z. L. (2009). Asian emissions in 2006 for the NASA INTEX-B mission. *Atmospheric Chemistry and Physics, 9*, 5131–5153. doi:10.5194/acp-9-5131-2009.

Zhao, Y., Nielsen, C. P., Lei, Y., McElroy, M. B., & Hao, J. (2011). Quantifying the uncertainties of a bottom-up emission inventory of anthropogenic atmospheric pollutants in China. *Atmospheric Chemistry and Physics, 11*, 2295–2308. doi:10.5194/acp-11-2295-2011.

Zhao, B., Wang, S. X., Liu, H., Xu, J. Y., Fu, K., Klimont, Z., Hao, J. M., He, K. B., Cofala, J., & Amann, M. (2013a). NO_x emissions in China: Historical trends and future perspectives. *Atmospheric Chemistry and Physics, 13*, 9869–9897. doi:10.5194/acp-13-9869-2013.

Zhao, Y., Zhang, J., & Nielsen, C. P. (2013b). The effects of recent control policies on trends in emissions of anthropogenic atmospheric pollutants and CO_2 in China. *Atmospheric Chemistry and Physics, 13*, 487–508. doi:10.5194/acp-13-487-2013.

Chapter 7
Biomass Burning Sources in China

Caiqing Yan, Jinting Yu, Yanan Zhao, and Mei Zheng

Abstract Biomass burning is an important source of pollution at both regional and global scales, with significant impacts on air quality, climate and human health. In China, it is also a major source of fine particulate matter. A better understanding of emissions, source contributions, transport, and transformation of biomass burning is essential for assessing its impacts on the environment. Here, studies related to biomass burning in China have been reviewed. Current understanding of its characteristics (e.g., biomass burning types, emission factors, and source profiles), emission amount and its contributions to ambient PM in China as well as its effects on climate and human health are summarized.

Keywords Biomass burning • Emissions in China • Source contribution • Health effect

7.1 Introduction

Biomass burning, including both open fires (e.g., forest fires, savanna fires, grassland fires, peat burning, and crop residue burning in fields) and domestic fires (residential biofuel burning), is one of the major sources of atmospheric aerosol at the global scale (van der Werf et al. 2010). With wide and intensive emissions of atmospheric pollutants, biomass burning sources have been recognized to have significant impacts on local and regional air quality, visibility, human health, global atmospheric chemistry, biogeochemical cycles, earth's radiative budget, and climate change (Andreae 1991; Reddington et al. 2015).

Biomass burning activities occur all around the world (e.g., in Asia, Africa, South America, and North America) with different scales and characteristics. To evaluate their roles in local, regional and global air quality and climate change, and

C. Yan • J. Yu • Y. Zhao • M. Zheng (✉)
SKL-ESPC and BIC-ESAT, College of Environmental Sciences and Engineering, Peking University, Beijing 100871, China
e-mail: cyan0325@pku.edu.cn; lingtingyjt@163.com; ynzhao0915@126.com; mzheng@pku.edu.cn

I. Bouarar et al. (eds.), *Air Pollution in Eastern Asia: An Integrated Perspective*, ISSI Scientific Report Series 16, DOI 10.1007/978-3-319-59489-7_7

to provide the scientific basis for policy making in air pollution control, it is essential to accurately quantify contributions of biomass burning to ambient air pollutants, and to better understand spatial and temporal variations of emission characteristics for each specific type of fires.

A number of studies have been conducted on biomass burning sources all over the world, as well as in China, where severe air pollution is of great concern, and biomass burning has been recognized as one of major sources of atmospheric fine particulate matter ($PM_{2.5}$). Studying biomass burning source is very important for reducing environmental and health impacts resulting from fine particulate pollution (Huang et al. 2016). However, a comprehensive understanding and assessment of biomass burning sources in China is still much needed. Therefore, the primary goals of this chapter are (1) to highlight the emission characteristics of biomass burning sources in China; (2) to summarize the current understanding of the biomass burning contributions to ambient fine particulate matter; and (3) to present current challenges and progress on biomass burning research and point out to future research needs.

7.2 Emission Characteristics of Biomass Burning in China

China is an important source region for atmospheric aerosol, with various origins including biomass burning, vehicular exhaust, coal combustion, dust storms, industrial and residential emissions (Fu et al. 2012a; Huang et al. 2014). Current understanding on biomass burning in China is summarized and presented below.

7.2.1 *Types of Fuel*

There are several fire types related to biomass burning in the world, including fires from savannas, grasslands, tropical forest, extratropical forest, domestic biofuel burning, charcoal production and combustion, and agricultural residue burning (Andreae and Merlet 2001). Among these sources, forest, savanna or grassland fires, burning of agricultural crop residues and domestic biofuel represent the major biomass burning activities in China (Streets et al. 2003a). However, forest and savanna fire emissions do not provide significant contributions in China, in comparison with those from tropical America, Africa and Southeast Asia (van der Werf et al. 2006). These types only occur in certain areas and have decreased remarkably since the early 1980s (Yan et al. 2006).

As China is a large agriculture country and it has the world's top-ranked agricultural production, tremendous amounts of agricultural residues are produced (Huang et al. 2012a). Agricultural residues (sometimes called agricultural waste) are often burned in-situ, especially in relatively undeveloped areas, as it is generally the quickest, simplest and cheapest approach. In addition, this operation also provides some nutrients back into the soil prior to the next planting. Based on previous studies, up to 40% of crop residues (e.g., rice, wheat, corn, coarse cereals, cotton, legumes, peanut or rapeseed) were burned in fields every year in some provinces in China, but this percentage has decreased in recent years (Cao et al. 2006, 2008a; Huang et al. 2012a; Li et al. 2007; Yan et al. 2006).

Crop residue burning in the fields accounts for a major fraction of the total biomass burning in China (Huang et al. 2016; Li et al. 2016a; Streets et al. 2003a), and poses a serious threat to air quality and human health, which has drawn increasing attention in China (Li et al. 2016a). After a series of regulations and policies for limiting or prohibiting agricultural burning in the fields, such activities have been reduced in many areas (Yan et al. 2006). There are also state-sponsored efforts to collect and burn biomass materials in power stations or to recycle the residues. However, in the absence of tight monitoring, crop residues are still burned, especially in rural areas during nighttime (Li et al. 2014). Indeed, such field-based burning of crop residues remains very common in most areas of China (Huang et al. 2012a; Zhang et al. 2008a). Crop residues are also used as fuels for household heating and cooking. Based on the first Chinese Environmental Exposure-related Human Activity Patterns Survey (CEERHAPS, Duan et al. 2014), 32.1% of households in China use biomass fuels for cooking, and 12.8% use it for heating. Some studies indicate that agricultural crop residues burned (as domestic fuel) is higher than amounts burned in the field (Yan et al. 2006; Zhang et al. 2008a).

7.2.2 *Emission Characteristics*

7.2.2.1 Emission Factor

Emission factor (EF) is a very important parameter that quantifies air pollutants from various activities (Oros et al. 2006). Several studies have been carried out in China to derive EF for different pollutants emitted from biomass burning, based on the following measurements: (1) in laboratory using combustion tower, self-built burning stove or dilution chamber system (Cao et al. 2008a, b; Liu et al. 2011; Zhang et al. 2008a, 2013); (2) in cookstove and Chinese Kang used in households in rural China (Li et al. 2007, 2009; Wang et al. 2009a); and (3) in areas downwind of agricultural fires (Li et al. 2007).

Tables 7.1a and 7.1b list estimated EFs for pollutants resulting from burning of different types of straws in open fires and household stoves based on several previous studies. For example, Li et al. (2007) conducted emission testings of open biomass burning in China. Shen et al. (2010) investigated EFs of particulate matter (PM) and elemental carbon (EC) for crop residues burned in typical household stoves. Zhang et al. (2000) developed a database of EFs for carbon monoxide (CO), carbon dioxide (CO_2), methane (CH_4), nitrogen oxide (NO_x), sulfur dioxide (SO_2), non-methane hydrocarbons (NMHCs), and total suspended particles from

Table 7.1a Speciated emission factors of pollutants from biomass burning in China (g/kg)

References	Fuel type	CO_2	CO	CH_4	N_2O	NO_X	SO_2	OC	EC	$PM_{2.5}$
Liu et al. (2011)	Rice straw	656.27 ± 26.15	44.12 ± 0.36	2.19 ± 0.73	0.11 ± 0.01					
	Wheat straw	586.39 ± 20.25	22.19 ± 0.77	2.22 ± 0.12	0.05 ± 0.002					
	Maize straw	620.72 ± 47.56	43.25 ± 3.31	2.95 ± 0.17	0.12 ± 0.01					
	Rape stalk	795.71 ± 26.38	32.61 ± 1.08	3.40 ± 1.27	0.06 ± 0.002					
	Soybean straw	543.11 ± 57.86	29.41 ± 3.13	2.89 ± 0.65	0.09 ± 0.01					
	Cotton stalk	464.14 ± 2.96	24.88 ± 0.16	1.82 ± 0.58	0.05 ± 0.01					
Zhang et al. (2008a)	Rice straw	791.3 ± 12.5	64.2 ± 4.9			1.81 ± 0.09				
	Wheat straw	1557.9 ± 85.8	1,41.2 ± 14.8			1.12 ± 0.19				
	Corn stover	1261.5 ± 59.9	1,14.7 ± 12.4			1.28 ± 0.04				
Cao et al. (2008a)	Rice straw	1757.6	72.4			3.52	0.147	1.96	0.52	
	Wheat straw	1483.6	65.5			2.59	0.049	3.83	0.52	
	Corn stover	2200.2	70.2			3.36	0.026	2.21	0.78	
	Cotton stalk	1453.4	111.8			2.68	0.002	1.83	0.82	

Cao et al. (2008b)	Wheat straw	1377.72 ± 431.12	57.78 ± 24.75			2.28 ± 1.00	0.04 ± 0.04	3.46 ± 2.05	0.42 ± 0.23	
	Rice straw	1674.12 ± 452.26	67.98 ± 25.58			3.43 ± 1.08	0.18 ± 0.31	2.01 ± 0.67	0.49 ± 0.21	
	Corn stover	2327.14 ± 709.57	67.64 ± 13.01			3.60 ± 0.85	0.04 ± 0.04	2.25 ± 0.74	0.95 ± 1.08	
	Cotton stalk	1345.42 ± 108.02	105.82 ± 6.02			2.49 ± 0.23		1.83 ± 0.54	0.82 ± 0.20	
Zhang et al. (2013)	Rice straw_F	1105.2 ± 189.3	53.2 ± 17.9					10.53 ± 4.87	0.49 ± 0.22	12.1 ± 4.4
	Rice straw_S	1024.0 ± 207.9	110.6 ± 37.9					8.77 ± 4.81	0.37 ± 0.11	18.3 ± 13.5
	Sugarcane	1152.5 ± 258	40.1 ± 15.7					1.25 ± 0.67	1.22 ± 0.66	4.12 ± 1.10
Li et al. (2007)	Wheat straw	1470 ± 46	60 ± 23	3.4 ± 0.85	0.07 ± 0.02	3.3 ± 1.7	0.85 ± 0.57	2.7 ± 1.0	0.49 ± 0.12	7.6 ± 4.1
	Maize straw	1350 ± 16	53 ± 4.0	4.4 ± 0.97	0.14 ± 0.03	4.3 ± 1.8	0.44 ± 0.20	3.9 ± 1.7	0.35 ± 0.10	11.7 ± 1.0

“F” presents “flaming”, and “S” presents “smoldering”

Table 7.1b Speciated emission factors of pollutants from biomass burning in China (g/kg)

References	Fuel type	CO_2	CO	CH_4	VOC	NO_X	OC	EC	$PM_{2.5}$
Li et al. (2009)	Crop residue						1.93 ± 1.00	0.43 ± 0.32	4.43 ± 1.87
	Woody fuel						1.12 ± 0.40	1.49 ± 0.69	3.08 ± 0.82
Wang et al. (2009a)	Brushwood_Sep	1568.4 ± 55.6	39.6 ± 3.4	2.3 ± 1.4	1.77 ± 0.42	1.71 ± 0.24			
	Maize straw_Sep	1499.7 ± 33.4	41.5 ± 1.7	1.4 ± 0.1	1.07 ± 0.06	1.66 ± 0.79			
	Sorghum stalk_Sep	1600.5 ± 17.1	31.5 ± 1.5	1.2 ± 0.1	1.01 ± 0.34	1.58 ± 0.82			
	Maize straw_Sep	1544.2 ± 6.4	75.7 ± 5.6	0.4 ± 0.1	0.18 ± 0.06	1.52 ± 0.52			
Wang et al. (2009a)	Brushwood_Apr	1171.9 ± 87.3	60.3 ± 13.2	2.2 ± 0.4	2.50 ± 0.76	1.11 ± 0.28			
	Maize straw_Apr	704.3 ± 80.2	135.3 ± 6.6	10.8 ± 0.8	19.69 ± 3.29	0.55 ± 0.17			
	Wheat straw_ Apr	675.7 ± 87.3	167.7 ± 35.8	15.4 ± 3.5	26.57 ± 6.0	1.01 ± 0.34			
	Rice straw_ Apr	976.8 ± 58.5	104.5 ± 18.4	4.8 ± 1.0	7.36 ± 1.5	0.42 ± 0.13			
	Brushwood_ Jan	1490.9 ± 23.5	102.7 ± 30.4	2.2 ± 0.8	0.92 ± 0.49	1.03 ± 0.47			
	Maize straw_Jan	1313.7 ± 11.0	183.1 ± 34.7	3.6 ± 1.4	1.48 ± 0.36	0.92 ± 0.34			
	Wheat straw_Jan	1147.6 ± 123	215.4 ± 51	1.2 ± 0.7	0.91 ± 0.52	0.71 ± 0.17			
	Sorghum stalk_Jan	1438.6 ± 71.5	142.8 ± 28.4	0.7 ± 0.4	0.23 ± 0.11	0.90 ± 0.11			

wood and crop residue burning in different types of household stoves in China. Field-measurement of the emission factor for air pollutants represents a different approach compared to the analysis of burned material in laboratory combustion chambers or in household stoves. This approach can be combined with fuel consumption estimates to provide crop residue fire emission information in atmospheric models (Zhang et al. 2015). However, it should be noted that field-based EF measured in one region using mainly automatic harvesting methods, may differ from those in farmed areas with manual harvesting, even if the actual crop type (e.g., wheat) is identical (Akagi et al. 2011).

7.2.2.2 $PM_{2.5}$ Source Profile

Chemical speciation of $PM_{2.5}$ provides important information for determining aerosol properties, quantification of biomass burning source contribution, as well as its impacts on climate change and human health. $PM_{2.5}$ source profiles for biomass burning emissions, which refer to the fractional mass abundance of measured chemical species relative to primary $PM_{2.5}$ mass in biomass burning source emissions, have been established in some previous studies and are summarized in Table 7.2. It is clear that carbonaceous species (i.e., organic carbon (OC) and EC) are dominant in $PM_{2.5}$ from biomass burning. The percentage of OC in $PM_{2.5}$ ranges from 30 to 87% for different types of fuels under different burning conditions. EC mainly ranges from 2 to 18%, with higher fraction when considering stove-burning emissions (Tang et al. 2014). The OC/EC ratio varies from 0.6 to 24, with higher values in crop burning emissions compared to wood burning emissions (Shen et al. 2014; Wei et al. 2014; Zhang et al. 2013). Further, $PM_{2.5}$ from field burning is mainly composed of Cl^-, K^+, K, NH_4^+ and SO_4^{2-} (Cao et al. 2008b; Li et al. 2007; Wang et al. 2016; Zhang et al. 2007, 2012a, 2013). Specific ratios from some biomass burning sources in China, which are useful for quantifying contribution of biomass burning to local and regional fine particulate matter, are listed in Table 7.3.

There are still a limited number of studies focusing on the development of $PM_{2.5}$ speciation profiles for biomass burning in the case of organic species in China. Based on current limited studies, sugars and phenols have been recognized as the dominant organic species in biomass burning smoke. Additionally, PAHs, *n*-alkanes and *n*-alkanols could be also detected in biomass burning emissions, but not in large amount (Wang et al. 2009b; Zhang et al. 2007). Zhang et al. (2007) analyzed 12 classes of organic compounds in $PM_{2.5}$ produced from straw burning, and found that sugars and methyloxylated phenols were the most abundant species, and that levoglucosan had the highest concentration (about 4.5% of $PM_{2.5}$). Wang et al. (2009b) characterized organic compositions from burning in the case of three kinds of monsoon evergreen broad-leaf trees and of three kinds of shrubs, which constitute the predominant species of the natural vegetation in South China. They found that the most abundant organics were saccharides and phenols, followed by carboxylic acids or wax esters, PAHs, *n*-alkanes and *n*-alkanols.

Table 7.2 $PM_{2.5}$ speciation profiles from biomass burning in China (unit: % of $PM_{2.5}$)

References	Li et al. (2007)		Cao et al. (2008b)				Zhang et al. (2012a)	Zhang et al. (2013)*			Wang et al. (2016)*		
Fuel types	Wheat straw	Maize stover	Wheat straw	Rice straw	Corn stover	Cotton stalk	Wood	Rice straw	Rice straw	Sugarcane leaves	Maize straw	Wheat straw1	Wheat straw2
Burning state	Opening burning		Combustion tower				Local cooking practices	Flaming	Smold-ering		Simulation burning		
OC	38.5 ± 16.0	33.6 ± 13.8	39.54	32.01	42.37	40.40	37.7 ± 14.9	87.0	47.9	30.3	54.0	46.9	43.9
EC	7.65 ± 3.97	2.98 ± 0.68	4.80	7.80	17.89	18.10	3.5 ± 1.1	4.0	2.0	29.6	1.9	2.6	2.4
Na	0.17 ± 0.01	0.17 ± 0.14											
Mg	0.03 ± 0.01	0.05 ± 0.01											
Al	0.14 ± 0.15	0.04 ± 0.02					0.11 ± 0.05						
Si	0.06 ± 0.07	0.05 ± 0.03					0.17 ± 0.07						
S	0.62 ± 0.13	0.67 ± 0.34					1.5 ± 0.5						
K	7.26 ± 4.28	6.88 ± 3.49					1.65 ± 1.4						
Ca	0.13 ± 0.06	0.06 ± 0.03					0.03 ± 0.01						
Fe	0.02 ± 0.01	0.01 ± 0.01					0.04 ± 0.01						
Ti	0.0033 ± 0.0023	0.0012 ± 0.0014											
V	0.0004 ± 0.0004	0.0009 ± 0.0007					0.0006 ± 0.0001						
Cr	0.0041 ± 0.0071						0.008 ± 0.0002						
Mn	0.0008 ± 0.0002	0.0011 ± 0.0002					0.007 ± 0.001						
Co	0.0001 ± 0.0001	0.0000 ± 0.0000					0.0005 ± 0.0001						
Ni	0.0046 ± 0.0045	0.0003 ± 0.0003					0.0005 ± 0.0001						
Cu							0.006 ± 0.001						
Zn		0.0003 ± 0.0005					0.09 ± 0.03						
Ga	0.0002 ± 0.0003	0.0000 ± 0.0001											
As	0.0007 ± 0.0002	0.0006 ± 0.0001					0.0003 ± 0.0001						
Se	0.0002 ± 0.0001	0.0005 ± 0.0001											
Ag	0.0000 ± 0.0001	0.0001 ± 0.0003											

Cd	0.0004 ± 0.0002	0.0006 ± 0.0002					0.0001 ± 0.0001						
Hg	0.0001 ± 0.0003												
Pb	0.0088 ± 0.0093	0.0095 ± 0.0037											
Cl^-	13.8 ± 14.6	23.0 ± 7.05	10.85	13.61	7	16.02	0.45 ± 0.32	6.86	6.23	9.71	5.98	6.08	7.21
F^-			0.03	0.01	0.06	0.01	0.11 ± 0.04	0.50	0.66	0.00			
NO_2^-			0.08	0.09	0.13								
NO_3^-	0.24 ± 0.21	0.6 ± 0.23		0.74	0.32		1.45 ± 1.56	0.25	0.11	0.00	0.04	0.06	0.11
SO_4^{2-}	1.54 ± 1.25	1.86 ± 1.08	2.06	2.89	2.53	4.77	6.44 ± 2.95	1.49	1.20	2.43	0.38	0.56	1.23
NH_4^+	3.69 ± 3.33	9.97 ± 2.3	0.12	0.05	0.01		10.78 ± 2.38	3.64	1.48	2.43	1.03	0.54	0.51
K^+	9.94 ± 11.8	8.51 ± 4.77	9.56	11.39	3.39	16.23	2.67 ± 1.03	5.29	4.48	7.28	2.81	4.69	5.78
Na^+			1.12	1.22	1.78	2.26	0.17 ± 0.03	0.25	0.27	0.00	0.01	0.02	0.30
Mg^{2+}			0.06	0.1	0.02		0.04 ± 0.01	0.00	0.00	0.00	0.00	ND	ND
Ca^{2+}			0.77	0.97	1.35	1.03	0.15 ± 0.08	0.33	0.27	0.00	0.00	ND	0.00

*presents calculated data based on the reference

Table 7.3 Ratios of chemical components in source emissions of biomass burning from different studies

References	Fuel type	Burning state	$K^+/PM_{2.5}$	K^+/OC	EC/K^+	Levoglucosan/OC	$BC/PM_{2.5}$	$TC/PM_{2.5}$	OC/EC
Shen et al. (2014)	Fuel wood log						0.33	61 ± 24	0.83 ± 1.14
	Brushwood/branches						0.58	70 ± 29	0.59 ± 0.88
	Crop residue						0.16	49 ± 14	1.54 ± 1.27
Ni et al. (2015)	Wheat straw	Field open burning					0.021		24.7 ± 13.4
	Rice straw						0.025		15.6 ± 5.9
	Corn stalk						0.023		21.4 ± 6.0
Wei et al. (2014)	Wheat straw						0.047		5.7
	Rape straw						0.13		1.2
	Rice straw						0.098		2.2
	Cotton straw						0.21		2.6
Zhang et al. (2013)	Rice straw	Open/flaming	0.0529	0.061	0.77		0.04		21.5
	Rice straw	Open/smoldering	0.0448	0.094	0.45		0.02		23.7
	Sugarcane leaves	Open/flaming	0.0730	0.24	4.1		0.3		1
	Maize straw	Open/flaming					0.059		4
	Maize straw	Stoves					0.078		6.4
	Soybean/oil crops straw	Stoves					0.22		1.6
	Firewood	Stoves					0.47		0.75
	Forest	Open					0.043		16.1
Li et al. (2007)	Wheat straw	Open burning	0.0763	0.21	0.84		0.064		5.5
	Maize stover	Open burning	0.0855	0.26	0.35		0.03		11.1
Li et al. (2009)	Wheat straw	Household stove							6.3–12.5
	Corn stalk								4–4.8

Cao et al. (2008b)	Wheat straw		0.0952	0.024	0.5		0.048		8.2
	Rice straw		0.1136	0.36	0.69		0.078		4.1
	Corn stover		0.0339	0.08	5.3		0.18		2.4
	Cotton stalk		0.1623	0.4	1.1		0.18		2.2
Zhang et al. (2007)	Cereal straw	Dilution chamber				0.037			

It should be noted that, biomass types, burning conditions (e.g., in the field or household stoves) and burning phases (smoldering or flaming phase) are major factors influencing biomass burning emission characteristics. The difference between different burning conditions and phases should be considered when choosing appropriate source profiles and ratios for estimating biomass burning source contribution.

7.2.2.3 Total Emission

Emissions from open burning including forest fires, grassland fires and field burning of crop residues, as well as residential burning of crop residues and woods in China have been estimated and reported in previous publications (Cao et al. 2005; Huang et al. 2012a; Li et al. 2016a; Lu et al. 2011a; Tian et al. 2011; Wang and Zhang 2008; Yan et al. 2006; Zhang et al. 2008a). Estimates of emissions in some of the studies are shown in Table 7.4.

Table 7.4 suggests that emissions estimated in different studies could be different. The differences might be primarily associated with the adopted method. For example, Li et al. (2016a) and Huang et al. (2012a) used data from moderate resolution imaging spectroradiometer (MODIS) Thermal Anomalies/Fire Daily Level 3 Global Product (MOD/MYD14A1), whereas other studies listed in Table 7.4 were based on the EFs and consumption values for biomass fuels. Large variations in activity data and EFs, as well as difference in parameters used such as the proportion of crop residues burned in fields, and grain-to-straw ratios could lead to large differences in final estimates (Li et al. 2016a; Yan et al. 2006). EFs used in previous studies were usually obtained from studies in other countries or using combined emission factors from different types of crops due to the lack of local estimates, or simply by applying the same emission factor for different crops. For example, Streets et al. (2003b) and Yan et al. (2006) have estimated emissions from crop burning in China using the emission factors from Andrea and Merlet (2001), which combined information of emissions from various types of biomass burning.

7.2.2.4 Emission Inventory

It is essential to establish accurate estimates of aerosol and gaseous pollutants from emissions of open field burning and residential burning for evaluating their roles in local air quality and climate change (Cao et al. 2008a), providing required information for air quality simulation and forecasting, and providing helpful information to policy makers for pollution control (Street et al. 2003a; Zhang et al. 2008a). Although there are a few studies available for developing emission inventories of open burning of crop residues in China, data are still scarce (Li et al. 2007). Zhang et al. (2008a) developed an emission inventory of gaseous pollutants from burning of rice, wheat and corn straws, but did not provide data for $PM_{2.5}$. Huang et al.

Table 7.4 Emission estimates of biomass burning in China (unit: Gg)

Type	References	Year	CO_2	CO	CH_4	$NMVOC_S$	NH_3	SO_2	NO_X	$PM_{2.5}$	OC	EC
Crop residue burning	Li et al. (2016)	2012	140,000	11,000	330	870	80	70	170	1220	360	60
	Huang et al. (2012)	2006	67,882	4378	249	2202	93	15	226	268	99	33
	Wang et al. (2008)	2006	154,502	7306	374	873	84	60	360	2167	480	49
Open burning and biofuel	Lu et al. (2011b)	2007	743,743	76,580	3333	6050	775	335	951	4044	784	268
	Tian et al. (2011)	2007	946,635	59,297	2202	3026	–	173	1085	3620	–	427
	Cao et al. (2005)	2000	883,884	39,338	1931	7340	793	175	1125	–	1309	3160

(2012a) developed a detailed and highly-spatially (1 km) and temporally (10 days) resolved emission inventory for agricultural crop residue resulting from field burning, based on daily MODIS Thermal Anomalies/Fire products and the Global Land Cover for China (GLC-China, Xu et al. 2005). Li et al. (2016a) established an improved crop residue open burning emission inventory with a spatial resolution of $0.25° \times 0.25°$ and a temporal resolution of 1 month, based on the MOD/MYD14A1 and on county-level agricultural statistical data, using locally measured EFs.

Biomass burning emission inventories can be improved in the following ways, which include (1) Uncertainty reduction: There are still large uncertainties in estimating the amount of biomass burned and the associated total emissions. The large uncertainties associated with burned biomass estimates, in particular those related to open burning, are mainly due to the uncertainty in estimating burned areas, fuel loads, combustion efficiency and EFs (Huang et al. 2012a; Ito and Penner 2004); (2) Speciated emission inventory: Most of the current estimates of biomass burning emissions are only focusing on several gaseous and particulate pollutants such as CO_2, CO, CH_4, N_2O, NO_X, SO_2, VOC, OC, EC, and $PM_{2.5}$. Only a few studies provide information of other pollutants from biomass burning emissions in China including mercury (Huang et al. 2011) and PAHs (Zhang and Tao 2008); and (3) High spatiotemporal resolution: High spatiotemporal emission inventories are required to meet the need of global and regional air quality simulations and to improve the accuracy of model results.

7.2.3 *Spatial and Seasonal Variations*

Biomass burning in China exhibits significant spatial and seasonal variations, due to spatial and seasonal variations in biomass types and economic levels. For example, crop residue burning is strongly correlated with agricultural practices (harvesting cycles, types of crops, etc.). Figure 7.1 presents the distribution of different provinces and regions in China defined in the present report.

7.2.3.1 Spatial Variation

Agricultural open fires with large amount of emissions per unit area are located in some provinces of eastern and northeastern China (Cao et al. 2008a), and are mainly distributed in three regions, i.e., the Northern China Plain (NCP, e.g., Henan, Shandong, Northern Anhui and Northern Jiangsu provinces), Northeastern China (e.g., Heilongjiang, Jilin and Liaoning provinces), and Southern China (e.g., Guangdong, Guangxi and Yunnan provinces). The regions account for approximately 75% of fires over the whole country. Most fire emissions originate in NCP (Huang et al. 2012a,b), which is the largest agricultural zone in China, with about 34% of the national rural population, 27% of the cultivated land and 35% of the crop yield according to National Bureau of Statistics of China (NBSC, http://www.stats.gov.cn/). On the contrary, fires are sparsely scattered in the western part of

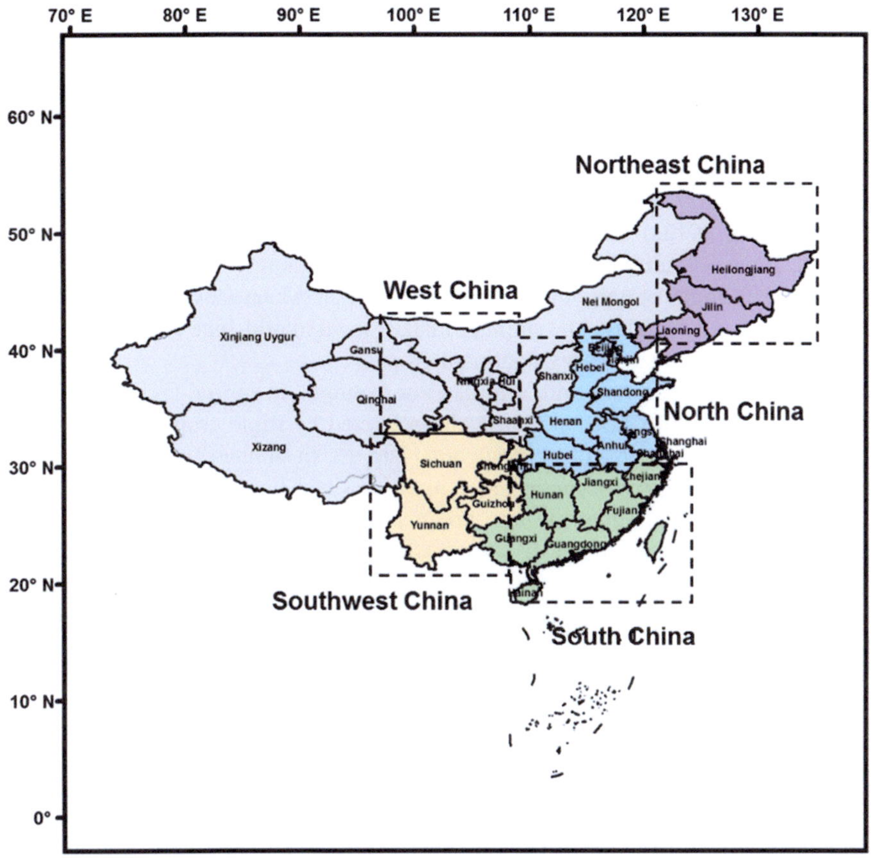

Fig. 7.1 Major regions in China defined in this study

China (e.g., Tibet, Qinghai and Gansu provinces) and in Inner Mongolia due to their lower rural population density and lower economic levels (Cao et al. 2008a). In general, the five provinces with the highest burning levels are Shandong, Henan, Anhui, Heilongjiang, and Hebei, where farmers' income and population density are higher than elsewhere in China (Cao et al. 2008a). The five provinces/cities with the lowest levels are Tibet, Tianjin, Beijing, Shanghai, and Hainan.

There are geographical diversities in both fuel types and total emission. Crop types burned are closely correlated with crop distributions in China. As corn, wheat, and rice are extensively planted in northeast China, NCP, and south China, respectively. The major biomass burned from north to south in China varies from maiz straw, to wheat straw and rice straw (Li et al. 2016a). Compared to eastern areas, forest and savanna fires are more important in the western regions, rather than agricultural burning. Compared to northern regions, broad-leaf trees and shrubs are burned in larger quantities in southern China. As some of plant residues are used as fuels for cooking in rural areas (Duan et al. 2014; Tao et al. 2013), residential biomass burning is more prominent in rural regions than in urban areas.

7.2.3.2 Seasonal Variation

Fires are affected by both natural and anthropogenic factors, and crop residue burning is associated with local sowing and harvest timing. Therefore, biomass burning types and emissions are not only variable spatially but also temporally in China. The highest fire count is observed in June (about 37% of total fires), followed by early October, and March to May (contributing 9–10% on average) (Huang et al. 2012a). Fire counts are lowest during November to January (only 2–3%). Biomass burning contribution is therefore highest in summer and fall, and follows the timing of agricultural activities. Winter wheat (sown in mid-October and reaped in end of May) and summer maize (sown in mid-June and reaped in end of September) are the two key dominant crops.

Fire locations also vary with seasons. For example, in spring, fires are active in the southern part of China, which is characterized by triple cropping cultivation. March to April is the time of the first-round harvest. In summer (mainly in early- to late-June), fires mostly occur in the central area of China. According to NBSC, intensive agricultural fires in June are mainly related to wheat straw burning because of cultivation mode of winter wheat and summer corn/rice in eastern China. A considerable number of fires in southern China is related to the second-round harvest. In autumn, crop residue burning uniformly scatters over most of agricultural zones since autumn is the main harvest season for many types of crops. In winter, fires are sparsely distributed (Huang et al. 2012a, 2016). Residential biomass burning becomes important in rural areas in winter (especially in northern China) due to intensive heating activities in cold season.

7.3 Biomass Burning Source Contribution in China

Previous studies have indicated that biomass burning could be one of the major sources leading to severe regional air pollution and haze in the Beijing-Tianjin-Hebei (BTH), the Pearl River Delta (PRD), the Yangtze River delta (YRD) regions (Cheng et al. 2013, 2014; Li et al. 2010; Wang et al. 2007; Zhang et al. 2010). Therefore, it is essential to quantify contributions of biomass burning to air pollution.

7.3.1 Methods for Quantifying Biomass Burning Source

7.3.1.1 The Satellite-Based Method

Recent developments of satellite monitoring systems, such as the Along-Track Scanning Radiometer (ATSR), advanced Very High Resolution Radiometer, and MODIS have provided useful information for identifying fire events and the duration of the burning (Lin et al. 2014; Morton et al. 2008). The Fire Information for Resource Management System (FIRMS) is a NASA-funded application for disseminating MODIS Active Fire Data. It provides fire locations and MODIS images (http://maps.geog.umd.edu/firms).

Emissions of trace gases and aerosol from open biomass burning could be estimated from burned areas, fuel load maps, combustion factors, and EFs (Ito and Penner 2004). Burned area can be quantified by using satellite data combined with fractional vegetation cover map. Fire counts for various land cover types along with MODIS Burned Area Product (MCD45A1) have been used to investigate contributions of biomass burning emissions and its spatiotemporal allocations (Boschetti and Roy 2009). Biomass burning emissions retrieved from satellite data provide spatiotemporal patterns and can serve as input to chemistry transport models for air quality forecasts in near-real time (Zhang et al. 2008b). The combination of satellite-detected fire spots with backward trajectory analysis can be used to identify source regions of fires and transport pathways of air pollutants (Chan et al. 2003, 2006; Cheng et al. 2014; Sang et al. 2011; Zhang et al. 2012b). However, it should be noted that, there are several limitations to this method: (1) Only large-scale biomass burning can be identified by satellite-detected fire spots. Local anthropogenic activities such as domestic usage of biofuels and small size fires are not easily detected (Huang et al. 2012a); (2) Satellites can only capture a snapshot of fire energy instead of a time-integrated measure of fire activity (Morton et al. 2008). For example, MODIS satellites AQUA and TERRA pass over eastern China at a fixed local time (i.e., Terra: 10:30 AM and PM; Aqua: 1:30 PM and AM) every day. Thus, fires outside these time periods might be missed (Roy et al. 2008); and (3) Types of fire fuel cannot be identified with satellite-remote sensing.

To improve and optimize this method, MODIS satellite data should be combined with county-level agricultural statistical data, to reduce the influence of missing fire counts due to the small fire size and resulting from cloud cover (Li et al. 2016a). Additionally, local estimates of EFs should be used when calculating primary pollutant emissions.

7.3.1.2 The Receptor Model-Based Method

Gaseous carbonaceous compounds (e.g., CO, CH_4, C_2H_4, HCN, HCO_2H, and CH_3COH) can be emitted in large amount from biomass burning, and emission of CO, CO_2, NO_x, water-soluble potassium (K^+), total particulate matter and black carbon (BC) are typically measured to trace biomass burning source (Simoneit 2002).

K^+ is widely used as an inorganic biomass burning tracer (Zhang et al. 2013), although it could be emitted by other sources such as fireworks, cooking and sea salt (Cheng et al. 2013; Simoneit 2002). Compared to K^+, levoglucosan, a major pyrolysis product of cellulose, is more specific to biomass burning sources and is recognized as a useful tracer for quantitatively assessing the contribution of biomass burning smoke to aerosol mass (Tao et al. 2013; Zhang et al. 2007, 2008c, 2010). Although some studies have suggested that levoglucosan might decay during transport under certain conditions (Hennigan et al. 2010), more direct evidence is still needed, especially from ambient measurements. Therefore, levoglucosan is still widely used as a biomass burning tracer (Duan et al. 2004; Wang et al. 2007, 2009b, c; Zhang et al. 2008c). Isomers of levoglucosan, including β-sitosterol, 4-hydroxybenzoic acid, and lignin/resin acids can also be used to identify the influence of biomass burning (Fu et al. 2012b). Other molecular tracers that provide signature of biomass burning have also been used in previous studies, including methoxyphenols (Simpson et al. 2005), resin acids (Simoneit et al. 1993) and retene (Ramdahl 1983). Some organic compounds (e.g., wax esters, phenols, monosaccharide, and associated thermal alteration products) and the relative abundances of key biomarkers have been used to trace and distinguish the original biomass fuel types (Fine et al. 2001, 2002; Wang et al. 2009b).

Tracers of biomass burning can be applied in receptor models (e.g., Chemical Mass Balance Model, CMB; Positive Matrix Factorization, PMF) for source apportionment work. The application of CMB models requires source profiles of local sources (e.g., biomass burning, coal combustion, vehicular emission, dust, industrial sources) and ambient concentrations of individual chemical tracers at the receptor site to apportion contributions of different sources. By contrast, PMF models do not require source profiles, but with a large dataset, it can identify aerosol sources based on tracers and source profiles. The receptor model-based method has been widely used in the current $PM_{2.5}$ source apportionment research in China (Huang et al. 2014; Tao et al. 2013).

Some studies simply apply the ratios of biomass burning tracers (e.g., levoglucosan) to $PM_{2.5}$ to roughly assess contribution of biomass burning based on Eq. 7.1, with the assumption that the ratio of levoglucosan to $PM_{2.5}$ concentration in biomass plume does not change during atmospheric transport:

$$\text{Contribution of biomass burning} = \frac{([Levoglucosan]/[PM_{2.5}])_{ambient}}{([Levoglucosan]/[PM_{2.5}])_{source}} \tag{7.1}$$

where the square bracket represents concentration. Levoglucosan in $PM_{2.5}$ resulting from cereal straw burning, a major source of biomass burning in China, is reported as 4.5% by Zhang et al. (2007), while the mass fraction of OC in $PM_{2.5}$ is equal to 0.546 ± 0.060 (mg/mg). Wang et al. (2007) and Zhang et al. (2010) have estimated the contribution of biomass burning in Guangzhou and Beijing with this method. Using the simplified receptor-based approach, biomass smoke aerosol is estimated to account for 16–28% of OC in the background atmosphere of Hainan and 4.9% of OC at the rural site of Hong Kong during air pollution episodes (Zhang et al. 2012b).

It should be noted that the relative abundance of various tracers in biomass burning smoke plumes is highly variable and dependent on combustion temperature, flame aeration and moisture content of fuels (Wang et al. 2009b). Therefore, biomass burning estimates based on tracer method should take the fuel types and combustion conditions into consideration (Yan et al. 2015).

7.3.1.3 The Carbon Isotope-Based Method

Radiocarbon ($\Delta^{14}C$) has been recognized as a powerful technique for differentiating between biomass/biogenic emissions and fossil fuel sources (Kirillova et al. 2014), providing direct and quantitative information on fossil/biomass carbon contribution. This is because ^{14}C is free from fossil source but it is enriched from biomass burning source. Stable carbon isotopic signature ($\delta^{13}C$) varies with fuel types (Kawashima and Haneishi 2012), and its measured value can be used to distinguish between fuel types (Kawashima and Haneishi 2012). In general, $\delta^{13}C$ values from open burning of C_3 plants (e.g., rice, pines and conifers, −34.7 ‰ to −28.0 ‰) are lighter compared to values resulting from C_4 plants (e.g., corn and grass, −19.3 ‰ to −16.1 ‰) (Kawashima and Haneishi 2012; O'leary 1981). Several carbon isotope studies to identify relative contributions of carbonaceous aerosols from fossil fuels versus biomass sources have been conducted in China (Li et al. 2016b; Sun et al. 2012).

7.3.1.4 The Air Quality Model-Based Method

As the satellite-based method could not provide detailed information on aerosol composition and since the receptor model-based method is often based on a few receptor sites, each method is associated with its own limitations. The 3-D air quality models, however, remain an effective approach to estimate source contributions of biomass burning with a wide spatial coverage. Their ability to simulate biomass burning at both regional and global scale can help assess its impacts on radiative forcing and climate (Lin et al. 2014).

A number of modeling studies have been performed to simulate the impact of crop burning on air quality during harvest season (Cheng et al. 2014; Huang et al. 2012b; Yamaji et al. 2010). For example, Cheng et al. (2014) used both ambient monitoring data and WRF/CMAQ (weather research and forecasting, WRF and community multi-scale air quality, CMAQ) model simulations to quantify the contribution of biomass burning. They found that open biomass burning could contribute by 37%, 70%, and by 61% to $PM_{2.5}$, OC and EC concentrations, respectively. Yamaji et al. (2010) evaluated the influence of open crop residual burning on ozone, CO, BC and OC concentrations in North China Plain using a regional chemical transport model. Air quality model results are however associated with uncertainties which are linked to uncertainties in emission inventories. Therefore, there is a great need to develop improved local emission inventories to improve model results.

7.3.2 Contribution of Biomass Burning in China

Table 7.5 lists results of biomass burning contribution to $PM_{2.5}$ mass and carbonaceous aerosol in different areas of China. It can be seen that the contribution of biomass burning to the ambient $PM_{2.5}$ mass concentration in China varies with location, season and depends on the adopted method. Results from these studies show that biomass burning contribute to 15–40% of $PM_{2.5}$ in Beijing (Cheng et al. 2013; Wang et al. 2009c; Zhang et al. 2008c), and to around 3–20% in Guangzhou (Huang et al. 2014; Wang et al. 2007; Zhang et al. 2010). The contribution of biomass burning to $PM_{2.5}$ mass concentration is higher in summer with a contribution of around 20% (Cheng et al. 2013, 2014; Li et al. 2014).

The contribution of biomass burning to carbonaceous aerosol was also found to be much higher during the summer harvest season, which could account for around 50–70% of the total OC or EC mass in Beijing, Anhui and PRD region (Cheng et al. 2013, 2014; Li et al. 2014).

7.4 Other Issues Related to Biomass Burning Study in China

As aforementioned, biomass burning is one of the major sources of airborne particulate matter and gaseous pollutants all over the world, influencing regional air quality, earth's radiative budget, ecosystem, and human health (Andreae and Merlet 2001; Penner et al. 1992). Although the effects of biomass burning on air quality have been relatively well understood, knowledge on its effects on meteorology, climate, and human health, is still very limited in China (Huang et al. 2016).

7.4.1 Trans-boundary Transport

In addition to its local effects, biomass burning plumes can be uplifted to high altitudes such as 2000–4000 m (Lin et al. 2014), and can be further transported over long distances. They can affect air quality in many remote regions and even other countries (Galanter et al. 2000). Table 7.6 lists several studies that investigated the long-range transport of biomass burning smoke.

7.4.1.1 Long-Range Transport of Biomass Burning Emissions to China from Southeast Asian Countries

Previous studies have shown that biomass burning emissions from Southeastern Asia have significant impacts on air pollutants in the Pacific Ocean region and

Table 7.5 Contributions of biomass burning in different cities in China

Region	Season	Source contribution				Method	References
		$PM_{2.5}$	PM_{10}	OC	EC		
Beijing	Summer (2005, 2006)			11%		CMB receptor model	Wang et al. (2009c)
	Winter (2006)			26%			
	Annual average (2005–2006)	17%					
	Annual average (2002–2003)	18–38%	14–32%			Tracer-based methods	Zhang et al. (2008c)
	Summer (2011)	24%[a]		50%	50%	PMF receptor model	Cheng et al. (2013)
	Winter (2013)	5.6%				PMF receptor model	Huang et al. (2014)
Mount Tai	Summer (2006)	17%				Tracer-based methods	Fu et al. (2012b)
Anhui	Summer (2013)	40.6%		67.5%	68.5%	PMF receptor model	Li et al. (2014)
Shanghai	Winter (2013)	4.8%				PMF receptor model	Huang et al. (2014)
PRD (five cities)	Summer (2011)	37%[a]		70%	61%	Tracer-based methods and air quality model	Cheng et al. (2014)
Guangzhou	Autumn_downtown (2004)	4–19%				Tracer-based methods	Wang et al. (2007))
	Autumn_suburban (2004)	3–16.8%					
	Summer_daytime (2006)			7%[b]		Tracer-based methods	Zhang et al. (2010)
	Summer_nighttime (2006)			14%[b]			
	Winter (2013)	6.7%				PMF receptor model	Huang et al. (2014)
Chengdu	Spring (2009, 2010)	33%				PMF receptor model	Tao et al. (2013)
Xi'an	Winter (2013)	6.4%				PMF receptor model	Huang et al. (2014)

[a]Presents reconstructed $PM_{2.5}$
[b]Presents primary contribution only

Table 7.6 Studies on long-range transport of biomass burning between China and other countries

Emission region	Influenced region	Influences	Methods	Season	References
Southeast Asian countries (e.g., Burma and Vietnam region)	Hong Kong	Elevated ozone enhancement in the lower troposphere	Satellite image and MAPS	Spring	Chan et al. (2000)
Southeast Asian countries (e.g., Burma, Laos, Indochina and northern Thailand, Nepal of the India subcontinent)	Hong Kong	Tropospheric ozone enhancement	Satellite imagery, air trajectory and trace gas data	Spring	Chan et al. (2003)
Southeast Asia and South Asia	Tibetan Plateau of Southwest China	Increase of pollutants (e.g., ozone, trace gases and aerosol) in the lower troposphere	Fire maps from satellite data, back trajectories	Spring	Chan et al. (2006)
Southeast Asian countries (north of Vietnam)	PRD region	(1) UV intensity reduction; (2) Decrease the formation of ozone in Guangzhou	Satellite data, ground measurement, model	Spring	Deng et al. (2008)
Philippines	Southeast coast of China (through western Pacific Ocean, South China Sea)	Noticeable enhancements in ambient levoglucosan and OC	Field study, satellite and backward trajectory model	Spring, prevalent monsoon	Zhang et al. (2012b)
Southeast Asian countries (e.g., Indochina and southern China)	Central Taiwan	Enhancement of $PM_{2.5}$ mass	Backward air trajectories fire products of satellite images, surface and aloft isobaric maps	Spring	Lee et al. (2011)
Northern Philippines	Hong Kong	Contribution to 7.5% and 2.9% of OC and $PM_{2.5}$	Meteorological data, backward trajectories, fire counts and aerosol index	Spring	Sang et al. (2011)

(continued)

Table 7.6 (continued)

Emission region	Influenced region	Influences	Methods	Season	References
			from the Earth Probe satellite		
Indochinese peninsula of Southeast Asia (e.g., Thailand, Laos, Cambodia, Myanmar, Vietnam, and the mainland territory of Malaysia)	Central Taiwan	Reduction in shortwave radiation of about 20 W/m^2 at ground surface	Satellite measurement and AERONET ground observation, WRF-Chem model	Spring	Lin et al. (2014)
Northeast China/North China	Japan	Effects on the abundance of the biomass burning tracers such as anhydrosugars in Okinawa aerosol	Back trajectories, field observation	Fall and winter	Zhu et al. (2015)

China. For example, Zhang et al. (2012b) illustrated that biomass-burning emissions from the Philippines could be transported over the western Pacific Ocean and the southeast coast of China under the influence of prevalent monsoon winds. Chan et al. (2000, 2003) indicated that biomass burning emissions from Southeast Asian countries (e.g., Burma, Laos, Indochina and northern Thailand) could be transported to Hong Kong and result in ozone enhancement in the lower troposphere in springtime. Deng et al. (2008) indicated that biomass burning in Southeast Asia has important contributions to atmospheric aerosol and ozone levels in the PRD region. Lin et al. (2014) reported that biomass-burning aerosol from Indochinese peninsula (e.g., Thailand, Laos, Cambodia, Myanmar, Vietnam, and the mainland territory of Malaysia) could be long-distance transported to Taiwan. Huang et al. (2013) reported that Asian spring monsoon facilitated the impact of biomass burning from peninsular Southeast Asia to the South China Sea, the Taiwan Strait, and some provinces in southern China.

It is recognized that the transport of biomass burning emissions from Southeastern Asian countries, including India, Bangladesh, Myanmar, Thailand, Laos, Cambodia and Vietnam to southern China is common in springtime especially in dry season (Chan et al. 2003, 2006).

7.4.1.2 Long-Range Transport of Biomass Burning Emissions from China to Other Regions

Biomass burning emissions in China may have some effects on other downwind areas such as Japan, Korea and even the west coast of the United States (Zhu et al. 2015). With aircraft campaign, de Gouw et al. (2004) showed that air masses from Southeast Asia and China, where biomass-burning emissions were relatively high, were typically transported across the Pacific Ocean at a higher altitude than air masses from Japan and Korea, implying the influence of long-range transport of pollutants from Asia.

7.4.2 Effects of Biomass Burning

7.4.2.1 Effects on Air Quality

As biomass burning could emit significant amount of atmospheric pollutants, it influences atmospheric aerosol loading in the troposphere, changes $PM_{2.5}$ chemical composition and leads to poor visibility (Du et al. 2011; Fu et al. 2012b). Agricultural field burning from northern to southern China could lead to dramatic increase of $PM_{2.5}$ mass concentrations during the summer harvest season, resulting in poor regional air quality in large areas of eastern China (Huang et al. 2012b, 2016; Zhang et al. 2010),

A number of studies have been conducted to investigate the impacts of biomass burning on local and regional air quality in China. For example, Li et al. (2008) observed a significant degradation of air quality in Beijing in relation to wheat straw burning, together with a significant increase in CO concentrations. Li et al. (2010) indicated that agricultural biomass burning aerosol is a major contributor to regional haze in northern China and prominent in Beijing during June. Wang et al. (2007) showed that biomass burning could significantly influence air quality in Guangzhou, and explain more than 50% of air pollution episodes in different areas of Guangzhou. Du et al. (2011) found that biomass burning activities lead to haze events in Shanghai, with biomass burning source tracers (e.g., K^+, Cl^-) increasing by nearly 20 times compared to clear days. Cheng et al. (2014) also reported a significant impact of biomass burning on haze pollution in the YRD region during the summer harvest season, and indicated that biomass open burning could lead to low visibility (2.9–9.8 km) and contribute to 37% of $PM_{2.5}$ during certain periods. Based on the integrated analysis of satellite-detected fire spots, back-trajectory analysis and air quality model simulations, Cheng et al. (2014) found that the impact of biomass burning was regional, in response to substantial inter-province transport of air pollutants.

Besides local sources, biomass burning emissions from other regions through long-range/regional transport could be also a major factor impacting local air quality (Huang et al. 2012a). Most of the current studies are focusing however on the air quality impacts of open biomass burning, while impacts of residential (indoor) biomass burning for household cooking and heating have received less attention.

7.4.2.2 Effects on Climate Change

Challenge on Radiative Forcing of Light-Absorbing Carbonaceous Aerosol

Biomass burning has been recognized to have significant impacts on atmospheric chemistry and global climate, as it could not only emit greenhouse gases such as CO_2 and CH_4, other gases like CO and NMHCs, but also particulate matter, such as BC and OC. According to emission factor and $PM_{2.5}$ source profiles, OC and BC are two major carbonaceous components emitted from biomass burning sources. BC, the strongest particulate absorber of visible solar radiation, ranks as the second most important climate-warming agent after CO_2 (Bond et al. 2013). However, accurately determining radiative forcing of BC still remains a challenge (Gustafsson and Ramanathan 2016).

In most climate models, BC has been assumed to be the only significant light absorbing carbonaceous aerosol, while OC mainly has cooling effects. However, current studies have found that biomass burning aerosol could contain significant amounts of light-absorbing organic matter (named "brown carbon"). Biomass burning is identified as the dominant source of brown carbon in most areas of the world (Laskin et al. 2015), including China (Du et al. 2014). Du et al. (2014), for example, indicated that biomass burning is the most important source of water-soluble brown carbon, contributing to 58% of total light absorption at 365 nm. Yan et al. (2015) demonstrated that biomass burning derived WSOC could contribute to $17 \pm 4\%$ and $19 \pm 5\%$ of the light absorption by total WSOC in winter and summer in 2013, respectively. They pointed out that residential biomass burning was a non-negligible source of brown carbon in winter. However, light-absorbing properties of brown carbon are still not well understood and this effect is still not well accounted for in current climate models. A better understanding of light-absorbing properties and radiative forcing of brown carbon from different types of biomass burning activities is needed.

Generally, light-absorbing aerosol heats the air, alters regional atmospheric stability and vertical motions, and affects the large scale circulation and hydrologic cycle with significant regional climate effects (Menon et al. 2002). Biomass burning emissions, with a high emission rate of light-absorbing carbonaceous aerosol, could heat the atmosphere and cool the ground surface through both direct solar radiation attenuation (direct radiative forcing) and cloud redistribution (semi-direct radiative forcing) (Feng et al. 2013). Quantifying changes in properties of black and brown carbon in response to the evolution of biomass burning smoke from different types of biomass fuels is required to properly model the climate effect of these compounds.

A growing number of studies have been carried out to investigate weather-air quality interactions and to understand the effects of air pollution on synoptic scale weather (e.g., Ding et al. 2013). Some studies have indicated that biomass burning smoke can significantly change synoptic weather patterns, inhibit cloud formation and suppress precipitation (Huang et al. 2016). However, there is a need to develop interdisciplinary approaches for investigating environmental, weather and climate impacts of biomass burning emissions in China. An important tool to conduct such study could be provided by a fully coupled meteorology-air quality models (Ding et al. 2013).

7.4.2.3 Effects on Human Health

Different methods are used to investigate health effects of biomass burning aerosol including epidemiological study, exposure experiments, and toxicological research. Some studies have provided evidence to demonstrate adverse health outcomes from biomass burning. For example, biomass burning smoke has been shown to be associated with several diseases, such as acute respiratory infection (ARI, pneumonia, bronchitis, etc.) and chronic obstructive pulmonary disease (COPD) (Ezzati and Kammen 2001; Hu and Guo 2007; Kurmi et al. 2010). Further, emissions from biomass fuel combustion have also been found to be correlated with adult lung cancer (Sarigiannis et al. 2015). Additional studies should investigate health effects of biomass burning emissions from different fuel types and burning conditions (e.g., stoves, flaming or smoldering burning, open or residential burning).

7.5 Summary

This chapter provides a summary of investigations focusing on the release of aerosol from biomass burning in China. Different perspectives are adopted, specifically emission characteristics (e.g., $PM_{2.5}$ chemical profiles and source-specific ratios, emission factor, emission inventory), source quantification, and effects on air quality. Significant amount of air pollutants can be emitted from open field burning and domestic burning for heating and cooking, which exhibit clear seasonal and spatial variations. Biomass burning aerosols can also be transported to China over long distances (from Southeast Asia regions especially in spring).

This chapter highlights also some limitations of current studies and suggests future prospects: (1) Localized emission characteristics of biomass burning sources for different fuel types and burning conditions need to be further studied and knowledge needs to be updated; (2) It is essential to develop a more complete database of emission inventories, emission factor estimates and source profiles for both residential and open biomass burning with higher spatial and temporal resolution and detailed speciation information; (3) More efforts should be made in the future to investigate the effects of biomass burning on human health and climate.

Acknowledgements This study was supported by funding from the National Natural Science Foundation of China (Grant No. 41375125, 41571130033), and partly supported by funding from China Postdoctoral Science Foundation (No. 2016M590022). We thank Yue Liu for her kind assistance in this work.

References

Akagi, S., Yokelson, R., Wiedinmyer, C., Alvarado, M., Reid, J., Karl, T., Crounse, J., & Wennberg, P. (2011). Emission factors for open and domestic biomass burning for use in atmospheric models. *Atmospheric Chemistry and Physics, 11*, 4039–4072.

Andreae, M.O. (1991). Biomass burning: Its history, use and distribution and its impact on environmental quality and global climate. In Global biomass burning: Atmospheric, climatic and biospheric implications. Cambridge, MA: MIT Press, 3–21.

Andreae, M. O., & Merlet, P. (2001). Emission of trace gases and aerosols from biomass burning. *Global Biogeochemistry Cycles, 15*, 955–966.

Bond, T. C., Doherty, S. J., Fahey, D. W., Forster, P. M., Berntsen, T., DeAngelo, B. J., Flanner, M. G., Ghan, S., Kärcher, B., Koch, D., Kinne, S., Kondo, Y., Quinn, P. K., Sarofim, M. C., Schultz, M. G., Schulz, M., Venkataraman, C., Zhang, H., Zhang, S., Bellouin, N., Guttikunda, S. K., Hopke, P. K., Jacobson, M. Z., Kaiser, J. W., Klimont, Z., Lohmann, U., Schwarz, J. P., Shindell, D., Storelvmo, T., Warren, S. G., & Zender, C. S. (2013). Bounding the role of black carbon in the climate system: A scientific assessment. *Journal of Geophysical Research, [Atmospheres], 118*, 5380–5552.

Boschetti, L., & Roy, D. P. (2009). Strategies for the fusion of satellite fire radiative power with burned area data for fire radiative energy derivation. *Journal of Geophysical Research, 114*, D20302. doi:10.1029/2008JD011645.

Cao, G. L., Zhang, X. Y., Wang, D., & Zheng, F. C. (2005). Inventory of atmospheric pollutants discharged from biomass burning in China continent (in Chinese). *China Environmental Science, 25*(4), 389–393.

Cao, G. L., Zhang, X. Y., Zheng, F. C., & Wang, Y. Q. (2006). Estimating the quantity of crop residues burnt in open field in China (in Chinese). *Resource Sciences, 28*(1), 9–13.

Cao, G. L., Zhang, X. Y., Wang, Y. Q., & Zheng, F. C. (2008a). Estimation of emissions from field burning of crop straw in China. *Chinese Science Bulletin, 53*(5), 784–790.

Cao, G. L., Zhang, X. Y., Gong, S. L., & Zheng, F. C. (2008b). Investigation on emission factors of particulate matter and gaseous pollutants from crop residue burning. *Journal of Environmental Sciences, 20*(1), 50–55.

Chan, L. Y., Chan, C. Y., Liu, H. Y., Christopher, S., Oltmans, S. J., & Harris, J. M. (2000). A case study on the biomass burning in Southeast Asia and enhancement of tropospheric ozone over Hong Kong. *Geophysical Research Letters, 27*(10), 1479–1482.

Chan, C. Y., Chan, L. Y., Harris, J. M., Oltmans, S. J., Blake, D. R., Qin, Y., Zheng, Y. G., & Zheng, X. D. (2003). Characteristics of biomass burning emission sources, transport, and chemical speciation in enhanced springtime tropospheric ozone profiles over Hong Kong. *Journal of Geophysical Research, 108*(D1), 4015. doi:10.1029/2001JD001555.

Chan, C. Y., Wong, K. H., Li, Y. S., Chan, L. Y., & Zheng, X. D. (2006). The effects of Southeast Asia fire activities on tropospheric ozone, trace gases and aerosols at a remote site over the Tibetan Plateau of Southwest China. *Tellus, 58B*, 310–318.

Cheng, Y., Engling, G., He, K.-B., Duan, F.-K., Ma, Y.-L., Du, Z.-Y., Liu, J.-M., Zheng, M., & Weber, R. J. (2013). Biomass burning contribution to Beijing aerosol. *Atmospheric Chemistry and Physics, 13*, 7765–7781.

Cheng, Z., Wang, S., Fu, X., Watson, J. G., Jiang, J., Fu, Q., Chen, C., Xu, B., Yu, J., Chow, J. C., & Hao, J. (2014). Impact of biomass burning on haze pollution in the Yangtze River delta, China: A case study in summer 2011. *Atmospheric Chemistry and Physics, 14*, 4573–4585.

Deng, X. J., Tie, X. X., Zhou, X. J., Wu, D., Zhong, L. J., Tan, H. B., Li, F., Huang, X. Y., Bi, X. Y., & Deng, T. (2008). Effects of Southeast Asia biomass burning on aerosols and ozone concentrations over the Pearl River Delta (PRD) region. *Atmospheric Environment, 42*, 8493–8501.

Ding, A. J., Fu, C. B., Yang, X. Q., Sun, J. N., Petäjä, T., Kerminen, V. M., Wang, T., Xie, Y., Herrmann, E., Zheng, L. F., Nie, W., Liu, Q., Wei, X. L., & Kulmala, M. (2013). Intense atmospheric pollution modifies weather: A case of mixed biomass burning with fossil fuel combustion pollution in eastern China. *Atmospheric Chemistry and Physics, 13*, 10545–10554.

Du, H. H., Kong, L. D., Cheng, T. T., Chen, J. M., Du, J. F., Li, J., Xia, X. G., Leng, C. P., & Huang, G. H. (2011). Insights into summertime haze pollution events over Shanghai based on online water-soluble ionic composition of aerosols. *Atmospheric Environment, 45*, 5131–5137.

Du, Z. Y., He, K. B., Cheng, Y., Duan, F. K., Ma, Y. L., Liu, J. M., Zhang, X. L., Zheng, M., & Weber, R. (2014). A yearlong study of water-soluble organic carbon in Beijing II: Light absorption properties. *Atmospheric Environment, 89*, 235–241.

Duan, F. K., Liu, X. D., Yu, T., & Cachier, H. (2004). Identification and estimate of biomass burning contribution to the urban aerosol organic carbon concentrations in Beijing. *Atmospheric Environment, 38*, 1275–1282.

Duan, X. L., Jiang, Y., Wang, B. B., Zhao, X. G., Shen, G. F., Cao, S. Z., Huang, N., Qian, Y., Chen, Y. T., & Wang, L. M. (2014). Household fuel use for cooking and heating in China: Results from the first Chinese Environmental Exposure-Related Human Activity Patterns Survey (CEERHAPS). *Applied Energy, 136*, 692–703.

Ezzati, M., & Kammen, D. M. (2001). Indoor air pollution from biomass combustion and acute respiratory infections in Kenya: An exposure-response study. *Lancet, 358*(9282), 619–624.

Feng, Y., Ramanathan, V., & Kotamarthi, V. R. (2013). Brown carbon: A significant atmospheric absorber of solar radiation? *Atmospheric Chemistry and Physics, 13*, 8607–8621.

Fine, P. M., Cass, G. R., & Simoneit, B. R. (2001). Chemical characterization of fine particle emissions from the fireplace combustion of woods grown in the southern United States. *Environmental Science & Technology, 35*(13), 2665–2675.

Fine, P. M., Cass, G. R., & Simoneit, B. R. T. (2002). Chemical characterization of fine particle emissions from the fire place combustion of woods grown in the Southern United States. *Environmental Science & Technology, 36*(7), 1442–1451.

Fu, T. M., Cao, J. J., Zhang, X. Y., Lee, S. C., Zhang, Q., Han, Y. M., Qu, W. J., Han, Z., Zhang, R., Wang, Y. X., Chen, D., & Henze, D. K. (2012a). Carbonaceous aerosols in China: Top-down constraints on primary sources and estimation of secondary contribution. *Atmospheric Chemistry and Physics, 12*(5), 2725–2746.

Fu, P. Q., Kawamura, K., Chen, J., Li, J., Sun, Y. L., Liu, Y., Tachibana, E., Aggarwal, S. G., Okuzawas, K., Tanimoto, H., Kanaya, Y., & Wang, Z. F. (2012b). Diurnal variations of organic molecular tracers and stable carbon isotopic composition in atmospheric aerosols over Mt. Tai in the North China Plain: An influence of biomass burning. *Atmospheric Chemistry and Physics, 12*, 8359–8375.

Galanter, M., Levy, H., & Carmichael, G. R. (2000). Impacts of biomass burning on tropospheric CO, NOx, and O_3. *Journal of Geophysical Research, 105*(D5), 6633–6653.

de Gouw, J. A., Cooper, O. R., Warneke, C., Hudson, P. K., Fehsenfeld, F. C., Holloway, J. S., Hübler, G., Nicks, D. K., Nowak, J. B., Parrish, D. D., Ryerson, T. B., Atlas, E. L., Donnelly, S. G., Schauffler, S. M., Stroud, V., Johnson, K., Carmichael, G. R., & Streets, D. G. (2004). Chemical composition of air masses transported from Asia to the U.S. West Coast during ITCT 2K2: Fossil fuel combustion versus biomass-burning signatures. *Journal of Geophysical Research, 109*, D23S20. doi:10.1029/2003JD004202.

Gustafsson, Ö., & Ramathan, V. (2016). Convergence on climate warming by black carbon aerosols. *Proceedings of the National Academy of Sciences of the United States of America, 113*(16), 4243–4245.

Hennigan, C. J., Sullivan, A. P., & Collett, J. L. (2010). Levoglucosan stability in biomass burning particles exposed to hydroxyl radicals. *Geophysical Research Letters, 37*, L09806. doi:10.1029/2010GL043088.

Hu, J., & Guo, X. (2007). Research progress in indoor air pollution and health effects caused by combustion of biomass fuels (In Chinese). *Journal of Environmental Health, 24*(10), 827–829.

Huang, X., Li, M. M., Friedli, H. R., Chang, D., & Zhu, L. (2011). Mercury emissions from biomass burning in China. *Environmental Science & Technology, 45*(21), 9442–9448.

Huang, X., Li, M. M., Li, J. F., & Song, Y. (2012a). A high-resolution emission inventory of crop burning in fields in China based on MODIS Thermal Anomalies/Fire products. *Atmospheric Environment, 50*, 9–15.

Huang, X., Song, Y., Li, M. M., Li, J. F., & Zhu, T. (2012b). Harvest season, high polluted season in East China. *Environmental Research Letters, 7*, 044033. doi:10.1088/1748-9326/7/4/044033.

Huang, K., Fu, J. S., Hsu, N. C., Gao, Y., Dong, X. Y., Tsay, S. C., & Lam, Y. F. (2013). Impact assessment of biomass burning on air quality in Southeast and East Asia during BASE-ASIA. *Atmospheric Environment, 78*, 291–302.

Huang, R. J., Zhang, Y. L., Bozzetti, C., Ho, K. F., Cao, J. J., Han, Y. M., Daellenbach, K. R., Slowik, J. G., Platt, S. M., Canonaco, F., Zotter, P., Wolf, R., Pieber, S. M., Bruns, E. A., Crippa, M., Ciarelli, G., Piazzalunga, A., Schwikowski, M., Abbaszade, G., Schnelle-Kreis, J. S., Zimmermann, R., An, Z. S., Szidat, S., Baltensperger, U., Haddad, I. E., & Prévôt, A. S. H. (2014). High secondary aerosol contribution to particulate pollution during haze events in China. *Nature, 514*, 218–222.

Huang, X., Ding, A. J., Liu, L. X., Liu, Q., Ding, K., Niu, X. R., Nie, W., Xu, Z., Chi, X. G., Wang, M. H., Sun, J. N., Guo, W. D., & Fu, C. B. (2016). Effects of aerosol-radiation interaction on precipitation during biomass-burning season in East China. *Atmospheric Chemistry and Physics, 16*, 10063–10082.

Ito, A., & Penner, J. E. (2004). Global estimates of biomass burning emissions based on satellite imagery for the year 2000. *Journal of Geophysical Research, 109*, D14S05. doi:10.1029/2003JD004423.

Kawashima, H., & Haneishi, Y. Y. (2012). Effects of combustion emissions from the Eurasian continent in winter on seasonal $\delta^{13}C$ of elemental carbon in aerosols in Japan. *Atmospheric Environment, 46*, 568–579.

Kirillova, E. N., Andersson, A., Han, J., Lee, M., & Gustafsson, Ö. (2014). Sources and light absorption of water-soluble organic carbon aerosols in the outflow from northern China. *Atmospheric Chemistry and Physics, 14*, 1413–1422.

Kurmi, O. P., Semple, S., Simkhada, P., Smith, W. C. S., & Ayres, J. G. (2010). COPD and chronic bronchitis risk of indoor air pollution from solid fuel: A systematic review and meta-analysis. *Thorax, 65*(3), 221–228.

Laskin, A., Laskin, J., & Nizkorodov, S. A. (2015). Chemistry of atmospheric brown carbon. *Chemical Reviews, 115*, 433–4382.

Lee, C. T., Chuang, M. T., Lin, N. H., Wang, J. L., Sheu, G. R., Chang, S. C., Wang, S. H., Huang, H., Chen, H. W., Liu, Y. L., Weng, G. H., Lai, H. Y., & Hsu, S. P. (2011). The enhancement of $PM_{2.5}$ mass and waters-soluble ions of biosmoke transported from Southeast Asia over the Mountain Lulin site in Taiwan. *Atmospheric Environment, 45*, 5784–5794.

Li, X. H., Wang, S. X., Duan, L., Hao, J. M., Li, C., Chen, Y. S., & Yang, L. (2007). Particulate and trace gas emissions from open burning of wheat straw and corn stover in China. *Environmental Science & Technology, 41*, 6052–6058.

Li, L. J., Wang, Y., Zhang, Q., Li, J. X., Yang, X. G., & Jin, J. (2008). Wheat straw burning and its associated impacts on Beijing air quality. *Science in China Series D-Earth Sciences, 51*(3), 403–414.

Li, X. H., Wang, S. X., Duan, L., Hao, J. M., & Nie, Y. (2009). Carbonaceous aerosol emissions from household biofuel combustion in China. *Environmental Science & Technology, 43*(15), 6076–6081.

Li, W. J., Shao, L. Y., & Buseck, P. R. (2010). Haze types in Beijing and the influence of agricultural biomass burning. *Atmospheric Chemistry and Physics, 10*, 8119–8130.

Li, J., Song, Y., Mao, Y., Mao, Z. C., Wu, Y. S., Li, M. M., Huang, X., He, Q. C., & Hu, M. (2014). Chemical characteristics and source apportionment of $PM_{2.5}$ during the harvest season in eastern China's agricultural regions. *Atmospheric Environment, 92*, 442–448.

Li, J., Bo, Y., & Xie, S. (2016a). Estimating emissions from crop residue open burning in China based on statistics and MODIS fire products. *Journal of Environmental Sciences, 44*, 158–170.

Li, C. L., Bosch, C., Kang, S. C., Andersson, A., Chen, P. F., Zhang, Q. G., Cong, Z. Y., Chen, B., Qin, D. H., & Gustafsson, Ö. (2016b). Source of black carbon to the Himalayan-Tibetan Plateau glaciers. *Nature Communications, 7*, 12574. doi:10.1038/ncomms12574.

Lin, C. Y., Zhao, C., Liu, X. H., Lin, N. H., & Chen, W. N. (2014). Modelling of long-range transport of Southeast Asia biomass-burning aerosols to Taiwan and their radiative forcings over East Asia. *Tellus B, 66*, 23733.

Liu, L. H., Jiang, J. Y., & Zong, L. G. (2011). Emission inventory of greenhouse gases from agricultural residues combustion: A case study of Jiangsu province (in Chinese). *Environmental Sciences, 32*(5), 1242–1248.

Lu, B., Kong, S. F., Han, B., Wang, X. Y., & Bai, Z. P. (2011a). Inventory of atmospheric pollutants discharged from biomass burning in China continent in 2007 (in Chinese). *China Environmental Science, 31*(2), 186–194.

Lu, Z., Zhang, Q., & Streets, D. G. (2011b). Sulfur dioxide and primary carbonaceous aerosol emissions in China and India, 1996–2010. *Atmospheric Chemistry and Physics, 11*, 9839–9864.

Menon, S., Hansen, J., Nazarenko, L., & Luo, Y. F. (2002). Climate effects of black carbon aerosol in China and India. *Science, 297*, 2250. doi:10.1126/science.1075159.

Morton, D. C., Defries, R. S., Randerson, J. T., Giglio, L., Schroeder, W., & van der Werf, G. R. (2008). Agricultural intensification increase deforestation fire activity in Amazonia. *Global Change Biology, 14*, 2262–2275.

Ni, H.Y., Han, Y.M., Cao, J.J., Chen, L.A., Tian, J., Wang, X.L., Chow, J.C., Watson, J.G., Wang, Q.Y., Wang, P., Li, H., Huang, R.J. (2015). Emission characteristics of carbonaceous particles and trace gases from open burning of crop residues in China. *Atmospheric Environment*, 1–8. http://dx.doi.org/10.1016/j.atmosenv.2015.05.007.

O'Leary, M. H. (1981). Carbon isotope fractionation in plants. *Phytochemistry, 20*, 553–567.

Oros, D. R., Abas, M. R., Omar, N. Y. M. J., Rahman, N. A., & Simoneit, B. R. T. (2006). Identification and emission factors of molecular tracers in organic aerosols from biomass burning: Part 3. *Grasses Applied Geochemistry, 21*, 919–940.

Penner, J. E., Dickinson, R. E., & O'Neill, C. A. (1992). Effects of aerosol from biomass burning on the global radiation budget. *Science, 256*(5062), 1432–1434.

Ramdahl, T. (1983). Retenee a molecular marker of wood combustion in ambient air. *Nature, 306*, 580–582.

Reddington, C. L., Butt, E. W., Ridley, D. A., Artaxo, P., Morgan, W. T., Coe, H., & Spracklen, D. V. (2015). Air quality and human health improvements from reductions in deforestation-related fire in Brazil. *Nature Geoscience, 8*, 768–771.

Roy, D. P., Boschetti, L., Justice, C. O., & Ju, J. (2008). The collection 5 MODIS burned area product e global evaluation by comparison with the MODIS active fire product. *Remote Sensing of Environment, 112*, 3690–3707.

Sang, X. F., Chan, C. Y., Engling, G., Chan, L. Y., Wang, X. M., Zhang, Y. N., Shi, S., Zhang, Z. S., Zhang, T., & Hu, M. (2011). Levoglucosan enhancement in ambient aerosol during springtime transport events of biomass burning smoke to Southeast China. *Tellus, 63B*, 129–139.

Sarigiannis, D. A., Karakitsios, S. P., Zikopoulos, D., Nikolaki, S., & Kermenidou, M. (2015). Lung cancer risk from PAHs emitted from biomass combustion. *Environmental Research, 137*, 147–156.

Shen, G. F., Yang, Y. F., Wang, W., Tao, S., Zhu, C., Min, Y. J., Xue, M., Ding, J. N., Wang, B., Wang, R., Shen, H. Z., Li, W., Wang, X. L., & Russell, A. G. (2010). Emission factors of particulate matter and elemental carbon for crop residues and coals burned in typical household stoves in China. *Environmental Science & Technology, 44*(18), 7157–7162.

Shen, G. F., Xue, M., Chen, Y. C., Yang, C. L., Li, W., Shen, H. Z., Huang, Y., Zhang, Y. Y., Chen, H., Zhu, Y., Wu, H. S., Ding, A. J., & Tao, S. (2014). Comparison of carbonaceous particulate matter emissions factors among different solid fuels burned in residential stoves. *Atmospheric Environment, 89*, 337–345.

Simoneit, B. R. T. (2002). Biomass burning- a review of organic tracers for smoke from incomplete combustion. *Applied Geochemistry, 17*, 129–162.

Simoneit, B. R. T., Rogge, W. F., Mazurek, M. A., Standley, L. J., Hlldemann, L. M., & Cass, G. R. (1993). Lignin pyrolysis products, lignans and resin acids as specific tracers of plant classes in emissions from biomass combustion. *Environmental Science and Technology, 12* (12), 2533–2541.

Simpson, C. D., Paulsen, M., Dills, R. L., Liu, S., & Kalman, D. A. (2005). Determination of methoxyphenols in ambient atmospheric particulate matter: Tracers for wood combustion. *Environmental Science and Technology, 39*, 631–637.

Streets, D.G., Yarber, K.F., Woo, J.H., Carmichael, G.R., (2003a). Biomass burning in Asia: Annual and seasonal estimates and atmospheric emissions. *Global Biogeochemical Cycles*, 17 (4). http://dx.doi.org/10.1029/2003gb002040.

Streets, D. G., Bond, T. C., Carmichael, G. R., Fernandes, S. D., Fu, Q., He, D., et al. (2003b). An inventory of gaseous and primary aerosol emissions in Asia in the year 2000. *Journal of Geophysical Research, 108*(D21), 8809. doi:10.1029/2002JD003093.

Sun, X. S., Hu, M., Guo, S., Liu, K. X., & Zhou, L. P. (2012). ^{14}C-based source assessment of carbonaceous aerosols at a rural site. *Atmospheric Environment, 50*, 36–40.

Tang, X. B., Huang, C., Lou, S. R., Qiao, L. P., Wang, H. L., Zhou, M., Chen, M. H., Chen, C. H., Wang, Q., Li, G. L., Li, L., Huang, H. Y., & Zhang, G. F. (2014). Emission factors and pm chemical composition study of biomass burning in the Yangtze River Delta region (in Chinese). *Environmental Sciences, 35*(5), 1623–1632.

Tao, J., Zhang, L. M., Engling, G., Zhang, R. J., Yang, Y. H., Cao, J. J., Zhu, C. S., Wang, Q. Y., & Luo, L. (2013). Chemical composition of $PM_{2.5}$ in an urban environment in Chengdu, China: Importance of springtime dust storms and biomass burning. *Atmospheric Research, 122*, 270–283.

Tian, H. Z., Zhao, D., & Wang, Y. (2011). Emission inventories of atmospheric pollutants discharged from biomass burning in China (in Chineses). *Acta Scientiae Circumstantiae, 31* (2), 349–357.

Van der Werf, G. R., Randerson, J. T., Giglio, L., Collatz, G. J., Kasibhatla, P. S., & Arellano Jr., A. F. (2006). Interannual variability in global biomass burning emissions from 1997 to 2004. *Atmospheric Chemistry and Physics, 6*, 3423–3441.

Van der Werf, G. R., Randerson, J. T., Giglio, L., Collatz, G. J., Mu, M., Kasibhatla, P. S., Morton, D. C., DeFries, R. S., Jin, Y., & van Leeuwen, T. T. (2010). Global fire emissions and the contribution of deforestation, savanna, forest, agricultural, and peat fires (1997–2009). *Atmospheric Chemistry and Physics, 10*(23), 11707–11735.

Wang, S. X., & Zhang, C. Y. (2008). Spatial and temporal distribution of air pollutant emissions from open burning of crop residues in China. *Science Paper Online, 3*(5), 329–333.

Wang, Q. Q., Shao, M., Liu, Y., William, K., Paul, G., Li, X., Liu, Y., & Lu, S. (2007). Impact of biomass burning on urban air quality estimated by organic tracers: Guangzhou and Beijing as cases. *Atmospheric Environment, 41*, 8380–8390.

Wang, S. X., Wei, W., Du, L., Li, G. H., & Hao, J. M. (2009a). Characteristics of gaseous pollutants from biofuel-stoves in rural China. *Atmospheric Environment, 43*(27), 4148–4154.

Wang, Z. Z., Bi, X. H., Sheng, G. Y., & Fu, J. M. (2009b). Characterization of organic compounds and molecular tracers from biomass burning smoke in South China I: Broad-leaf trees and shrubs. *Atmospheric Environment, 43*, 3096–3102.

Wang, Q. Q., Shao, M., Zhang, Y., Wei, Y., Hu, M., & Guo, S. (2009c). Source apportionment of fine organic aerosols in Beijing. *Atmospheric Chemistry and Physics, 9*, 8573–8585.

Wang, Y. J., Hu, M., Wang, Y., Qin, Y. H., Chen, H. Y., Zeng, L. M., Lei, J. R., Huang, X. F., He, L. Y., Zhang, R. Q., & Wu, Z. J. (2016). Characterization and influence factors of $PM_{2.5}$ emitted from crop straw burning (in Chinese). *Acta Chimica Sinica (Chinese Edition), 74*(4), 356–362.

Wei, S. Y., Shen, G. F., Zhang, Y. Y., Xue, M., Xie, H., Lin, P. C., Chen, Y. C., Wang, X. L., & Tao, S. (2014). Field measurement on the emissions of PM, OC, EC and PAHs from indoor crop straw burning in rural China. *Environmental Pollution, 184*, 18–24.

Xu, W., Wu, B., Yan, C., & Huang, H. (2005). China land cover 2000 using SPOT VGT S10 data (in Chinese). *Journal of Remote Sensing, 9*, 204–213.

Yamaji, K., Li, J., Uno, I., Kanaya, Y., Irie, H., Takigawa, M., Komazaki, Y., Pochanart, P., Liu, Y., Tanimoto, H., Ohara, T., Yan, X., Wang, Z., & Akimoto, H. (2010). Impact of open crop residual burning on air quality over Central Eastern China during the Mount Tai experiment 2006 (MTX2006). *Atmospheric Chemistry and Physics, 10*(15), 7353–7368.

Yan, X. Y., Ohara, T., & Akimoto, H. (2006). Bottom-up estimate of biomass burning in mainland China. *Atmospheric Environment, 40*(27), 5262–5273.

Yan, C. Q., Zheng, M., Sullivan, A. P., Bosch, C., Desyaterik, Y., Andersson, A., Li, X. Y., Guo, X. S., Zhou, T., Gustafsson, Ö., & Collett Jr., J. L. (2015). Chemical characteristics and light-absorbing property of water-soluble organic carbon in Beijing: Biomass burning contributions. *Atmospheric Environment, 121*, 4–12.

Zhang, Y. X., & Tao, S. (2008). Seasonal variation of polycyclic aromatic hydrocarbons (PAHs) emissions in China. *Environmental Pollution, 156*(3), 657–663.

Zhang, J., Smith, K. R., Ma, Y., Ye, S., Jiang, F., Qi, W., Liu, P., Khalil, M. A. K., Rasmussen, R. A., & Thorneloe, S. A. (2000). Greenhouse gases and other airborne pollutants from household stoves in China: A data base for emission factors. *Atmospheric Environment, 34*, 4537–4549.

Zhang, Y. X., Shao, M., Zhang, Y. H., Zeng, L. M., He, L. Y., Zhu, B., Wei, Y. J., & Zhu, X. L. (2007). Source profiles of particulate organic matters emitted from cereal straw burnings. *Journal of Environmental Sciences, 19*(2), 167–175.

Zhang, H. F., Ye, X. N., Cheng, T. T., Chen, J. M., Yang, X., Wang, L., & Zhang, R. Y. (2008a). A laboratory study of agricultural crop residue combustion in China: Emission factors and emission inventory. *Atmospheric Environment, 42*(36), 8432–8441.

Zhang, X. Y., Kondragunta, S., Schmidt, C., & Kogan, F. (2008b). Near real time monitoring of biomass burning particulate emissions (PM2.5) across contiguous United States using multiple satellite instruments. *Atmospheric Environment, 42*(29), 6959–6972.

Zhang, T., Claeys, M., Cachier, H., Dong, S. P., Wang, W., Maenhaut, W., & Liu, X. D. (2008c). Identification and estimation of the biomass burning contribution to Beijing aerosols using levoglucosan as a molecular marker. *Atmospheric Environment, 42*, 7013–7021.

Zhang, Z. S., Engling, G., Lin, C. Y., Chou, C. K., Lung, S. C., Chang, S. Y., Fan, S. J., Chan, C. Y., & Zhang, Y. H. (2010). Chemical speciation, transport and contribution of biomass burning smoke to ambient aerosol in Guangzhou, a mega city of China. *Atmospheric Environment, 44*, 3187–3195.

Zhang, H. F., Wang, S. X., Hao, J. M., Wan, L., Jiang, J. K., Zhang, M., Mestl, H. E. S., Alnes, L. W. H., Aunan, K., & Mellouki, A. W. (2012a). Chemical and size characterization of particles emitted from the burning of coal and wood in rural households in Guizhou, China. *Atmospheric Environment, 51*(5), 94–99.

Zhang, Y. N., Zhang, Z. S., Chan, C. Y., Engling, G., Sang, X. F., Shi, S., & Wang, X. M. (2012b). Levoglucosan and carbonaceous species in the background aerosol of southeast China: Case study on transport of biomass burning smoke from the Philippines. *Environmental Science and Pollution Research, 19*, 244–255.

Zhang, Y. S., Shao, M., Lin, Y., Luan, S. J., Mao, N., Chen, W. T., & Wang, M. (2013). Emission inventory of carbonaceous pollutants from biomass burning in the Pearl River Delta region, China. *Atmospheric Environment, 76*(5), 189–199.

Zhang, T. R., Wooster, M. J., Green, D. C., & Main, B. (2015). New field-based agricultural biomass burning trace gas, $PM_{2.5}$ and black carbon emission ratios and factors measured in situ at crop residue fires in Eastern China. *Atmospheric Environment, 121*, 22–34.

Zhu, C., Kawamura, K., & Kunwar, B. (2015). Effect of biomass burning over the western North Pacific Rim: Wintertime maxima of anhydrosugars in ambient. *Atmospheric Chemistry and Physics, 15*, 1959–1973.

Chapter 8
Sources and Long-Term Trends of Ozone Precursors to Asian Pollution

Trissevgeni Stavrakou, Jean-François Müller, Maite Bauwens, and Isabelle De Smedt

Abstract Due to its fast economic development, China's emissions are in the spotlight of efforts to mitigate climate change and improve regional and city-scale air quality. Despite growing efforts to better quantify China's emissions, the current estimates are often poor or inadequate. Bottom-up inventories are generally based on sectoral statistical information and therefore rely strongly on the accuracy of the input data. Complementary to bottom-up methodologies, inverse modeling of fluxes has the potential to improve those estimates through the use of atmospheric observations of trace gas compounds. Here we present comparisons of key pollutant emissions from different bottom-up inventories, and perform 20-year model simulations of the atmospheric composition over China using either the EDGARv4.2 or the MACCity bottom-up emission databases. The skill of the model to capture the observed variability and trends is assessed through comparisons with satellite NO_2 observations retrieved from GOME, SCIAMACHY and OMI sensors through 1997–2008 and HCHO columns observed by OMI over 2005–2010. Next, we use a decade (2005–2014) of OMI HCHO columns to constrain the VOC emissions over China in a flux inversion framework built on the IMAGESv2 chemistry-transport model, and adjust the emissions of VOC precursors of HCHO in the model in order to reduce the discrepancy between the model predictions and the HCHO observations. The interannual and seasonal variability of the resulting top-down VOC fluxes (anthropogenic, pyrogenic and biogenic) is presented and confronted to past studies.

Keywords Top-down Chinese emissions • Inverse modeling • Ozone trends • Satellite formaldehyde columns

T. Stavrakou (✉) • J.-F. Müller • M. Bauwens • I. De Smedt
Royal Belgian Institute for Space Aeronomy, 3, Avenue Circulaire, 1180 Brussels, Belgium
e-mail: trissevgeni.stavrakou@aeronomie.be; jfm@aeronomie.be; maite.bauwens@aeronomie.be; isad@aeronomie.be

I. Bouarar et al. (eds.), *Air Pollution in Eastern Asia: An Integrated Perspective*, ISSI Scientific Report Series 16, DOI 10.1007/978-3-319-59489-7_8

8.1 Introduction

Air quality is a serious environmental problem in China. China is the most populated country with a population that has grown by 43% since the 1980s, to reach 1.38 billion people as of May 2016 (Worldometers 2016). The gross domestic product increase during the last 35 years has been tremendous, with today's GDP being almost about 36 times higher than in 1980, 28 times than in 1990, and 9 times compared to 2000, making China the driving force of the global economy (World Development Indicators 2016). This fast economic development gave rise to several environmental concerns including a severe degradation of the air quality, which has been recognized to have numerous adverse effects on human health and ecosystems. It is estimated that currently 55% of China's population lives in cities, but the situation might change drastically in the future, since by 2030 the urban population is expected to rise to more than one billion people (United Nations 2014). As the number of Chinese megacities is steadily increasing, air pollution issues are brought at the forefront of public awareness. In Shanghai and Beijing, ambient $PM_{2.5}$ concentrations often exceed by a large factor the upper threshold recommended by the European and US Air Quality Standards, testifying that development in China has come at a serious cost. On the other hand, emerging rural regions transform quickly into centers of sustained industrialization, and because of lacking regulation policy of environmental resources, they become increasingly vulnerable. In the aftermath of the astounding progress of the Chinese economy and social welfare, the emissions of air pollutants estimated using the traditional statistical approach become quickly outdated and do not reflect the real air pollution status in many regions (Mijling and van der A 2012; Ding et al. 2015), whereas large discrepancies between national emission estimates for the entire China and the sum of the regional estimates of its provinces (Guan et al. 2012) underline the need for alternative ways of emission estimation.

Besides anthropogenic impacts, the warming climate is another important factor contributing to the current atmospheric composition change. Long-term analysis of surface temperature (Dee et al. 2011) reveals an upward trend of about 0.37 °C per decade since the eighties for the entire country, almost three times faster than the global average. This trend is even higher for big cities like Beijing. The temperature trend could lead to increased biogenic emission fluxes, which are important actors for air quality. The effect of the combined influence in anthropogenic and biogenic emissions on the long-term levels of surface ozone is crucial and warrants investigation.

This study presents Chinese emissions of key pollutants from well-established bottom-up inventories, and examines the evolution of atmospheric composition over China during a period of 20 years, based on model simulations and comparisons with multi-year satellite observations. The bottom-up VOC emission inventories over China are updated relying on the source inversion technique constrained by a decade of formaldehyde observations retrieved from the OMI instrument, and the resulting top-down anthropogenic, biogenic and pyrogenic fluxes are analyzed.

8.2 Bottom-Up Estimates for Chinese Emissions

Three bottom-up anthropogenic emission inventories are examined here. MACCity and EDGARv4.2 are the most widely used global anthropogenic emission inventories. The MACCity emissions inventory (Granier et al. 2011) builds on the ACCMIP historical emission dataset (Lamarque et al. 2010) until 2000 and the 2005 and 2010 emissions provided by one of the future emissions scenarios (RCP8.5, Riahi et al. 2011). It covers the emissions of 19 chemical species from 1980 to 2010. Emissions in the EDGARv4.2 inventory (EC-JRC/PBL 2011) are calculated based on energy and industrial balance statistics and cover direct greenhouse gases, ozone precursors and acidifying gases for all years between 1970 and 2008. On the other hand, the REASv2 anthropogenic inventory (Kurokawa et al. 2013) is based on activity data for Asian countries, includes most major air pollutants and greenhouse gases, and covers the period between 2000 and 2008. However, owing to the lack of detailed statistical data and emission factor measurements, the anthropogenic bottom-up inventories over China are believed to be highly uncertain. This is illustrated in the comparison between MACCity, EDGARv4.2 and REASv2 emission estimates for China for three key tropospheric pollutants, CO, NOx and VOCs (Fig. 8.1, upper panels). The MACCity emission estimates are approximately twice higher than in EDGARv4.2 for CO, and by 60% higher for VOCs, whereas REASv2 trends are much stronger with respect to the other inventories for all three cases.

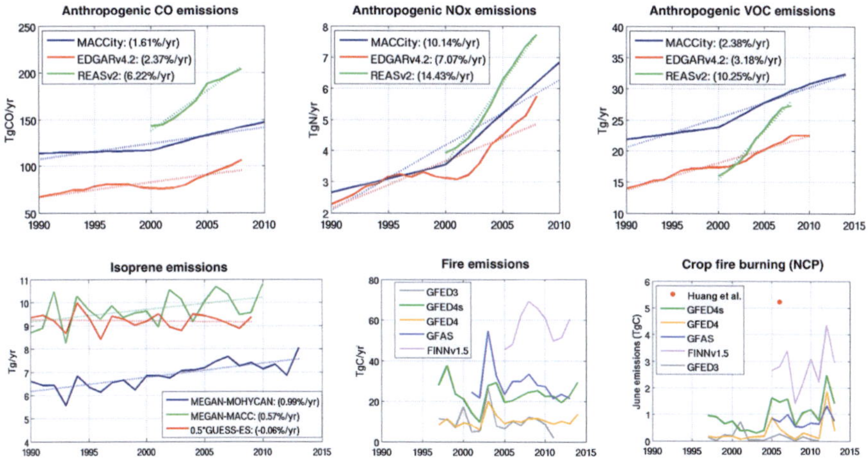

Fig. 8.1 Anthropogenic CO, NOx and VOC emission estimates for China according to the MACCity, EDGARv4.2 and REASv2 inventories (*upper panels*). Bottom-up Chinese isoprene fluxes obtained from the MEGAN-MOHYCAN, MEGAN-MACC and GUESS-ES inventories are shown in the lower *left panel*. Bottom-up estimates of emissions from biomass burning over China (*lower middle*) and from crop fires in the North China Plain (*lower right*) are taken from GFED3, GFED4, GFAS, FINNv1.5 fire databases, and the Huang et al. (2012) inventory

Biogenic emissions are an important source of hydrocarbons in the atmosphere, with isoprene accounting for almost half of the biogenic emission on the global scale, and about 20% of the total VOC emissions in China. According to MEGAN (Guenther et al. 2006, 2012), the most widely used model for estimating emissions from vegetation, isoprene emissions depend on the nature and abundance of plants and are modulated by meteorological and phenological (leaf age) parameters. Those emissions are therefore affected by land use changes and by the changing climate. Due to the scarcity of isoprene flux measurements and to the fast environmental changes in China, isoprene emissions predicted by models in this region are expected to be highly uncertain. Figure 8.1 (left lower panel) illustrates a comparison between three emission inventories for isoprene emission in China, two of them based on the MEGAN model, MEGAN-MOHYCAN (Stavrakou et al. 2014) and MEGAN-MACC (Sindelarova et al. 2014), and the third from the process-based GUESS-ES model (Arneth et al. 2007). MEGAN-MACC is by about 40% higher than MEGAN-MOHYCAN, whereas the GUESS-ES inventory is about factor of 2 higher than MEGAN-MACC. Note also that the isoprene emissions exhibit an increasing trend due to the warming temperatures in both MEGAN-based inventories, but do not show any trend in GUESS-ES inventory. A more detailed comparison of these inventories can be found in Bauwens et al. (2016).

Biomass burning emissions exhibit strong seasonal features and interannual variability. Bottom-up estimates depend on assumptions in fire models regarding fuel loading and consumption efficiency, and on the quality of satellite land cover maps and proxies, and are therefore subject to large uncertainties. In China, this emission category is only a minor contributor to the total emission, accounting for ca. 3%, 1% and 6% of the total annual CO, NOx and VOC emissions, respectively. Figure 8.1 (lower middle and right panels) shows a comparison between estimates from different fire inventories over China. GFAS (Kaiser et al. 2012) provides daily global fire emission estimates based on assimilation of fire radiative power observed by the MODIS satellite instruments and FINNv1.5 (Wiedinmyer et al. 2011) uses a fire emission model and satellite products for active fires. Three versions of the GFED emission databases are compared, GFED3 (van der Werf et al. 2010), GFED4, which adopted lower fuel consumption rates than GFED3 in line with field observations (van Leeuwen et al. 2014), and GFED4s which accounts for burnt area from small fires. The inventory of Huang et al. (2012) for agricultural crop residue burning in China relies on province-level statistical data and satellite fire proxies for spatial allocation. The large discrepancies (factor of 6) between fire estimates over China, and in particular for crop residue burning emissions (factor of ~10), underscore the urgent need for narrowing down these uncertainties in order to better address air quality issues.

According to REASv2, MEGAN-MACC and GFED4s inventories for 2010, the annual Chinese anthropogenic emissions are estimated at 200 Tg CO, 7.6 Tg NOx-N, and 25.6 Tg VOC, isoprene emissions at 7.1 Tg, and emissions from vegetation fires at 7 Tg CO, 0.1 Tg NOx-N and 2. Tg VOC. Overall, anthropogenic emissions contribute about 90% to the total CO and NOx emissions, and 74% to the total VOC emission.

8.3 Long-Term Model Simulations of Atmospheric Composition Over China

The IMAGESv2 global chemistry-transport model (Stavrakou et al. 2015; Müller and Stavrakou 2005) is used to perform long-term simulations of the atmospheric composition for China. The model calculates the concentrations of 172 short and long-lived trace gases with a time step of 6 h at a horizontal resolution of $2° \times 2.5°$ and at 40 vertical levels between the surface and the lower stratosphere (44 hPa). The effect of diurnal variations is accounted for through correction factors on the photolysis and kinetic rates obtained from model simulations with a time step of 20 min, which are also used to calculate the diurnal shapes of formaldehyde columns required for the comparison with satellite data. A detailed model description is provided in Stavrakou et al. (2015). Meteorological fields are obtained from ERA-Interim analyses of the European Centre for Medium-range Weather Forecasts (ECMWF, Dee et al. 2010). Two model simulations are performed using anthropogenic emissions of CO, NOx, VOC, SO_2, and NH_3 either from the MACCity inventory (Granier et al. 2011) over 1990–2010 or the EDGARv4.2 inventory (EC-JRC/PBL 2011) over 1990–2008. The former simulation uses vegetation fires from the MACCity inventory, which are derived from modified RETRO carbon emission data (Schultz et al. 2008) over 1960–1996 and from GFEDv2 carbon emission data over 1997–2008 (van der Werf et al. 2006). In the EDGARv4.2 simulation, fires are obtained and from the Global Fire Emission Database (GFED4s, Giglio et al. 2013). Both simulations use isoprene emissions from the MEGAN-MOHYCAN inventory (Müller et al. 2008; Stavrakou et al. 2014), which is available for all years of the target period at 0.5° resolution (http://emissions.aeronomie.be). Other biogenic emissions are obtained from the MEGAN model (Guenther et al. 2006, 2012).

The simulated annually averaged surface mixing ratios over China for key trace gases are illustrated in Fig. 8.2. The concentration levels and the simulated trends are due to the combined effect of CO, NOx, and VOC emission levels and their trends in the MACCity and EDGARv4.2 bottom-up inventories and of the chemical interactions of the CO-VOC-NOx-Ox-HOx system. The higher CO emission levels in the MACCity inventory (Fig. 8.1) lead to significantly higher surface CO concentrations (by ca. 30%) in comparison to the EDGARv4.2 simulation, whereas the combination of higher CO and VOC emissions in MACCity results in lower OH levels. Owing to the stronger CO and VOC emission trends in EDGARv4.2 inventory, the OH trend is higher in this simulation (0.71%/year vs. 0.43%/year). The higher MACCity surface HCHO levels is a consequence of the higher CO and VOC emissions, but the difference between MACCity and EDGARv4.2 concentrations is generally lower than 10%. The simulated surface ozone is found to be weakly sensitive to the large differences in emissions and trends between MACCity and EDGARv4.2. Overall, due to compensating chemical effects and despite the large discrepancies of the bottom-up emission fluxes, the levels of atmospheric pollutants in China grow at much slower pace than the emissions themselves. The ozone trends are similar in both simulations.

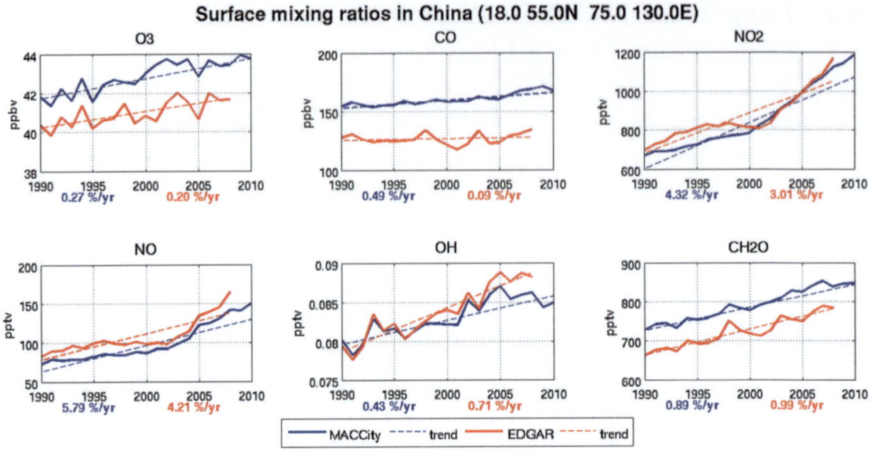

Fig. 8.2 Time series of annual surface O_3, CO, NO_2, NO, OH and CH_2O concentrations modeled using either the MACCity inventory between 1990 and 2010 (*blue line*), or the EDGARv4.2 inventory between 1990 and 2008 (*red line*) over China. Trend lines over the respective simulation periods are shown as *dashed lines* and trend estimates are shown at the *bottom* of each panel

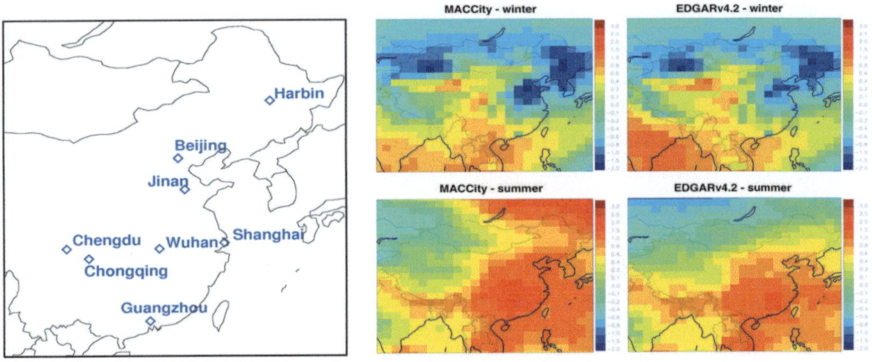

Fig. 8.3 *Left* locations of Chinese cities discussed in this section and in Table 8.1 *Right* Summer and winter ozone trend maps between 1990 and 2008 (in % per year) simulated with the model using either MACCity or EDGARv4.2 inventories for anthropogenic emissions

The summer and winter ozone trend distributions between 1990 and 2008 using either the MACCity or the EDGARv4.2 inventory are very similar (Fig. 8.3). The higher CO emission in MACCity (Fig. 8.1) leads to a more efficient ozone production per NOx molecule in the presence of light and explains the slightly higher ozone summertime trend in MACCity. In winter the ozone trend is negative (ca. −1.5%/year) in regions with significant upward NOx emission trend (e.g. Beijing and surrounding regions). This is due to the fact that higher NOx emission in winter suppresses photochemical activity which is necessary for ozone formation, as opposed to the summertime conditions where the ozone production is favored, primarily because of the reaction of NO with HO_2 leading to NO_2 formation and subsequent photolysis ($NO + HO_2 \rightarrow NO_2 + OH$, and $NO_2 + h\nu \rightarrow NO + O$). Both simulations lead to the same annual ozone trend in Beijing (0.21%/year), and to similar but higher trends in Shanghai (0.44%/year in MACCity and 0.31%/year in EDGARv4.2), and especially in Guangzhou (0.74%/year in MACCity and 058%/year in EDGARv4.2), due to their lower latitudes and subsequently warmer conditions and higher radiation levels.

In a next step we examine whether the model concentrations and trends are supported by direct comparisons with monthly averaged NO_2 and HCHO column observations retrieved from GOME (1997–2002, Boersma et al. 2004; De Smedt et al. 2008), SCIAMACHY (2003–2008, Boersma et al. 2011; De Smedt et al. 2008) and OMI (2005–2010, De Smedt et al. 2015; Levelt et al. 2006; Boersma et al. 2011) satellite instruments. The calculation of the monthly modeled NO_2 columns accounts for the averaging kernels, for the sampling times of observations at each location, and for the satellite overpass times (morning for GOME and SCIAMACHY, early afternoon for OMI). The marked observed NO_2 seasonal cycle with maxima in winter and minima in summer, is due to the difference in NOx lifetime throughout the seasons, and is reasonably well reproduced by the model (Fig. 8.4). The observed NO_2 column trends, 5.7%/year in Northern China

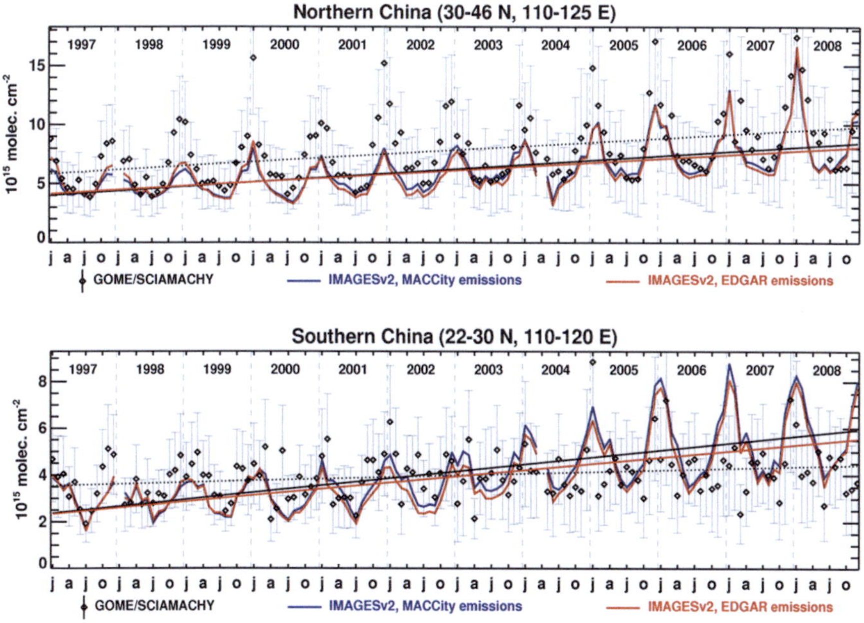

Fig. 8.4 NO_2 columns observed by GOME and SCIAMACHY instruments over 1997–2008 in Northern and Southern China and simulated with IMAGESv2 using anthropogenic emissions either from MACCity (*blue*) or from EDGARv4.2 (in *red*) inventory. The observed columns are shown as *black diamonds* with their error bars. The *dotted black line* represents the trend line of the observations. The blue and red trend straight lines represent the model trend

Table 8.1 Comparison between observed GOME and SCIAMACHY and modeled NO_2 column trends (in %/year) over 1997–2008

Region	Observed NO_2 trend	Model NO_2 trend MACCity	Model NO_2 trend EDGARv4.2	Correlation of observed vs. MACCity NO_2	Correlation of observed vs. EDGAR NO_2
Beijing	7.4	16	11.7	0.61	0.64
Jinan	7.8	17.2	12.9	0.68	0.70
Wuhan	12.1	17.1	14.3	0.85	0.85
Guangzhou	7.4	14.8	15.3	0.43	0.38
Harbin	13.8	9.3	8.9	0.77	0.79
Shanghai	19.8	11.1	12.2	0.69	0.66
Northern China	5.7	9.2	8.0	0.86	0.87
Southern China	2.1	12.7	11.5	0.50	0.50

and 2.1%/year in Southern China, are, however, lower than the model trends from both simulations, 8%/year in Northern China, and 12%/year in Southern China, suggesting that the NOx emission trends in the bottom-up inventories are most likely overestimated. In addition, as summarized in Table 8.1, the observed trends are generally lower than in the model (except in Shanghai and Harbin), and large differences are found over Beijing, Jinan and Guangzhou. Correlations are strong in Northern China (>0.8), but weaker in Southern China, independently of the inventory used, likely due to the lower observed NO_2 levels and therefore larger relative errors, and also to the larger share of the very uncertain natural emissions in Southern China.

Similar comparisons between model and HCHO columns observed by OMI between 2005 and 2010 are shown in Fig. 8.5. In this case the agreement with the observations is very satisfactory in both regions, and similar correlations (0.9) are calculated in both simulations. The trend in the observations is higher in the Northern (2.8%/year) than in Southern China (1.4%/year) over this period due to the dominance of anthropogenic emissions in the north. The summertime columns in Southern China are generally overestimated, indicating that the emissions of HCHO precursors from the biosphere might be too high in the model. In the next sections, we will investigate in detail how the HCHO columns can inform us about the sources of the volatile organic precursors of HCHO in China.

8.4 Contribution of VOCs to HCHO Formation in China

As formaldehyde (HCHO) is a high-yield secondary product in the oxidation of a large majority of volatile organic compounds (VOCs) emitted by fires, vegetation and anthropogenic activities, the satellite observations of this compound are well-

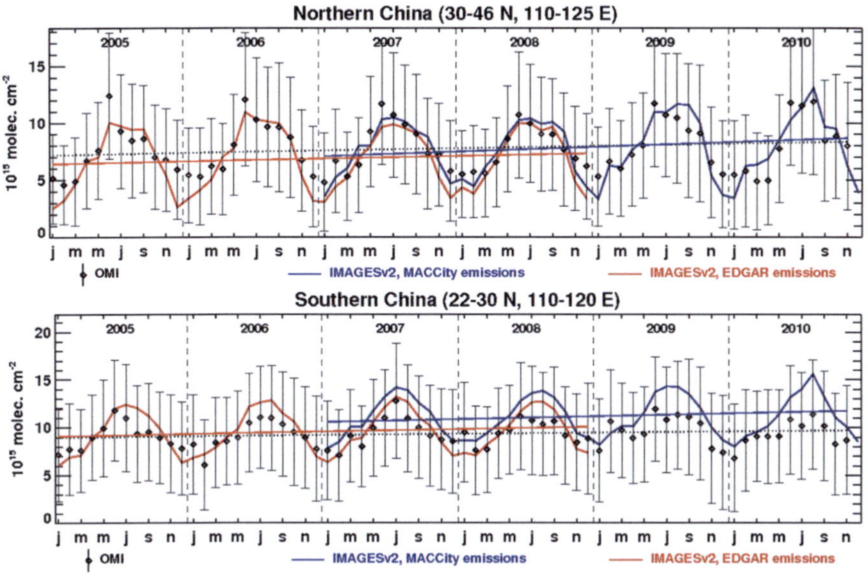

Fig. 8.5 HCHO columns observed by the OMI instrument between 2005 and 2010 in Northern and Southern China and simulated with IMAGESv2 using anthropogenic emissions either from MACCity (*blue*) or from EDGARv4.2 (in *red*). The observed columns are shown as *black diamonds* with their error bars. The *dotted black line* represents the observation trend line. The blue and red straight lines represent the model trend

suited to inform us on the spatial and temporal variability of the underlying VOC sources. In this section, we take advantage of HCHO column observations from the Ozone Monitoring Instrument (OMI) to infer emission flux estimates from VOCs emitted by different source categories over China over a complete decade of measurements (2005–2014). This is realized through source inverse modeling, which consists in varying and optimizing the emission in a chemistry-transport model (CTM) in order to minimize the discrepancy between the observed and modeled HCHO columns. The top-down fluxes are derived in the global CTM IMAGESv2 by an iterative minimization algorithm based on the adjoint of IMAGESv2, starting from a priori emission estimates provided by bottom-up inventories (cf. next section). Due to the large number and variety of VOCs which are precursors of HCHO, a required preliminary step is to update the chemical mechanism of the model so as to reproduce the production of HCHO from these precursors based on up-to-date mechanistic data and to a quasi-explicit chemical mechanism (Master Chemical Mechanism v3.2, http://mcm.leeds.ac.uk/MCM/).

Using box model calculations, we have calculated the relative contribution of the different HCHO precursors considered in the model over Northern and Southern China. This calculation relies on a priori emission estimates from REASv2 for anthropogenic VOCs, GFED4s for vegetation fires, and MEGAN-MOHYCAN for isoprene emissions (Sect. 8.3). Note that biogenic methanol emissions are obtained from an inversion study constrained by IASI methanol columns (Stavrakou et al. 2011). As summarized in Table 8.2, the main biogenic precursors of HCHO, isoprene, methanol and α-pinene, account for 27% and 41% of the total HCHO production in Northern China (34–46 N) and Southern China (18–34 N), respectively. Owing to the high HCHO formation yield in the oxidation of C_2H_4 (1.8 mol/mol), 10% of the HCHO production in Northern China and 7% in Southern China is due to this compound alone. The oxidation of the predominantly anthropogenic C2-C3 alkanes and alkenes contributes to the total HCHO production by 26% in Northern and by 16% in Southern China. The most common aromatics, benzene, toluene and xylenes, have only a small contribution to the total HCHO production, due to their low HCHO yields from these compounds. Acetic acid and small carbonyls contribute for ~10% of HCHO production.

Other hydrocarbons, mainly butanes and higher alkanes, butane and higher alkenes, and higher aromatic hydrocarbons of anthropogenic origin, represent a significant part of the total HCHO production, estimated at 31.5% in Northern and at 25.6% in Southern China. Overall, according to IMAGESv2 model simulations (Stavrakou et al. 2015), the HCHO production from the above precursors in China is estimated at 34 Tg (44% from Northern China).

Table 8.2 Percentage contribution to HCHO production in Northern and Southern China from individual VOCs included in the model

Contributors to HCHO production	Northern China	Southern China
Isoprene	11.87	24.74
α-pinene	0.32	1.21
CH_3OH	15.13	15.18
C_2H_6	7.66	4.35
C_3H_8	4.80	2.39
C_2H_4	10.54	7.14
C_3H_6	2.60	2.00
HCHO	1.22	1.39
CH_3CHO	2.33	3.42
Methyl ethyl ketone	0.61	0.57
CH_3COOH	4.09	3.58
CH_3COCH_3	2.18	3.80
Benzene	0.27	0.16
Toluene	3.12	3.08
Xylenes	1.70	1.34
Other HCs	31.54	25.64

8.5 HCHO Satellite Column Abundances and Model Predictions

We use formaldehyde observations retrieved from the OMI spectrometer aboard the Aura mission (De Smedt et al. 2015) as top-down constraints in an inverse modeling system based on the adjoint of the IMAGESv2 model (Müller and Stavrakou 2005). The adjoint method is a powerful tool to calculate the sensitivities (i.e. partial derivatives) of the model output with respect to the control parameters (here a priori emission fluxes) and is suitable for addressing non-linear large-scale problems. The satellite retrievals are based on an improved DOAS algorithm which was designed to reduce the effect of interferences between the observed species. The current data version (v14) uses an iterative algorithm for spike removal in the slant column residuals and a procedure based on the background normalization to remove striping artefacts due to calibration problems (Boersma et al. 2011; De Smedt et al. 2015). In addition to de-striping, the effects of the OMI row anomaly issue affecting the spectra after 2007 were taken into account by systematically removing the OMI rows presenting higher levels of noise and fitting residuals than the average. This filtering results in a more appropriate dataset for addressing trend studies, as discussed in De Smedt et al. (2015). The observed HCHO columns are averaged on a monthly basis, whereas scenes with cloud fractions higher than 40% and land fractions lower than 20%, as well as data with a retrieval error higher than 100% are excluded from the averages. The satellite columns are accessible at the BIRA-IASB website (http://h2co.aeronomie.be) (Fig. 8.6).

The model simulations are performed with the IMAGESv2 global CTM (cf. Sect. 8.3) over all years between 2005 and 2014 following a 4-month spin-up

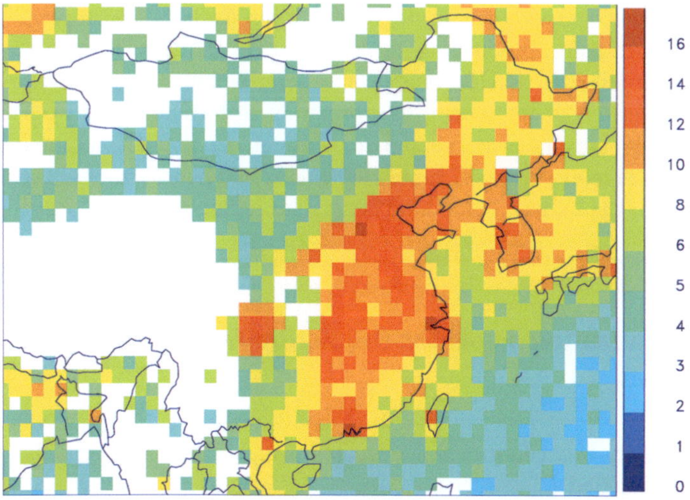

Fig. 8.6 OMI HCHO column distribution over East Asia in July 2005 at a resolution of 1° × 1°. Units are 10^{15} molec.cm^{-2}

time. The a priori anthropogenic fluxes in IMAGESv2 are obtained from the REASv2 inventory for Asia until 2008 (Kurokawa et al. 2013), and the 2008 inventory is used for all years between 2009 and 2014. The NMVOC emission speciation is also taken from REASv2 over Asia. Agricultural fire burning in China is obtained from Huang et al. (2012) inventory compiled based on statistical data for 2006 (and is kept the same for all years), while biomass burning from other emission categories is taken from the GFED4s inventory (Randerson et al. 2012; Giglio et al. 2013) for all years of the target period. Biogenic emissions of isoprene are obtained from the MEGAN-MOHYCAN global inventory driven by the ECMWF ERA-Interim meteorology (Stavrakou et al. 2014). The 2005–2014 mean annual isoprene flux in China is estimated at 7.4 Tg, with the highest emission calculated in 2013 (8 Tg) and the lowest in 2012 (6.8 Tg).

The chemical degradation mechanism of pyrogenic NMVOCs is largely described in Stavrakou et al. (2009), with only minor modifications. This mechanism includes an explicit treatment for 16 pyrogenic formaldehyde precursors. The emissions of other pyrogenic compounds is represented through a lumped compound (OTHC) with a simplified oxidation mechanism designed in order to reproduce the overall formaldehyde yield of the explicit NMVOC mix it represents. The oxidation mechanism for isoprene is based on Stavrakou et al. (2010), but was modified to account for the revised kinetics of isoprene peroxy radicals according to the Leuven Isoprene Mechanism version 1 (LIM1) (Peeters et al. 2014), as well as for the chemistry of the isoprene epoxides (IEPOX) following the Master Chemical Mechanism MCMv3.2 (http://mcm.leeds.ac.uk/MCMv3.2/). The HCHO yield in isoprene oxidation by OH is calculated using a box model to be 2.4 mol/mol under high NOx (1 ppbv NO_2, after 2 months of simulation) and 1.9 mol/mol at 0.1 ppbv NO_2. However, it should be acknowledged that important uncertainties persist in the isoprene mechanism in low NOx conditions, due to unresolved issues in the degradation mechanisms of both the oxidation products of the isoprene epoxides and the isomerization products of isoprene peroxy radicals (Peeters et al. 2014; Bates et al. 2016). The HCHO simulation is described in Bauwens et al. (2016). In the current version of the model, the formation of formaldehyde from anthropogenic VOC precursors is accounted for through 17 explicit compounds and a lumped species representing 32 higher anthropogenic hydrocarbons. The oxidation mechanism of the lumped compound is detailed in Stavrakou et al. (2015).

8.6 Optimization Tools for the Estimation of Top-Down Fluxes

The discrepancy between the model predictions and the observations is quantified by the cost function J,

$$J(\mathrm{f}) = 1/2((\mathrm{H}(\mathrm{f}) - y)^T \mathrm{E}^{-1}(\mathrm{H}(\mathrm{f}) - y) + \mathrm{f}^T \mathrm{B}^{-1}\mathrm{f})$$

and is minimized using an iterative optimization algorithm, adapted for large-scale problems, which is based on the calculation of the partial derivatives of J with respect to the input variables. In our case the input variables are scalar variables f = ($\mathrm{f_j}$), such that the optimized flux can be expressed as

$$\Phi_{\mathrm{i}}^{\mathrm{opt}}(x,t) = \sum_{\mathrm{j}=1}^{\mathrm{m}} e^{\mathrm{f_j}} \Phi_{\mathrm{i}}(x,t),$$

where $\Phi_{\mathrm{i}}(x, t)$ is the initial flux depending on spatial coordinates (latitude, longitude) and time (month). H(f) is the model operator, y the observation vector (monthly HCHO columns in our case), E and B are the covariance matrices of the errors on the observations and on the a priori parameters f, respectively, and ^{T} denotes the transpose of the matrix. The partial derivatives of J with respect to the input variables f are calculated using the discrete adjoint of IMAGESv2 chemistry-transport model (CTM) (Müller and Stavrakou 2005; Stavrakou et al. 2009). The derivation of monthly top-down (anthropogenic, biogenic and pyrogenic) VOC fluxes is conducted at the model resolution (2° × 2.5°), as discussed in Stavrakou et al. (2015). The inversions are performed separately for all years of the target decade. The covariance matrix of the observational errors (E) is assumed diagonal. The errors are defined as the squared sum of the retrieval error (De Smedt et al. 2015) and a representativity error set to 2×10^{15} molec.cm^{-2}. Convergence of the minimization algorithm is reached when the gradient of the cost function is strictly zero, but in large-scale problems this criterion is rarely attained. For that reason, we assume that convergence is reached when the gradient of the cost function is reduced by a factor of 1000 with respect to its initial value. About 50 iterations are generally needed to meet this requirement.

Figure 8.7 illustrates a comparison over China between observed monthly mean HCHO column densities over 2005–2014, monthly columns simulated by IMAGESv2 sampled at the time and location of the satellite measurements, and optimized columns inferred after inversion. In Northern China the observed seasonal cycle exhibits a large amplitude between a minimum in winter (ca. 5×10^{15} molec.cm^{-2}) and a maximum in summer (ca. 10×10^{15} molec.cm^{-2}), and in particular in June (ca. 12×10^{15} molec.cm^{-2}). This variability is the result of higher photochemical activity and to biogenic emission fluxes in summertime, whereas in winter, formaldehyde production from anthropogenic hydrocarbon oxidation is slowed down due to lower oxidant and radiation levels. The peak in HCHO columns in June is the result of the intense agricultural burning in the North China Plain and will be discussed in detail in Sect. 8.9. In Southern China, the seasonal amplitude is less pronounced due to the higher radiation levels present in this region all year round, and ranges between ca. 8×10^{15} molec.cm^{-2} in wintertime and a summer maximum of ca. 12×10^{15} molec.cm^{-2}. The a priori model simulation reproduces well the magnitude of the columns and the observed seasonal cycle, but underestimates the winter columns in Northern China and overestimates

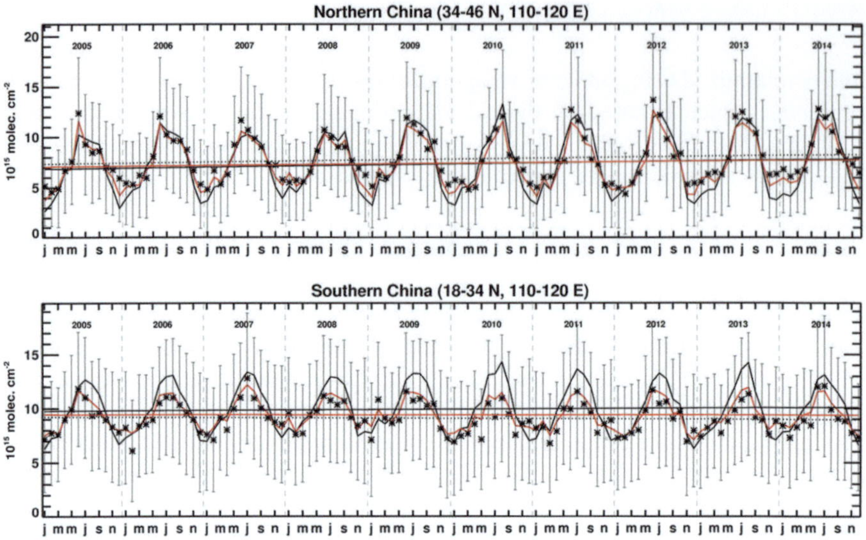

Fig. 8.7 A priori (in *black*), optimized (in *red*) modeled HCHO columns, and observed columns from the OMI sensor between 2005 and 2014 over Northern and Southern China. The observed columns are shown as *black diamonds* with their error bars. The *dotted black line* represents the observation trend line. The *black* and *red straight lines* represent the model trend in the a priori and a posteriori simulation

the summer columns in Southern China throughout the decade of measurements (Fig. 8.7).

As expected, the inversion successfully brings the model much closer to the observations over all seasons and years. This is achieved by increasing the anthropogenic VOC precursor emissions in Northern China and by decreasing the anthropogenic VOC and biogenic emission fluxes in Southern China. The observations in Northern China exhibit a positive trend of 1.3%/year over 2005–2014, which is somewhat lower than the trend in the a priori model columns (1.5%/year), suggesting that the a priori REASv2 VOC emission trends emissions might be slightly overestimated. A similar conclusion is also drawn for Southern China, where the a priori model trend is close to zero but positive (+0.3%/year), whereas the observed trend is slightly negative (−0.3%/year). It should be noted, however, that the observed trend in Southern China is not significant because of the relatively high uncertainties of the observed columns in this region (De Smedt et al. 2015). After optimization, the model columns show a trend of 1.2%/year and −0.24%/year in Northern and Southern China, respectively, much closer to the observed trends. The seasonal and interannual variability of the inferred top-down emissions over China is discussed in the following sections.

8.7 Anthropogenic VOC Emissions Based on OMI HCHO Columns

Figure 8.8 illustrates the distribution of the mean 2005–2014 anthropogenic VOC emission update inferred by the flux inversion of HCHO observations. The top-down emissions are generally higher than in the REASv2 inventory, with the strongest updates (up to 80%) derived in the provinces of Liaoning, Jilin, and Heilongjiang (Northeast). The satellite observations of HCHO suggest moderate positive emission updates in the Beijing region (up to 10–30%), and in the rapidly developing Chengdu-Chongqing region in southwest China (up to 40%). A substantial reduction of the emissions is found in Shanghai (up to 50%), and along the Yangtze River, whereas moderate flux decreases (ca. 20%) are deduced in the Guangzhou region.

Overall, the a posteriori annual emissions in China remain close to the a priori between 2005 and 2008, slightly decline between 2009 and 2012, likely related to the aftermath of the 2008–2009 economic recession, and increase again in 2013 and 2014 (Fig. 8.9). The EDGARv4.2 inventory between 2005 and 2008 is by ca. 20% lower than both REASv2 and the top-down fluxes, and in good agreement with the MEIC inventory for 2008 (20.2 TgVOC, Li et al. 2014), and close to Wei et al. (2008) estimate for 2005 (20.1 TgVOC), whereas Bo et al. (2008) reported a much lower flux (16.5 TgVOC). Although the total top-down anthropogenic VOC fluxes in China show little interannual variability, the top-down emissions in Chinese megacities often present important year-to-year variability, which is generally linked to the implementation of emission control measures. Bottom-up estimates during short-term events, like the 2008 Beijing Olympic Games, or the 2010 Guangzhou Asian Games, are rarely available, and therefore top-down estimates

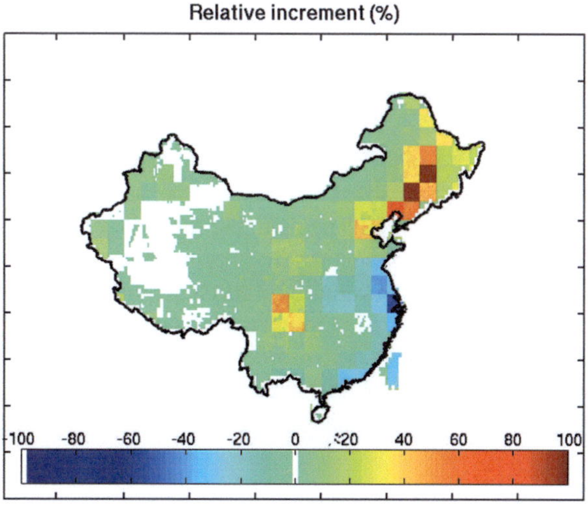

Fig. 8.8 Percentage flux difference between the optimized anthropogenic fluxes averaged over 2005–2014 and the a priori REASv2 inventory. Note that REASv2 data are available until 2008, and the 2008 values are used as a priori in the model for all years between 2009 and 2014

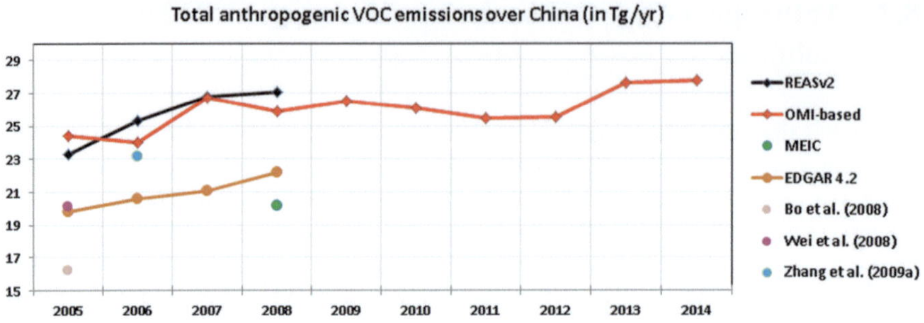

Fig. 8.9 Annual anthropogenic VOC emissions in China according to bottom-up and OMI-based estimates

are particularly useful to evaluate the effectiveness of anti-pollution measures and define future strategies.

In Beijing, although the top-down emissions are higher with respect to the bottom-up inventory, they suggest an emission drop by 14% in 2008, compared to 2007 and 2009, consecutive to the emission restrictions implemented in order to improve air quality during the Beijing Olympics. In comparison, based on GOME-2 and OMI satellite observations, a 60% reduction of NO_2 columns was reported in Beijing area during the Olympic Games (Mijling et al. 2009), and MOPITT CO observations showed an average 32% reduction in CO columns for 2008 with respect to 2007 (Worden et al. 2012), whereas the corresponding reductions of CO and NOx emissions were estimated at 28% (Yang et al. 2011) and at 38% (Worden et al. 2012) respectively. Ground observations at a rural site near Beijing reported reductions in daytime SO_2, CO, and NOy of 61%, 25%, and 21%, respectively, and the mean daytime mixing ratio of O_3 was decreased by ca. 23% in August 2008 compared to the same month of the previous year (Wang et al. 2009).

Emission control practices from the 2008 Beijing Olympics were implemented during the 2010 World Exposition in Shanghai (May–October 2010) in order to ensure good air quality (Huang et al. 2013). The OMI HCHO columns suggest an emission drop in VOC estimates by 36% in 2010 relative to 2009, stronger than the NO_2 emission reduction (7.5–14.5%) inferred by ground-based and OMI observations (Chan et al. 2015).

In the region of Guangzhou, the top-down emissions are lower than the a priori REASv2, likely reflecting the city's acclaimed efforts to efficiently reduce the emissions of air pollutants by implementing strict measures for air remediation. Moreover, the inversion infers a significant emission drop (48%) in 2010, related to control measures taken in order to ensure good air conditions during the Asian Games (November 2010, Li et al. 2012).

8.8 Isoprene Top-Down Fluxes

The satellite observations of HCHO infer a reduction of the Chinese isoprene fluxes, from 7.3 Tg/year in MEGAN-MOHYCAN to 5.8 Tg/year on average over 2005–2014. The decrease is more significant in Southern China and ranges between 27 and 45% depending on the year. The emissions peak in summertime and exhibit weak interannual variability (Fig. 8.11), with a maximum in 2007 (2.6 Tg/year) and a minimum in 2010 (1.7 Tg/year). Both MEGAN-MACC and GUESS-ES estimates are higher than the optimized results, by 60% and by factor of 2.5, respectively. The OMI-based emission in 2010 is in good agreement with an earlier estimate inferred from an inversion using GOME-2 HCHO observations (2.4 Tg/year, Stavrakou et al. 2015).

In South China, the negative trend (−0.53%/year) in the observed HCHO columns leads to a stronger negative trend in the top-down fluxes, from −0.12%/year in the a priori to −1.05%/year after optimization (Fig. 8.10). The negligible trend in the MEGAN-MOHYCAN emissions can be explained by a compensation between the declining trend in temperature (−0.03 °C/year) and in the ERA-Interim photochemically active radiation (PAR, −0.33%/year), and the increasing trend in the MODIS leaf area index data in this region (1%/year, Bauwens et al. 2016). Furthermore, the model accounts for the effects of land use changes, and more specifically the reduction of crop extent in South China, as a result of crop

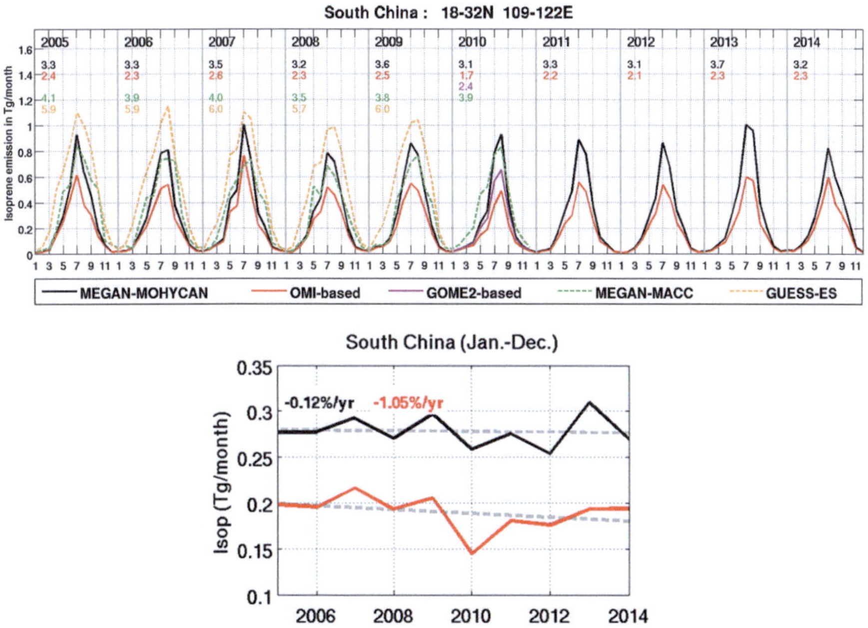

Fig. 8.10 A priori annual isoprene fluxes based on MEGAN-MOHYCAN (in *black*) and a posteriori fluxes (in *red*) over Southern China

abandonment, as suggested by the database of Ramankutty and Foley (1999) and supported by MODIS land cover data (Friedl et al. 2010). However, a recent land cover database suggests that the extent of crops increased in eastern China between 1979 and 2005 (Hurtt et al. 2011). Moreover, an alternative surface radiation dataset, based on ERA-Interim data but accounting for the effects of aerosols, estimated a declining trend in PAR in this region (Weedon et al. 2014). The crop expansion and declining PAR were proposed to cause a negative isoprene trend in South China (Yue et al. 2015), and could explain the observed negative trend in HCHO column observations (Bauwens et al. 2016).

8.9 Crop Residue Burning in the North China Plain: The View from OMI

Crop residue burning is an important contributor to biomass burning fluxes in the North China Plain, where post-harvest burning is a well-established agricultural management practice, allowing for land clearing from residual straw and preparation for the subsequent crop cultivation. The most extensive crop fires in the North China Plain occur in June after the winter wheat comes to maturity, and are blamed for air pollution spikes causing serious health problems. Estimating harvest season burning emissions is therefore of primary importance to assess air quality and define best policies for its improvement in this sensitive region, home to more than 300 million people. Bottom-up approaches, either based on crop production and emission factors, or on satellite burned area and fire radiative power products, have been adopted so far. However, these methods crucially depend, among other assumptions, on the satellite skill to detect small fires, and generally tend to underestimate the magnitude of the actual emissions. The OMI HCHO observations reveal a strong peak over the North China Plain in June throughout the decade of measurements, resulting from enhanced pyrogenic emissions of a large suite of volatile organic compounds (VOCs), precursors of HCHO. The observed columns are higher in June compared with the neighboring months, by about 50% and 15% compared to May and July, respectively. This behavior indicates a significant source from agricultural burning alongside the anthropogenic and biogenic source in this region.

The satellite observations suggest a substantial increase (Fig. 8.11) of the agricultural waste burning source in June by about a factor of 2 on average for 2005–2014 in comparison with the a priori inventory of Huang et al. (2012), which is based on a statistical approach and is much higher than other bottom-up inventories (Fig. 8.1, lower right panel). The top-down crop emissions in the North China Plain are found to be by 14, 7 and 4 times higher than GFAS (Kaiser et al. 2012), GFED4s (Giglio et al. 2013), and FINNv1.5 (Wiedinmyer et al. 2011) independent estimates, respectively, whereas the GFED3 (van der Werf et al. 2010) emissions are even lower (Fig. 8.1). The interannual variability of the top-down source is

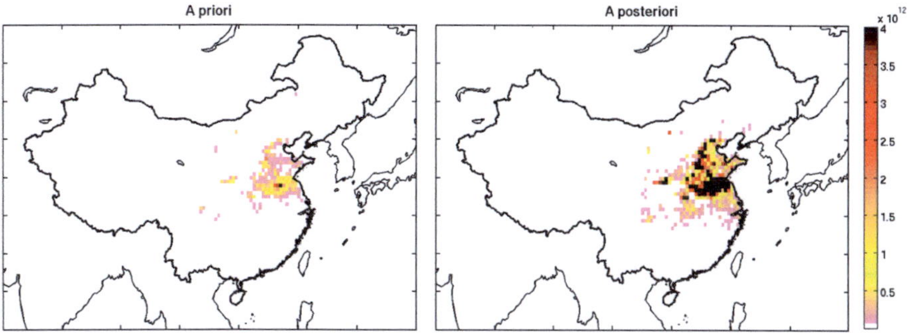

Fig. 8.11 A priori (Huang et al. 2012) and optimized crop waste burning fluxes over the North China Plain in June averaged over 2005–2014. Units are in molec.cm^{-2} s^{-1}

pronounced, with lower crop fluxes in 2008 and 2009, likely due to anti-pollution regulation measures during the Beijing Olympics, followed by higher emissions in the subsequent years, in consistency with an increase in MODIS fire counts after 2008 (Stavrakou et al. 2016).

8.10 Conclusions

Understanding and quantifying the sources and sinks of pollutants represents a major challenge for society in the face of air quality and climate change. To evaluate the impact of bottom-up inventories on the atmospheric composition over China we performed two long-term (1990–2008) model simulations using emissions from either the EDGARv4.2 or the MACCity bottom-up inventories. Interestingly, despite the large discrepancies between the emission flux estimates from these inventories, the averaged surface ozone trends over China calculated in these two simulations were found to be similar over the entire country (ca. 2% per decade), due to compensating chemical effects, whereas the trends of surface CO, NO_2 and OH exhibit larger deviation between the two simulations. Evaluation against satellite NO_2 column trends suggests that the NOx emission trends in the bottom-up inventories are most likely overestimated over 1997–2008, while comparison between observed and modeled HCHO columns indicates that the biogenic emissions might be too high in the model and that the anthropogenic VOC emission trends are only slightly overestimated (if at all) in the REASv2 inventory. These results highlight the value of alternative methods for emission estimation, complementary to the bottom-up emission approach.

Inverse modelling offers an enticing framework for using satellite observations to constrain the fluxes of key pollutants. The detection of formaldehyde columns from satellites opens new avenues for the derivation of fluxes of volatile organic compounds, a wide class of formaldehyde precursors emitted by vegetation, fires and anthropogenic activities. In this work we have used an atmospheric model constrained by one decade (2005–2014) of HCHO columns retrieved from the OMI sounder and an advanced inversion framework in order to infer emissions of VOC compounds over China. With respect to state-of-art bottom-up inventories, the satellite observations of HCHO

- Suggest moderate anthropogenic emission increases in Beijing, and Chengdu-Chongqing, but decreased emissions in Shanghai, and Guangzhou broad regions for all years
- Are able to capture the lower emission fluxes due to the implementation of pollution control measures during important events
- Suggest a reduction of the isoprene fluxes in Southern China, and a negative trend, possibly due to crop expansion and declining solar radiation fluxes
- Indicate large underestimation of the agricultural crop burning source in the North China Plain every year in June.

New challenges are closing in, with the wealth of data expected from upcoming enhanced resolution satellites which should, if properly exploited, lead to major advances in our understanding and characterization of emission sources, at spatial and temporal resolutions far superior to previous flux inversion exercises.

References

Arneth, A., Niinemets, Ü., Pressley, S., et al. (2007). Process-based estimates of terrestrial ecosystem isoprene emissions: Incorporating the effects of a direct CO_2-isoprene interaction. *Atmospheric Chemistry and Physics, 7*(1), 31–53.

Bates, K. H., Nguyen, T. B., Teng, A. P., Crounse, J. D., Kjaergaard, H. G., Stoltz, B. M., Seinfeld, J. H., & Wennberg, P. O. (2016). Production and fate of C4 dihydroxycarbonyl compounds from isoprene oxidation. *The Journal of Physical Chemistry. A, 120*(1), 106–117.

Bauwens, M., Stavrakou, T., Müller, J.-F., De Smedt, I., Van Roozendael, M., van der Werf, G. R., Wiedinmyer, C., Kaiser, J. W., Sindelarova, K., & Guenther, A. (2016). Nine years of global hydrocarbon emissions based on source inversion of OMI formaldehyde observations. *Atmospheric Chemistry and Physics, 16*, 10133–10158.

Bo, Y., Cai, H., & Xie, S. D. (2008). Spatial and temporal variation of historical anthropogenic NMVOCs emission inventories in China. *Atmospheric Chemistry and Physics, 8*, 7297–7316.

Boersma, K. F., Eskes, H. J., & Brinksma, E. J. (2004). Error analysis for tropospheric NO_2 retrieval from space. *Journal of Geophysical Research, 109*, D04311. doi:10.1029/2003JD003962.

Boersma, K. F., et al. (2011). An improved tropospheric NO2 column retrieval algorithm for the Ozone Monitoring Instrument. *Atmospheric Measurement Techniques, 4*, 1905–1928.

Chan, K. L., Hartl, A., Lam, Y. F., Xie, P. H., Liu, W. Q., Cheung, H. M., Lampl, J., Pöhler, D., Li, A., Xu, J., Zhou, H. J., Ning, Z., & Wenig, M. O. (2015). Observations of tropospheric NO_2 using ground based MAX-DOAS and OMI measurements during the Shanghai World Expo 2010. *Atmospheric Environment, 119*, 45–58.

De Smedt, I., Müller, J.-F., Stavrakou, T., van der A, R. J., Eskes, H., & Van Roozendael, M. (2008). Twelve years of global observation of formaldehyde in the troposphere using GOME and SCIAMACHY sensors. *Atmospheric Chemistry and Physics, 8*, 4947–4963.

De Smedt, I., Stavrakou, T., Hendrick, F., et al. (2015). Diurnal, seasonal, and long-term variation of global formaldehyde columns inferred from combined OMI and GOME-2 observations. *Atmospheric Chemistry and Physics, 15*, 12519–12545.

Dee, D. P., et al. (2011). The ERA-Interim reanalysis: Configuration and performance of the data assimilation system. *Quarterly Journal of the Royal Meteorological Society, 137*, 553–597.

Ding, J., van der A, R. J., Mijling, B., Levelt, P. F., & Hao, N. (2015). NOx emission estimates during the 2014 Youth Olympic Games in Nanjing. *Atmospheric Chemistry and Physics, 15*, 9399–9412.

EC-JRC/PBL European Commission. (2011). Joint Research Center/Netherlands Environmental Assessment Agency. Emission Database for Global Atmospheric Research version 4.2. http://edgar.jrc.ec.europa.eu. Accessed 20 Sept 2016.

Friedl, M.A., Sulla-Menashe, D., Tan, B., Schneider, A., Ramankutty, N., Sibley, A., Huang, X. (2010). MODIS collection 5 global land cover: Algorithm refinements and characterization of new datasets 2001–2012, Collection 5.1 IGBP Land Cover. Boston: University.

Giglio, L., Randerson, J. T., & Werf, G. R. (2013). Analysis of daily, monthly, and annual burned area using the fourth-generation global fire emissions database (GFED4). *Journal of Geophysical Research, 118*(1), 317–328.

Granier, C., Bessagnet, B., Bond, T., D'Angiola, A., Denier van der Gon, H., Frost, G. J., et al. (2011). Evolution of anthropogenic and biomass burning emissions of air pollutants at global and regional scales during the 1980–2010 period. *Climatic Change, 109*, 163–190.

Guan, D., Liu, Z., Geng, Y., Lindner, S., & Hubacek, K. (2012). The gigatonne gap in China's carbon dioxide inventories. *Nature Climate Change, 2*, 672–675.

Guenther, A. B., Jiang, X., Heald, C. L., Sakulyanontvittaya, T., Duhl, T., Emmons, L. K., & Wang, X. (2012). The model of emissions of gases and aerosols from nature version 2.1 (MEGAN2.1): An extended and updated framework for modeling biogenic emissions. *Geoscientific Model Development, 5*, 1471–1492.

Guenther, A. B., Karl, T., Harley, P., Wiedinmyer, C., Palmer, P. I., & Geron, C. (2006). Estimates of global terrestrial isoprene emissions using MEGAN (Model of Emissions of Gases and Aerosols from Nature). *Atmospheric Chemistry and Physics, 6*, 3181–3210.

Huang, X., Li, M., Li, J., & Song, Y. (2012). A high-resolution emission inventory of crop burning in fields in China based on MODIS Thermal Anomalies/Fire products. *Atmospheric Environment, 50*, 9–15.

Huang, K., Zhuang, G., Lin, Y., Wang, Q., Fu, J. S., Fu, Q., Liu, T., & Deng, C. (2013). How to improve the air quality over megacities in China: Pollution characterization and source analysis in Shanghai before, during, and after the 2010 World Expo. *Atmospheric Chemistry and Physics, 13*, 5927–5942.

Hurtt, G. C., Chini, L. P., Frolking, S., Betts, R. A., Feddema, J., Fischer, G., et al. (2011). Harmonization of land-use scenarios for the period 1500–2100: 600 years of global annual land-use transitions, wood harvest, and resulting secondary lands. *Climatic Change, 109*, 117–161.

Kaiser, J., Heil, A., Andreae, M., Benedetti, A., Chubarova, N., Jones, L., Morcrette, J.-J., Razinger, M., Schultz, M., Suttie, M., et al. (2012). Biomass burning emissions estimated with a global fire assimilation system based on observed fire radiative power. *Biogeosciences, 9* (1), 527–554.

Kurokawa, J., Ohara, T., Morikawa, T., Hanayama, S., Greet, J.-M., Fukui, T., Kawashima, K., et al. (2013). Emissions of air pollutants and greenhouse gases over Asian regions during 2000–2008: Regional Emission inventory in ASia (REAS) version 2. *Atmospheric Chemistry and Physics, 13*, 11019–11058.

Lamarque, J.-F., et al. (2010). Historical (1850–2000) gridded anthropogenic and biomass burning emissions of reactive gases and aerosols: Methodology and application. *Atmospheric Chemistry and Physics, 10*, 7017–7039.

Levelt, P. F., Hilsenrath, E., Leppelmeier, G. W., van Den Oord, G. H. J., Bhartia, P. K., Tamminen, J., De Haan, J. F., & Veefkind, P. (2006). Science objectives of the ozone monitoring instrument. *Geoscience and Remote Sensing, 44*, 1199–1208.

Li, T. Y., Deng, X. J., Fan, S. J., Wu, D., Li, F., Deng, T., Tan, H. B., & Jiang, D. H. (2012). Study on air quality and pollution meteorology conditions of Guangzhou during the 2010 Asian games. *Huan Jing Ke Xue, 33*(9), 2932–2938.

Li, M., Zhang, Q., Streets, D. G., He, K. B., Cheng, Y. F., Emmons, L. K., Huo, H., Kang, S. C., Lu, Z., Shao, M., Su, H., Yu, X., & Zhang, Y. (2014). Mapping Asian anthropogenic emissions of non-methane volatile organic compounds to multiple chemical mechanisms. *Atmospheric Chemistry and Physics, 14*, 5617–5638.

Mijling, B., & van Der A, R. J. (2012). Using daily satellite observations to estimate emissions of short-lived air pollutants on a mesoscopic scale. *Journal of Geophysical Research*. doi:10.1029/2012JD017817.

Mijling, B., van Der A, R. J., Boersma, K. F., Van Roozendael, M., De Smedt, I., & Kelder, H. M. (2009). Reductions of NO_2 detected from space during the 2008 Beijing Olympic Games. *Geophysical Research Letters*. doi:10.1029/2009GL038943.

Müller, J.-F., & Stavrakou, T. (2005). Inversion of CO and NOx emissions using the adjoint of the IMAGES model. *Atmospheric Chemistry and Physics, 5*, 1157–1186.

Müller, J.-F., Stavrakou, T., Wallens, S., De Smedt, I., Van Roozendael, M., Rinne, J., Munger, B., Goldstein, A., & Guenther, A. (2008). Global isoprene emissions estimated using MEGAN, ECMWF analyses and a detailed canopy environmental model. *Atmospheric Chemistry and Physics, 8*, 1329–1341.

Peeters, J., Müller, J.-F., Stavrakou, T., & Nguyen, S. V. (2014). Hydroxyl radical recycling in isoprene oxidation driven by hydrogen bonding and hydrogen tunneling: The upgraded LIM1 mechanism. *The Journal of Physical Chemistry A, 118*, 8625–8643.

Ramankutty, N., & Foley, J. A. (1999). Estimating historical changes in global land cover. croplands from 1700 to 1992. *Global Biogeochem. Cy., 13*, 997–1027.

Randerson, J., Chen, Y., van der Werf, G. R., Rogers, B., & Morton, D. (2012). Global burned area and biomass burning emissions from small fires. *Journal of Geophysical Research*. doi:10.1029/2012JG002128.

Riahi, K., Rao, S., Krey, V., et al. (2011). RCP8.5 – a scenario of comparatively high greenhouse gas emissions. *Climatic Change, 109*, 33. doi:10.1007/s10584-011-0149-y.

Schultz, M. G., Heil, A., Hoelzemann, J. J., Spessa, A., Thonicke, K., Goldammer, J. G., Held, A. C., Pereira, J. M. C., & van het Bolscher, M. (2008). Global wildland fire emissions from 1960 to 2000. *Global Biogeochemical Cycles, 22*, GB2002. doi:10.1029/2007GB003031.

Sindelarova, K., Granier, C., Bouarar, I., Guenther, A., et al. (2014). Global dataset of biogenic VOC emissions calculated by the MEGAN model over the last 30 years. *Atmospheric Chemistry and Physics, 14*, 9317–9341.

Stavrakou, T., Müller, J.-F., De Smedt, I., Van Roozendael, M., van der Werf, G. R., Giglio, L., & Guenther, A. (2009). Global emissions of non-methane hydrocarbons deduced from SCIAMACHY formaldehyde columns through 2003–2006. *Atmospheric Chemistry and Physics, 9*, 3663–3679.

Stavrakou, T., Müller, J.-F., De Smedt, I., Van Roozendael, M., van der Werf, G. R., Giglio, L., & Guenther, A. (2009a). Evaluating the performance of pyrogenic and biogenic emission inventories against one decade of space-based formaldehyde columns. *Atmospheric Chemistry and Physics, 9*, 1037–1060.

Stavrakou, T., Peeters, J., & Müller, J.-F. (2010). Improved global modelling of HOx recycling in isoprene oxidation: Evaluation against the GABRIEL and INTEX-A aircraft campaign measurements. *Atmospheric Chemistry and Physics, 10*, 9863–9878.

Stavrakou, T., Guenther, A., Razavi, A., Clarisse, L., Clerbaux, C., Coheur, P. F., et al. (2011). First space-based derivation of the global atmospheric methanol emission fluxes. *Atmospheric Chemistry and Physics, 11*, 4873–4898.

Stavrakou, T., Müller, J.-F., Bauwens, M., De Smedt, I., Van Roozendael, M., Guenther, A., Wild, M., & Xia, X. (2014). Isoprene emissions over Asia 1979–2012: Impact of climate and land use changes. *Atmospheric Chemistry and Physics, 14*, 4587–4605.

Stavrakou, T., Müller, J.-F., Bauwens, M., De Smedt, I., Van Roozendael, M., et al. (2015). How consistent are top-down hydrocarbon emissions based on formaldehyde observations from GOME-2 and OMI? *Atmospheric Chemistry and Physics, 15*, 12007–12067.

Stavrakou, T., Müller, J.-F., Bauwens, M., De Smedt, I., Lerot, C., Van Roozendael, M., Coheur, P. F., Clerbaux, C., et al. (2016). Substantial underestimation of post-harvest burning emissions in the North China Plain revealed by multi-species space observations. *Scientific Reports, 6*, 32307. doi:10.1038/srep323072016.

United Nations, Department of Economic and Social Affairs, Population Division, World Urbanization Prospects. (2014). The 2014 revision, highlights (ST/ESA/SER.A/352). https://esa.un.org/unpd/wup/Publications/Files/WUP2014-Highlights.pdf. Accessed 20 Sept 2016.

van der Werf, G. R., Randerson, J. T., Giglio, L., Collatz, G. J., Kasibhatla, P. S., & Arellano Jr., A. F. (2006). Interannual variability in global biomass burning emissions from 1997 to 2004. *Atmospheric Chemistry and Physics, 6*, 3423–3441. doi:10.5194/acp-6-3423-2006.

van der Werf, G. R., Randerson, J. T., Giglio, L., Collatz, G. J., Mu, M., Kasibhatla, P. S., Morton, D. C., DeFries, R. S., Jin, Y., & van Leeuwen, T. T. (2010). Global fire emissions and the contribution of deforestation, savanna, forest, agricultural, and peat fires (1997–2009). *Atmospheric Chemistry and Physics, 10*(23), 11707–11735.

van Leeuwen, T. T., van der Werf, G. R., Hoffmann, A. A., Detmers, R. G., Rücker, G., French, N. H. F., Archibald, S., et al. (2014). Biomass burning fuel consumption rates: A field measurement database. *Biogeosciences, 11*(24), 7305–7329.

Wang, Y., Hao, J., McElroy, M. B., William, J., et al. (2009). Ozone air quality during the 2008 Beijing Olympics: Effectiveness of emission restrictions. *Atmospheric Chemistry and Physics, 9*, 5237–5251.

Weedon, G. P., Balsano, G., Bellouin, N., Gomes, S., Best, M. J., & Viterbo, P. (2014). The WFDEI meteorological forcing data set: WATCH Forcing Data methodology applied to ERA-Interim reanalysis data. *Water Resources Research, 50*, 7505–7514.

Wei, W., Wang, S., Chatani, S., & Klimont, Z. (2008). Emission and speciation of non-methane volatile organic compounds from anthropogenic sources in China. *Atmospheric Environment, 42*, 4976–4988.

Wiedinmyer, C., Akagi, S., Yokelson, R., Emmons, L., Al-Saadi, J., Orlando, J., & Soja, A. (2011). The Fire INventory from NCAR (FINN): A high resolution global model to estimate the emissions from open burning. *Geoscientific Model Development, 4*(3), 625–641.

Worden, H. M., Cheng, Y., Pfister, G., Carmichael, G. R., Zhang, Q., Streets, D. G., et al. (2012). Satellite-based estimates of reduced CO and CO_2 emissions due to traffic restrictions during the 2008 Beijing Olympics. *Geophysical Research Letters, 39*. doi:10.1029/2012GL052395.

World Development Indicators. (n.d.) http://en.wikipedia.org/wiki/Historical_GDP_of_China. Accessed 20 Sept 2016.

Worldometers. (n.d.) http://www.worldometers.info/world-population/china-population. Accessed 20 Sept 2016.

Yang, Q., Wang, Y., Zhao, C., Liu, Z., Gustafson, W. I., & Shao, M. (2011). NOx emission reduction and its effects on ozone during the 2008 Olympic Games. *Environmental Science & Technology, 45*, 6404–6410.

Yue, X., Unger, N., & Zheng, Y. (2015). Distinguishing the drivers of trends in land carbon fluxes and plant volatile emissions over the past 3 decades. *Atmospheric Chemistry and Physics, 15*, 11931–11948.

Chapter 9
Source Apportionment of Tropospheric Ozone by Chemical Transport Model: From Global to City Cluster

Bin Zhu, Jinhui Gao, Xuewei Hou, Hanqing Kang, and Chen Pan

Abstract Quantifying the ozone contributions from all sources to a given location can help understanding ozone source-receptor relationships and interpreting the mechanism of distributions and variations of ozone and its precursors. In this Chapter, we introduce the methods of ozone source apportionment, with a special focus on the source apportionment of ozone by tracer tagging methods in chemical transport model (air quality model). To some extent, the tagging methods can decrease the error caused by the chemical nonlinearity in comparison to traditional perturbing emission method. Two source apportionment case studies, one with a modified global scale chemical transport model (MOZART4) used in the East Asia Pacific rim and another with a modified regional air quality model (WRF-Chem) applied to the Yangtze River Delta, china. Two tagging methods, tagging ozone production regions and tagging ozone precursors (NO_x and VOCs), will be compared and their feasibility in global and regional model will be discussed.

Keywords Ozone • Ozone precursors • Source apportionment • Chemical transport model

9.1 Introduction

Tropospheric ozone (O_3) is a major air pollutant formed in the troposphere through a series of physical processes and complex photochemical reactions (Fishman and Crutzen 1978). The formation mechanism is non-linearly related to carbon monoxide (CO), methane (CH_4), nitrogen oxide ($NO_x = NO + NO_2$), and volatile organic compounds (VOCs) with the presence of solar radiation. The surface ozone problem is often governed by local and regional emission factors, such as O_3 lifetime and transport mechanism (Li et al. 2007; Geng et al. 2011), or even by

B. Zhu (✉) • J. Gao • X. Hou • H. Kang • C. Pan
Collaborative Innovation Center on Forecast and Evaluation of Meteorological Disasters, Nanjing University of Information Science and Technology, Nanjing 210044, China
e-mail: binzhu@nuist.edu.cn; yaweihappy2004@gmail.com; houxw@nuist.ed.cn; hanqingkang@aliyun.com; chen-pan_jsnj@qq.com

I. Bouarar et al. (eds.), *Air Pollution in Eastern Asia: An Integrated Perspective*, ISSI Scientific Report Series 16, DOI 10.1007/978-3-319-59489-7_9

intercontinental long-range transport (Li et al. 2002; Sudo and Akimoto 2007; Pfister et al. 2013). From the intercontinental, continental to city cluster scales, there are still gaps in our understanding of the distributions and variations of ozone and its precursors. Emission changes in one region can alter the atmospheric components in other regions, and the feasibility of further emissions control shows important variations among different regions. Quantifying the ozone contribution from each source to a certain place or region can help in filling these gaps of ozone source-receptor relationships and interpreting the – distributions and variations of ozone and its precursors.

Since ozone has a relatively short but versatile atmospheric lifetime, its transport scales are highly variable, from between cities, to between continents. For example, in polluted urban regions where concentrations of its precursors are high, its lifetime is of the order of hours, however, in the free troposphere where concentrations of its precursors are low, its lifetime is of the order of several weeks (Stevenson et al. 2006; Young et al. 2013). If we consider the ozone precursors, with many more species of varied atmospheric lifetimes, their impact on ozone chemical production and transport scales are of great variety and complexity. In addition to its role as a priority pollutant on urban scale, ozone may influence air quality on a hemispheric scale (Akimoto 2003; Dentener et al. 2010). Long-term measurements of surface ozone and of vertical profiles by background and ozonesonde stations can help to interpret the mechanism of ozone distribution, variation, trends and origins. Long-term trends and the possible origins of surface ozone in Asia were explored based on the analysis of observations at several background stations (Naja and Akimoto 2004; Tanimoto et al. 2005; Xu et al. 2008; Cooper et al. 2014; Oltmans et al. 2006). Combining observations with back trajectory analysis is a good method to identify the air mass origins and to give the semi-quantitative contributions of ozone from diverse sources. Trajectory-based analysis of long-term ozonesonde data in Japan indicated a positive trend in boundary layer ozone from regionally polluted outflow (Lee et al. 1998), and this is attributed to the large increase in NO_x emissions over China (Naja and Akimoto 2004). However, there is evidence of intercontinental transport of ozone from Europe/North America to Sapporo, northeast of Japan. Using a residence time approach with trajectory data over central Europe, the photochemically built boundary layer ozone showed a decreasing trend, consistent with the NO_x emissions trends, however, long-term changes of ozone suggest that intercontinental transport has a significant effect on ozone levels in the lower free troposphere and boundary layer over central Europe (Naja et al. 2003).

Since measurements are lacking in many regions of the world, there are several gaps in our knowledge (Cooper et al. 2014). We therefore rely on chemical transport models in global and regional scales to not only fill in these observational knowledge gaps, but also to interpret the observations, to identify key processes and variables that determine ozone distributions and variability, and to project the relations between sources and receptors pairs in difference regions.

A number of studies have applied chemical transport models to quantify intercontinental source-receptor relationships for ozone pollution. They have applied either of two approaches: sensitivity studies perturbing the sources (Jacob et al. 1999; Yienger et al. 2000; Wild and Akimoto 2001; Fiore et al. 2009) or source attribution using ozone (or its precursors) tagging by production region (or emission region) (Wang et al. 1998; Auvray and Bey 2005; Emmons et al. 2012; Hou et al. 2014; Gao et al. 2015). Traditionally, sensitivity studies with perturbed emissions are of most policy relevance to quantify the impacts of changes in emissions scenarios, or past/future evaluation for addressing emission changes on air qualities. Most studies switch off completely the anthropogenic emissions in the source region (Berntsen et al. 1999; Fiore et al. 2002; Auvray and Bey 2005). Others increase emissions twofold or more to examine the impact of industrialization in the future (Jacob et al. 1999; Zhang et al. 2008). Wild and Akimoto (2001) investigate the effects of small increases (10%) of the base emission to mitigate non-linearity. Wu et al. (2009) indicated that estimates of O_3 contributions are significantly different when estimated by removing all of the emissions in a source region, in contrast to making small perturbations, which is due to the nonlinearity of O_3 photochemistry. Emmons et al. (2012) and Brown-Steiner and Hess (2011) compared the ozone-tagging method with the emission-perturbation method and found a large difference between the two approaches, especially near the emission regions of ozone precursors.

To some extent, the tagging methods can decrease the error caused by the non-linearity. The tagging method is a mass balance technique and it treats a tagged chemical species emitted or chemically produced in a certain region as a separate tracer and calculates its transport, chemical transform and surface deposition, exactly as its template species (see details in Sect. 9.2).

In this Chapter, we focus on the source apportionments of ozone by chemical transport model (air quality model) using tagging methods. Firstly, the tagging methodology is presented. Then we present two source apportionment studies, one with a modified global scale chemical transport model (MOZART4) used in the East Asia Pacific rim and another with a modified regional air quality model (WRF-Chem) applied to the Yangtze River Delta. Two tagging methods, tagging ozone production regions and tagging ozone precursors, will be compared and their feasibility in global and regional model will be discussed.

9.2 Introduction to Ozone Tagging in a Chemical Transport Model

As a typical secondary atmospheric pollutant, tropospheric ozone is not directly emitted but rather formed through a series of chemical reactions, involving relevant precursors (NO_x and VOCs, CO and CH_4). When the latter are emitted to the atmosphere, ozone may, or may not, be formed, dependent on the solar radiation level, the relative concentration, and lifetime of NO_x, VOCs, CO and CH_4, etc. Hence, the ozone accumulated in any region corresponds to a mixture of locally derived ozone with molecules transported from other locations, resulting in the possibility of inconsistent understanding of ozone sources.

To isolate the ozone contributions from individual source regions, generally two tagging methods were employed in chemical transport models. In the first approach, ozone is taggedbased on its production regions. Ozone was tagged with a label based on the region of its chemical production, and it is treated as a separate tracer with its transport, chemical loss and surface deposition, being calculated exactly, as is calculated in the model. Here, we follow the ozone tagging method of Sudo and Akimoto (2007) which was first introduced by Wang et al. (1998).

$$\frac{dQ_i(x,y,z)}{dt} = P_i(x,y,z) - \beta(x,y,z)Q_i(x,y,z); \tag{9.1}$$

with Q_i the mixing ratio of O_3 tagged by the region i, β the chemical loss rate constant (S^{-1}), and P_i the gross chemical production within the region i.

$$Pi(x,y,z) = \begin{cases} P(x,y,z) & : \text{ inside region } i \\ 0 & : \text{ outside region } i \end{cases}; \tag{9.2}$$

For P(x, y, z), we use chemical production of the conventionally defined odd oxygen family Ox ($= O_3 + O + O(1D) + NO_2 + 2NO_3 + 3N_2O_5 +$ PANs). In addition to the chemical tendency, the model calculates the tendency due to physical processes, e.g. transport and dry deposition at the surface for each O_3 tracer.

The second approach for ozone tagging is based on the precursor emission region. Ozone was tagged with a label based on where its precursors (NO_x and VOCs) are emitted. This means that the approach requires not only tagging ozone itself, but also tracking relevant precursors. Since the ozone formation mechanism involves both NO_x and VOCs, either can be followed for source apportionment, depending on the intended goals. For large-scale research in global models, tagging NO embodies a more appropriate choice (Emmons et al. 2012). On the other hand, tagging both NO_x and VOCs better serves the purposes of air quality regional-scale modeling.

The tagging method by Emmons et al. (2012) is based on tagging NO emissions and its resulting products, following them to the production of O_3. Synthetic tracers are added to the chemical mechanism that do not modify the original chemistry but make use of the mixing ratios and loss rates of the full, standard chemistry. Essentially a duplicate set of tracers for all compounds containing N are added to the mechanism The NO and its resulting products include: NO, NO_2, NO_3, HNO_3, HNO_4, N_2O_5, $CH_3CO_3NO_2$, $CH_3COCH_2ONO_2$, $CH_2CCH_3CO_3NO_2$, $CH_2CHCCH_3OOCH_2ONO_2$, $CH_2CCH_3CHONO_2CH_2OH$, NH_4NO_3, O, O_3. Tagged O_3 is then destroyed at the same rate as the total ozone. Emmons et al. (2012) presented the details of the mechanism and illustrated the additive equalities of the technique and comparisons to other attribution techniques. A test was also shown by Emmons et al. (2012) to illustrate that the tagging method is sound.

At city cluster and regional scales, photochemical ozone production is generally VOC-limited condition in urban air and NO_x-limited under rural and remote conditions (Kley 1997). To account for such highly non-linear behavior, one tagging method consists in tracking both ozone and its precursor emissions (VOCs and NO_x) from each geographical source region and calculating all ozone tracers in a unique simulation (Yarwood et al. 1996; Gao et al. 2016). In this method, the simulation domain is divided into several source regions. Each species of the relevant pollutants (ozone and its precursors) from each source region will be tracked as an independent compound. These tracers undergo the effects of physical processes (advection, vertical mixing and convection etc.) as in the standard simulation, but without perturbing the normal calculations. The change rate of ozone with identification i in any grid cell is treated as:

$$\frac{dQ_i}{dt} = P_i - D \bullet \frac{Q_i}{\sum Q_i}; \tag{9.3}$$

where Q_i represents the ozone concentration identified with i at this grid cell, P_i is the chemical production of ozone formed by the precursors emitted at region i, and D is the total chemical destruction of ozone in this grid cell. Different ratios of NO_x/VOCs cause the formation of tropospheric ozone in a certain region to be controlled either by NO_x–limited or VOCs-limited condition. We use the ratio of HCHO/NOy (Sillman 1995; Xie et al. 2014) to assess the occurrence of NO_x- or VOC-limited ozone production. Ozone production from region i is estimated as follows:

(a) If the ozone production process is NO_x-limited

$$P_i = PO_3 \times \frac{NO_{xi}}{\sum NO_{xi}}; \tag{9.4}$$

(b) If the ozone production process is VOC-limited

$$P_i = PO_3 \times \frac{VOC_{ij} \times kOH_j \times MIR_j}{\sum VOC_{ij} \times kOH_j \times MIR_j}; \tag{9.5}$$

$$PO_3 = k4[NO][HO_2] + k5[NO][CH_3O_2] + k6[NO][RO_2] \tag{9.6}$$

where PO_3 represents the total ozone production in each grid cell, calculated as described by Davis et al. (2003) in Eq. 9.6, and *k4*, *k5* and *k6* represent the reaction rate constants of the three reactions; NO_{xi} and VOC_{ij} are the concentrations of NO_x and VOCs from region i in that grid cell, respectively; because the reactivity of each VOC species can be different, a weighting factor base on the OH-reactivity rate of each VOC species j (kOH_j) is introduced; MIR_j is the maximum incremental reactivity factor of species j of VOCs emitted at region i (Carter 1994), corresponding to its ozone generating potential.

9.3 Case Studies of Ozone Tagging Method

In this section, we present two case studies employing the tagging methods to investigate the source and reception relations of ozone. In the first case, two tagging methods (ozone tagging by production region and ozone tagging by NO emission region) were implemented into a global chemistry model, the Model of Ozone and Related Chemical Tracers (MOZART4), and the seasonal ozone budget and spring ozone latitudinal gradient variation in the boundary layer of the Asia-Pacific region were interpreted using the modified model. In the second case, ozone tagging by both NO_x and VOCs emission region were implemented into a regional air quality model WRF-Chem, and a high surface ozone episode under frequent shifting wind conditions over the Yangtze River Delta, China was investigated using the modified model.

9.3.1 Ozone Tagging in Global Chemistry Model: Seasonal Ozone Budget and Spring Ozone Latitudinal Gradient Variation in the Boundary Layer of the Asia-Pacific Rim

9.3.1.1 Ozone Tagging Methods in MOZART4 and Used Data

MOZART is a global tropospheric CTM that has been extensively evaluated by comparing results with observations from ozonesonde, aircraft, and surface monitoring stations. The output data have been shown to simulate the concentrations of tropospheric O_3 and its precursors reasonably well (Horowitz et al. 2003; Pfister et al. 2008; Emmons et al. 2010). In this study, we used MOZART-4, which is described and evaluated in great detail in Emmons et al. (2010). MOZART-4 was run with the standard chemical mechanism in this study. The Synoz (synthetic ozone) scheme of McLinden et al. (2000) was used as an upper boundary condition for O_3 flux in the stratosphere, yielding a cross-tropopause O_3 flux of 500 Tg/year. This was driven by the NCEP/NCAR reanalysis meteorological fields, having a horizontal resolution of approximately $2.8 \times 2.8°$, with 28 vertical levels from the surface to approximate 2 hPa. The initial conditions and emissions were based on NCAR Community Data Portal (Emmons et al. 2010). The model was run with a time step of 20 min from 1 Jan 2000 to 31 Dec 2007, and the first year was discarded as spin-up. It is noted that the modeled results used for analyses are based on the mean values in the boundary layer (the six lower-most layers in the model, surface to ~2 km) and that only the surface values were used for model evaluation.

In this study, two tagging experiments (listed in Table 9.1) were performed to quantify the source of the latitudinal gradient in O_3 during the spring. To isolate the contributions from individual source regions, we used a tagging method, which was introduced by Sudo and Akimoto (2007). We classify O_3 by ten source regions as

Table 9.1 The sets of model experiments

Experiment	Year	Description
Tagging experiment	2001–2007	Tagging the O_3 source region
Tracer ID: JK-BL		BL[a]: Japan and Korean Peninsula[b]
CN-BL		BL: China[b]
NA-BL		BL: North America[c]
ER-BL		BL: Europe[c]
JK-FT		FT[d]: Japan and Korean Peninsula
CN-FT		FT: China
NA-FT		FT: North America
ER-FT		FT: Europe
Strato		Stratosphere
REST		The rest region
Tagged NO experiment	2001–2007	Tagging ozone produced from NO sources

From Hou et al. (2014)
[a]The boundary layer (BL) is defined as the six lower-most layers in the model (surface to ~2 km)
[b]The regions are displayed in Fig. 9.3b
[c]The North America region extends over 30–65°N, 50–130°W and the Europe region extends over 30–70°N, −12–60°E
[d]Free troposphere (FT) is defined here to extend to a pressure of 100 hPa

shown in Table 9.1. In this table, Strato denotes the contribution of the stratosphere (O_3S). Ozone produced in other regions is tagged as REST. Another tagging experiment quantified the source contribution from Eastern China. In this experiment, NO emissions and the resulting products are tagged, following them down to the production of O_3, as proposed by Emmons et al. (2012).

We use monthly O_3 observations from eight regional stations of the Acid Deposition Monitoring Network in East Asia (EANET) from 2001 to 2007: Cheju, Rishiri, Tappi, Sado-Seki, Oki, Yusuhara, Hedo, and Ogasawara. The descriptions of these sites can be found at http://www.eanet.asia/site/index.html. In addition, we use data from the WMO-World Data Centre for Greenhouse Gases (WDCGG, http://ds.data.jma.go.jp/gmd/wdcgg/) from 2001 to 2007, which includes the four sites of Ryori, Tae-Ahn Peninsula, Yonagunijima, and Minamitorishima. Finally, monthly data for Beijing, Tianjin, Mt. Tai, Mt. Huang, Lin'an and Hong Kong are obtained from references (Hou et al. 2014, and references therein). The coordinates of all sites is provided in Figs. 9.1 and 9.2.

9.3.1.2 Validation of Simulated O_3

Figures 9.1 and 9.2 show the comparisons of the modeled monthly O_3 with observations over Japan, the Korean Peninsula and Eastern China. Seasonal variations of O_3 are generally well reproduced by the model. The model also properly reproduces the different variability of O_3 at six sites over Eastern China (Fig. 9.2). Ozone peaks in May in Tianjin, Lin'an and Mt. Huang; in June in Beijing and

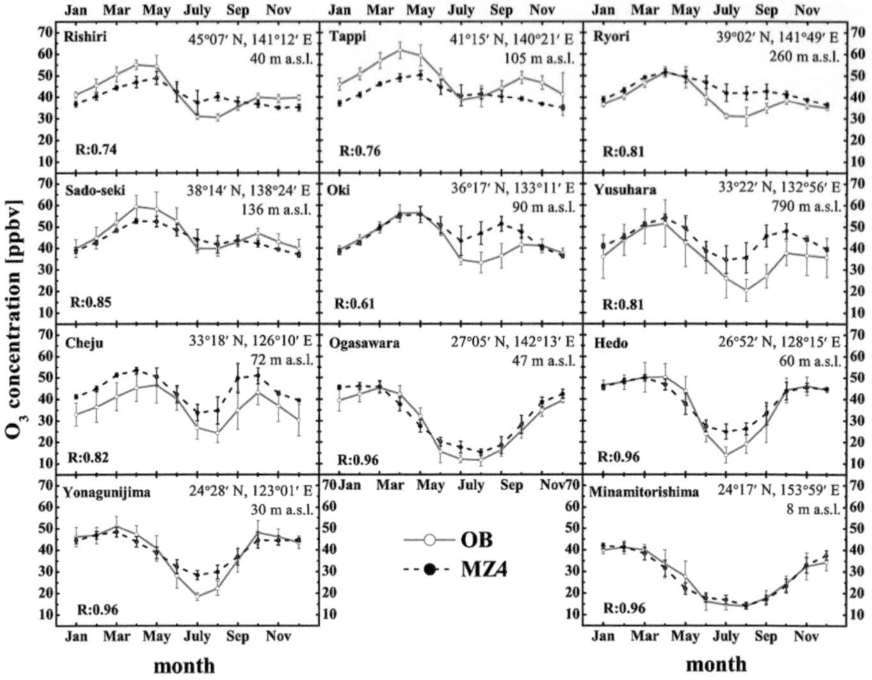

Fig. 9.1 Comparison of modeled O_3 (ppbv) (*dashed line*) and observations (*solid line*) with their mean biases at selected sites over Japan and the Korean Peninsula during 2001–2007. *R* is the correlation coefficient (From Hou et al. 2014)

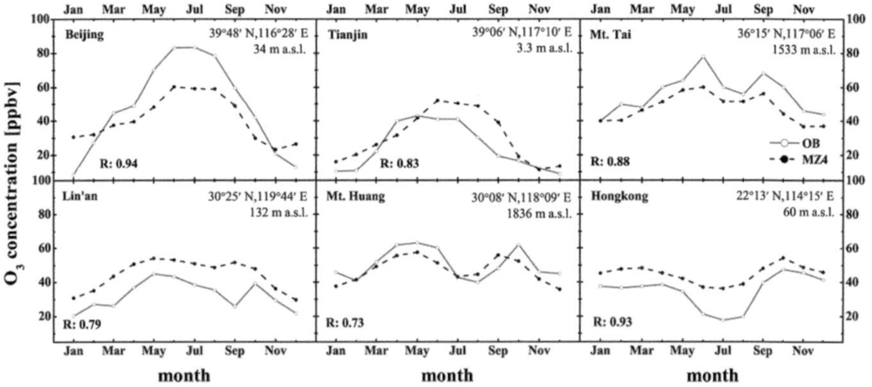

Fig. 9.2 Comparison of the modeled O_3 (ppbv; *dashed line*) and observations (*solid line*) at selected sites over Eastern China (From Hou et al. 2014)

Mt. Tai; and in autumn in Hong Kong. The O_3 values over Southeastern China (represented by Lin'an, Mt. Huang and Hong Kong) are characterized by a double-peak pattern in the late spring and autumn and have a single peak in late spring or early summer in Northeastern China (Tianjin and Beijing). The correlation coefficient (R) between the simulated and observed passed the t-test under the 95% confidence level. The high correlation of 0.96 at four oceanic sites is primarily attributed to their remote locations with less direct impact of continental emissions. Most land sites at mid latitude may be affected by local anthropogenic emissions and complex topography, which may be responsible for the relatively low correlation for mid-latitude land sites.

9.3.1.3 Seasonal Cycles and Budget Of Boundary-Layer O_3 Over the Asia-Pacific Region

Based on the seasonal O_3 patterns and the NO emission rate in East Asia, we define four regions over the Asia-Pacific as shown in Fig. 9.3. In region I (18–28°N, 122–135°E) a high correlation coefficient of 0.96 is found between observed and simulated O_3 due to a less direct effect of continental emissions. In Region II (28–46°N, 122–144°E), including Japan and the Korean Peninsula, correlation coefficients range from 0.61 to 0.85. Region III (33–43°N, 112–122°E) and Region IV (22–33°N, 112–122°E) are located in Eastern China (divided by the latitude of 33°N) and have relatively stronger emissions of anthropogenic pollutants. The four regions are all affected by the Asian monsoon. We define two regions in Eastern China due to their differences in pollutant emission rates and climate (Ding and Chan 2005).

In this section, the controlling factors of O_3concentration (transport, net photochemical production and deposition) in the four regions are discussed. Here, every term affecting the O_3 budget is evaluated quantitatively from the surface to

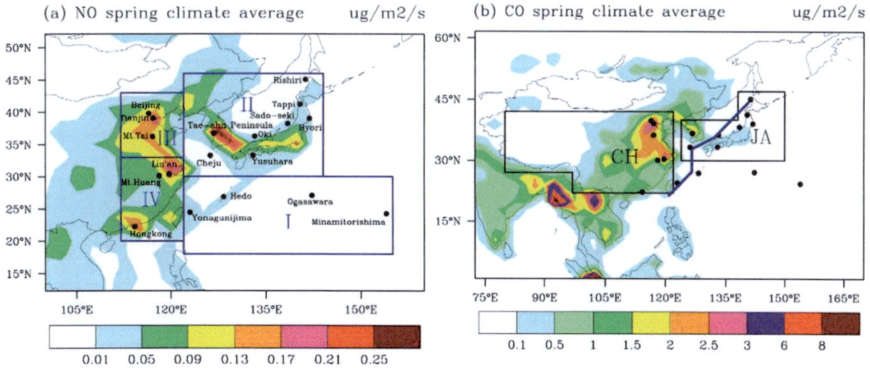

Fig. 9.3 Average emission rate (μg/m^2/s) during 2001–2007 of (**a**) NO (including anthropic, biomass burning and soil emissions) and (**b**) CO e in spring (*shaded*). The four regions for the budget analyses are indicated by *blue* boxes in (**a**). The *solid blue line* in (**b**) indicates the diagnostic transect. The *black* boxes in (**b**) mark two of the ten tagged regions. The locations of monitoring sites are indicated by *solid dots* (From Hou et al. 2014)

approximately 2 km (boundary layer) in each region. The rate of change of O_3 can be expressed as:

$$\frac{dO_3}{dt} = Chem + Adv + Con + Dif - Dep; \tag{9.7}$$

where Chem represents the net chemical production; Adv, Con and Dif are ozone tendencies due to advection, convection and diffusion, respectively; and Dep is the dry deposition rate. Figure 9.4 shows each of these components of the O_3 budget for the four regions: Net chemistry, Advection, Convection, Diffusion, and Dry deposition.

As shown in Fig. 9.4, O_3S has a February–March maximum and a summer minimum in the boundary layer over the study regions. This is similar to the results of Danielsen (1968) and Langford (1999). The seasonal variability of O_3S is inconsistent with that of O_3, except in Region I where the peaks of O_3 and O_3S

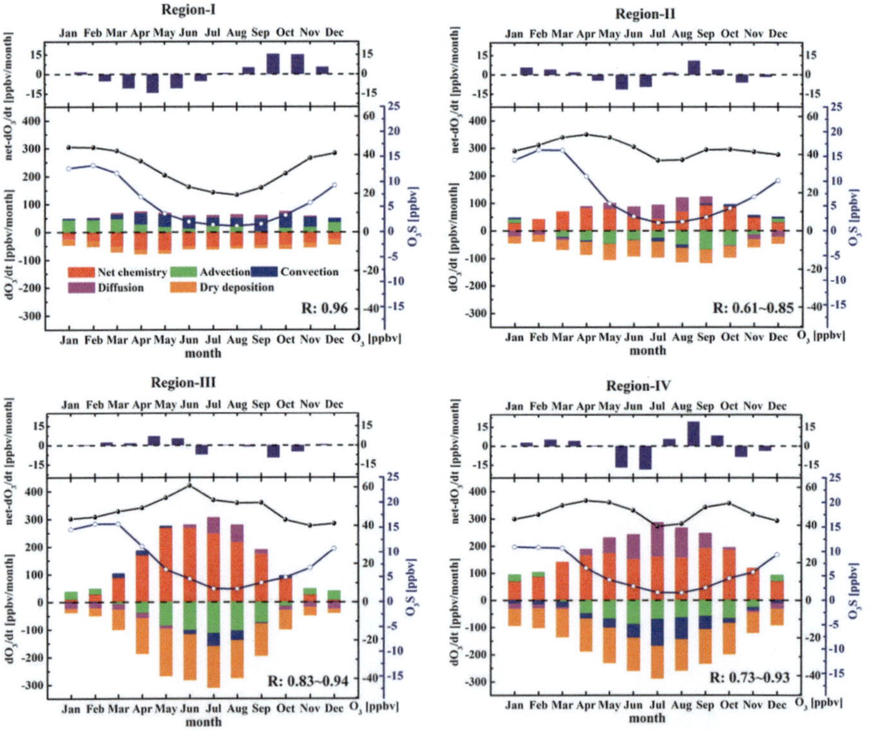

Fig. 9.4 Monthly O_3 mixing ratio (units: ppbv), O_3S (units: ppbv) and budget (units: ppbv/month) calculated by MOZART-4 over the four regions. Advection, convection, diffusion, net chemistry, and dry deposition represent the tendencies due to advection, convection, diffusion, photochemical production, and dry deposition, respectively. O_3S is the stratospheric O_3 tracer mixing ratio in the boundary layer. *Net-dO$_3$/dt* represents the net monthly rate of change in the O_3 mixing ratio between consecutive months (units: *ppbv/month*) (From Hou et al. 2014)

(30% of O_3) coincide (early spring). This result implies that the seasonal cycle of O_3 is weakly influenced by stratospheric downward transport, except in Region I during the spring.

In Region I, O_3 exhibits an early-spring maximum (approximately 44 ppbv) and a summer minimum (approximately 19 ppbv). The net chemistry, advection, vertical transport, and deposition are relatively small compared to those of the other regions. The net monthly rates of change are obviously negative from February to August (net-dO_3/dt in Fig. 9.4), in agreement with the decreasing tendency of O_3 during that period. Net chemistry shows a loss of O_3, and the rate of loss reaches its maximum in the spring. Advection, convection and diffusion have positive values with distinct seasonal variations, indicating that Region I is an O_3 inflow region. Because of the subtropical high, southeasterly wind prevails during the summer. The lower O_3 mixing ratios in the upwind remote ocean lead to weaker advection in summer in Region I. Weaker dry deposition results from lower local O_3 mixing ratios in summer. We can see that the inflow resulting from advection, convection and diffusion is the main source of O_3 in Region I, while chemistry is the sink. Advection and stratospheric contribution may play critical roles in the seasonal cycle of O_3 due to their similar seasonal variations.

In Region II, O_3 shows a seasonal pattern with two peaks at 50.51 and 43 ppbv in spring and autumn, respectively. Net chemistry is positive throughout the year and has two peaks of 80 and 110 ppbv/month in the spring and autumn, respectively. Advection is an ozone sink during the whole year, and diffusion is a net source only in summer, which reveals that Region II is generally an ozone outflow region. As such, the main factor causing O_3 seasonal cycle in Region II is photochemistry.

In Region III, O_3 has a summer maximum (61 ppbv) and a late fall-early winter minimum (40 ppbv). Net chemistry, advection and deposition are relatively large. Net chemistry reaches a maximum of 280 ppbv/month in summer, with a minimum of 5 ppbv/month in December. Northeastern China has less precipitation and is relatively drier than Region II and Region IV in the summer (Yang et al. 2003), resulting in a summer maximum of photochemistry in Region III due to lower humidity. The main cause for the summer maximum in Region III is the high level of O_3 precursors. The advection tendency is negative (approximately -105 ppbv/month) in summer due to the outflow of polluted air from Region III and the inflow of clean air from the upwind region. Thus, the O_3 seasonal cycle in Region III mainly results from photochemistry.

In Region IV, the double-peak pattern of O_3 is distinctly different from that in Region I where O_3 is low from March to November. The first peak (53 ppbv) occurs in spring. After spring, the O_3 mixing ratio begins to decrease until July, with a minimum of 39 ppbv. In autumn, there is a second peak, which is slightly lower than the springtime peak. The net chemistry presents two peaks in the spring and autumn, similar to the seasonal variation of O_3. Region IV is an O_3 outflow region as seen from the negative tendencies associated with advection, convection and diffusion during the whole year. In summer, this region is located at the bottom part of subtropical high, and the southeasterly wind prevails, leading to the transport of oceanic air masses with low levels of O_3 and O_3 precursors to Region IV, as inferred from advection. However, low levels of O_3 precursors, high humidity and cloudiness (Yang et al. 2003) lead to weaker photochemistry during the summer. Thus, the seasonal variation of O_3 is principally affected by chemistry, and the summer minimum in Region IV results from weak net chemical production and strong outflow.

9.3.1.4 Causes for the Spring Latitudinal O_3 Gradient

Tanimoto et al. (2005) indicated that exchanges between continental and maritime air masses driven by the Asian monsoon play a central role in producing O_3-latitudinal in homogeneity over East Asia, in addition to regional photochemistry. The authors also estimated that the regional build-up of O_3 over Japan is primarily due to anthropogenic emissions in Eastern China and the Korean Peninsula. As discussed in Sect. 9.3.1.3, the O_3-latitudinal gradient crosses Region I and Region II where the controlling factors of O_3 concentration differ significantly. What is the factor controlling the O_3-latitudinal gradient? What are the implications of Chinese pollution export to the latitudinal gradient? The following experiments quantified the impact of the physical and chemical processes on the O_3-latitudinal gradient by analyzing the ozone budget using tagging experiments.

Figure 9.5a shows that the spring O_3 concentration increases with latitude, reaching a maximum (52 ppbv) around 35°N, and then decreases from 35 to 45°N, in agreement with the result of Tanimoto et al. (2005). The stratospheric tracer is weakest at low latitudes (approximately 5 ppbv) and increases continuously between 22 and 32°N (to approximately 10 ppbv). It is less than one-third of the total mean O_3 mixing ratio, in accordance with the results of Monks (2000). This finding indicates that the stratospheric input is not the dominant reason leading

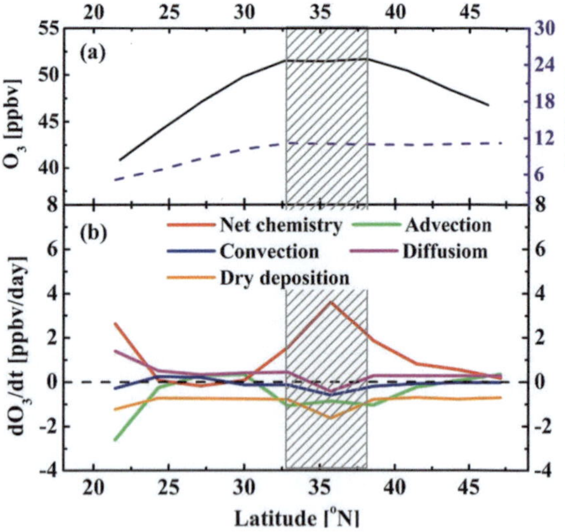

Fig. 9.5 Latitude variation of O_3 and O_3S (**a**) (units: ppbv) and budget (**b**) (units: ppbv/day) along the diagnostic transect in the spring from 2001 to 2007. The *shaded* region marks a relatively high ozone concentration (above 51 ppbv) (From Hou et al. 2014)

to the O_3-latitudinal gradient. Figure 9.5b shows that convection has little impact on the latitudinal gradient. Positive tendencies associated to net chemistry and diffusion and the negative tendency due to advection in the relatively low latitudes (20–24°N) reveal that this is a region of net O_3 outflow and that the spring maximum at 20–24°N is mainly due to net chemistry. The area between 24 and 28°N is the O_3 inflow region due to the positive roles of advection and diffusion. The net chemical tendency increases with latitude from 28°N to its peak at 35°N. The advection, convection and diffusion terms become negative (outflow of O_3), and advection plays a main role in O_3 outflow. At 42°N, the effect is reversed. It is concluded that the spring O_3-latitudinal gradient is mainly due to net chemistry, and advection makes the gradient weaker.

The results of the tagged experiments are shown in Fig. 9.6. The stratospheric contribution increases with latitude and accounts for 20% of the total at 35°N. The contribution of O_3 produced in the boundary layer over Japan and the Korean Peninsula (JK-BL) increases with latitude and reaches a peak (approximately 25%) at 35°N. The O_3 export from China's boundary layer (CN-BL) is highest in low latitudes, approximately 35%, and decreases with latitude. It is less than 15% at 35°N. Although each contribution of O_3 production over North America and Europe is relatively low, together they contribute roughly 20% of the total O_3 at 35°N. At higher latitudes, the contribution reaches 25%. The contribution of REST is relatively low at 35°N, only 14%. As such, the key contributor is CN-BL (30%) south of 32°N, and Strato and JK-BL are ranked second and third. Around 35°N, the contributions to the high- O_3 values are arranged in a decreasing order, except for REST, as follows: JK-BL (local, 25%), Strato (20%), CN-BL (15%), ER-BL (7%), ER-FT (6%), NA-BL (5%), NA-FT (4%), CN-FT (3%), and JK-FT(2%). North of 38°N, JK-BL is still the key contributor, but it slightly decreases with latitude while ER-BL, NA-BL and REST increase.

O_3 can be produced downwind from the emission regions of its precursors. The export of China's emissions leads to secondary O_3 production at mid-latitudes. Jaeglé et al. (2003) found that the influence of secondary O_3 production increased

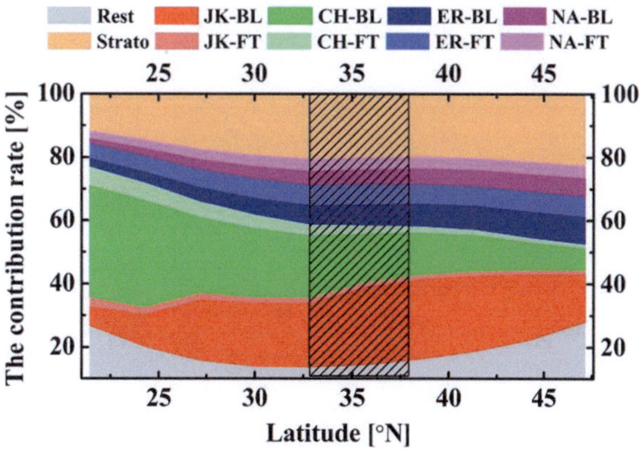

Fig. 9.6 Mean contribution (units: %) of tagged species to ozone in the boundary layer along the diagnostic transect in spring from 2001 to 2007. The abbreviations of tagged species can be found in Table 9.1 (From Hou et al. 2014)

O_3 by 2–4 ppbv on average in the Northeast Pacific for the 0–6 km column using GEOS-CHEM. The ozone precursors are obviously more abundant over China and Japan than in the Northeast Pacific (Fig. 9.3a, b). Because most of the atmosphere is NO_x-limited, we use a tagged NO and related tracers to identify the contribution of NO and its products in Eastern China to O_3 over Japan, as previously by Emmons et al. (2012). The simulation result shows that the contribution of NO and its resulting products in Eastern China decrease with latitude and account for 7.8 ppbv (15.4%) at 35°N (not shown), which is consistent with the tagged experiment in Fig. 9.6.

9.3.2 *Ozone Tagging in a Regional Chemistry Model: High Ozone Episode Under Shifting Wind Over the Yangtze River Delta, China*

9.3.2.1 Ozone Tagging Methods in WRF-Chem and Used Data

The version 3.4 of WRF-Chem model (Weather Research and Forecasting with Chemistry) was used in the study and a detailed description of this approach can be consulted in Skamarock et al. (2008). The chemistry component, fully coupled online with the WRF model, is also defined in Grell et al. (2005). The model was set up considering two domains (Figure 9.7a) with horizontal resolutions of 36 and 12 km, and 140 × 140 and 126 × 126 grid cells, for domains d01 and d02, respectively. Domain d01 covers most of East Asia and the second domain covers middle and east China, so as to address the high concentration of ozone in this region. The vertical structure of the atmosphere is represented with 38 sigma levels between the

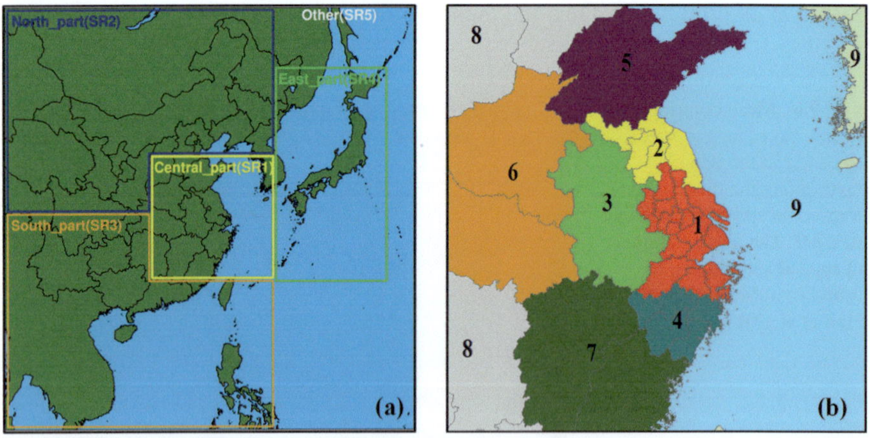

Fig. 9.7 Source regions setting for the two domains (**a**) sub-division of d01 into five source regions; (**b**) sub-division of d02 in nine source regions (From Gao et al. 2016)

surface and 50 hPa, including 12 levels within the lowest 2 km. Simulation time spanned from 1 to 27 May 2013. Initial and boundary meteorological conditions were taken from the National Centers for Environmental Prediction (NCEP) final (FNL) Operational Global Analysis data files with $1 \times 1°$ and 6 h of horizontal and temporal resolution, respectively. Global model simulations obtained from MOZART4 (Emmons et al. 2010) provided initial and spatial/temporal chemical boundary conditions. The gas-phase chemical mechanism Carbon Bond Mechanism version Z (CBM-Z) (Zaveri and Peters 1999) was used in this work.

The initial estimations for anthropogenic emissions relied on the Asian emission inventory INTEX-B 2006 (Zhang et al. 2009), with a horizontal spatial resolution of $0.5 \times 0.5°$. This database includes both primary gas phase pollutants (NO_x, SO_2, CO and VOCs) and particle matter ($PM_{2.5}$ and PM_{10}). Emission data has been modified in the past, incorporating results of several previous works. Thus, the INTEX-B data was interpolated into emissions with a horizontal resolution of $0.1 \times 0.1°$ (Zhu et al. 2015), in agreement with TRACE-P information (Street et al. 2003). In addition, the biogenic emissions generated by the Model of Emissions of Gas and Aerosols from Nature (MEGAN) (Guenther et al. 2006).

In this case study, an ozone tagging method both by NO_x and VOCs was applied to quantify contributions due to NO_x and VOC emissions in different geographical source regions to the surface ozone, over the YRD region. The tagging procedure within WRF-Chem is very similar to the Ozone Source Apportionment Technology (OSAT) (Yarwood et al. 1996) developed in the Comprehensive Air quality Model with extensions (ENVIRON 2011), while incorporating several modifications to suit the model and the chemical mechanism requirements. See details in Sect. 9.2.

Figure 9.7b depicts domain d02 divided into nine source regions: (1) Yangtze River Delta (YRD) containing 15 cities belonging to Jiangsu province, Zhejiang province and Shanghai city; (2) North of Jiangsu province (NJS), including the remaining cities of Jiangsu; (3) Anhui province (AH); (4) South of Zhejiang province (SZJ), including the cities from Zhejiang province outside the YRD region; (5) Shandong province (SD); (6) Henan and Hubei provinces (HN&HB); (7) Jiangxi and Fujian provinces (JX&FJ); (8) other areas of China in domain d02 (other1); (9) the remaining areas in domain d02 (other2) mainly including sea, part of North and South Korea, and the islands. Such setting of source regions enables analyzing YRD surface ozone source apportionment, by determining both local and remotely derived contributions. The contribution outside domain d02 was set as super region, which includes the ozone contribution from domain d01 and the chemical boundaries provided by the global model. Domain d01 was divided into five source regions (Fig. 9.7a): (1) the central part (SR1), corresponding to the area covered by domain d02; (2) the northern part (SR2) mostly comprising the northern part of China; (3) the southern part (SR3) covering the southern part of China and other areas in Southeast Asia; (4) the eastern part (SR4) including Japan and the surrounding areas; and (5) the remaining zones in domain d01 (SR5). The chemical boundaries provided by MOZART-4 also defined as an independent source region (BC).

9.3.2.2 Validation of Simulated O_3

Here, we used hourly ozone time series (18–27 May 2013) and Meteorological Information Comprehensive Analysis and Process System (MICAPS) surface meteorological (temperature, wind speed, and direction) data (1–27 May 2013), with a temporal resolution of 3 h, for assessing performance of the model system.

The simulated and observed data of temperature (°C), wind speed (m/s), and direction (°) data are compared at two sites, one is in Nanjing (NJ, 118.70°E, 32.20°N), the other one is in Kunshan (KS, 120.95°E, 31.39°N), Suzhou city, during May 2013. The model successfully captured the synoptic features, especially the shift in wind direction from west to south and then to easterly, during the measurement period. Moreover, there is a good level of agreement between measured and modeled temperature and wind speed values for both locations (not shown, see details in Gao et al. 2016). Table 9.2 displays the statistical metrics of the meteorological variables. The Index of Agreement (IOA) for wind direction (IOA_{wdir}) was calculated separately, owing to the vectorial nature of wind, following the procedure described in Kwok et al. (2010). The Mean Bias (MB, −0.47 and 0.63 at NJ and KS site, respectively), Normalized Mean Bias (NMB, −13.9 and 22.1 at NJ and KS site, respectively), and correlation coefficient (r, 0.75 and 0.72 at NJ and KS site, respectively) are shown. Regarding wind direction, the obtained IOA_{widr} were 0.96 and 0.94 at NJ and KS site, respectively. The relatively uniform gathered IOA values reflect the good agreement between model simulations and measured wind directions.

Figure 9.8 depicts comparisons between simulated and observed surface O_3 concentrations at three sites, one in Nanjing and the other two (Kunshan and Suzhou) in Suzhou city. Generally, the simulated time series for ozone concentration behave satisfactorily. The obtained correlation coefficients between simulations and observations were 0.85, 0.83 and 0.67, and IOAs are 0.91, 0.87 and 0.78, for Nanjing, Kunshan, and Suzhou, respectively. All aforementioned statistical metrics point towards a good agreement between simulation and observations.

Table 9.2 Summary of statistical comparisons between simulated and measured variables at Nanjing (NJ) and Kunshan (KS) locations. Parameters for ozone in Suzhou site are also included

Site	VAR	r	IOA	MB	RMSE	NMB(%)
Nanjing(NJ)	Ozone(ppb)	0.85	0.91	8.34	22.33	22.9
	Temp.(°C)	0.97	0.98	0.68	1.31	3.1
	Wspd(m s^{-1})	0.75	0.84	−0.47	1.19	−13.9
	Wdir(degree)	0.75	0.96	6.57	51.05	4.8
Kunshan(KS)	Ozone(ppb)	0.83	0.87	−5.77	21.81	−13.8
	Temp.(°C)	0.93	0.94	−0.71	2.13	−3.3
	Wspd(m s^{-1})	0.72	0.78	0.63	1.23	22.1
	Wdir(degree)	0.62	0.94	−12.95	62.71	−8.1
Suzhou(SZ)	Ozone(ppb)	0.67	0.78	−3.27	22.67	−9.6

From Gao et al. (2016)

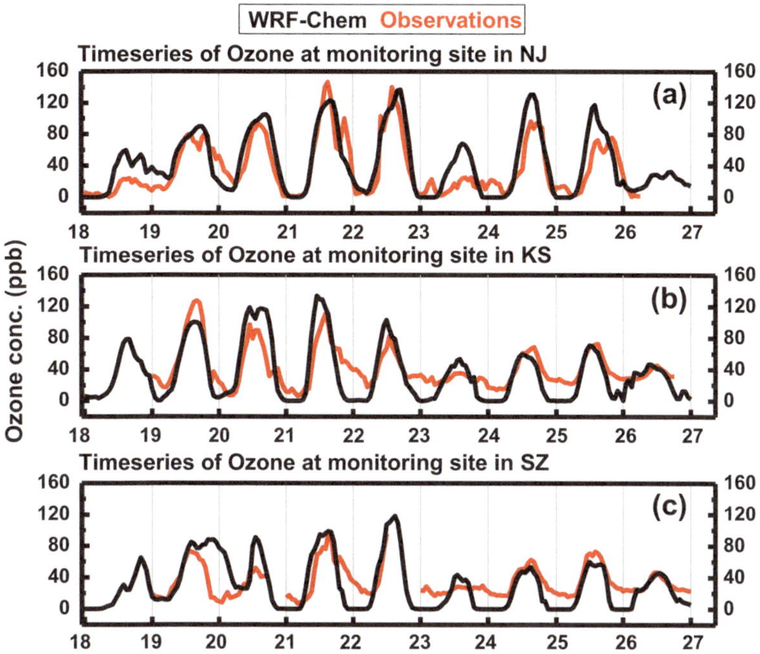

Fig. 9.8 Measured and modeled time series of ozone concentrations (ppb) in (**a**) Nanjing, (**b**) Kunshan, and (**c**) Suzhou (From Gao et al. 2016)

The MB (8.34, −5.77 and −3.27) and the NMB (22.9, −13.8 and −9.6) values reflect surface ozone over- and under-estimation in Nanjing and Suzhou, respectively.

9.3.2.3 Surface Ozone Distribution and Source Regions Over YRD

The simulation outputs from domain d02, ranging from 19 to 26 May 2013, were used to conduct ozone source apportionment analysis. We selected this time period since it include a high ozone episode and wind direction shifted from west to south, then to easterly, influencing the ozone source contribution patterns in the YRD region. The simulation was started on 1st May, such that the impact of initial conditions was dissipated in the analyzed time period (Napelenok et al. 2008).

The transport of surface ozone was highly affected by wind speed and direction. As showed in Fig. 9.9a–c, the wind direction shifted from west to south and then to easterly, during the studied period. Depending on shifting wind conditions, this time lapse can be divided into three stages: stage I from 00:00 LST (Local Standard Time) 19 to 17:00 20 May, stage II from 22:00 20 to 00:00 22 May, and stage III from 06:00 22 to 08:00 26 May, during which the YRD was affected by westerly,

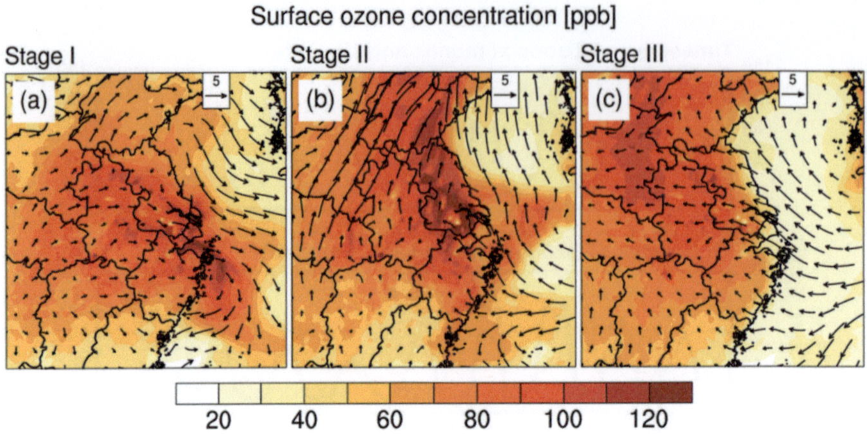

Fig. 9.9 Average surface ozone concentration at time stages (**a**) I, (**b**) II, and (**c**) III (From Gao et al. 2016)

southerly, and easterly wind, respectively. Figure 9.9a–c present the average spatial distribution of surface ozone during the three time stages.

During stage I, the YRD was affected by a westerly wind, driving air masses from western zones to the YRD region. Areas of highest concentrations (80 ppb, corresponding to Grade I level of Chinese National Ambient Air Quality Standards, is selected as a threshold for high ozone concentration determination) were distributed over Anhui, Jiangsu and Zhejiang provinces, alongside a central high value on the YRD region (Fig. 9.9a). When the wind direction changed to south (stage II), a high ozone episode occurred further expanding northwards the affected areas. When the YRD region was dominated by easterly wind (stage III), fresh air was driven from sea to mainland. The ozone content over the YRD area and other coastal regions was therefore diluted by incoming ozone-depleted air masses, and O_3 concentrations decreased to values below 80 ppb.

Upwind source regions of the YRD changed as a function of shifting wind, increasing the complexity of the ozone contribution problem in this area. The ozone tagging method can be expanded from the mere geographical identification of sources, by also including contribution quantification. In this section, the whole YRD region and two prefecture-level cities (Nanjing and Suzhou) are envisaged as ozone receptors. A high ozone episode occurred from 22 to 23 May, reflected by measurements in Nanjing.

Figure 9.10 displays O_3 contributions from different geographical source regions to the surface ozone at Nanjing, Suzhou, and the whole YRD region, during 19–26 May 2013. Nanjing stands at the west of the YRD, where ozone is easily affected by western and southern source regions, especially under influence of the westerly and southerly winds (stages I and II). When the wind direction shifted to easterly (stage III), the contribution from the YRD became more significant (Fig. 9.10a). Suzhou is in the central portion of the YRD, receiving a

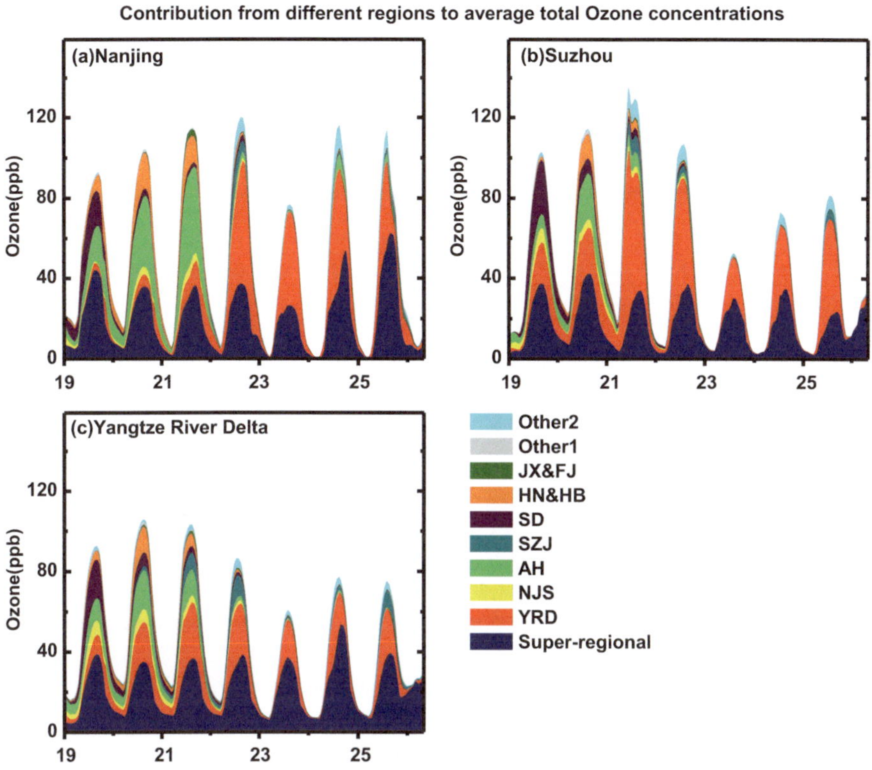

Fig. 9.10 Contributions from different source regions to the average surface ozone at the three receptors (**a**) Nanjing (**b**) Suzhou (**c**) Yangtze River Delta (From Gao et al. 2016)

correspondingly higher contribution from YRD region than in the case of Nanjing (Fig. 9.10b). The contributions to the whole YRD region were shown in Fig. 9.10c. Besides the super-region and the YRD region contributions, it was affected by AH, NJS, SZJ, SD, and HN&HB source regions. These results reflect the regional nature of the ozone problem during this period. Surface ozone in the YRD was attributed to different source regions, especially those located upwind where ozone and precursors concentrations were high. Moreover, it is worth noting the significant super-region contribution, which remained relatively stable, regardless of wind direction.

Table 9.3 shows the quantified impacts to the three receptors from the different source regions, in terms of averaged maximum daytime ozone for the three stages. During stage I, the whole YRD region was mostly affected by source regions YRD, AH, SD, and HN&HB, involving average contributions of 13.6 ppb/15.0%, 14.7 ppb/16.2%, 12.4 ppb/13.6%, and 8.2 ppb/9.0%, respectively. Throughout stage II, the contributions from YRD and SZJ became noticeably higher. Finally, the whole YRD region was under the prevailing influence of easterly wind during time stage III, with an increase in YRD region contribution up to 20.6 ppb/31.3%.

Table 9.3 Contributions from source regions to the maximum daytime averaged ozone from 10:00 to 17:00, during the three time stages mentioned in Fig. 9.9

	I: Westerly(ppb)			II: Southerly(ppb)			III: Easterly(ppb)		
Source regions	NJ	SZ	YRD	NJ	SZ	YRD	NJ	SZ	YRD
YRD	4.0	20.0	13.6	10.2	64.4	26.8	39.6	36.3	20.6
NJS	2.5	4.9	6.3	4.3	2.7	3.4	0.6	0.3	0.6
AH	23.4	14.5	14.7	42.5	6.2	12.3	3.1	0.4	0.9
SZJ	0.1	0.1	1.1	0.1	7.0	7.9	2.8	1.8	4.8
SD	11.3	16.1	12.4	2.1	3.1	3.2	0.7	0.4	0.5
HB&HN	10.8	6.5	8.2	13.9	3.7	6.0	0.7	0.2	0.4
JX&FJ	0.2	0.04	0.6	3.0	0.7	1.5	0.4	0.3	0.4
Other1	0.6	0.6	0.5	0.4	0.2	0.3	0.1	0.1	0.1
Other2	0.5	1.4	1.7	0.2	9.0	2.6	5.7	4.7	3.1
SR	34.3	33.2	31.8	29.5	26.6	32.2	35.2	25.8	34.4

From Gao et al. (2016)

Ozone source contributions varied during the study time interval, taking different values depending on location within the YRD. Nanjing received a relatively larger contribution from AH province (23.4 ppb/26.7%) in case of westerly winds. Since Suzhou is located in the central portion of the YRD, and this area received a relatively larger contribution from the YRD region (20.0 ppb/20.6%) and a smaller one from AH province (14.5 ppb/14.9%). Shandong province, in the northeast section of YRD region, contributed to both Nanjing and Suzhou in 19 May (11.3 ppb/12.9% and 16.1 ppb/16.5% in Nanjing and Suzhou, respectively). This reflects a southwest transport of ozone and precursors, following emissions in SD prior to 19 May, and subsequent transfer to the YRD when wind direction shifted from northeast to northwest. Throughout stage II, Nanjing was highly affected by AH region (42.5 ppb/40.0%), while the YRD region strongly influenced Suzhou (64.4 ppb/52.1%). When the whole YRD was under east wind conditions, both Nanjing and Suzhou received a relatively larger contribution from the YRD region of 39.6 ppb/44.5% and 36.3 ppb/51.7%, respectively. It is worth mentioning the high importance of super-region contribution to the three receptors.

The mean ozone concentrations in the three receptors expressed as the daytime maximum averaged for the whole period were 34.2 ppb (Nanjing), 28.0 ppb (Suzhou) and 33.3 ppb (YRD). For the entire YRD region, the daytime maximum mean concentrations from the super-region contribution, regarding the three time stages, were 31.8 ppb/35.0%, 32.2 ppb/33.4% and 34.4 ppb/52.3%. The small differences among the mean values suggest that the super-region contributions were less sensitive to shifting wind conditions than the source regions in domain d02. Furthermore, the super-region contribution during high ozone episode was 32.2 ppb, and 34.4 ppb of which during low ozone concentrations days. Given the little difference between the two concentrations, the super-region contribution was not a key factor to the occurrence of high ozone events in the YRD region. Increasing local and regional contributions played a more important role in the formation of such concentration peaks.

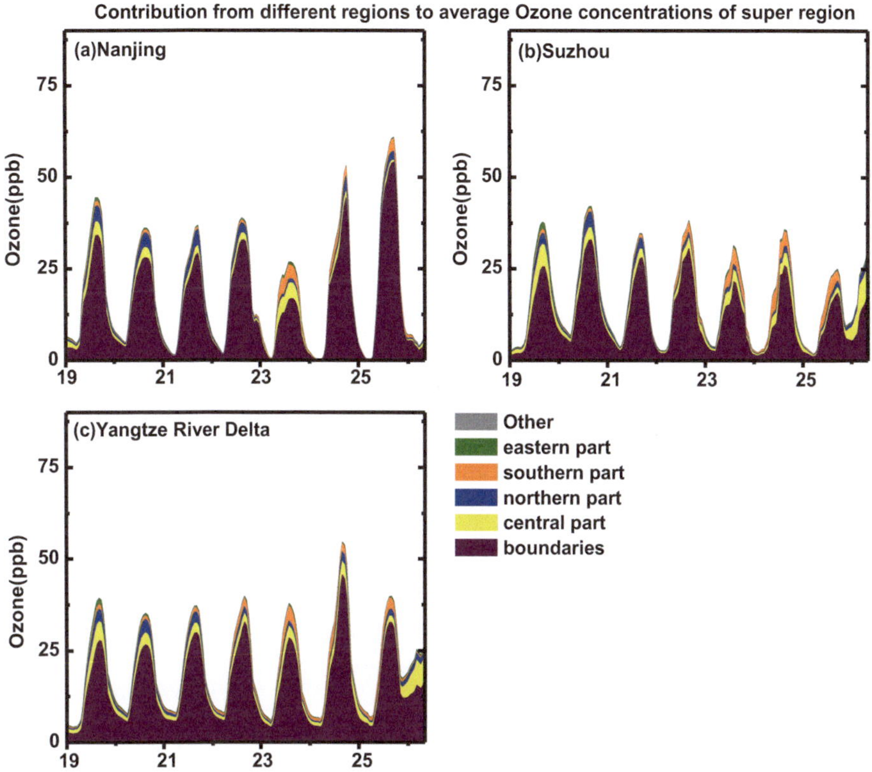

Fig. 9.11 Contributions from different super regions to the average surface ozone over the three receptors (**a**) Nanjing, (**b**) Suzhou, (**c**) Yangtze River Delta (From Gao et al. 2016)

The super-region contribution included: (i) anthropogenic and naturally emissions in domain d01; (ii) the ozone from outside the model domain d01 (BC, the chemical boundary conditions). The contributions from these source regions to the three receptors in domain d02 are shown in Fig. 9.11. Sources in domain d01 provided small contributions, but BC contributed with significant ozone to the three receptors. Table 9.4 displays the determined impacts, quantified as maximum daytime averaged ozone concentrations during the studied period. In Nanjing, the contributions from the central, northern, and southern parts accounted for 2.3, 2.4, and 1.8 ppb, respectively, in the overall averaged concentration. Contributions from other source regions (Eastern and other) were smaller (0.6 ppb and 0.1 ppb, respectively). Suzhou shows a similar profile to Nanjing, but with larger contributions from the central and southern parts (2.9 ppb and 2.5 ppb), and a smaller one from the northern. The contributions from BC to Nanjing and Suzhou were 27 and 20 ppb, respectively, and approximately three times higher than total contributions from domain d01. The large difference between these two contributions is partly related to the different ozone spatial distributions and formation mechanisms, discussed in the following section. For the whole YRD region, ozone from BC

Table 9.4 Contributions from super-region to the daytime averaged ozone maximum, from 10:00 to 17:00 during the study period

Receptors	BC (ppb)	Central (ppb)	Northern (ppb)	Southern (ppb)	Eastern (ppb)	Other (ppb)	DM1_total (ppb)
NJ	27.0	2.3	2.4	1.8	0.6	0.1	7.2
SZ	19.9	2.9	2.0	2.5	0.6	0.1	8.1
YRD	25.5	2.8	2.1	2.2	0.6	0.1	7.9

From Gao et al. (2016)

accounted for 25.5 ppb. The comparison of our results with the global background ozone study (Wang et al. 2011) reveals a high level of similarity between the BC contributions and Total Background Ozone (TBO) contributions in South China (SC) region (~30 ppb in May). The contribution to the ozone over the YRD region in domain d02, from domain d01 was ~8 ppb. The BC contribution, involving anthropogenic and biogenic sources outside China and East Asia, was an important contributor to the ozone over the YRD region in domain d02. Thus, the tagging method has the ability to clearly quantify the ozone source apportionment problem, especially when the receptors are controlled by complex wind fields.

In summary, the ozone source apportionment results showed that when the whole YRD region was under westerly wind conditions, the entire zone was highly affected by source regions of YRD, AH, SD, and HN&HB, with relative average contributions of 13.6 ppb/15.0%, 14.7 ppb/16.2%, 12.4 ppb/13.6%, and 8.2 ppb/ 9.0%, respectively. The super-region contribution was also very significant and maximum daytime mean concentrations from the super region during three time stages were 31.8 ppb/35.0%, 32.2 ppb/33.4%, and 34.4 ppb/52.3%, respectively. The results showed that surface ozone in the YRD region was attributed to different geographical source regions, especially from upwind source regions where the concentrations of ozone and precursors were high. In addition, the super-region contribution was significant and relatively stable regardless of wind direction. In high ozone events, the super-region was not the key factor, as increased local and regional contributions were the main governing factors in the formation of high ozone episodes.

9.4 Conclusion and Outlook

From scientific and application views, quantifying the ozone contributions from all sources to a given location can help understanding ozone source-receptor relationships and interpreting the mechanism of distributions and variations of ozone and its precursors. In this Chapter, we introduce the methods of ozone source apportionment, with a special focus on the source apportionment of ozone by a chemical transport model (air quality model). There are two approaches: sensitivity studies perturbing the emission sources and source attribution using the tagging of ozone (or its precursors) by production region (or emission region). To some extent, the tagging methods can decrease the error caused by the chemical nonlinearity. There are generally two main approaches to ozone tagging: by its production regions and by its precursor emission region. Regarding ozone tagging by precursor emission region, there are also two ideas: ozone tagging by NO emission region and ozone tagging by both NO_x and VOCs emission region.

We implemented two tagging approaches (ozone tagging by production region and ozone tagging by NO emission region) into the global chemistry model MOZART4. A case study was presented to illustrate the ozone tagging method and its application to ozone source apportionment over the Northwest Pacific rim using the modified MOZART4.The feasibility of the two tagging methods were compared. The ozone tagging methods can give reasonable seasonal source apportionments, and the two methods (ozone tagging by production region and ozone tagging by NO emission region) correspond very well. But in case of regions with complex chemical nonlinearities, such as the city cluster of Yangtze River Delta and more generally over East Asia, we consider that both ozone formation tagging and the tagging of its precursor (VOCs and NO_x) might provide useful information. From city cluster to regional scale, the method of ozone tagging by production region could induce great errors due to the large amount of precursors being transported among sub-regions. In another case, we employed a method of ozone tagging by both NO_x and VOCs emission region in a regional air quality model (WRF-Chem) with a focus on the Yangtze River Delta. The results showed that surface ozone in the YRD region can be attributed to different geographical source regions, especially from upwind source regions where the concentrations of ozone and precursors were high. Increased local and regional photochemical contributions were the main governing factors in the formation of high ozone episodes.

It is noteworthy that tagging method combined with process analysis (chemical production/destruction, advection, convection, vertical mixing, deposition etc.) can further unravel how ozone and its precursors are transported between different regions. Adopting a similar approach for primary and secondary aerosol tagging can be also valuable and actually has been performed by some researchers (Yarwood et al. 1996; Li et al. 2008).

Finally, the approach consisting in perturbing the emission sources remains valuable for investigating e.g. emissions scenarios, or past/future evaluation for addressing emission changes on air quality. Grewe et al. (2012) demonstrate that both methods are valid and have their area of application, but only tagging calculates contributions of emissions to concentrations, whereas the perturbation method identifies changes in the ozone concentrations due to emission changes. Emmons et al. (2012) indicate that in order to determine the effect of a change in emissions (due to policy controls or climate change), then the perturbation method is appropriate. However, the tagged ozone mechanism quantifies the contribution of a given source to a given state in the atmosphere, without any change in the chemical composition, and is useful in determining the ozone chemical budget. We think the perturbation method and tagging method could be integrated by alternate calculation, and an optimized option will be obtained for local and regional air pollution mitigation and control.

Acknowledgment This work was supported by the grants from European Union Seventh Framework Program (FP7/2007-2013, no. 606719, PANDA project), the National Natural Science Foundation of China (91544229), and from the National Key Research and Development Program of China (2016YFA0602003).

References

Akimoto, H. (2003). Global air quality and pollution. *Science, 302*(5651), 1716–1719.

Auvray, M., & Bey, I. (2005). Long-range transport to Europe: Seasonal variations and implications for the European ozone budget. *Journal of Geophysical Research, 110*, D11303. doi:10.1029/2004JD005503.

Berntsen, T. K., Karlsdo' ttir, S., & Jaffe, D. A. (1999). Influence of Asian emissions on the composition of air reaching the north western United States. *Geophysical Research Letters, 26*, 2171–2174.

Brown-Steiner, B., & Hess, P. (2011). Asian influence on surface ozone in the United States: A comparison of chemistry, seasonality, and transport mechanisms. *Journal of Geophysical Research, 116*, D17309. doi:10.1029/2011JD015846.

Carter, W. P. L. (1994). Development of ozone reactivity scales for volatile organic compounds. *Air & Waste, 44*, 881–899.

Cooper, O. R., Parrish, D., Ziemke, J., et al. (2014). Global distribution and trends of tropospheric ozone: An observation-based review. *Elementa Science of the Anthropocene, 2*.

Danielsen, E. F. (1968). Stratosphere-troposphere exchange based on radioactivity, ozone and potential vorticity. *Journal of the Atmospheric Sciences, 25*, 502–518.

Davis, D. D., Chen, G., Crawford, J. H., et al. (2003). An assessment of western North Pacific ozone photochemistry based on springtime observations from NASA's PEM-West B (1994) and TRACE-P (2001) field studies. *Journal of Geophysical Research, 108*(D21), 8829.

Dentener, F., Keating, T., & Akimoto, H. (2010). Hemispheric transport of air pollution 2010: Part A-ozone and particulate matter. In *Convention on long-range transboundary Air pollution*, United Nations.

Ding, Y., & Chan, J. C. L. (2005). The East Asian summer monsoon: An overview. *Meteorology and Atmospheric Physics, 89*, 117e142. http://dx.doi.org/10.1007/s00703-005-0125-z.

Emmons, L. K., Walters, S., Hess, P. G., et al. (2010). Description and evaluation of the model for ozone and related chemical tracers, version4 (MOZART-4). *Geoscientific Model Development, 3*(1), 43–67.

Emmons, L. K., Hess, P. G., Lamarque, J. F., & Pfister, G. G. (2012). Tagged ozone mechanism for MOZART-4CAM-chem and other chemical transport models. *Geoscientific Model Development, 5*, 1531–1542.

ENVIRON (2011), CAMx 5.4 manual. http://www.camx.com.

Fiore, A. M., Jacob, D. J., Bey, I., et al. (2002). Background ozone over the United States in summer: Origin, trend, and contribution to pollution episodes. *Journal of Geophysical Research, 107*(D15), 4275. doi:10.1029/2001JD000982.

Fiore, A. M., et al. (2009). Multi-model estimates of intercontinental source-receptor relationships for ozone pollution. *Journal of Geophysical Research, 114*, D04301. doi:10.1029/2008JD010816.

Fishman, J. & Crutzen, P. J. (1978). The origin of ozone in the troposphere. Nature, 274, 5, 9641–9668.

Gao, J., Zhu, B., Xiao, H., et al. (2015). A case study of surface ozone source apportionment during a high concentration episode, under frequent shifting wind conditions over the Yangtze River Delta, China.[J]. *Science of the Total Environment, 544*, 853–863.

Gao, J., Zhu, B., Kang, H., et al. (2016). A case study of surface ozone source apportionment during a high concentration episode, under frequent shifting wind conditions over the Yangtze River Delta, China. *Science of the Total Environment, 544*, 853–863.

Geng, F., Tie, X., Guenther, A., et al. (2011). Effect of isoprene emissions from major forests on ozone formation in the city of Shanghai, China. *Atmospheric Chemistry and Physics, 11*, 10449–10459.

Grell, G. A., Peckham, S. E., Schmitz, R., Mckeen, S. A., Frost, G., Skamarock, K., & Eder, B. (2005). Fully coupled "online" chemistry within the WRF model. *Atmospheric Environment, 39*(37), 6957–6975.

Grewe, V., Dahlmann, K., Matthes, S., et al. (2012). Attributing ozone to NO_x emissions: Implications for climate mitigation measures. *Atmospheric Environment, 59*(7), 102–107.

Guenther, A., Karl, T., Harley, P., Wiedinmyer, C., Palmer, P. I., & Geron, C. (2006). Estimates of global terrestrial isoprene emissions using MEGAN (Model of Emissions of Gases and Aerosols from Nature). *Atmospheric Chemistry and Physics, 6*, 3181–3210. doi:10.5194/acp-6-3181-2006.

Horowitz, L. W., Walters, S., Mauzerall, D. L., et al. (2003). A global simulation of tropospheric ozone and related tracers: Description and evaluation of MOZART, version 2. *Journal of Geophysical Research, 108*, 4784. http:// dx.doi.org/10.1029/2002JD002853. D2.

Hou, X. W., Zhu, B., Kang, H. Q., & Gao, J. H. (2014). Analysis of seasonal ozone budget and spring ozone latitudinal gradient variation in the boundary layer of the Asia-Pacific region. *Atmospheric Environment, 94*, 734–741.

Jacob, D. J., Logan, J. A., & Murti, P. P. (1999). Effect of rising Asian emissions on surface ozone in the United States. *Geophysical Research Letters, 26*, 2175–2178.

Jaegle, L., Jaffe, D. A., Price, H. U., et al. (2003). Sources and budgets for CO and O_3 in the northeastern Pacific during the spring of 2001: Results from the PHOBEA-II experiment. *Journal of Geophysical Research, 108*(D20), 8802. http://dx.doi.org/10.1029/2002JD003121.

Kley, D. (1997). Tropospheric chemistry and transport. *Science, 276*(5315), 1043–1044.

Kwok, R. H. F., Fung, J. C. H., Lau, A. K. H., et al. (2010). Numerical study on seasonal variations of gaseous pollutants and particulate matters in Hong Kong and Pearl River Delta Region. *Journal of Geophysical Research, 115*, D16308.

Langford, A. O. (1999). Stratosphere-troposphere exchange at the subtropical jet: Contribution to the tropospheric ozone budget at mid-latitudes. *Geophysical Research Letters, 26*(16), 2449e2452.

Lee, S. H., Akimoto, H., Nakane, H., et al. (1998). Lower tropospheric ozone trend observed in 1989–1997 at Okinawa, Japan[J]. *Geophysical Research Letters, 25*(10), 1637–1640.

Li, Q., Jacob, D. J., Bey, I., Simmonds, P. G., Oltmans, S. J., et al. (2002). Transatlantic transport of pollution and its effects on surface ozone in Europe and North America. *Journal of Geophysical Research, 107*(D13), 4166. doi:10.1029/2001JD001422.

Li, J., Wang, Z., Akimoto, H., Gao, C., Pochanart, P., and Wang, X. (2007). Modeling study of ozone seasonal cycle in lower troposphere over East Asia. *Journal of Geophysical Research, 112*(D22S25). doi:10.1029/2006JD008209.

Li, J., Wang, Z., Akimoto, H., et al. (2008). Nearground ozone source attributions and outflow in central eastern China during MTX2006. *Atmospheric Chemistry and Physics, 8*, 7335–7351.

McLinden, C. A., Olsen, S. C., Hannegan, B., Wild, O., & Prather, M. J. (2000). Stratospheric ozone in 3-D models: A simple chemistry and the cross-tropopause flux. *Journal of Geophysical Research, 105*, 14,653e–114,665.

Monks, P. S. (2000). A review of the observations and origins of the spring ozone maximum. *Atmospheric Environment, 34*, 3545–e3561.

Naja, M., & Akimoto, H. (2004). Contribution of regional pollution and long-range transport to the Asia-Pacific region: Analysis of long-term ozonesonde data over Japan. *Journal of Geophysical Research, 109*, D21306. doi:10.1029/2004JD004687.

Naja, M., Akimoto, H., & Staehelin, J. (2003). Ozone in background and photochemically aged air over central Europe: Analysis of long-term ozonesonde data from Hohenpeissenberg and Payerne. *Journal of Geophysical Research, 108*(D2), 49–65.

Napelenok, S. L., Cohan, D. S., Odman, M. T., & Tonse, S. (2008). Extension and evaluation of sensitivity analysis capabilities in a photochemical model. *Environmental Modeling and Software, 23*(8), 994–999.

Oltmans, S. J., Lefohn, A. S., Harris, J. M., et al. (2006). Long-term changes in tropospheric ozone. *Atmospheric Environment, 40*(17), 3156–3173.

Pfister, G. G., Emmons, L. K., Hess, P. G., et al. (2008). Analysis of the summer 2004 ozone budget over the United States using Intercontinental transport Experiment Ozonesonde Network Study (IONS) observations and Model of Ozone and Related Tracers (MOZART- 4) simulations. *Journal of Geophysical Research, 113*, D23306.

Pfister, G. G., Walters, S., Emmons, L. K., et al. (2013). Quantifying the contribution of inflow on surface ozone over California during summer 2008. *Journal of Geophysical Research – Atmospheres, 118*, 12,182–112299. doi:10.1002/2013JD020336.

Sillman, S. (1995). The use of NO_y, H_2O_2, and HNO_3 as indicators for ozone-NO_x-hydrocarbon sensitivity in urban locations. *Journal of Geophysical Research, 100*, 14175–14188.

Skamarock, W., Klemp, J. B., Dudhia, J., et al. (2008). A description of the advanced research WRF version 3, *NCAR technical note,* NCAR/TN/u2013475.

Stevenson, D. S., Dentener, F. J., Schultz, M. G., et al. (2006). Multimodel ensemble simulations of present-day and near-future tropospheric ozone. *Journal of Geophysical Research Atmospheres, 111*(D8), 263–269.

Street, D. G., Bond, T. C., Carmichael, G. R., et al. (2003). An inventory of gaseous and primary aerosol emissions in Asia in the year 2000. *Journal of Geophysical Research, 108*(D21), 8809. doi:10.1029/2002JD003093.

Sudo, K., & Akimoto, H. (2007). Global source attribution of tropospheric ozone: Long-range transport from various source regions. *Journal of Geophysical Research, 112*, D12302. doi:10. 1029/2006JD007992.

Tanimoto, H., Sawa, Y., Matsueda, H., et al. (2005). Significant latitudinal gradient in the surface ozone spring maximum over East Asia. *Geophysical Research Letters, 32*(L21805). doi:10. 1029/2005GL023514.

Wang, Y., Jacob, D. J., & Logan, J. A. (1998). Global simulation of tropospheric O3-NOx-hydrocarbon chemistry: 3. Origin of tropospheric ozone and effects of nonmethane hydrocarbons. *Journal of Geophysical Research, 103*, 10,757–10,767.

Wang, Y., Zhang, Y., Hao, J., & Luo, M. (2011). Seasonal and spatial variability of surface ozone over China: Contributions from background and domestic pollution. *Atmospheric Chemistry and Physics, 11*, 3511–3525. doi:10.5194/acp-11-3511-2011.

Wild, O., & Akimoto, H. (2001). Intercontinental transport of ozone and its precursors in a three dimensional globe CTM. *Journal of Geophysical Research, 106*, 27,729–27,744.

Wu, S., Duncan, B. N., Jacob, D. J., et al. (2009). Chemical nonlinearities in relating intercontinental ozone pollution to anthropogenic emissions. *Geophysical Research Letters, 36*(5), 277–291.

Xie, M., Zhu, K., Wang, T., et al. (2014). Application of photochemical indicators to evaluate ozone nonlinear chemistry and pollution control countermeasure in China. *Atmospheric Environment, 99*, 466–477.

Xu, X., Lin, W., Wang, T., et al. (2008). Long-term trend of surface ozone at a regional background station in eastern China 1991–2006: Enhanced variability. *Atmospheric Chemistry & Physics, 8*(1), 2595–2607.

Yang, J., Ding, Y., Chen, R., & Liu, L. (2003). The fluctuation of dry and wet climate boundary and its causal analysis in China. *Acta Meteorologica Sinica, 61*(3), 364e373. (in Chinese).

Yarwood, G., Morris, R. E., Yocke, M. A., et al. (1996). *Development of a methodology for source apportionment of ozone concentrations estimates from a photochemical grid model* (p. 15222). Pittsburgh: Air and Waste Management Association.

Yienger, J. J., Galanter, M., Holloway, T. A., et al. (2000). The episodic nature of air pollution transport from Asia to North America. *Journal of Geophysical Research, 105*, 26,931–26,945.

Young, P. J., Archibald, A. T., Bowman, K. W., et al. (2013). Pre-industrial to end 21st century projections of tropospheric ozone from the Atmospheric Chemistry and Climate Model Intercomparison Project (ACCMIP). *Atmospheric Chemistry & Physics, 13*(4), 2063–2090.

Zaveri, R. A., & Peters, L. K. (1999). A new lumped structure photochemical mechanism for large-scale applications. *Journal of Geophysical Research, 104*(D23), 30387–30415.

Zhang, L., et al. (2008). Transpacific transport of ozone pollution and the effect of recent Asian emission increases on air quality in North America: An integrated analysis using satellite, aircraft, ozonesonde, and surface observations. *Atmospheric Chemistry and Physics, 8*, 6117–6136.

Zhang, Q., Streets, D. G., Carmichael, G. R., et al. (2009). Asian emissions in 2006 for the NASA INTEX-B mission. *Atmospheric Chemistry and Physics, 9*, 5131–5153.

Zhu, B., Kang, H., Zhu, T., et al. (2015). Impact of Shanghai urban land surface forcing on downstream city ozone chemistry. *Journal of Geophysical Research – Atmospheres, 120*, 4340–4351. doi:10.1002/2014JD022859.

Part III
Analysis of In-Situ Measurements

Chapter 10
Real-Time Characterization of Aerosol Particle Composition During Winter High-Pollution Events in China

Qiyuan Wang, Miriam Elser, Ru-Jin Huang, Suixin Liu, Yichen Wang, Imad El Haddad, André S.H. Prévôt, and Junji Cao

Abstract China has faced severe air pollution issues in recent decades, and we deployed an aerosol mass spectrometer with a novel $PM_{2.5}$ lens to study fine particle pollutants from Beijing and Xi'an, China during winter. The average mass concentration of non-refractory (NR)-$PM_{2.5}$ was 87 $\pm$ 73 μg m^{-3} in Beijing and 262 $\pm$ 213 μg m^{-3} in Xi'an during the entire campaign periods. The NR-$PM_{2.5}$ is dominated by organic aerosol (OA) in both sites. The percent contribution of each chemical component to NR-$PM_{2.5}$ mass was rather stable at Beijing regardless of pollution levels, but secondary inorganic aerosol contributions increased by ~170% during polluted periods in Xi'an. Five components were identified by a multi-linear engine statistical factor analysis of the OA mass spectra, and in Beijing, primary aerosols from domestic coal combustion accounted for 57% and 50% of OA for non-polluted and polluted periods, respectively while biomass burning emissions dominated in Xi'an (32% of the OA for non-polluted and 42% for polluted periods). The percent contribution of oxygenated organic aerosol (OOA) to the OA mass increased from non-polluted to polluted periods, from 14 to 22% in Beijing and from 21 to 25% in Xi'an, and OOA was the second largest contributor to total OA mass in the polluted samples from both sites. Concentration-weighted trajectory analyses indicated that the central North China Plain was the most likely source

---

Q. Wang • S. Liu • Y. Wang
Key Laboratory of Aerosol Chemistry and Physics, SKLLQG, Chinese Academy of Sciences, Institute of Earth Environment, Xi'an 710061, China
e-mail: Wangqy@ieecas.cn; lsx@ieecas.cn; wangych@ieecas.cn

M. Elser • I. El Haddad • A.S.H. Prévôt (✉)
Laboratory of Atmospheric Chemistry, Paul Scherrer Institute, 5232 Villigen PSI, Switzerland
e-mail: miriam.elser@psi.ch; imad.el-haddad@psi.ch; andre.prevot@psi.ch

R.-J. Huang • J. Cao (✉)
Key Laboratory of Aerosol Chemistry and Physics, State Key Laboratory of Loess and Quaternary Geology, Institute of Earth Environment, Chinese Academy of Sciences, Xi'an 710061, China
e-mail: rujin.huang@ieecas.cn; cao@loess.llqg.ac.cn

I. Bouarar et al. (eds.), *Air Pollution in Eastern Asia: An Integrated Perspective*,
ISSI Scientific Report Series 16, DOI 10.1007/978-3-319-59489-7_10

region for the NR-$PM_{2.5}$ at Beijing, while the Guanzhong Basin and southern Shaanxi Province were the most important potential source areas for Xi'an.

Keywords Air pollution • Aerosol mass spectrometer • Organic aerosol • Source apportionment • Regional transport

10.1 Introduction

China has been facing severe air pollution problems in recent decades due to the country's rapid industrialization and urbanization (Chan and Yao 2008; Rohde and Muller 2015; Zhang and Cao 2015). Heavy pollution episodes have been all too frequent in recent winters, and these typically occur when stagnant meteorological conditions trap both gas-phase and particulate pollutants near ground level (Guo et al. 2014; Liang et al. 2015). Air pollution in China has received worldwide attention due to the pollutants' effects on air quality, visibility, climate, and human health (Kan et al. 2012; Xie et al. 2016). A key component of this pollution is fine particulate matter ($PM_{2.5}$, particulate matter with aerodynamic diameters $\leq$2.5 μm), which can be the result of regional transport or produced by a variety of sources, including motor vehicles, coal and biomass burning, road/soil dust, as well as secondary formation processes (Huang et al. 2014; Long et al. 2016).

A 2013 report from the Asian-development Bank suggests that only <1% of 500 cities in China could meet the air quality guidelines (10 μg m^{-3} for annual mean and 25 μg m^{-3} for 24-h mean) suggested by the World Health Organization, and seven Chinese cities are ranked among the ten most polluted cities in the world. China's State Council issued an Action Plan on Prevention and Control of Air Pollution on September 2013, with the goal of reducing $PM_{2.5}$ levels in Beijing/Tianjin/Hebei by 25%, 20% in the Yangtze River Delta (YRD), and 15% in the Pearl River Delta.

A crucial step in improving air-quality is identifying the major sources for atmospheric particulate matter and the processes that produce it. Most previous $PM_{2.5}$ studies in Chinese cities have relied on filter-based measurements (e.g., Shen et al. 2011; Tao et al. 2012; Cao et al. 2013; Wang et al. 2013a; Zhang et al. 2013; Gao et al. 2015; Ho et al. 2015). Organic matter, secondary water-soluble ions (SO_4^{2-}, NO_3^-, and NH_4^+), and mineral dust have been found to be the major components of $PM_{2.5}$ (Cao et al. 2012a; Huang et al. 2014). During haze periods, elevated concentrations of secondary water-soluble ions cause light scattering which in turn leads to visibility degradation (Cao et al. 2012b; Wang et al. 2013b). High relative humidity is an important meteorological factor involved in haze formation because this enhances water uptake by hygroscopic aerosol particles and promotes the formation of secondary aerosol particles (Sun et al. 2013a; Wang et al. 2015). Huang et al. (2014) investigated the chemical composition of $PM_{2.5}$ and its sources in four Chinese cities during January 2013, and they found that heavy pollution episodes were largely a result of secondary aerosol formation. That process accounted for 30–70% of the $PM_{2.5}$ mass and 44–71% of organic aerosol mass. Furthermore, their studies showed that on average, the contributions of secondary organic aerosol (SOA) and secondary inorganic aerosol (SIA) were relatively of similar; that is, the SOA/SIA mass ratios ranged from 0.6 to 1.4.

Although filter-based measurements do provide useful information on air pollution and haze formation, samples of this type have low time resolution (typically hours to days), and the methods are not well suited for investigating the rapid evolution of atmospheric aerosols, and they are less than optimal for distinguishing among emission sources. Furthermore, filter-based methods may suffer from sampling and/or measurement artifacts, such as losses of semi-volatile species due to evaporative processes during sampling or the sorption of gaseous species onto the collected particles or filter material (Nie et al. 2010; Chow et al. 2011; Pathak and Chan 2005).

Real-time measurements of aerosol particle composition (including organics, SO_4^{2-}, NO_3^-, NH_4^+, and Cl^-) with rapid time resolution have been conducted in China using online mass aerosol spectrometers, such as the Aerodyne High-Resolution Aerosol Mass Spectrometer (AMS) and Aerosol Chemical Speciation Monitor (ACSM) (e.g, Huang et al. 2011; Xu et al. 2014; Chen et al. 2015; Du et al. 2015; Jiang et al. 2015; Zhang et al. 2015; Elser et al. 2016; Hu et al. 2016). The contributions of several primary sources (e.g., traffic, coal combustion, biomass burning, and cooking emissions) to the organic aerosol (OA) have been determined using these instruments. For example, Sun et al. (2016) found that coal combustion emissions dominated the OA in Beijing during Dec. 2013–Jan. 2014, and on average accounted for 20% of the OA mass, followed by cooking emissions (18%), traffic (10%), and biomass burning (9%). In comparison, a higher contribution from coal combustion was found in measurements made in Beijing during Nov. 2011–Jan. 2012 (Sun et al. 2013b)–in that study, coal combustion emissions accounted for 33% of the total OA. In addition, several studies have investigated different types of oxygenated OA particles that result from secondary processes. For instance, Zhang et al. (2014) were able to distinguish between a highly-oxidized, low-volatility oxygenated organic aerosol (LV-OOA) and a less oxidized, semi-volatile oxygenated OA (SV-OOA) in Beijing in winter. On average LV-OOA accounted for 28% and SV-OOA 26% of OA mass, with higher impacts on more polluted days.

Although real-time measurements with AMS/ACSM are increasing in China, in cities such as Beijing (Han et al. 2015; Sun et al. 2016), Shanghai (Huang et al. 2012), Guangzhou (Xiao et al. 2011), Shenzhen (He et al. 2011), Nanjing (Zhang et al. 2015), Lanzhou (Xu et al. 2014), Xiamen (Yan et al. 2015), etc., those studies have only dealt with the submicron fraction (PM_1, particulate matter with aerodynamic diameters $\leq$1.0 μm). To address the air quality standards issued by China's State Council, it is necessary to take on-line measurements of the $PM_{2.5}$ composition. In this study, results of online high-resolution time-of-flight aerosol mass spectrometer (HR-ToF-AMS) measurements of the non-refractory (NR)-$PM_{2.5}$ fraction are presented and discussed for two large Chinese cities. These measurements were made during the winter and the multi-linear engine (ME-2) tool was used to identify and quantify the OA sources.

10.2 Experiments in Two Major Chinese Cities

10.2.1 Measurement Campaign

Two intensive measurement programs were conducted at the two cities as follows: (1) Xi'an–from 13 Dec. 2013 to 6 Jan. 2014 on the roof (~10 m above ground level) of the Institute of Earth Environment, Chinese Academy of Sciences building (34.23°N, 108.88°E, Fig. 10.1) and (2) Beijing–from 9 to 26 Jan. 2014 on the rooftop (~20 m above ground level) of the Institute of Remote Sensing Applications (40.00°N, 116.38°E, Fig. 10.1).

Beijing, the capital of China, is located in an area bordering the North China Plain, and it is surrounded by the Yanshan Mountains to the north and Taihang Mountains to the west. Beijing is one of the largest and most populated cities in the

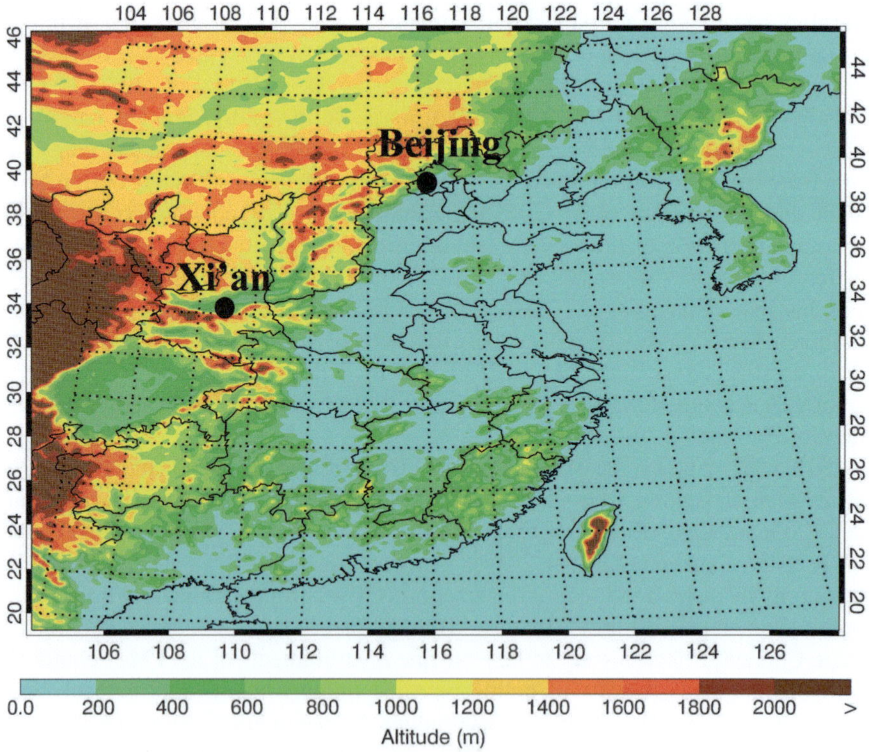

Fig. 10.1 Location of the Beijing and Xi'an sampling sites

world, and in 2013 it had a population of >20 million. As a result of rapid economic development, population expansion, and urbanization over the past decades, Beijing has experiencing severe $PM_{2.5}$ pollution (Tie et al. 2015; Yu et al. 2016). A study in Beijing by Zhang et al. (2013) showed that the annual mean concentration of $PM_{2.5}$ from Apr. 2009 to Jan. 2010 was 135 ± 63 μg m^{-3}. The largest contributor to $PM_{2.5}$ mass was secondary inorganic aerosol (26%), followed by industrial pollution (25%), coal combustion (18%), soil dust (15%), biomass burning (12%), and the combination of traffic and waste incineration emissions (4%). Air pollution in Beijing is not only a local issue but also a regional problem due to the mixing of pollutants from both local and surrounding areas (Li et al. 2015; Chen et al. 2016; Wang et al. 2016). The sampling site in Beijing was surrounded by educational, commercial, and residential districts with no major industrial activities.

Xi'an, located on the Guanzhong Plain at the southern edge of the Loess Plateau, is the largest city in Northwest China, and in 2013 it was the home to over eight million inhabitants. Due to the rapid urbanization over the past several decades, Xi'an often suffers from high $PM_{2.5}$ loadings, especially in winter (Shen et al. 2009; Zhang et al. 2011). Numerous severe pollution episodes occurred in the winter of 2012–2013. They were among the worst in the country (http://datacenter.mep.gov.cn). In addition to the local anthropogenic emissions, the region is often affected by the transport of dust particles from the Gobi and other arid and semi-arid regions (Wang et al. 2014). The sampling site was located in the high-tech area of the city, ~8 km southwest of downtown, which was surrounded by a residential/commercial area. Nearby streets were watered sporadically during heavy pollution periods to minimize the resuspension of fugitive dust.

10.2.2 Instrumental Analyses

An Aerodyne HR-ToF-AMS (Aerodyne Research, Inc., Billerica, MA, USA) was used to determine the concentrations of a suite of aerosol constituents, including OA, SO_4^{2-}, NO_3^-, NH_4^+, and Cl^-. The HR-ToF-AMS has three main sections (1) the particle inlet, (2) the particle sizing chamber, and (3) the particle composition detection unit. The operation of the HR-ToF-AMS has been described in detail elsewhere (Jayne et al. 2000; DeCarlo et al. 2006; Canagaratna et al. 2007). In contrast to the PM_1 lens that has been used most often with the instrument, we equipped it with a recently developed aerodynamic lens, which extends the measured particle size to the $PM_{2.5}$ fraction. This lens has been well characterized by Williams et al. (2013), and it efficiently transmits particles from 80 nm up to at least 3 μm in vacuum aerodynamic diameter. The atmospheric particles were dried with a Nafion® dryer (MD-110-48S; Perma Pure, Inc., Lakewood, NJ, USA) before entering the HR-ToF-AMS to reduce uncertainties in the bounce-related collection efficiency (CE_b) and possible transmission losses of large particles at high humidity.

The AMS data were analyzed with the use of Igor Pro 6.3 (WaveMetrics) and the standard ToF-AMS Data Analysis toolkit (SQUIRREL version 1.52L and PIKA version 1.11L). Standard relative ionization efficiencies (RIE) of 1.4, 1.1, and 1.3 were assumed for OA, NO_3^- and Cl^-, respectively, and the experimentally determined RIEs for SO_4^{2-} and NH_4^+ were 1.48 and 3.37, respectively. A composition-dependent collection efficiency algorithm developed by Middlebrook et al. (2012) was used to correct CE_b in the calculation of ambient mass concentrations. The source apportionment was performed on the organic AMS data using positive matrix factorization (PMF) as implemented by the multilinear engine (ME-2; Paatero 1997) and controlled via the interface SoFi code in Igor WaveMetrics (Source Finder; Canonaco et al. 2013).

10.3 Characterization of NR-$PM_{2.5}$ Chemical Composition in Two Major Chinese Cities in Winter

10.3.1 NR-$PM_{2.5}$ Mass Loadings

Figure 10.2 is a time-series of the NR-$PM_{2.5}$ mass concentrations (defined as the sums of the HR-ToF-AMS measured OA, SO_4^{2-}, NO_3^-, NH_4^+, and Cl^- masses) during the studies in Beijing and Xi'an. The hourly NR-$PM_{2.5}$ concentrations varied dramatically at both sites, with ranges of 3–324 μg m^{-3} for Beijing and 11–1070 μg m^{-3} for Xi'an. The NR-$PM_{2.5}$ concentrations for the entire observation period in Beijing showed an arithmetic mean and standard deviation of 87 ± 73 μg m^{-3}, which was ~15% higher than the grade II level of the China National Ambient Air Quality Standard (NAAQS) for daily $PM_{2.5}$ mass concentrations (75 μg m^{-3}, GB3095-2012). Based on the NAAQS, we defined polluted periods as times

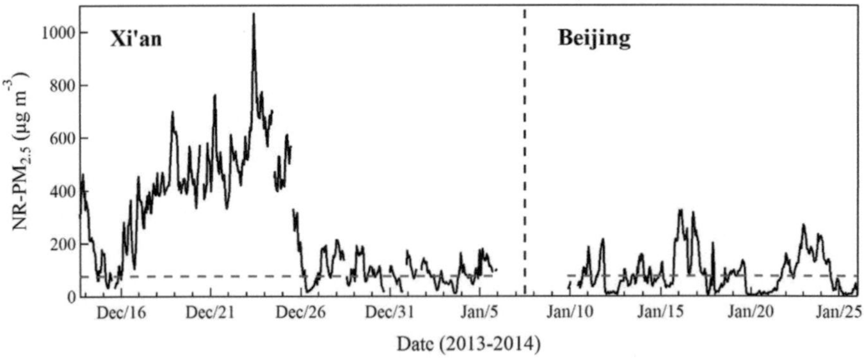

Fig. 10.2 Time-series plot of hourly-averaged NR-$PM_{2.5}$ mass concentrations during the entire campaign periods in Xi'an and Beijing. The horizontal *dashed red line* shows the Grade II standard value for daily $PM_{2.5}$ concentration (75 μg m^{-3}) promulgated as the China National Ambient Air Quality Standard (GB3095-2012)

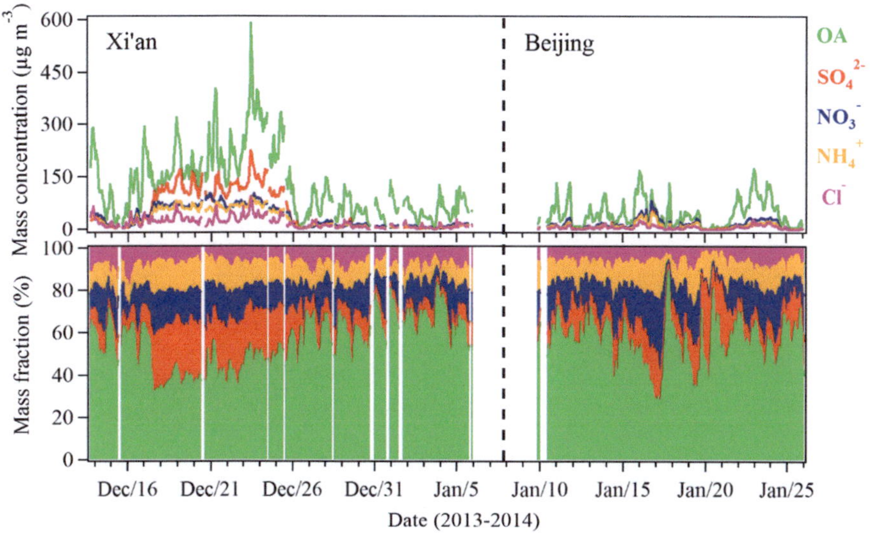

Fig. 10.3 Time-series of mass concentrations of NR-$PM_{2.5}$ composition and mass fraction of each species in NR-$PM_{2.5}$ for the entire campaign periods in Xi'an and Beijing

when the hourly $PM_{2.5}$ >75 μg m^{-3}, while all other times were classified as non-polluted periods. The average mass concentration of NR-$PM_{2.5}$ during polluted periods was 148 μg m^{-3}, which was higher than that during non-polluted periods (33 μg m^{-3}) by a factor of 4.5.

In comparison, the grand arithmetic mean NR-$PM_{2.5}$ concentration for the entire campaign in Xi'an was 262 ± 213 μg m^{-3}, which was roughly three times higher than that in Beijing. About 80% of the hourly NR-$PM_{2.5}$ concentrations at Xi'an exceeded the secondary class of NAAQS, demonstrating serious and pervasive air pollution in this city. The average mass concentration of NR-$PM_{2.5}$ was 320 μg m^{-3} during polluted periods, which was 6.6 times higher than that during non-polluted periods (49 μg m^{-3}). An extended interval of extremely high NR-$PM_{2.5}$ loadings occurred from 17 to 26 Dec. 2013, and during that event the NR-$PM_{2.5}$ average was 446 μg m^{-3}, and the peak concentrations exceeded 1000 μg m^{-3}.

10.3.2 NR-$PM_{2.5}$ Chemical Composition

Figure 10.3 shows time-series plots of the NR-$PM_{2.5}$ chemical species for the entire study at both Xi'an and Beijing. All NR-$PM_{2.5}$ species exhibited dynamic variations in mass concentrations due to changes in source emissions, meteorological conditions (such as wind speed, relative humidity, and planetary boundary layer height), photochemical reactions, and regional transport. At Beijing, the mass concentrations of OA, SO_4^{2-}, NO_3^-, NH_4^+, and Cl^- ranged from 1–174, 1–63,

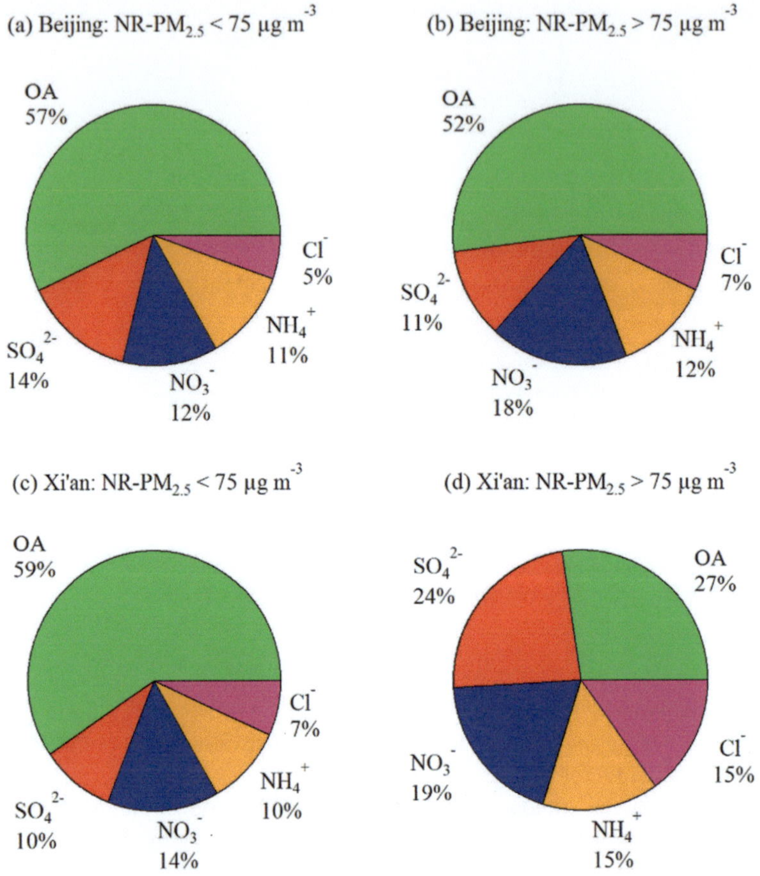

Fig. 10.4 Contributions of each chemical species to NR-$PM_{2.5}$ for non-polluted periods ($PM_{2.5}$ <75 µg m^{-3}) and polluted periods ($PM_{2.5}$ <75 µg m^{-3}) in **a**, **b** Beijing and **c**, **d** Xi'an

0–82, 1–50, and 0–29 µg m^{-3}, respectively, with corresponding average values of 46 ± 39, 10 ± 11, 15 ± 14, 10 ± 10, and 6 ± 6 µg m^{-3}. At Xi'an, the mass concentrations of OA, SO_4^{2-}, NO_3^-, NH_4^+, and Cl^- varied from 7–589, 1–224, 1–102, 1–94, and 0–93 µg m^{-3}, with averages of 128 ± 97, 51 ± 56, 35 ± 30, 30 ± 27, and 19 ± 15 µg m^{-3}, respectively.

Overall, OA was the most abundant component of the NR-$PM_{2.5}$ at both Beijing and Xi'an, with similar contributions of ~55% to the mass at the two sites (range: 27–90% for Beijing and 30–84% for Xi'an). NO_3^- was the second largest contributor to the NR-$PM_{2.5}$ at Beijing, accounting for 15% (range: 1–34%) of NR-$PM_{2.5}$, followed by SO_4^{2-} (13%, range: 3–40%), NH_4^+ (12%, range: 3–19%), and Cl^- (6%, range: 1–15%). In Xi'an, SO_4^{2-} was the second largest contributor, accounting for 15% (range: 3–33%) of NR-$PM_{2.5}$, followed by NO_3^- (13%, range: 2–24%), NH_4^+ (10%, range: 4–16%), and Cl^- (8%, range: 2–21%).

Figure 10.4 summarizes the average NR-$PM_{2.5}$ chemical composition during non-polluted and polluted periods. For Beijing, the relative contribution of each species to NR-$PM_{2.5}$ varied slightly even though each analyte increased dramatically in mass from non-polluted periods to polluted periods. For example, the mass concentration of OA increased fourfold from non-polluted periods to polluted periods (19 versus 77 μg m^{-3}), but the contribution of OA to NR-$PM_{2.5}$ only varied by ~10% (57% versus 52%). In comparison, the contributions of the chemical species to NR-$PM_{2.5}$ were more variable (~20–55%) in the samples from the different pollution conditions in Xi'an. For example, SO_4^{2-}, NO_3^-, NH_4^+, and Cl^- showed enhanced contributions to NR-$PM_{2.5}$ during polluted periods, but OA showed a decrease. Indeed, the percent contribution of SIA (sum of SO_4^{2-}, NO_3^-, and NH_4^+) to the NR-$PM_{2.5}$ mass increased from 34% during non-polluted periods to 58% during polluted periods while the OA percent contribution to NR-$PM_{2.5}$ decreased by a factor of two (59% versus 27%). This indicates that increases in the SIA were the main cause of air pollution in Xi'an.

10.3.3 Diurnal Variations of NR-$PM_{2.5}$ Mass and Chemical Composition

The diurnal variations of NR-$PM_{2.5}$ mass loadings and chemical composition in Beijing and Xi'an are shown in Fig. 10.5. The NR-$PM_{2.5}$ mass showed a pronounced diurnal cycle, with the concentrations ranging from 59 to 114 μg m^{-3} in Beijing and from 200 to 296 μg m^{-3} in Xi'an, and these variations can be explained by the combined influences of human activities and meteorological conditions. In Beijing, the NR-$PM_{2.5}$ showed a clear peak in the morning between 10:00 and 11:00 local time (LT–all time references that follow are given in LT), and that was followed by a gradual decrease to a minimum value in the afternoon around 14:00. The NR-$PM_{2.5}$ mass concentrations then increased again to a maximum value at night around 20:00 and then remained high values until 02:00. After that, the NR-$PM_{2.5}$ decreased continuously until 06:00.

The morning peak in NR-$PM_{2.5}$ in Beijing was 1–2 h later than the maximum in local rush-hour traffic, and the timing of the peak may be explained by the effect of enhanced solar radiation resulting in chemical production of secondary aerosol. As the day progressed, the depth of the planetary boundary layer increased with solar heating, and in the afternoon, the aerosol particles mixed into a larger volume leading to low aerosol loadings. After sunset, the mixing layer became shallower, a nocturnal layer formed due to radiative cooling, which trapped particles near the surface, resulting in enhanced nighttime NR-$PM_{2.5}$ loadings. The decreasing trend after midnight can be explained by reductions in anthropogenic emissions, possibly coupled with the removal of particles dry deposition.

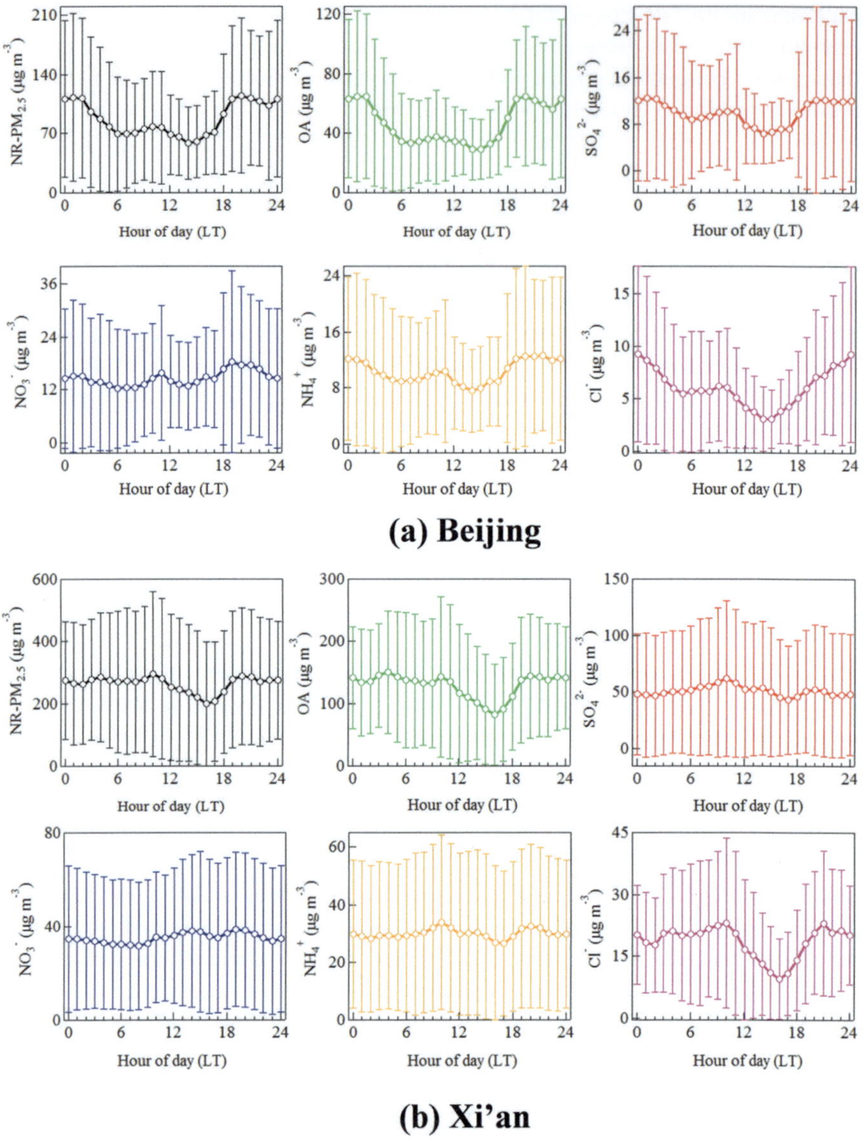

Fig. 10.5 Diurnal variations of $NR\text{-}PM_{2.5}$ and its chemical components averaged over the campaign in (**a**) Beijing and (**b**) Xi'an. *Vertical lines* indicate one standard deviation. LT represents local time

Compared with Beijing, the $NR\text{-}PM_{2.5}$ mass concentrations at Xi'an exhibited some similar trends; that is, increasing trends in the morning and after sunset, but in contrast to Beijing, they remained relatively stable during the night in Xi'an until around 07:00 in the morning. This nighttime difference between sites may be the result of enhanced liquid-phase reactions at Xi'an that lead to the formation of greater quantities of sulfate. Evidence for this can be seen in the increases in nighttime SO_4^{2-} loadings in Xi'an while decreasing trends were observed during the night in Beijing.

As shown in Fig. 10.5a, all measured aerosol species (including OA, SO_4^{2-}, NO_3^-, NH_4^+, and Cl^-) in Beijing showed diurnal cycles similar to the daily pattern in NR-$PM_{2.5}$ mass loadings; that is, increasing trends in the morning and at night as well as decreasing trends after 02:00. In Xi'an, the diurnal variations of OA, NH_4^+, and Cl^- exhibited patterns similar to NR-$PM_{2.5}$ (Fig. 10.5b). However, SO_4^{2-} showed an increasing trend after midnight and a decreasing trend in the afternoon while NO_3^- decreased after midnight and increased in the afternoon. These patterns indicate that liquid-phase reactions are especially important for SO_4^{2-} formation in Xi'an, and they suggest that photochemical reactions are important for NO_3^- formation.

10.4 Organic Aerosol Source Apportionment

The ME-2 analysis of OA spectra resolved five OA components, (1) hydrocarbon-like OA (HOA), (2) cooking OA (COA), (3) coal combustion OA (CCOA), (4) biomass burning OA (BBOA), and (5) oxygenated OA (OOA), and the mass spectra for each these components are shown in Fig. 10.6. The absolute concentrations and the relative percent contributions of the OA sources over time are shown in Fig. 10.7, and the average contributions of OA sources during non-polluted and polluted periods are shown in Fig. 10.8. The diurnal variations of the absolute concentrations of the identified five OA components are shown in Fig. 10.9.

10.4.1 HOA

The HOA profile is characterized by the presence of alkyl fragment signatures of $C_nH_{2n-1}^+$ (*m*/*z* 29, 43, 57, 71, 85, etc.) and $C_nH_{2n+1}^+$ (*m*/*z* 27, 41, 55, 69, 83, etc.), and prominent contributions of non-oxygenated species at *m*/*z* 43 ($C_3H_7^+$), *m*/*z* 55 ($C_4H_7^+$), and *m*/*z* 57 ($C_4H_9^+$) (Ng et al. 2011). Despite the larger number density of motor vehicles per km^2 in Beijing, higher mass concentrations of HOA were observed in Xi'an, possibly because of greater emissions from older cars in Xi'an. In Xi'an, HOA ranks as the third most important source for the OA mass, explaining ~15% overall, with 10% for non-polluted periods and 16% for polluted periods, and this can be explained by the accumulation of primary pollutants under stagnant conditions during polluted periods. In contrast, HOA contributes relatively little to the OA mass in Beijing and does not vary with pollution level (~9% both for non-polluted and polluted periods). The similar contributions for different levels of pollution may be due to the traffic restriction policy which aims to reduce motor vehicle emissions during polluted periods. The HOA diurnal plots show small

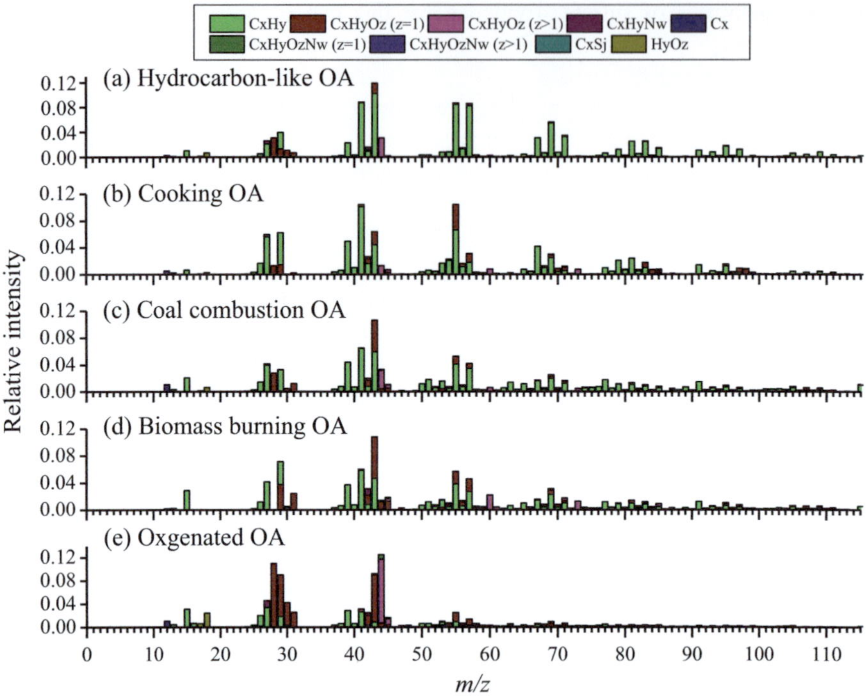

Fig. 10.6 Mass spectra of the five OA components for the entire study

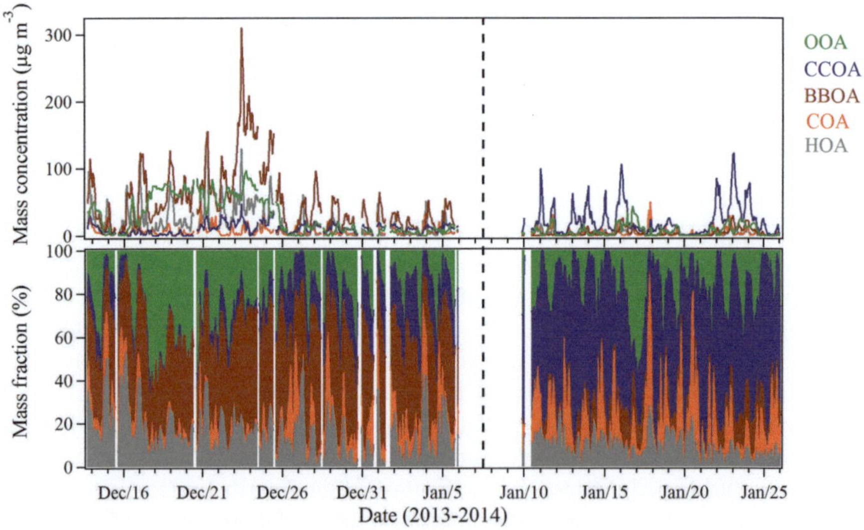

Fig. 10.7 Time series of the five identified OA sources and their relative contributions to OA

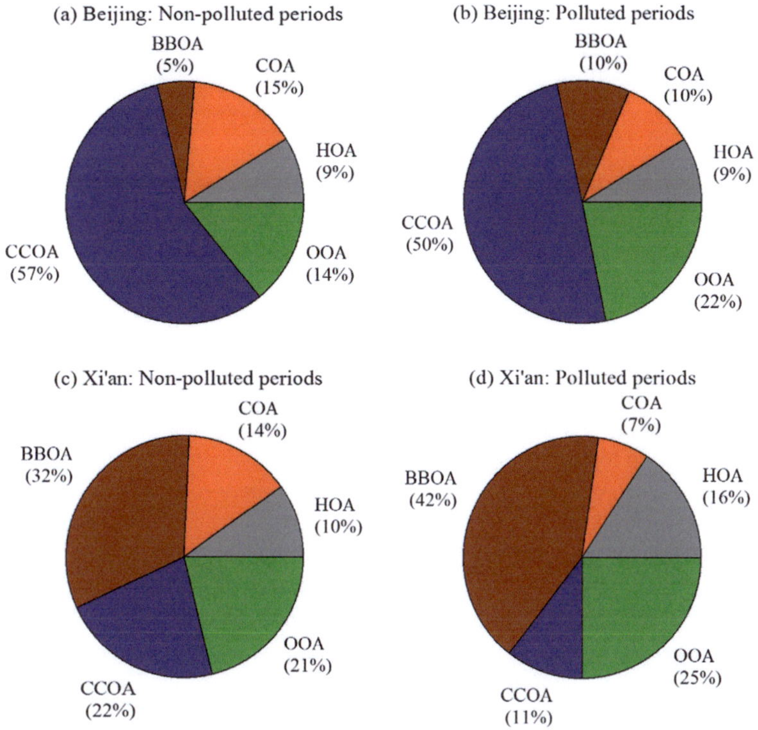

Fig. 10.8 Contributions of each chemical species to OA for non-polluted periods ($PM_{2.5} < 75\ \mu g\ m^{-3}$) and polluted periods ($PM_{2.5} < 75\ \mu g\ m^{-3}$) in (**a**, **b**) Beijing and (**c**, **d**) Xi'an

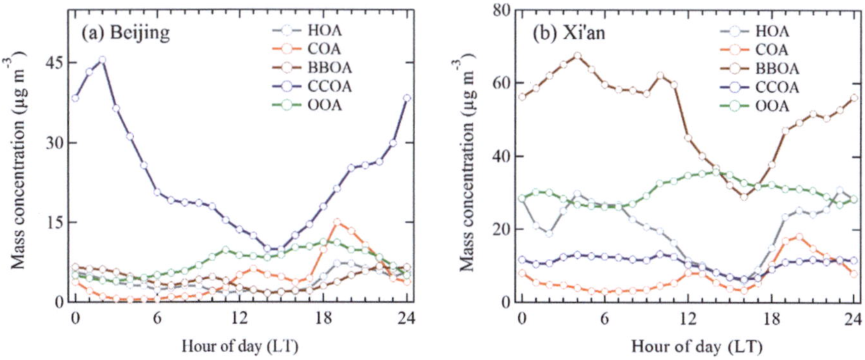

Fig. 10.9 Diurnal variations of the five identified OA sources averaged over the campaign in (**a**) Beijing and (**b**) Xi'an

peaks during morning at both locations due to the effects of rush-hour traffic. Additional increasing trends were observed in the HOA after midnight (e.g., 00:00–01:00 for Beijing and 03:00–05:00 for Xi'an). These peaks may be related to truck activity, which is enhanced during the night hours because in both cities, trucks are restricted during the day.

10.4.2 COA

The COA profile is very similar to that of HOA, but the COA spectra have higher contributions of the oxygenated ions at *m/z* 55 ($C_3H_3O^+$) and *m/z* 57 ($C_3H_5O^+$). The average COA relative percent contribution to total OA is generally low for polluted periods (10% in Beijing and 7% in Xi'an) and around 15% (15% in Beijing and 14% in Xi'an) for the non-polluted periods. COA shows similar diurnal patterns in Beijing and Xi'an, with strong peaks around noon (between 12:00 and 13:00) and evening hours (19:00–21:00). The fragment ion $C_6H_{10}O^+$ at *m/z* 98 has been suggested as a marker ion for the COA factor (Sun et al. 2011; Crippa et al. 2013), but COA correlated weakly with $C_6H_{10}O^+$. This low correlation was mostly due to enhanced concentrations of $C_6H_{10}O^+$ during the night when COA decreased to back-ground levels. Those nighttime peaks in COA are consistent with the enhancement of BBOA (see below), which explains on average ~40% of the mass of the $C_6H_{10}O^+$ fragment.

10.4.3 CCOA

The CCOA profile is characterized by unsaturated hydrocarbons (Dall'Osto et al. 2013), with greater proportions of explained variability for the unsaturated fragments at higher *m/z*. Coal burning is the largest source for OA in Beijing, on average accounting for 54% of the OA mass, with 57% and 50% contributions for non-polluted and polluted periods, respectively. In comparison, CCOA in Xi'an is of lesser importance, accounting for 13% of OA mass. Although, the relative contribution of CCOA to OA decreased by a factor of ~2 from non-polluted to polluted periods in Xi'an, the CCOA mass increased twofold during polluted periods, presumably due to the accumulation of primary emissions during the stagnant conditions. CCOA concentrations decreased substantially in the afternoon and increased dramatically at night in both cities. The decreasing trend in the afternoon can be explained by deepening of the boundary layers, which led to dilution of the atmospheric aerosols and as result lower CCOA loadings. The increasing trend at night most likely was due to enhanced anthropogenic emissions and that was exacerbated by shallow boundary layers, which led to the near-surface accumulation of pollutants. CCOA was relative stable in the early morning between 04:00 and 07:00 in Xi'an, but this OA fraction typically started to decrease at ~04:00 in Beijing.

10.4.4 BBOA

The BBOA is characterized by the prominent of signals at *m*/*z* 60 ($C_2H_4O_2^+$) and *m*/*z* 73 ($C_3H_5O_2^+$); these are known fragments produced from anhydrous sugars present in biomass smoke (Alfarra et al. 2007). Unlike Beijing, BBOA–not CCOA–is the largest contributor to OA in Xi'an, composing ~40% of the OA mass on average. The contribution of BBOA to OA increased by ~30% (from 33 to 42%) for non-polluted to polluted periods, and the BBOA mass increased by a factor of ~6 during polluted periods (10 versus 63 μg m^{-3}). In contrast, at Beijing, BBOA represented a much smaller percentage (~8%) of the total OA, averaging 5% for non-polluted periods to 10% for polluted periods. Although combustion emissions (e.g., CCOA and BBOA) from domestic heating and cooking activities dominate the organic aerosol mass at both locations, our results highlight clear differences in the types of fuels used in the two cities; that is a larger impact from coal burning in Beijing compared with a stronger influence from biomass burning in Xi'an. The diurnal cycles in BBOA loadings are similar in both cities because the dilution of the particles in the deeper daytime boundary layers causes a general decrease in the BBOA concentrations from 11:00 to 16:00 LT. Small peaks between 11:00 and 12:00 are probably related to the residential cooking activities. Peaks in BBOA starting at ~18:00 LT can be ascribed to emissions from residential heating combined with decreases in boundary layer depths.

10.4.5 OOA

The OOA profile is characterized by very high *m*/*z* 44 (CO_2^+). Sharp increases in the OOA mass concentrations were observed during polluted periods in Beijing (from 2 to 13 μg m^{-3}) and Xi'an (from 5 to 37 μg m^{-3}), and these events highlight the impact of secondary organic aerosol formation on air quality. These increases in mass also are reflected in greater contributions of OOA to total OA during the polluted periods (from 14 to 22% in Beijing and from 21 to 25% in Xi'an). In Beijing, a plot of the diurnal variations in OOA (Fig. 10.9) shows a peak around noon, which can be attributed to enhanced photochemical activity, then a decreasing trend from 13:00 to 16:00. Although strong solar heating in the afternoon can intensify chemical production of OOA, deeper boundary layers and increases in wind speeds at that time of day would tend to dilute the aerosols, and these effects can explain the lower OOA loadings observed in the afternoon. Higher concentrations of OOA were observed at night between 17:00 and 20:00, which may be associated with aqueous related SOA production at high humidity at night. In comparison, OOA showed a broad increased from 11:00 to 17:00 in Xi'an (Fig. 10.10b), which indicates that the photochemical production of OOA more than offset the dilution caused by the deepening of the daytime boundary layers.

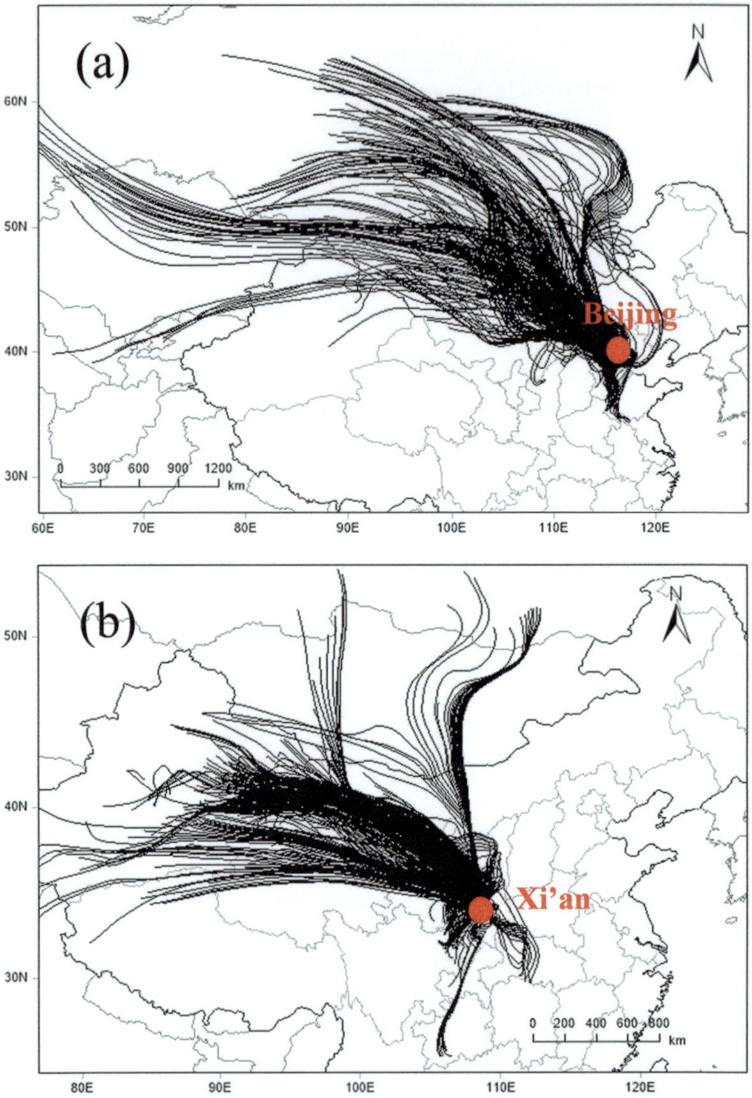

Fig. 10.10 Three-day backward air mass trajectories reaching at (**a**) Beijing and (**b**) Xi'an at 500 m above ground every hour during the entire campaign period

10.5 Regional Influences

A concentration-weighted trajectory (CWT) model was used to identify the likely regional transport source areas for the OA fractions and the NR-$PM_{2.5}$ chemical analytes sampled at Beijing and Xi'an. The CWT was based on 3-day air mass trajectories, calculated hourly and backward in time using the hybrid single-particle

Lagrangian integrated (HYSPLIT) model (Draxler and Roloh 2003). The HYSPLIT model was driven by full vertical dynamics and gridded meteorological data (Global Data Assimilation System; GDAS1, http://ready.arl.noaa.gov/HYSPLIT.php). The backward trajectories were calculated for an arrival height of 500 m above ground level, and totals of 360 and 552 trajectories were retrieved for Beijing and Xi'an, respectively (Fig. 10.10). To calculate the CWT, the entire geographic region covered by the trajectories was divided into an array of grid cells that were 0.5° latitude × 0.5 longitude (defined by the cell indices i and j). Each grid cell was assigned a residence-time weighted concentration from the aerosol sample matched with the trajectories that crossed that grid cell as follows:

$$C_{ij} = \frac{\sum_{l=1}^{M} C_l \tau_{ijl}}{\sum_{l=1}^{M} \tau_{ijl}} \tag{10.1}$$

where, C_{ij} is the average weighted-concentration in the ij^{th} cell, l is the index of the trajectory, M is the total number of trajectories, C_l is the concentration of each chemical composition observed on the arrival of trajectory l, and τ_{ijl} is the time spent in the ij^{th} cell by trajectory l. The amount of time a trajectory spent in a cell was represented by the number of trajectory segments located in that cell. A high value for C_{ij} implies that air parcels traveling over the ij^{th} cell would be, on average, associated with high concentration at the receptor with the implicit understanding that the receptor site itself is not included as a source.

Maps of the concentration-weighted trajectory model results for the major NR-$PM_{2.5}$ chemical species (e.g., SO_4^{2-}, NO_3^-, NH_4^+, and three OA species) during sampling campaign in Beijing and Xi'an are shown in Fig. 10.11. In Beijing, the regional transport source areas were similar for SO_4^{2-}, NO_3^-, NH_4^+, CCOA, BBOA, and OOA. Even though most of the trajectories to Beijing were from the northwest (see Fig. 10.10a), their effects on this suite of analytes were comparatively small. Rather, the CWT concentration gradients reveal that the most probable sources for the chemical species in NR-$PM_{2.5}$ during the Beijing campaign were in the central portion of the North China Plain, including Shijiazhuang and Baoding, where aerosol emissions are known to be high (Zhang et al. 2009).

For Xi'an, the CWT results were broadly similar for SO_4^{2-}, NO_3^-, NH_4^+, CCOA, BBOA, and OOA, and the model indicates that the most important potential source region was to the southwest of Xi'an, including an area in the west of the Guanzhong Basin where it is nearly always heavily pollution in wintertime. This is due to the strong wintertime emissions combined with the topography which tends to trap the pollutants in the area (Zhao et al. 2015). Another likely potential source region for Xi'an was found to the north area that includes the northeastern part of the Guanzhong Basin. The Guanzhong region is bordered on the south by the Qinling Mountains whose peak rise to 3000 m or more. Although these mountains impose a "blocking effect", which tends to inhibit the transport of pollutants from south, the CWT results indicate that pollutants from south may cross over the Qinling Mountains and impact Xi'an during winter.

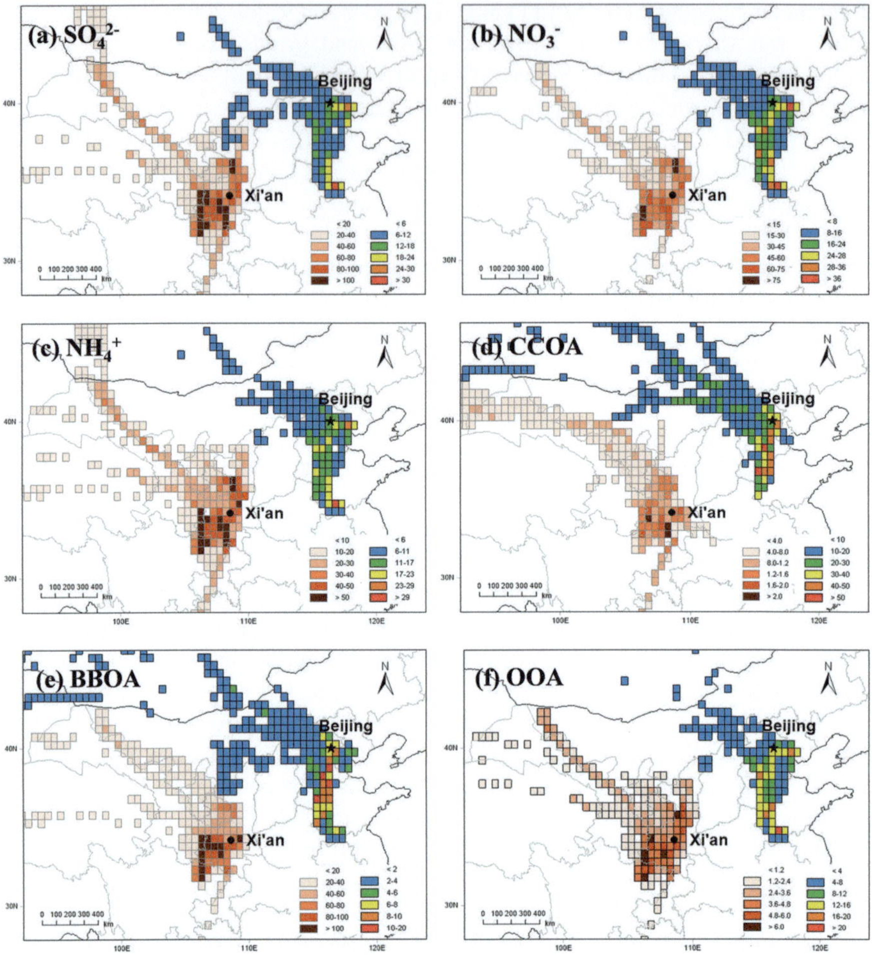

Fig. 10.11 The maps of the concentration weighted trajectory (μg m^{-3}) for the mass concentrations of (**a**) SO_4^{2-}, (**b**) NO_3^-, (**c**) NH_4^+, (**d**) CCOA, (**e**) BBOA, and (**f**) OOA during the sampling period in Beijing and Xi'an

10.6 Summary and Conclusions

The aerosol particle composition was investigated during Dec. 2013–Jan. 2014 in Xi'an and Beijing, China using an online high-resolution time-of-flight aerosol mass spectrometer (HR-ToF-AMS). The average mass concentration of non-refractory (NR)-$PM_{2.5}$ was 87 ± 73 μg m^{-3} in Beijing, which was ~15% higher than the grade II of China National Ambient Air Quality Standard

(NAAQS) for daily $PM_{2.5}$ mass (75 μg m^{-3}). In Xi'an, the average mass concentration of NR-$PM_{2.5}$ was 262 ± 213 μg m^{-3}, which was three times larger than that in Beijing, and ~80% of the hourly NR-$PM_{2.5}$ concentration exceeded the grade II of NAAQS. Similar diurnal cycles in NR-$PM_{2.5}$ was observed in both sites: the concentrations peaked during the morning rush hour, fell to minimum in the afternoon, and then rose again during the evening. Organic aerosol (OA) constituted a major fraction of NR-$PM_{2.5}$ in Beijing and Xi'an, with a similar contribution of ~55%. NO_3^- was the secondary largest contributor to NR-$PM_{2.5}$ (15%), followed by SO_4^{2-} (13%), NH_4^+ (12%), and Cl^- (6%). While in Xi'an, SO_4^{2-} was the second largest contributor to NR-$PM_{2.5}$ (15%), followed by NO_3^- (13%), NH_4^+ (10%), and Cl^- (8%). The contribution of each chemical component to NR-$PM_{2.5}$ was rather stable regardless of pollution levels in Beijing with some increases of nitrate and chloride and slight decrease of sulphate, while secondary inorganic aerosol contributions increased by ~170% during polluted periods in Xi'an. All aerosol species in Beijing present similar diurnal cycles to NR-$PM_{2.5}$ loadings, while only OA, NH_4^+, and Cl^- exhibit similar patters to NR-$PM_{2.5}$ in Xi'an. The different diurnal variations of SO_4^{2-} and NO_3^- in Xi'an indicate that liquid-phase reactions are important for SO_4^{2-} formation while photochemical reactions are important for NO_3^- formation.

Five OA components were identified by multi-linear engine (ME-2) of the OA mass spectra. In Beijing, coal combustion OA (CCOA) was the largest contributor to OA (54%) during the entire campaign period, followed by oxygenated organic aerosol (OOA: 18%), cooking OA (COA: 13%), hydrocarbon-like OA (HOA: 9%), and biomass burning OA (BBOA: 8%). Although CCOA was the largest contributor for both non-polluted and polluted periods, an increase of OOA contribution was observed, and it becomes the second largest contributor to OA mass. In Xi'an, BBOA was the largest contributor to OA (40%), followed by OOA (24%), HOA (15%), CCOA (13%), and COA (9%). A markedly increase of the BBOA contribution during polluted periods was found, suggesting the biomass burning activities in Guanzhong Basin were important factor for pollution formation in Xi'an. Concentration-weighted trajectory analyses indicate that the central North China Plain were the most likely source region for the NR-$PM_{2.5}$ chemical composition that impact Beijing, while Guanzhong Basin and Southern parts of the Shaanxi Province were the potential source areas that affect Xi'an. Those results suggest that it would be misleading to tackle urban pollution with a common approach between different cities. Instead, it is crucial to handle from observations the multitude of factors which make their differences: from emission to physical and chemical processes.

Acknowledgements This work was supported by the projects from the "Strategic Priority Research Program" of the Chinese Academy of Sciences (Grant No. XDB05000000) and National Natural Science Foundation of China (Grant No. 41403110).

References

Alfarra, M. R., Prevot, A. S. H., Szidat, S., Sandradewi, J., Weimer, S., Lanz, V. A., Schreiber, D., Mohr, M., & Baltensperger, U. (2007). Identification of the mass spectral signature of organic aerosols from wood burning emissions. *Environmental Science & Technology, 41*(16), 5770–5777.

Canagaratna, M. R., Jayne, J. T., Jimenez, J. L., Allan, J. D., Alfarra, M. R., Zhang, Q., Onasch, T. B., Drewnick, F., Coe, H., Middlebrook, A., Delia, A., Williams, L. R., Trimborn, A. M., Northway, M. J., DeCarlo, P. F., Kolb, C. E., Davidovits, P., & Worsnop, D. R. (2007). Chemical and microphysical characterization of ambient aerosols with the aerodyne aerosol mass spectrometer. *Mass Spectrometry Reviews, 26*(2), 185–222.

Canonaco, F., Crippa, M., Slowik, J. G., Baltensperger, U., & Prévôt, A. S. H. (2013). SoFi, an IGOR-based interface for the efficient use of the generalized multilinear engine (ME-2) for the source apportionment: ME-2 application to aerosol mass spectrometer data. *Atmospheric Measurement Techniques, 6*(12), 3649–3661.

Cao, J.-J., Shen, Z.-X., Chow, J. C., Watson, J. G., Lee, S.-C., Tie, X.-X., Ho, K.-F., Wang, G.-H., & Han, Y.-M. (2012a). Winter and Summer $PM_{2.5}$ Chemical Compositions in Fourteen Chinese Cities. *Journal of the Air & Waste Management Association, 62*(10), 1214–1226.

Cao, J.-J., Wang, Q.-Y., Chow, J. C., Watson, J. G., Tie, X.-X., Shen, Z.-X., Wang, P., & An, Z.-S. (2012b). Impacts of aerosol compositions on visibility impairment in Xi'an, China. *Atmospheric Environment, 59*, 559–566.

Cao, J. J., Zhu, C. S., Tie, X. X., Geng, F. H., Xu, H. M., Ho, S. S. H., Wang, G. H., Han, Y. M., & Ho, K. F. (2013). Characteristics and sources of carbonaceous aerosols from Shanghai, China. *Atmospheric Chemistry and Physics, 13*(2), 803–817.

Chan, C. K., & Yao, X. (2008). Air pollution in mega cities in China. *Atmospheric Environment, 42*(1), 1–42.

Chen, C., Sun, Y. L., Xu, W. Q., Du, W., Zhou, L. B., Han, T. T., Wang, Q. Q., Fu, P. Q., Wang, Z. F., Gao, Z. Q., Zhang, Q., & Worsnop, D. R. (2015). Characteristics and sources of submicron aerosols above the urban canopy (260 m) in Beijing, China, during the 2014 APEC summit. *Atmospheric Chemistry and Physics, 15*(22), 12879–12895.

Chen, Z., Xu, B., Cai, J., & Gao, B. (2016). Understanding temporal patterns and characteristics of air quality in Beijing: A local and regional perspective. *Atmospheric Environment, 127*, 303–315.

Chow, J. C., Watson, J. G., Robles, J., Wang, X., Chen, L.-W. A., Trimble, D. L., Kohl, S. D., Tropp, R. J., & Fung, K. K. (2011). Quality assurance and quality control for thermal/optical analysis of aerosol samples for organic and elemental carbon. *Analytical and Bioanalytical Chemistry, 401*(10), 3141–3152.

Crippa, M., El Haddad, I., Slowik, J. G., DeCarlo, P. F., Mohr, C., Heringa, M. F., Chirico, R., Marchand, N., Sciare, J., Baltensperger, U., & Prevot, A. S. H. (2013). Identification of marine and continental aerosol sources in Paris using high resolution aerosol mass spectrometry. *Journal of Geophysical Research-Atmospheres, 118*(4), 1950–1963.

Dall'Osto, M., Ovadnevaite, J., Ceburnis, D., Martin, D., Healy, R. M., O'Connor, I. P., Kourtchev, I., Sodeau, J. R., Wenger, J. C., & O'Dowd, C. (2013). Characterization of urban aerosol in Cork city (Ireland) using aerosol mass spectrometry. *Atmospheric Chemistry and Physics, 13* (9), 4997–5015.

DeCarlo, P. F., Kimmel, J. R., Trimborn, A., Northway, M. J., Jayne, J. T., Aiken, A. C., Gonin, M., Fuhrer, K., Horvath, T., Docherty, K. S., Worsnop, D. R., & Jimenez, J. L. (2006). Field-deployable, high-resolution, time-of-flight aerosol mass spectrometer. *Analytical Chemistry, 78*(24), 8281–8289.

Draxler, RR., Roloh, GD. (2003). HYSPLIT (HYbrid Single-Particle Lagrangian Integrated Trajectory) Model. *NOAA ARL READY Website*. http://www.arl.noaa.gov/ready/hysplit4.html. *NOAA Air Resources Laboratory,* Silver Spring, MD.

Du, W., Sun, Y. L., Xu, Y. S., Jiang, Q., Wang, Q. Q., Yang, W., Wang, F., Bai, Z. P., Zhao, X. D., & Yang, Y. C. (2015). Chemical characterization of submicron aerosol and particle growth events at a national background site (3295 m a.s.l.) on the Tibetan Plateau. *Atmospheric Chemistry and Physics, 15*(18), 10811–10824.

Elser, M., Huang, R. J., Wolf, R., Slowik, J. G., Wang, Q. Y., Canonaco, F., Li, G. H., Bozzetti, C., Daellenbach, K. R., Huang, Y., Zhang, R. J., Li, Z. Q., Cao, J. J., Baltensperger, U., El-Haddad, I., & Prevot, A. S. H. (2016). New insights into $PM_{2.5}$ chemical composition and sources in two major cities in China during extreme haze events using aerosol mass spectrometry. *Atmospheric Chemistry and Physics, 16*(5), 3207–3225.

Gao, J., Tian, H., Cheng, K., Lu, L., Zheng, M., Wang, S., Hao, J., Wang, K., Hua, S., Zhu, C., & Wang, Y. (2015). The variation of chemical characteristics of $PM_{2.5}$ and PM_{10} and formation causes during two haze pollution events in urban Beijing, China. *Atmospheric Environment, 107*, 1–8.

Guo, S., Hu, M., Zamora, M. L., Peng, J., Shang, D., Zheng, J., Du, Z., Wu, Z., Shao, M., Zeng, L., Molina, M. J., & Zhang, R. (2014). Elucidating severe urban haze formation in China. *Proceedings of the National Academy of Sciences of the United States of America, 111*(49), 17373–17378.

Han, T. T., Xu, W. Q., Chen, C., Liu, X. G., Wang, Q. Q., Li, J., Zhao, X. J., Du, W., Wang, Z. F., & Sun, Y. L. (2015). Chemical apportionment of aerosol optical properties during the Asia-Pacific Economic Cooperation summit in Beijing, China. *Journal of Geophysical Research-Atmospheres, 120*(23), 12281–12295.

He, L. Y., Huang, X. F., Xue, L., Hu, M., Lin, Y., Zheng, J., Zhang, R. Y., & Zhang, Y. H. (2011). Submicron aerosol analysis and organic source apportionment in an urban atmosphere in Pearl River Delta of China using high-resolution aerosol mass spectrometry. *Journal of Geophysical Research, 116*, D12304. doi:10.1029/2010jd014566.

Ho, K. F., Huang, R. J., Kawamura, K., Tachibana, E., Lee, S. C., Ho, S. S. H., Zhu, T., & Tian, L. (2015). Dicarboxylic acids, ketocarboxylic acids, α-dicarbonyls, fatty acids and benzoic acid in $PM_{2.5}$ aerosol collected during CAREBeijing-2007: An effect of traffic restriction on air quality. *Atmospheric Chemistry and Physics, 15*(6), 3111–3123.

Hu, W., Hu, M., Hu, W., Jimenez, J. L., Yuan, B., Chen, W., Wang, M., Wu, Y., Chen, C., Wang, Z., Peng, J., Zeng, L., & Shao, M. (2016). Chemical composition, sources, and aging process of submicron aerosols in Beijing: Contrast between summer and winter. *Journal of Geophysical Research, 121*(4), 1955–1977.

Huang, X. F., He, L. Y., Hu, M., Canagaratna, M. R., Kroll, J. H., Ng, N. L., Zhang, Y. H., Lin, Y., Xue, L., Sun, T. L., Liu, X. G., Shao, M., Jayne, J. T., & Worsnop, D. R. (2011). Characterization of submicron aerosols at a rural site in Pearl River Delta of China using an Aerodyne High-Resolution Aerosol Mass Spectrometer. *Atmospheric Chemistry and Physics, 11*(5), 1865–1877.

Huang, X. F., He, L. Y., Xue, L., Sun, T. L., Zeng, L. W., Gong, Z. H., Hu, M., & Zhu, T. (2012). Highly time-resolved chemical characterization of atmospheric fine particles during 2010 Shanghai World Expo. *Atmospheric Chemistry and Physics, 12*(11), 4897–4907.

Huang, R.-J., Zhang, Y., Bozzetti, C., Ho, K.-F., Cao, J.-J., Han, Y., Daellenbach, K. R., Slowik, J. G., Platt, S. M., & Canonaco, F. (2014). High secondary aerosol contribution to particulate pollution during haze events in China. *Nature, 514*(7521), 218–222.

Jayne, J. T., Leard, D. C., Zhang, X., Davidovits, P., Smith, K. A., Kolb, C. E., & Worsnop, D. R. (2000). Development of an aerosol mass spectrometer for size and composition analysis of submicron particles. *Aerosol Science and Technology, 33*(1–2), 49–70.

Jiang, Q., Sun, Y. L., Wang, Z., & Yin, Y. (2015). Aerosol composition and sources during the Chinese Spring Festival: Fireworks, secondary aerosol, and holiday effects. *Atmospheric Chemistry and Physics, 15*(11), 6023–6034.

Kan, H., Chen, R., & Tong, S. (2012). Ambient air pollution, climate change, and population health in China. *Environment International, 42*, 10–19.

Li, P., Yan, R., Yu, S., Wang, S., Liu, W., & Bao, H. (2015). Reinstate regional transport of $PM_{2.5}$ as a major cause of severe haze in Beijing. *Proceedings of the National Academy of Sciences, 112*(21), E2739–E2740.

Liang, X., Zou, T., Guo, B., Li, S., Zhang, H., Zhang, S., Huang, H., & Chen, S. X. (2015). Assessing Beijing's $PM_{2.5}$ pollution: Severity, weather impact, APEC and winter heating. *Proceedings of the Royal Society of London A: Mathematical, Physical and Engineering Sciences, 471*, 20150257.

Long, X., Tie, X., Cao, J., Huang, R., Feng, T., Li, N., Zhao, S., Tian, J., Li, G., & Zhang, Q. (2016). Impact of crop field burning and mountains on heavy haze in the North China Plain: A case study. *Atmospheric Chemistry and Physics, 16*(15), 9675–9691.

Middlebrook, A. M., Bahreini, R., Jimenez, J. L., & Canagaratna, M. R. (2012). Evaluation of composition-dependent collection efficiencies for the aerodyne aerosol mass spectrometer using field data. *Aerosol Science and Technology, 46*(3), 258–271.

Ng, N. L., Canagaratna, M. R., Jimenez, J. L., Zhang, Q., Ulbrich, I. M., & Worsnop, D. R. (2011). Real-time methods for estimating organic component mass concentrations from aerosol mass spectrometer data. *Environmental Science & Technology, 45*(3), 910–916.

Nie, W., Wang, T., Gao, X., Pathak, R. K., Wang, X., Gao, R., Zhang, Q., Yang, L., & Wang, W. (2010). Comparison among filter-based, impactor-based and continuous techniques for measuring atmospheric fine sulfate and nitrate. *Atmospheric Environment, 44*(35), 4396–4403.

Paatero, P. (1997). Least squares formulation of robust non-negative factor analysis. *Chemometrics and Intelligent Laboratory Systems, 37*(1), 23–35.

Pathak, R. K., & Chan, C. K. (2005). Inter-particle and gas-particle interactions in sampling artifacts of $PM_{2.5}$ in filter-based samplers. *Atmospheric Environment, 39*(9), 1597–1607.

Rohde, R. A., & Muller, R. A. (2015). Air pollution in China: Mapping of concentrations and sources. *PloS One, 10*(8), e0135749.

Shen, Z., Cao, J., Arimoto, R., Han, Z., Zhang, R., Han, Y., Liu, S., Okuda, T., Nakao, S., & Tanaka, S. (2009). Ionic composition of TSP and $PM_{2.5}$ during dust storms and air pollution episodes at Xi'an, China. *Atmospheric Environment, 43*(18), 2911–2918.

Shen, Z., Cao, J., Liu, S., Zhu, C., Wang, X., Zhang, T., Xu, H., & Hu, T. (2011). Chemical composition of PM_{10} and $PM_{2.5}$ collected at ground level and 100 meters during a strong winter-time pollution episode in Xi'an, China. *Journal of the Air & Waste Management Association, 61*(11), 1150–1159.

Sun, Y. L., Zhang, Q., Schwab, J. J., Demerjian, K. L., Chen, W. N., Bae, M. S., Hung, H. M., Hogrefe, O., Frank, B., Rattigan, O. V., & Lin, Y. C. (2011). Characterization of the sources and processes of organic and inorganic aerosols in New York city with a high-resolution time-of-flight aerosol mass apectrometer. *Atmospheric Chemistry and Physics, 11*(4), 1581–1602.

Sun, Y., Wang, Z., Fu, P., Jiang, Q., Yang, T., Li, J., & Ge, X. (2013a). The impact of relative humidity on aerosol composition and evolution processes during wintertime in Beijing, China. *Atmospheric Environment, 77*, 927–934.

Sun, Y. L., Wang, Z. F., Fu, P. Q., Yang, T., Jiang, Q., Dong, H. B., Li, J., & Jia, J. J. (2013b). Aerosol composition, sources and processes during wintertime in Beijing, China. *Atmospheric Chemistry and Physics, 13*(9), 4577–4592.

Sun, Y., Du, W., Fu, P., Wang, Q., Li, J., Ge, X., Zhang, Q., Zhu, C., Ren, L., Xu, W., Zhao, J., Han, T., Worsnop, D. R., & Wang, Z. (2016). Primary and secondary aerosols in Beijing in winter: Sources, variations and processes. *Atmospheric Chemistry and Physics, 16*(13), 8309–8329.

Tao, J., Cheng, T. T., & Zhang, R. J. (2012). Chemical composition of summertime $PM_{2.5}$ and its relationship to aerosol optical properties in Guangzhou, China. *Atmospheric and Oceanic Science Letters, 5*(2), 88–94.

Tie, X., Zhang, Q., He, H., Cao, J., Han, S., Gao, Y., Li, X., & Jia, X. C. (2015). A budget analysis of the formation of haze in Beijing. *Atmospheric Environment, 100*, 25–36.

Wang, Q., Cao, J., Shen, Z., Tao, J., Xiao, S., Luo, L., He, Q., & Tang, X. (2013a). Chemical characteristics of $PM_{2.5}$ during dust storms and air pollution events in Chengdu, China. *Particuology, 11*(1), 70–77.

Wang, Q., Cao, J., Tao, J., Li, N., Su, X., Chen, L. W. A., Wang, P., Shen, Z., Liu, S., & Dai, W. (2013b). Long-term trends in visibility and at Chengdu, China. *PloS One, 8*(7), e68894.

Wang, G. H., Cheng, C. L., Huang, Y., Tao, J., Ren, Y. Q., Wu, F., Meng, J. J., Li, J. J., Cheng, Y. T., Cao, J. J., Liu, S. X., Zhang, T., Zhang, R., & Chen, Y. B. (2014). Evolution of aerosol chemistry in Xi'an, inland China, during the dust storm period of 2013 – part 1: Sources, chemical forms and formation mechanisms of nitrate and sulfate. *Atmospheric Chemistry and Physics, 14*(21), 11571–11585.

Wang, Y. H., Liu, Z. R., Zhang, J. K., Hu, B., Ji, D. S., Yu, Y. C., & Wang, Y. S. (2015). Aerosol physicochemical properties and implications for visibility during an intense haze episode during winter in Beijing. *Atmospheric Chemistry and Physics, 15*(6), 3205–3215.

Wang, Q., Huang, R.-J., Cao, J., Tie, X., Shen, Z., Zhao, S., Han, Y., Li, G., Li, Z., Ni, H., Zhou, Y., Wang, M., Chen, Y., & Su, X. (2016). Contribution of regional transport to the black carbon aerosol during winter haze period in Beijing. *Atmospheric Environment, 132*, 11–18.

Williams, L. R., Gonzalez, L. A., Peck, J., Trimborn, D., McInnis, J., Farrar, M. R., Moore, K. D., Jayne, J. T., Robinson, W. A., Lewis, D. K., Onasch, T. B., Canagaratna, M. R., Trimborn, A., Timko, M. T., Magoon, G., Deng, R., Tang, D., de la Rosa Blanco, E., Prévôt, A. S. H., Smith, K. A., & Worsnop, D. R. (2013). Characterization of an aerodynamic lens for transmitting particles greater than 1 micrometer in diameter into the Aerodyne aerosol mass spectrometer. *Atmospheric Measurement Techniques, 6*(11), 3271–3280.

Xiao, R., Takegawa, N., Zheng, M., Kondo, Y., Miyazaki, Y., Miyakawa, T., Hu, M., Shao, M., Zeng, L., Gong, Y., Lu, K., Deng, Z., Zhao, Y., & Zhang, Y. H. (2011). Characterization and source apportionment of submicron aerosol with aerosol mass spectrometer during the PRIDE-PRD 2006 campaign. *Atmospheric Chemistry and Physics, 11*(14), 6911–6929.

Xie, Y., Dai, H., Dong, H., Hanaoka, T., & Masui, T. (2016). Economic impacts from $PM_{2.5}$ pollution-related health effects in China: A provincial-level analysis. *Environmental Science & Technology, 50*(9), 4836–4843.

Xu, J., Zhang, Q., Chen, M., Ge, X., Ren, J., & Qin, D. (2014). Chemical composition, sources, and processes of urban aerosols during summertime in northwest China: Insights from high-resolution aerosol mass spectrometry. *Atmospheric Chemistry and Physics, 14*(23), 12593–12611.

Yan, J. P., Chen, L. Q., Lin, Q., Li, Z., Chen, H. Y., & Zhao, S. H. (2015). Chemical characteristics of submicron aerosol particles during a long-lasting haze episode in Xiamen, China. *Atmospheric Environment, 113*, 118–126.

Yu, X., Kumar, K. R., Lü, R., & Ma, J. (2016). Changes in column aerosol optical properties during extreme haze-fog episodes in January 2013 over urban Beijing. *Environmental Pollution, 210*, 217–226.

Zhang, Y.-L., & Cao, F. (2015). Fine particulate matter ($PM_{2.5}$) in China at a city level. *Scientific Reports, 5*, 14884.

Zhang, Q., Streets, D. G., Carmichael, G. R., He, K. B., Huo, H., Kannari, A., Klimont, Z., Park, I. S., Reddy, S., Fu, J. S., Chen, D., Duan, L., Lei, Y., Wang, L. T., & Yao, Z. L. (2009). Asian emissions in 2006 for the NASA INTEX-B mission. *Atmospheric Chemistry and Physics, 9*(14), 5131–5153.

Zhang, T., Cao, J. J., Tie, X. X., Shen, Z. X., Liu, S. X., Ding, H., Han, Y. M., Wang, G. H., Ho, K. F., Qiang, J., & Li, W. T. (2011). Water-soluble ions in atmospheric aerosols measured in Xi'an, China: Seasonal variations and sources. *Atmospheric Research, 102*(1–2), 110–119.

Zhang, R., Jing, J., Tao, J., Hsu, S. C., Wang, G., Cao, J., Lee, C. S. L., Zhu, L., Chen, Z., Zhao, Y., & Shen, Z. (2013). Chemical characterization and source apportionment of $PM_{2.5}$ in Beijing: Seasonal perspective. *Atmospheric Chemistry and Physics, 13*(14), 7053–7074.

Zhang, J. K., Sun, Y., Liu, Z. R., Ji, D. S., Hu, B., Liu, Q., & Wang, Y. S. (2014). Characterization of submicron aerosols during a month of serious pollution in Beijing, 2013. *Atmospheric Chemistry and Physics, 14*(6), 2887–2903.

Zhang, Y. J., Tang, L. L., Wang, Z., Yu, H. X., Sun, Y. L., Liu, D., Qin, W., Canonaco, F., Prevot, A. S. H., Zhang, H. L., & Zhou, H. C. (2015). Insights into characteristics, sources, and evolution of submicron aerosols during harvest seasons in the Yangtze River delta region, China. *Atmospheric Chemistry and Physics, 15*(3), 1331–1349.

Zhao, S. Y., Tie, X. X., Cao, J. J., & Zhang, Q. (2015). Impacts of mountains on black carbon aerosol under different synoptic meteorology conditions in the Guanzhong region, China. *Atmospheric Research, 164*, 286–296.

Chapter 11
Chemical Composition During Severe Haze Events in Northern China

Lili Wang and Yuesi Wang

Abstract Haze pollution in North China during winter and autumn has been rather severe in recent years due to substantial emissions from anthropogenic activities. To optimize pollution control measures and explore the formation mechanism of haze pollution, a detailed analysis of aerosol chemical compositions during severe haze events is investigated by collecting results from many studies. Secondary inorganic aerosols (primarily SNA: sulfate, nitrate and ammonium) drive the severe haze pollution evolution because the relative contributions of SNA increase rapidly with elevated pollution levels through the rapid secondary transformation of primary gaseous pollutants to secondary aerosols. Remarkably, nitrate plays an increasingly important role in $PM_{2.5}$ pollution in Beijing. Organic matter (OM) is also very important for haze formation and accounts for the largest fraction. In addition, the size distribution shift is significant from the submicron to 1.1–2.1 μm due to heterogeneous formation mechanisms and hygroscopic growth.

Keywords Chemical composition • Secondary inorganic aerosols • Size distribution • Northern China • Severe haze

11.1 Introduction

Haze pollution in North China (NC) has increased over the past decades (Fu et al. 2014; Quan et al. 2011; Zhang et al. 2012; Wu et al. 2014) due to growth in the economy and energy consumption and rapid industrialization, urbanization and motorization (Han et al. 2015, 2016; Hu et al. 2014; Sheehan et al. 2014). In addition, the unique topography (mountains in the north and west, ocean in the east, plains in the south) and massive city clusters has contributed to poor diffusion conditions and more anthropogenic emissions. Frequently persistent severe pollution events, such as the heaviest pollution on record in January 2013 (Zhang et al.

L. Wang (✉) • Y. Wang
State Key Laboratory of Atmospheric Boundary Layer Physics and Atmospheric Chemistry (LAPC), Institute of Atmospheric Physics, Chinese Academy of Sciences, Beijing 100029, China
e-mail: wll@mail.iap.ac.cn; wys@mail.iap.ac.cn

I. Bouarar et al. (eds.), *Air Pollution in Eastern Asia: An Integrated Perspective*,
ISSI Scientific Report Series 16, DOI 10.1007/978-3-319-59489-7_11

2015c; Wang et al. 2014b) and the highest $PM_{2.5}$ (fine particulate matter, with aerodynamic diameters $\leq$2.5 μm) levels (Zhang and Cao 2015), have occurred in this region. High levels of $PM_{2.5}$ have resulted in impaired visibility; in particular, submicron particles (PM_1) primarily composed of carbonaceous and secondary species have induced approximately 80% of light scattering in Beijing during low visibility (less than 5 km) episodes (Bergin et al. 2001). $PM_{2.5}$ was not only the dominant PM (particle matter) category during episodes but also the major driving factor during episode evolution (Zheng et al. 2016a).These severe pollution and haze episodes have attracted extensive attention and contributed to serious influences on climate, environment, and public health. Aerosol chemical species are affected by local source emissions and meteorological conditions. Therefore, a detailed analysis of aerosol chemical compositions during severe haze events is critical for optimizing pollution control measures and promoting model simulation and validation.

Over the past decade, single-source pollution from coal combustion has remarkably converted to complex atmospheric pollution due to mixing pollutants from coal combustion, vehicular emissions, and even biomass burning and fugitive dust (Yang et al. 2011) in North China. The chemical compositions of aerosols have become more complex and include primary and secondary particles. Some references have summarized the chemical compositions in $PM_{2.5}$ or PM_{10} in China (Zhou et al. 2015; Zhang et al. 2012, 2015b; Yang et al. 2011; Lv et al. 2016), with a focus on average conditions. This chapter primarily focuses on the chemical compositions and characterization of aerosols during severe haze pollution episodes in North China for different seasons and different size particles to provide a scientific basis for pollution control strategies aimed at severe haze events.

11.2 Spatial and Temporal Distribution of Severe Haze Events in North China

A spatial distribution of $PM_{2.5}$ concentrations in China (Zhang and Cao 2015) revealed that haze events most frequently occurred in the North China which has the highest density of coal consumption and heavy industries (e.g., iron, steel and cement); in addition, the highest concentrations were found during winter and autumn due to enhanced anthropogenic emissions from fossil fuel combustion and biomass burning, unfavorable meteorological conditions for pollution dispersion, and secondary production of both inorganic aerosols and organic matters. A spatial evolution of regional severe haze pollution from Dec 19 to 21, 2015 (Fig. 11.1) indicated that the severe pollution formed in the south of NC and developed to encompass all NC, eventually expanding to Eastern China because of the influence of persistent southerly winds. Moreover, these severe haze events occurred more often along the eastern foot of the Taihang and Yan Mountains as well as in the south of NC.

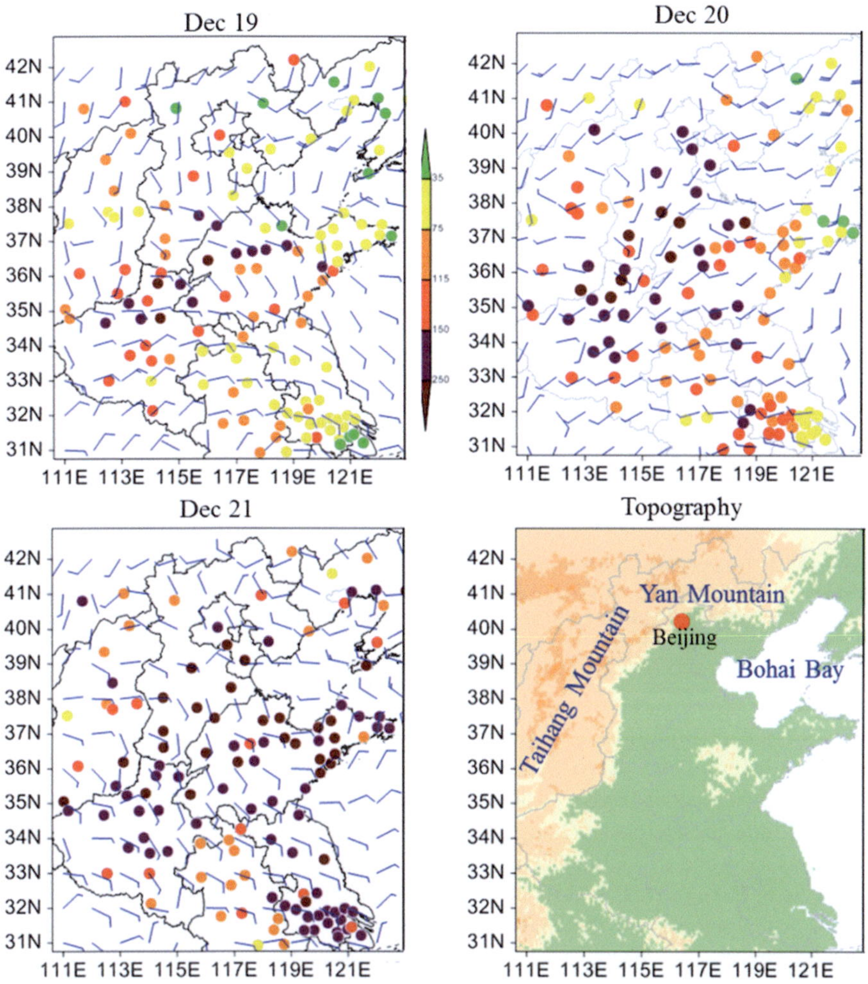

Fig. 11.1 Maps of the evolution of regional severe haze pollution (Dec 19–21, 2015, $PM_{2.5}$ concentrations μg m^{-3}) and the topography of Northern China

Overall, the frequency of severe haze days increased at a rate of 2.52% or 9 days/year based on 46 urban sites and primarily occurred in winter (a rate of 4.29%/year) and autumn (a rate of 3.21%/year) (Fig. 11.2). Notably, the increasing frequency of severe haze occurrences in 2014 and 2015 was significant, and the highest occurrence frequencies were in 2014 and 2015 at average 20% (winter) and 13% (autumn), respectively. Although a significant increased rate also appeared in spring, the increase of slight haze days was predominant and the haze events primarily occurred during early spring (March) when domestic heating was still occurring in this region.

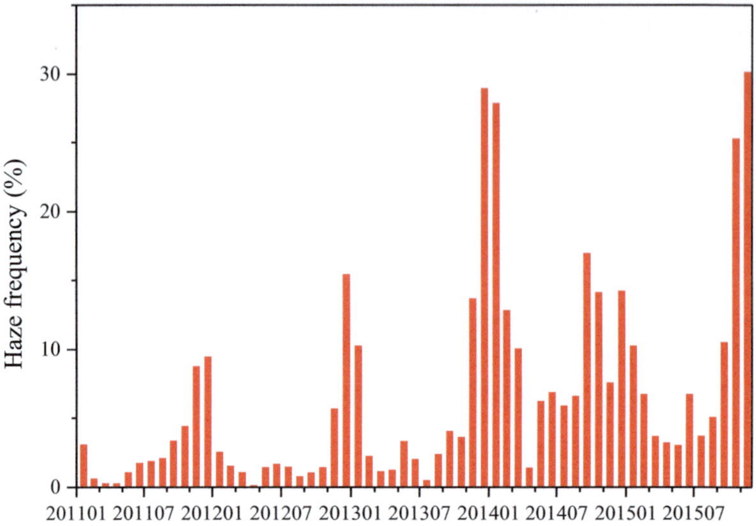

Fig. 11.2 The monthly variation trend of the average occurrence frequencies for severe haze events in NC from 2011 to 2015. (Severe haze events were defined using visibility and RH (relative humidity) at 14:00 LST as days with visibility ≤5 km and RH<90%, meanwhile excluding days with fog, precipitation, dust storms, smoke, or snow storms based on 46 urban sites in NC)

As for haze pollution in Beijing, 73 haze episodes (more than two successive days with $PM_{2.5} \geq 75\ \mu g\ m^{-3}$) and 189 polluted days were recorded in 2013; in addition, 54 episodes ($PM_{2.5} \geq 150\ \mu g\ m^{-3}$) reached severe haze levels, occurring in winter, June and October (Zheng et al. 2016a). The highest $PM_{2.5}$ concentration in NC was recorded in January 2013 with the maximum hourly $PM_{2.5}$ mass concentrations in Beijing (680 and 881 $\mu g\ m^{-3}$ on Jan 12 for the Beijing tower site and SQL, respectively) and the highest regional average concentration (250 $\mu g\ m^{-3}$ during Jan 10–14) (Wang et al. 2014a, b). The evolution of severe haze episodes was always accompanied with stagnant weather conditions such as enhanced atmospheric stability, decreased wind speeds, mixed layer height, and increased humidity (Zheng et al. 2016a; Wang et al. 2014a; Tang et al. 2016), all of which suppress pollution diffusion and facilitate particle hygroscopic growth. Two types of pollution evolution were found in NC and characterized by the "explosive growth" (EG) and "gradual growth" (GG) of $PM_{2.5}$. EG pollution episodes primarily occurred in Beijing in winter due to wind direction shearing from south to north and then south due to recirculation and regional transport along with secondary transformation (Liu et al. 2016; Wang et al. 2014a; Zheng et al. 2015b). Many studies have demonstrated the important role of regional transport, largely from the southwest but also from the east, and of coal combustion emissions for winter haze formation in Beijing.(Sun et al. 2016a; Li et al. 2015; Wang et al. 2014a, 2015a).

11.3 Chemical Characteristics of Aerosols During Severe Haze Events

In general, secondary inorganic ions (i.e., SNA: sulfate, nitrate and ammonium), the carbonaceous species (organic carbon (OC) and elemental carbon (EC)), and crustal species (i.e., Al, Si, Ca and Fe) have comprised the major components of aerosols in China (Fig. 11.3). However, few simultaneous measurements of these species have been taken in North China. In this section, we focus on a discussion of the major species that contribute to PM mass (SNA, OC, and EC). Although the dominant components were the same in Northern China, the fractions of each compound varied between different sites and seasons.

In this chapter the results from the offline filter based analysis method (Sun et al. 2006; Tian et al. 2014) and online analysis such as AMS (aerosol mass spectrometer) (Zhang et al. 2014; Elser et al. 2016) or ACMS (Aerodyne Aerosol Chemical Speciation Monitor) (Sun et al. 2013) were collected through a large number of references aimed at haze pollution episodes. Most studies have indicated that absolute concentrations of SNA and OC in PM_1 and $PM_{2.1}$, together with the relative contribution of SNA, increased with the elevated pollution levels; whereas, the relative contribution of OC decreased (Zhang et al. 2014; Liu et al. 2016; Elser et al. 2016; Pan et al. 2016b; Zheng et al. 2016a, b). In addition, aerosol compositions showed substantial changes during the formation and evolution of haze

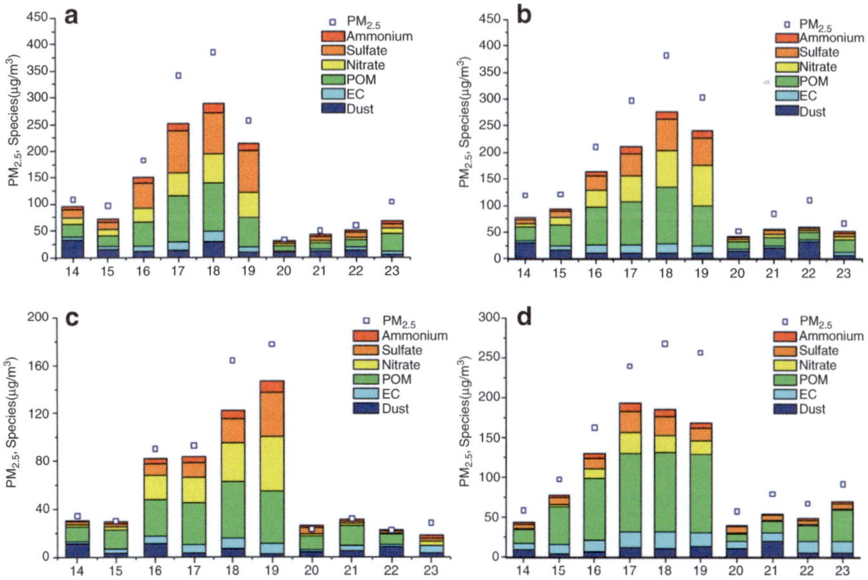

Fig. 11.3 Variation of concentrations of $PM_{2.5}$ and chemical species at (**a**) TJ, (**b**) BJ, (**c**) SDZ and (**d**) CD from 14 to 23 January 2010 (BJ and TJ are the urban sites, SDZ is a rural site in the northeast of BJ, CD is an urban site in the north of NC) (from Zhao et al. 2013)

episodes but were generally dominated by secondary aerosols (53–67% in PM_1 and 81% in $PM_{2.5}$) (Sun et al. 2016a; Huang et al. 2014), which were also significantly affected by regional transport from the heavily populated, urbanized and industrialized surrounding areas.

The concentration level of secondary pollutants was significantly higher in large cities (BJ and TJ) than in the regional background area (SDZ) and the cities in the northern area of Hebei province (CD): meanwhile, the percentage was comparable at three sites in NC and evidently higher than at the CD site (Fig. 11.3). In addition, an increase of secondary aerosol species was observed simultaneously in aerosols at the urban and rural sites, indicating the greater regional impact of a secondary formation process during the haze episode (Zhao et al. 2013; Zheng et al. 2016b).

11.3.1 Inorganic Species

Secondary inorganic species played enhanced roles in haze formation, as suggested by their rapid growth and elevated contributions during haze episodes under stagnant weather conditions, high humidity, and regional transport. SNA was dominant in aerosol inorganic species, which are secondary particles formed from gaseous pollutants (SO_2, NOx and NH_3). The sum of SNA accounted for more than 90% of total water soluble inorganic ion species and contributed to 34–50% of $PM_{2.5}$ in China on average (Zhou et al. 2015; Zheng et al. 2016b). The increasing sulfur and nitrogen oxidation ratios (SOR and NOR) in the late haze period or severe haze events indicated the enhanced production of secondary inorganic species, which is was likely due to heterogeneous aqueous reactions under weak solar radiation; thus, the photochemical capacity was reduced by the increases in aerosols and RH (relative humidity) (Quan et al. 2015; Zheng et al. 2016a). SNA were elevated synchronously at both urban and rural sites during heavy haze episodes, which is indicative of the characteristic of regional fine particulate pollution in NC (Zheng et al. 2016b; Sun et al. 2016b). The heaviest global concentrations of SO_2 and NO_2 were found in NC, a finding that provides abundant precursors for SNA formation (Krotkov et al. 2016).

The average values of SNA at TJ, BJ, SDZ and CD were 127.0 (accounting for 44.8% of $PM_{2.5}$), 113.5 (37.7%), 56.9 (41.9%) and 46.2 (20.1%) $\mu g\ m^{-3}$, respectively, which were approximately 4.3–9.1 times those on non-haze days (Zhao et al. 2013). When $PM_{2.5}$ concentrations reached peak values ($\geq 120\ \mu g\ m^{-3}$), the fractions of SNA were approximately 60–80%, indicating the importance of secondary formation in the pollution episodes (Zheng et al. 2016b). Moreover, during harvest season, the formation of secondary inorganic particles was intensified due to the interactions of smoke plumes and urban/industrial pollutants in the Beijing urban environment, resulting in higher concentrations and percentages (79%) of sulfate, nitrate, ammonium, and organic carbon in the fine particles under high relative humidity, and southerly winds conditions contributed to deteriorated urban visibility (Wang et al. 2015b).

11.3.1.1 Sulfate

Three major formation mechanisms for sulfate are (1) homogeneous gas phase reactions of SO_2 with OH and H_2O_2 radicals, (2) heterogeneous reactions on aerosol surfaces, and (3) in-cloud processes. Sulfate is the main species of SNA in PM_1 and $PM_{2.5}$ in winter in NC and often shows the largest increase and highest concentration during haze episodes due to coal combustion for heating, especially in extreme haze pollution (Sun et al. 2014), as shown in Fig. 11.3. During Jan and Feb 2013, the sulfate concentration was 33.3 μg m^{-3} in $PM_{2.5}$ during heavy haze days in Beijing (Tian et al. 2014).

The high concentration of sulfate in winter was formed through secondary formation (increasing sulfate oxidation ratio (SOR)) and primarily through the heterogeneous aqueous reaction process rather than gas-phase oxidation. This result was due to the weak photochemical activity associated with very low levels of O_3 (less than 5 ppb) during the haze episodes and the high levels of NO_2 being unable to produce sufficient conversion oxidants (OH and H_2O_2 radicals) for the gas-phase oxidation of SO_2 (Poppe et al. 1993; Hua et al. 2008). Additionally, the increased RH on haze days was favorable for the aqueous phase oxidation of SO_2 (Zhao et al. 2013; Quan et al. 2015). The most polluted stage in a severe winter haze episode was associated with fog processing and significantly enhanced sulfate levels (from 10–16% to 21–26% in PM_1) due to the aqueous-phase production of sulfate at high RH levels (Sun et al. 2016a; Zhang et al. 2015a; Liu et al. 2016). The results were similar for sulfate formation in $PM_{2.5}$ (Elser et al. 2016). In addition, high NOx promoted the formation of sulfate from SO_2 on the surface of dust particles (He et al. 2014; Wang et al. 2014b).

Statistically, the eastern region consumes 43% of gross coal while comprising only 11.1% of the total land area of China, so the eastern SO_2 and NOx emissions per land area are 3.3 and 4.4 times the corresponding national averages (Yang et al. 2011). In addition, briquettes and honeycomb coal for cooking and refueling and coal-burning boilers for domestic and commercial heating enhance the SO_2 emissions in winter (Xue et al. 2016). Thus, implementing strict coal-fired boilers, using low-sulfur (coal content <0.5%) coal, and improving the combustion efficiency of boilers are necessary to reduce and relieve haze pollution in NC in winter.

11.3.1.2 Nitrate

Nitrate was the primary species in SNA in summer in NC and played an important role in haze formation through the aqueous-phase process such as the hydrolysis of N_2O_5 on aerosol surfaces during the nighttime under high humidity; in addition, a homogeneous (gas-phase reactions of NO_2+OH, NO_2 with OH or O_3 reported to be positively correlated with temperature) reaction was involved in the formation of nitric acid during the daytime (Sun et al. 2015b, 2016b; Pathak et al. 2009). NH_4NO_3 is a semi-volatile species that is affected by formation processes and the gas-aerosol equilibrium, unstable and partially volatile at temperatures over 20 °C. Thus, it possesses a marked seasonality, with maximum values in winter and minimum values in summer (Calvo et al. 2013). However, in NC, the maximum values were in autumn and summer and the minimum values were in winter and spring (Huang et al. 2016). One explanation for this outcome is that high humidity facilitates nitrate formation under abundant NOx and NH_3 emissions.

In autumn, the absolute and relative concentrations of nitrate increased with visibility degradation and increasing RH for PM_1 and $PM_{2.5}$, whereas the relative contribution of organics decreased. In addition, a larger fraction of nitrate (25% in PM_1 and 35.6% in $PM_{2.5}$) than of sulfate and ammonium was found during haze events, and the increased contribution of nitrate to PM was the most pronounced. This rapid increase in the relative contribution of nitrate to PM_1 observed during the early stages of haze pollution (from 11% to 23%) was due to new particle formation, whereas the nitrate formed in $PM_{1-2.5}$ during the latter stages was due to heterogeneous formation and hygroscopic growth (Pan et al. 2016b; Yang et al. 2015). These results indicated that nitrate formation in $PM_{2.5}$ was enhanced by the development of haze pollution compared with nitrate formation in PM_1; moreover, water vapor played a more important role in promoting the formation of nitrate than in promoting the formation of organics during the development of haze pollution.

In winter, nitrate also played a significant role during haze episodes, even though the temperature was very low. The ratio of nitrate in PM_1 increased from 14% to 18% with the elevated haze levels and NOR (nitrogen oxidation ratio) was lower than SOR. Nitrate is formed via the heterogeneous hydrolysis of N_2O_5 on the surface of moist PM, especially during humid and hazy weather because of the higher NO_2 concentration and lower O_3 concentration resulting in weak photochemical activity (Zhang et al. 2015a; Ji et al. 2014). Furthermore, a higher concentration and contribution (15%) of nitrate to PM_1 was observed at 260 m (Fig. 11.4c, d) and was likely due to the favorable gas-particle partitioning under lower temperature and higher humidity conditions (Sun et al. 2015a).

No matter how complicated the conversion process, nitrate is closely related to high emissions of NOx from traffic in Beijing. Thus, the control of vehicular emissions is crucial in Beijing during severe haze episodes.

11.3.1.3 Ammonium

Ammonium is associated with nitrate and sulfate in particles. Ammonium in the fine mode might originate from the reaction of NH_3 vapor with acidic gases such as H_2SO_4, HNO_3, and HCl or the reaction/condensation of NH_3 vapor on the acidic particle surfaces of anthropogenic origin that accumulate in the fine mode. The ratio of nitrate and sulfate to ammonium was approximately 1:2, demonstrating that sulfate aerosols can be completely neutralized. As the main cation, ammonium is very important for neutralizing sulfate and nitrate. The increasing concentration and relative contribution of ammonium has been documented during haze events by many studies (Zhang et al. 2015a; Pan et al. 2016b; Tian et al. 2014).

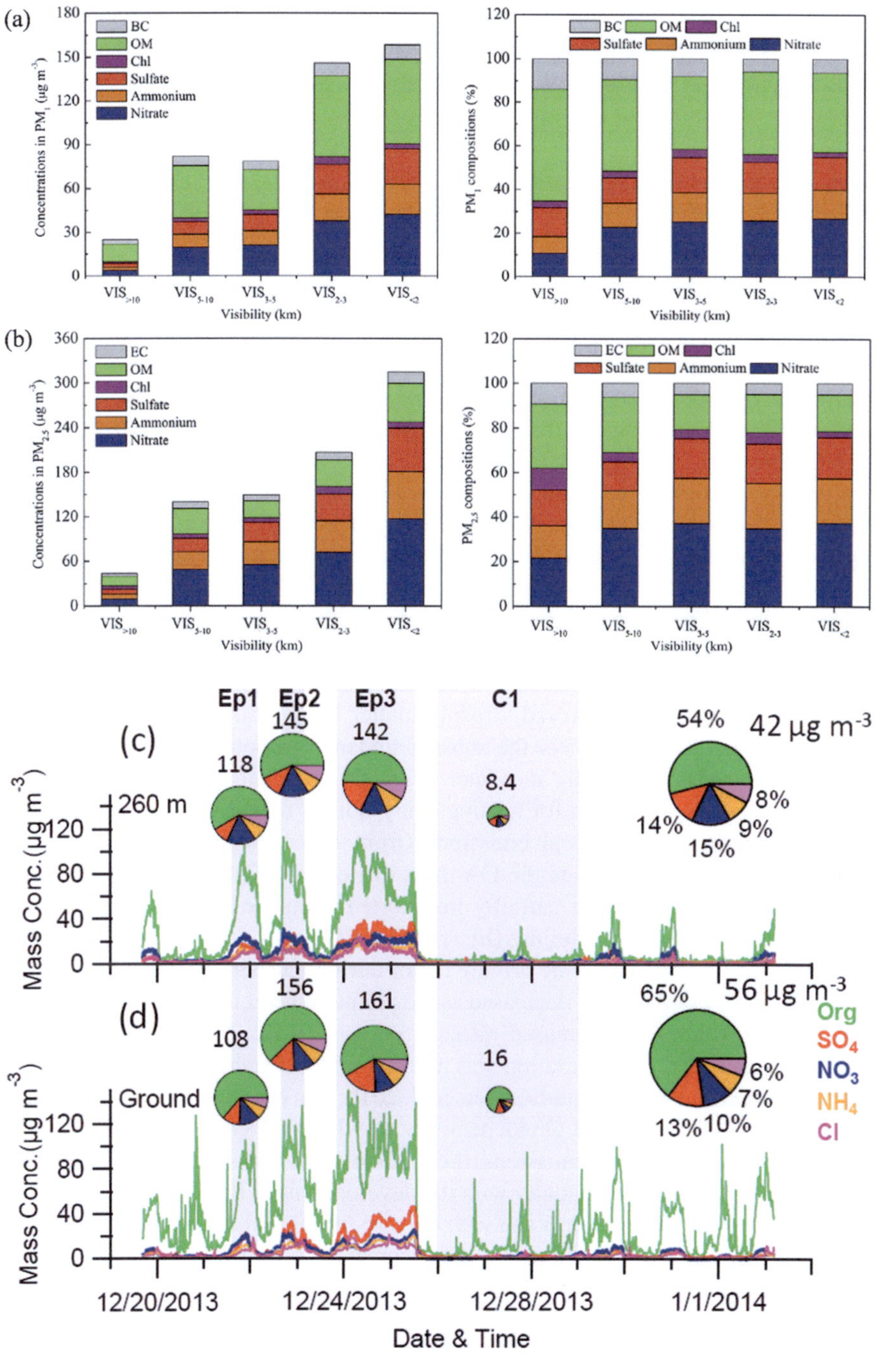

Fig. 11.4 Chemical concentrations and compositions of (**a**) PM_1 and (**b**) filter-based $PM_{2.5}$ at different visibility levels. Time series of (**c**) PM_1 species at 260 m Beijing meteorological tower and (**d**) the ground level. The pie charts of (**c**) and (**d**) show the average composition for the entire period, and the other pie charts show the average composition during three pollution episodes and one clean event (from Pan et al. 2016b; Sun et al. 2015a)

The highest NH_3 concentration in NC was revealed by the IASI satellite (Van Damme et al. 2014), which provides abundant NH_3 to convert into ammonium. NH_3 emissions primarily result from agricultural activities. However, fossil fuel emissions and NH_3 slip from power plants are more important than agricultural sources during haze pollution periods in urban Beijing, contributing 90% of the total NH_3 (Pan et al. 2016a). Thus, the reduction of ammonia emissions is urgently needed due to their role in aerosol nucleation and growth, causing haze formation.

11.3.2 Carbonaceous Species

Carbonaceous species include organic carbon (OC) and black carbon (or elemental carbon, EC), which comprise the major fraction in the locally formed haze episodes in NC (Sun et al. 2016b) or on clear days (Zheng et al. 2016b). EC and parts of OC are primary particles, and organic aerosol is also comprised of secondary particles. Based on results from AMS or ACMS, primary organic aerosols ((POA), including hydrocarbon-like OA (HOA) from traffic emissions, cooking OA (COA), coal combustion OA (CCOA) and biomass burning OA (BBOA)) and oxygenated OA ((OOA), including a highly oxidized, low-volatility oxygenated organic aerosol (LV-OOA) and a less oxidized, semi-volatile oxygenated OA (SV-OOA)) are differentiated using positive matrix factorization (PMF).

In all $PM_{2.5}$ species observed, organic matter (OM, scaling OC by a constant factor of 1.4–1.8) was always the most abundant component (18–46%) during pollution episodes, especially in winter. High OC concentrations in winter were ascribed to coal combustion for heating using boilers or stoves using low-quality bunker fuel. In addition, coal emissions (from domestic heating, cooking and industrial processes) dominate the OA mass in Beijing (47–55%), and these coal-related particles are at least partially transported from the industrial province of Hebei to highly populated Beijing (Elser et al. 2016).

However, OC was rarely the driving factor during episode evolution because its relative contribution usually decreased as the pollution level increased, although the absolute concentration increased (Zheng et al. 2015b, 2016a), as shown in Figs. 11.5 and 11.6. This variation was a rather universal phenomenon in NC that has been observed in many studies (Liu et al. 2016; Pan et al. 2016b) and that might be caused by the weakened photochemistry in winter (Zheng et al. 2015a, b). In autumn, due to decreasing emissions, the relative contributions of organics to PM_1 and $PM_{2.5}$ decreased substantially with the development of haze, i.e., from 51% on clear days to 36% on haze days and with only 16.5% of OM to $PM_{2.5}$ during severe haze periods, although their concentrations increased slightly as visibility decreased (Pan et al. 2016b).

For primary organic aerosols, the contribution of BBOA increased to 21% during harvest season in late June, indicating a large impact of agricultural burning on air pollution in summer (Sun et al. 2016b). In winter, CCOA was the dominant contributor to POA or OA, accounting for 33–37% of PM_1, because coal

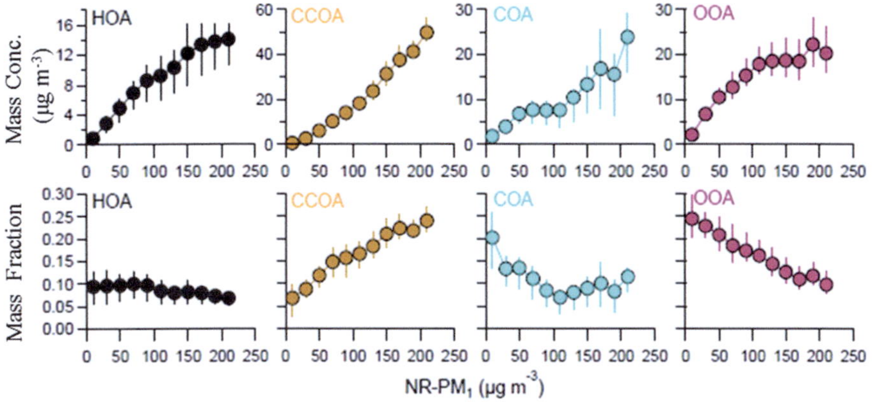

Fig. 11.5 Variations of mass concentrations (*top*) and mass fractions (*bottom*) of PM_1 OA components as a function of PM_1 mass loadings. The data were binned according to the total PM_1 mass, and the vertical lines are 25th (*bottom*) and 75th percentiles (*top*) (from Sun et al. 2013)

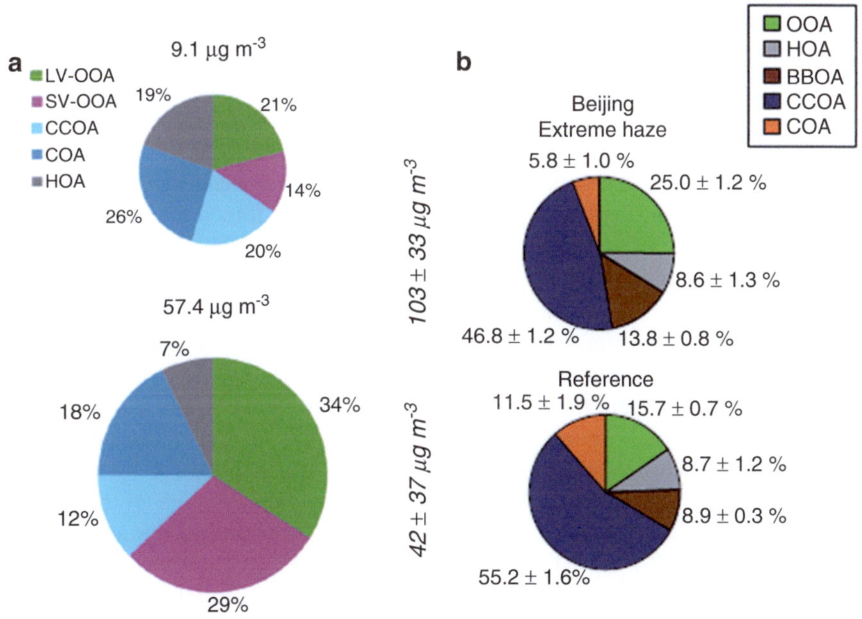

Fig. 11.6 Mean relative contributions of the OA sources for different pollution levels in Beijing in winter. (**a**) OA in PM_1 taken from Zhang et al. 2014, (**b**) OA in $PM_{2.5}$ (from Elser et al. 2016)

combustion was the dominant primary emission source in winter (Sun et al. 2016a). Figure 11.5 shows the increasing absolute and relative contributions of CCOA for PM_1 with the increasing PM_1 mass loading; CCOA increased from 0.4 µg m^{-3} (PM_1 is 13.8 µg m^{-3}) to 43.6 µg m^{-3} (PM_1 is 299.5 µg m^{-3}) from clear days to severe haze days (Sun et al. 2014) in Beijing. This finding demonstrated that CCOA, as the largest primary source of OA (33% in OA and 17% in PM_1), played a significant role. In general, POA (54–62%) was the dominant component in OA in PM_1 on the ground, but the relative contribution of SOA increased with the elevated height (Sun et al. 2015a). However, as shown in Fig. 11.6, the relative contribution of POA (CCOA, HOA and COA) to PM_1 or $PM_{2.5}$ decreased with the elevated haze levels in Beijing (Zhang et al. 2015a).

Secondary organic species (SOC) concentration was ascribed to the photochemical formation of SOC at high temperatures. Condensation and absorption are the main processes for SOC formation under low temperatures and high RH. However, in winter with very low temperatures, chemical transformation was rather important, as secondary organic aerosols (SOA) made comparable contributions to secondary inorganic aerosol (Huang et al. 2014). The high level of SOA in winter is associated with OH-radical-initiated oxidation, or perhaps aqueous-phase oxidation and NO3-radical-initiated nocturnal chemistry. As shown in Fig. 11.6, the increasing contribution of OOA during the haze episode reveals that SOA formation is very important during haze events. However, a decreasing relative contribution of OM was also found (Tian et al. 2014; Pan et al. 2016a), indicating the limited SOA formation at high haze pollution levels during the heating season. SOA sources are highly uncertain and fossil SOA is important for particulate pollution in Beijing (Huang et al. 2014). Overall, the formation mechanism of SOA in winter is very complex and further study is needed during severe haze episodes.

BC (or EC) is one of the most health-detrimental components of $PM_{2.5}$ (Janssen et al. 2011, 2012) and generates positive radiation that forces the atmosphere to warm (Jacobson 2001; Menon et al. 2002). Residential coal contributes significantly to BC in China; additionally, biomass burning during harvest seasons also emits abundant BC. The annual mean concentrations of EC in Beijing were $4.1 \pm 3.2\ \mu g/m^3$ from Mar 2013 to Feb 2014, which were lower than those over the past 10 years but higher than those from most of the megacities in North America and Europe (Ji et al. 2016). The high contribution of EC was from coal combustion, biomass burning, petroleum products, and vehicle exhaust in Beijing (Ji et al. 2016). Based on the dual carbon isotope constrained ($\Delta^{14}C$ and $\delta^{13}C$) source apportionment method, coal combustion accounted for 66% of the EC and biomass combustion accounted for approximately 30% during the severe haze events in January 2013 in NC (Andersson et al. 2015). BC under polluted urban environments in the surface and upper air could play an essential role in pollution development and could significantly contribute to large positive radiative forcing, more stable boundary layers and intensive haze pollution, especially in more polluted developing countries (Peng et al. 2016; Zheng et al. 2016a). During the haze episode, BC (or EC) concentration increased with elevated haze levels while the relative contribution decreased substantially (Pan et al. 2016b).

11.4 Size Distribution and Evolution of Aerosol Chemical Species During Severe Haze Events

The size distribution of aerosol chemical species changed with the elevated haze pollution levels. Two autumn Beijing haze episodes began from extensive nucleation processes and subsequently developed through continuous aerosol growth (Guo et al. 2014). During the most polluted haze periods in Jan. 2013, the percentage of PM_1 in $PM_{2.5}$ was <50%, probably because of the large contributions of droplet-mode aerosols. The peak mass concentration of particles (also including species such as organic carbon, elemental carbon, NH_4^+, SO_4^{2-}, and NO_3^-) shifted from 0.43–0.65 μm on clear days to 0.65–1.1 μm on lightly polluted days to 1.1–2.1 μm on heavily polluted days (Tian et al. 2014), as shown in Fig. 11.7. In addition to fine particles, anthropogenic chemical species also substantially accumulated in the coarse mode, suggesting that particles with aerodynamic diameters greater than 2.1 m cannot be neglected during severe haze events (Tian et al. 2014). The size distribution shift was attributed to (1) the high RH on hazy days so that the aqueous phase on the aerosol surface provided a means for the rapid heterogeneous gas–liquid conversion of gaseous precursors to produce secondary inorganic aerosols in $PM_{1.1-2.1}$; and (2) the formation of secondary aerosols due to abundant gaseous precursors. A significant size distribution shift for nitrate was also found in autumn (Pan et al. 2016b), indicating that the rapid increase of nitrate in $PM_{1-2.5}$ during severe haze episodes was attributable to heterogeneous formation mechanisms.

As shown in Fig. 11.8, based on the AMS observation for PM_1, an accumulation mode peaking between 500 and 700 nm was consistently observed for both organic and inorganic species during the growth and peak stages of haze pollution evolution. Secondary transformation was the primary reason for the pollution episodes: SNA accounted for approximately 42–60% of the 100–1,000 nm particle mass increase, with a higher increase rate of sulfate than of nitrate (Liu et al. 2016). However, during the extreme haze events, the AMS observation for $PM_{2.5}$ showed much larger particles, with a volume weighted mode at approximately 800–1,000 nm, in contrast to approximately 400 nm during reference periods. These large particle sizes made the use of the $PM_{2.5}$ inlet crucial, especially during the severe haze events, where 39% of the mass would have been lost in the conventional PM_1 inlet. The growth of the particles was associated with high secondary aerosol fractions and condensation of semi-volatile compounds on preexisting particles (Elser et al. 2016).

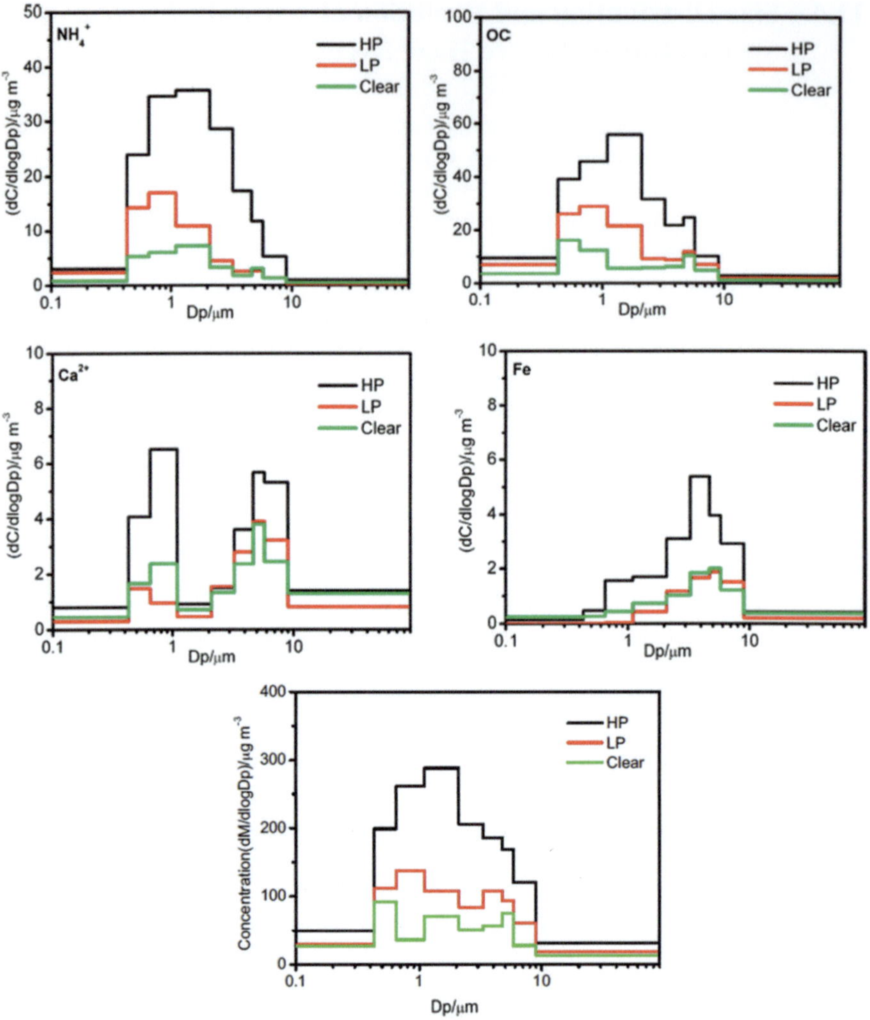

Fig. 11.7 Size distributions of typical chemical species in different categories based on nine-stage samplers (Andersen Series 20–800, USA) (from Tian et al. 2014)

11.5 Conclusions

Haze pollution in Northern China has been rather severe in recent years due to substantial emissions from anthropogenic activities. These severe haze events have primarily occurred in winter and autumn and the highest concentrations have often appeared along the eastern foot of the Taihang and Yan Mountains and in the south of NC. These severe haze episodes usually persist for several days with an extremely high $PM_{2.5}$ concentration. However, these haze events differed in

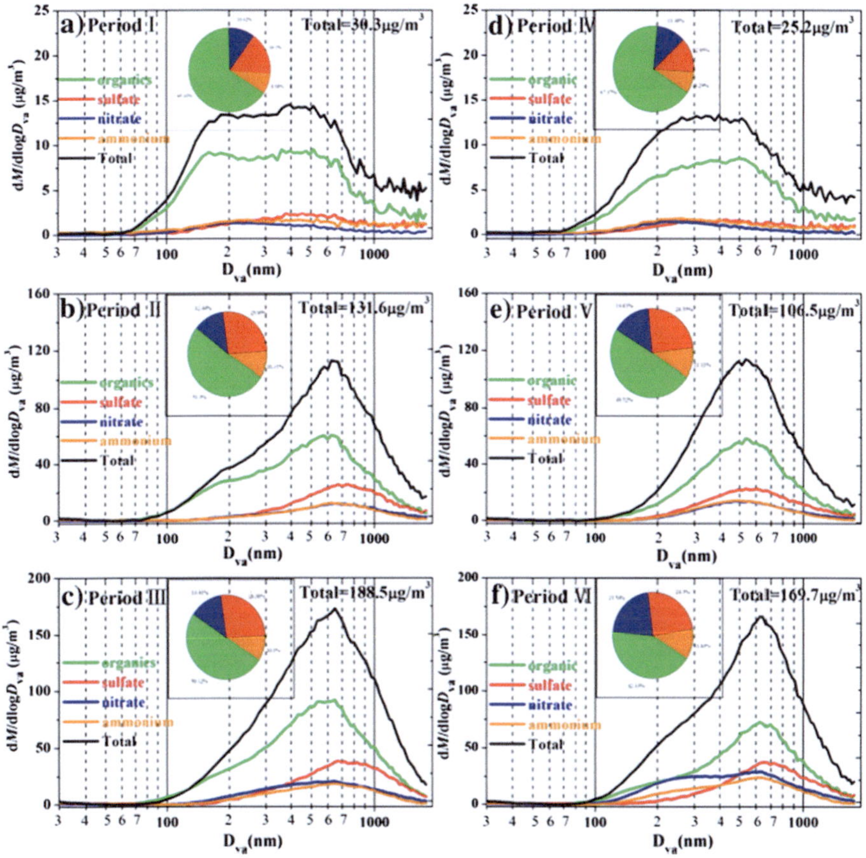

Fig. 11.8 (**a–f**) Average size distribution and chemical composition of particle chemical species in PM_1 during six time periods in the pollution episodes (**a** and (**d**) for clean period, (**b**) and (**e**) for growth period, and (**c**) and (**f**) for peak period). Taken from Liu et al. 2016. (**g**) Diurnal trends, size distributions, mean concentrations for $PM_{2.5}$ and relative contributions of the AMS species and BC for the four periods. Note: size distributions only available for AMS species (from Elser et al. 2016)

terms of sources, formation processes, and chemical composition, thus requiring different control policies to resolve this heavy haze pollution.

Results from many studies have indicated that fine particles were largely composed of organic matter and secondary inorganic species. The relative contributions of SNA increased rapidly with elevated pollution levels, revealing a secondary inorganic aerosol-driven trend for severe haze pollution evolution due to the rapid secondary transformation of primary gaseous pollutants to secondary aerosols, primarily through heterogeneous reactions under high humidity. Remarkably, a sulfate-driven trend was found in winter and a nitrate-driven trend was found in

Beijing, especially during warm seasons. Due to a continuous effort to reduce SO_2 emissions, NOx emissions increased rapidly, resulting in an increasingly more important role played by nitrate in PM pollution in Beijing. In addition, OM was also very important for haze formation and accounted for the largest fraction; CCOA and OOA were the main components in winter, although the relative contribution to PM_1 or $PM_{2.5}$ usually decreased. Therefore, mitigating severe winter haze in NC requires a priority control of secondary aerosol precursors and coal combustion emissions on a regional scale.

However, to date, most studies for haze pollution have focused on Beijing's pollution problem. Simultaneously, observations across multiple sites are rare. Thus large-scale investigation is required to adequately characterize the regional features of haze pollution in NC.

Acknowledgments This work was supported by National Natural Science Foundation of China (No. 41505133 and 41230642), the European Union Seventh Framework Program ([FP7/2007-2013]) under grant agreement No. 606719 (PANDA project), and the CAS Strategic Priority Research Program Grant (No. XDB05020203).

References

Andersson, A., Deng, J., Du, K., Zheng, M., Yan, C., Sköld, M., & Gustafsson, O. (2015). Regionally-varying combustion sources of the January 2013 severe haze events over Eastern China. *Environmental Science and Technology, 49*(4), 2038–2043.

Bergin, M., Cass, G., Xu, J., Fang, C., Zeng, L., Yu, T., Salmon, L., Kiang, C., Tang, X., & Zhang, Y. (2001). Aerosol radiative, physical, and chemical properties in Beijing during June 1999. *Journal of Geophysical Research D, 106*(D16), 17969–17980.

Calvo, A. I., Alves, C., Castro, A., Pont, V., Vicente, A. M., & Fraile, R. (2013). Research on aerosol sources and chemical composition: Past, current and emerging issues. *Atmospheric Research, 120*, 1–28. doi:10.1016/j.atmosres.2012.09.021.

Elser, M., Huang, R.-J., Wolf, R., Slowik, J. G., Wang, Q., Canonaco, F., Li, G., Bozzetti, C., Daellenbach, K. R., & Huang, Y. (2016). New insights into PM 2.5 chemical composition and sources in two major cities in China during extreme haze events using aerosol mass spectrometry. *Atmospheric Chemistry and Physics, 16*(5), 3207–3225.

Fu, G. Q., Xu, W. Y., Yang, R. F., Li, J. B., & Zhao, C. S. (2014). The distribution and trends of fog and haze in the North China Plain over the past 30 years. *Atmospheric Chemistry and Physics, 14*(21), 11949–11958.

Guo, S., Hu, M., Zamora, M. L., Peng, J., Shang, D., Zheng, J., Du, Z., Wu, Z., Shao, M., & Zeng, L. (2014). Elucidating severe urban haze formation in China. *Proceedings of the National Academy of Sciences, 111*(49), 17373–17378.

Han, L., Zhou, W., & Li, W. (2015). Increasing impact of urban fine particles (PM2.5), on areas surrounding Chinese cities. *Scientific Reports, 5*, 12467. doi:10.1038/srep12467. http://www.nature.com/articles/srep12467#supplementary-information.

Han, L., Zhou, W., & Li, W. (2016). Fine particulate (PM2.5) dynamics during rapid urbanization in Beijing, 1973–2013. *Scientific Reports, 6*, 23604. doi:10.1038/srep23604.

He, H., Wang, Y. S., Ma, Q. X., Ma, J. Z., Chu, B. W., Ji, D. S., Tang, G. Q., Liu, C., Zhang, H. X., & Hao, J. M. (2014). Mineral dust and NOx promote the conversion of SO2 to sulfate in heavy pollution days. *Scientific Reports, 4*. doi:10.1038/srep04172.

Hu, H., Zhang, X.-H., & Lin, L.-L. (2014). The interactions between China's economic growth, energy production and consumption and the related air emissions during 2000–2011. *Ecological Indicators, 46*, 38–51.

Hua, W., Chen, Z. M., Jie, C. Y., Kondo, Y., Hofzumahaus, A., Takegawa, N., Chang, C. C., Lu, K. D., Miyazaki, Y., Kita, K., Wang, H. L., Zhang, Y. H., & Hu, M. (2008). Atmospheric hydrogen peroxide and organic hydroperoxides during PRIDE-PRD'06, China: Their concentration, formation mechanism and contribution to secondary aerosols. *Atmospheric Chemistry and Physics, 8*(22), 6755–6773. doi:10.5194/acp-8-6755-2008.

Huang, R.-J., Zhang, Y., Bozzetti, C., Ho, K.-F., Cao, J.-J., Han, Y., Daellenbach, K. R., Slowik, J. G., Platt, S. M., Canonaco, F., Zotter, P., Wolf, R., Pieber, S. M., Bruns, E. A., Crippa, M., Ciarelli, G., Piazzalunga, A., Schwikowski, M., Abbaszade, G., Schnelle-Kreis, J., Zimmermann, R., An, Z., Szidat, S., Baltensperger, U., Haddad, I. E., & Prevot, A. S. H. (2014). High secondary aerosol contribution to particulate pollution during haze events in China. *Nature, 514*(7521), 218–222. doi:10.1038/nature13774. http://www.nature.com/nature/journal/v514/n7521/abs/nature13774.html#supplementary-information.

Huang, X., Liu, Z., Zhang, J., Wen, T., Ji, D., & Wang, Y. (2016). Seasonal variation and secondary formation of size-segregated aerosol water-soluble inorganic ions during pollution episodes in Beijing. *Atmospheric Research, 168*, 70–79. doi:10.1016/j.atmosres.2015.08.021.

Jacobson, M. Z. (2001). Strong radiative heating due to the mixing state of black carbon in atmospheric aerosols. *Nature, 409*(6821), 695–697.

Janssen, N. A., Hoek, G., Simic-Lawson, M., Fischer, P., Van Bree, L., Ten Brink, H., Keuken, M., Atkinson, R. W., Anderson, H. R., & Brunekreef, B. (2011). Black carbon as an additional indicator of the adverse health effects of airborne particles compared with PM10 and PM2.5. *Environmental Health Perspectives, 119*(12), 1691.

Janssen, N. A., Gerlofs-Nijland, M. E., Lanki, T., Salonen, R. O., Cassee, F., Hoek, G., Fischer, P., Brunekreef, B., & Krzyzanowski, M. (2012). *Health effects of black carbon*. Copenhagen: WHO Regional Office for Europe.

Ji, D., Li, L., Wang, Y., Zhang, J., Cheng, M., Sun, Y., Liu, Z., Wang, L., Tang, G., Hu, B., Chao, N., Wen, T., & Miao, H. (2014). The heaviest particulate air-pollution episodes occurred in northern China in January, 2013: Insights gained from observation. *Atmospheric Environment, 92*, 546–556. doi:10.1016/j.atmosenv.2014.04.048.

Ji, D., Zhang, J., He, J., Wang, X., Pang, B., Liu, Z., Wang, L., & Wang, Y. (2016). Characteristics of atmospheric organic and elemental carbon aerosols in urban Beijing, China. *Atmospheric Environment, 125*(Part A), 293–306. http://dx.doi.org/10.1016/j.atmosenv.2015.11.020.

Krotkov, N. A., McLinden, C. A., Li, C., Lamsal, L. N., Celarier, E. A., Marchenko, S. V., Swartz, W. H., Bucsela, E. J., Joiner, J., Duncan, B. N., Boersma, K. F., Veefkind, J. P., Levelt, P. F., Fioletov, V. E., Dickerson, R. R., He, H., Lu, Z., & Streets, D. G. (2016). Aura OMI observations of regional SO2 and NO2 pollution changes from 2005 to 2015. *Atmospheric Chemistry and Physics, 16*(7), 4605–4629. doi:10.5194/acp-16-4605-2016.

Li, P., Yan, R., Yu, S., Wang, S., Liu, W., & Bao, H. (2015). Reinstate regional transport of PM2. 5 as a major cause of severe haze in Beijing. *Proceedings of the National Academy of Sciences, 112*(21), E2739–E2740. doi:10.1073/pnas.1502596112.

Liu, Z., Hu, B., Zhang, J., Yu, Y., & Wang, Y. (2016). Characteristics of aerosol size distributions and chemical compositions during wintertime pollution episodes in Beijing. *Atmospheric Research, 168*, 1–12. http://dx.doi.org/10.1016/j.atmosres.2015.08.013.

Lv, B., Zhang, B., & Bai, Y. (2016). A systematic analysis of PM 2.5 in Beijing and its sources from 2000 to 2012. *Atmospheric Environment, 124*, 98–108.

Menon, S., Hansen, J., Nazarenko, L., & Luo, Y. (2002). Climate effects of black carbon aerosols in China and India. *Science, 297*(5590), 2250–2253.

Pan, Y., Tian, S., Liu, D., Fang, Y., Zhu, X., Zhang, Q., Zheng, B., Michalski, G., & Wang, Y. (2016a). Fossil fuel combustion-related emissions dominate atmospheric ammonia sources during severe haze episodes: Evidence from 15N-stable isotope in size-resolved aerosol ammonium. *Environmental Science and Technology, 50*, 8049.

Pan, Y., Wang, Y., Zhang, J., Liu, Z., Wang, L., Tian, S., Tang, G., Gao, W., Ji, D., Song, T., & Wang, Y. (2016b). Redefining the importance of nitrate during haze pollution to help optimize an emission control strategy. *Atmospheric Environment, 141*, 197–202. doi:10.1016/j.atmosenv.2016.06.035.

Pathak, R., Wu, W., & Wang, T. (2009). Summertime PM2. 5 ionic species in four major cities of China: Nitrate formation in an ammonia-deficient atmosphere. *Atmospheric Chemistry and Physics, 9*, 1711–1722.

Peng, J., Hu, M., Guo, S., Du, Z., Zheng, J., Shang, D., Zamora, M. L., Zeng, L., Shao, M., & Wu, Y.-S. (2016). Markedly enhanced absorption and direct radiative forcing of black carbon under polluted urban environments. *Proceedings of the National Academy of Sciences, 113*, 4266–4271.

Poppe, D., Wallasch, M., & Zimmermann, J. (1993). The dependence of the concentration of OH on its precursors under moderately polluted conditions: A model study. *Journal of Atmospheric Chemistry, 16*(1), 61–78. doi:10.1007/bf00696625.

Quan, J., Zhang, Q., He, H., Liu, J., Huang, M., & Jin, H. (2011). Analysis of the formation of fog and haze in North China Plain (NCP). *Atmospheric Chemistry and Physics, 11*(15), 8205–8214. doi:10.5194/acp-11-8205-2011.

Quan, J., Liu, Q., Li, X., Gao, Y., Jia, X., Sheng, J., & Liu, Y. (2015). Effect of heterogeneous aqueous reactions on the secondary formation of inorganic aerosols during haze events. *Atmospheric Environment, 122*, 306–312. doi:10.1016/j.atmosenv.2015.09.068.

Sheehan, P., Cheng, E., English, A., & Sun, F. (2014). China's response to the air pollution shock. *Nature Climate Change, 4*(5), 306–309.

Sun, Y., Zhuang, G., Tang, A., Wang, Y., & Ans, Z. (2006). Chemical Characteristics of PM2.5 and PM10 in Haze- Fog Episodes in Beijing. *Environmental Science and Technology, 40*(10), 3148–3155.

Sun, Y. L., Wang, Z. F., Fu, P. Q., Yang, T., Jiang, Q., Dong, H. B., Li, J., & Jia, J. J. (2013). Aerosol composition, sources and processes during wintertime in Beijing, China. *Atmospheric Chemistry and Physics, 13*(9), 4577–4592. doi:10.5194/acp-13-4577-2013.

Sun, Y. L., Jiang, Q., Wang, Z. F., Fu, P. Q., Li, J., Yang, T., & Yin, Y. (2014). Investigation of the sources and evolution processes of severe haze pollution in Beijing in January 2013. *Journal of Geophysical Research-Atmospheres, 119*(7), 4380–4398.

Sun, Y., Du, W., Wang, Q., Zhang, Q., Chen, C., Chen, Y., Chen, Z., Fu, P., Wang, Z., & Gao, Z. (2015a). Real-time characterization of aerosol particle composition above the urban canopy in Beijing: Insights into the interactions between the atmospheric boundary layer and aerosol chemistry. *Environmental Science and Technology, 49*(19), 11340–11347.

Sun, Y. L., Wang, Z. F., Du, W., Zhang, Q., Wang, Q. Q., Fu, P. Q., Pan, X. L., Li, J., Jayne, J., & Worsnop, D. R. (2015b). Long-term real-time measurements of aerosol particle composition in Beijing, China: Seasonal variations, meteorological effects, and source analysis. *Atmospheric Chemistry and Physics, 15*(17), 10149–10165. doi:10.5194/acp-15-10149-2015.

Sun, Y., Chen, C., Zhang, Y., Xu, W., Zhou, L., Cheng, X., Zheng, H., Ji, D., Li, J., Tang, X., Fu, P., & Wang, Z. (2016a). Rapid formation and evolution of an extreme haze episode in Northern China during winter 2015. *Scientific Reports, 6*, 27151. doi:10.1038/srep27151.

Sun, Y., Jiang, Q., Xu, Y., Ma, Y., Zhang, Y., Liu, X., Li, W., Wang, F., Li, J., Wang, P., & Li, Z. (2016b). Aerosol characterization over the North China Plain: Haze life cycle and biomass burning impacts in summer. *Journal of Geophysical Research: Atmospheres, 121*(5), 2508–2521. doi:10.1002/2015jd024261.

Tang, G., Zhang, J., Zhu, X., Song, T., Münkel, C., Hu, B., Schäfer, K., Liu, Z., Zhang, J., Wang, L., Xin, J., Suppan, P., & Wang, Y. (2016). Mixing layer height and its implications for air pollution over Beijing, China. *Atmospheric Chemistry and Physics, 16*(4), 2459–2475. doi:10.5194/acp-16-2459-2016.

Tian, S., Pan, Y., Liu, Z., Wen, T., & Wang, Y. (2014). Size-resolved aerosol chemical analysis of extreme haze pollution events during early 2013 in urban Beijing, China. *Journal of Hazardous Materials, 279*, 452–460. doi:10.1016/j.jhazmat.2014.07.023.

Van Damme, M., Clarisse, L., Heald, C., Hurtmans, D., Ngadi, Y., Clerbaux, C., Dolman, A., Erisman, J. W., & Coheur, P.-F. (2014). Global distributions, time series and error characterization of atmospheric ammonia (NH3) from IASI satellite observations. *Atmospheric Chemistry and Physics, 14*(6), 2905–2922.

Wang, L., Zhang, N., Liu, Z., Sun, Y., Ji, D., & Wang, Y. (2014a). The influence of climate factors, meteorological conditions, and boundary-layer structure on severe haze pollution in the Beijing-Tianjin-Hebei Region during January 2013. *Advances in Meteorology, 2014*, 1–14. doi:10.1155/2014/685971.

Wang, Y., Yao, L., Wang, L., Liu, Z., Ji, D., Tang, G., Zhang, J., Sun, Y., Hu, B., & Xin, J. (2014b). Mechanism for the formation of the January 2013 heavy haze pollution episode over central and eastern China. *Science China Earth Sciences, 57*(1), 14–25. doi:10.1007/s11430-013-4773-4.

Wang, L., Liu, Z., Sun, Y., Ji, D., & Wang, Y. (2015a). Long-range transport and regional sources of PM2.5 in Beijing based on long-term observations from 2005 to 2010. *Atmospheric Research, 157*, 37–48. doi:10.1016/j.atmosres.2014.12.003.

Wang, L., Xin, J., Li, X., & Wang, Y. (2015b). The variability of biomass burning and its influence on regional aerosol properties during the wheat harvest season in North China. *Atmospheric Research, 157*, 153–163. http://dx.doi.org/10.1016/j.atmosres.2015.01.009.

Wu, J., Luo, J., Zhang, L., Xia, L., Zhao, D., & Tang, J. (2014). Improvement of aerosol optical depth retrieval using visibility data in China during the past 50 years. *Journal of Geophysics Research: Atmosphere, 119*(23).

Xue, Y., Tian, H., Yan, J., Zhou, Z., Wang, J., Nie, L., Pan, T., Zhou, J., Hua, S., & Wang, Y. (2016). Temporal trends and spatial variation characteristics of primary air pollutants emissions from coal-fired industrial boilers in Beijing, China. *Environmental Pollution, 213*, 717–726.

Yang, F., Tan, J., Zhao, Q., Du, Z., He, K., Ma, Y., Duan, F., & Chen, G. (2011). Characteristics of PM 2.5 speciation in representative megacities and across China. *Atmospheric Chemistry and Physics, 11*(11), 5207–5219.

Yang, Y. R., Liu, X. G., Qu, Y., An, J. L., Jiang, R., Zhang, Y. H., Sun, Y. L., Wu, Z. J., Zhang, F., Xu, W. Q., & Ma, Q. X. (2015). Characteristics and formation mechanism of continuous hazes in China: A case study during the autumn of 2014 in the North China Plain. *Atmospheric Chemistry and Physics, 15*(14), 8165–8178. doi:10.5194/acp-15-8165-2015.

Zhang, Y., & Cao, F. (2015). Fine particulate matter (PM2. 5) in China at a city level. *Scientific Reports, 5*, 14884. doi:10.1038/srep14884.

Zhang, X., Wang, Y., Niu, T., Zhang, X., Gong, S., Zhang, Y., & Sun, J. (2012). Atmospheric aerosol compositions in China: spatial/temporal variability, chemical signature, regional haze distribution and comparisons with global aerosols. *Atmospheric Chemistry and Physics, 12*(2), 779–799.

Zhang, J. K., Sun, Y., Liu, Z. R., Ji, D. S., Hu, B., Liu, Q., & Wang, Y. S. (2014). Characterization of submicron aerosols during a month of serious pollution in Beijing, 2013. *Atmospheric Chemistry and Physics, 14*(6), 2887–2903. doi:10.5194/acp-14-2887-2014.

Zhang, Q., Quan, J., Tie, X., Li, X., Liu, Q., Gao, Y., & Zhao, D. (2015a). Effects of meteorology and secondary particle formation on visibility during heavy haze events in Beijing, China. *Science of the Total Environment, 502*(0), 578–584. http://dx.doi.org/10.1016/j.scitotenv.2014.09.079.

Zhang, Y.-L., Huang, R.-J., El Haddad, I., Ho, K.-F., Cao, J.-J., Han, Y., Zotter, P., Bozzetti, C., Daellenbach, K., & Canonaco, F. (2015b). Fossil vs. non-fossil sources of fine carbonaceous aerosols in four Chinese cities during the extreme winter haze episode of 2013. *Atmospheric Chemistry and Physics, 15*(3), 1299–1312.

Zhang, X. Y., Wang, J. Z., Wang, Y. Q., Liu, H. L., Sun, J. Y., & Zhang, Y. M. (2015c). Changes in chemical components of aerosol particles in different haze regions in China from 2006 to 2013 and contribution of meteorological factors. *Atmospheric Chemistry and Physics, 15*(22), 12935–12952. doi:10.5194/acp-15-12935-2015.

Zhao, X., Zhao, P., Xu, J., Meng, W., Pu, W., Dong, F., He, D., & Shi, Q. (2013). Analysis of a winter regional haze event and its formation mechanism in the North China Plain. *Atmospheric Chemistry and Physics, 13*(11), 5685–5696.

Zheng, B., Zhang, Q., Zhang, Y., He, K., Wang, K., Zheng, G., Duan, F., Ma, Y., & Kimoto, T. (2015a). Heterogeneous chemistry: A mechanism missing in current models to explain secondary inorganic aerosol formation during the January 2013 haze episode in North China. *Atmospheric Chemistry and Physics, 15*(4), 2031–2049.

Zheng, G., Duan, F., Su, H., Ma, Y., Cheng, Y., Zheng, B., Zhang, Q., Huang, T., Kimoto, T., & Chang, D. (2015b). Exploring the severe winter haze in Beijing: The impact of synoptic weather, regional transport and heterogeneous reactions. *Atmospheric Chemistry and Physics, 15*(6), 2969–2983.

Zheng, G., Duan, F., Ma, Y., Zhang, Q., Huang, T., Kimoto, T., Cheng, Y., Su, H., & He, K. (2016a). Episode-based evolution pattern analysis of haze pollution: Method development and results from Beijing, China. *Environmental Science & Technology, 50*(9), 4632–4641.

Zheng, J., Hu, M., Peng, J., Wu, Z., Kumar, P., Li, M., Wang, Y., & Guo, S. (2016b). Spatial distributions and chemical properties of PM 2.5 based on 21 field campaigns at 17 sites in China. *Chemosphere, 159*, 480–487.

Zhou, X., Cao, Z., Ma, Y., Wang, L., Wu, R., & Wang, W. (2015). Concentrations, correlations and chemical species of PM/PM based on published data in China: Potential implications for the revised particulate standard. *Chemosphere, 144*, 518–526. doi:10.1016/j.chemosphere. 2015.09.003.

Chapter 12
Spatial Distributions, Chemical Properties, and Sources of Ambient Particulate Matters in China

Song Guo, Jing Zheng, and Min Hu

Abstract Severe air pollution and its associated health impacts have become one of the major concerns in China. Among all the air pollutants, ambient particles are of great interest to the scientific community and policy makers because of their ability to carry health risks, as well as influence air quality and global climate. The researches on chemical characteristics of particles remain key priorities since the mass and ratios of different components imply the various sources and formation mechanisms. It is crucial to develop detailed chemical databases of particles in China to precisely conduct source apportionment and evaluate environmental impacts. This paper elucidates chemical properties of $PM_{2.5}$ at the particular sites and seasons in major regions of China. Characteristics of regional air pollutions are also analyzed by urban-regional pair studies. The formation of heavy pollution episodes is also discussed. Contributions of primary and secondary sources to the organic aerosols are investigated as well. Secondary organic aerosol (SOA) presents a major fraction of organic aerosol. Different from most of other cities in the world, anthropogenic SOA is dominant, suggesting reducing the anthropogenic VOCs is an efficient way to reduce organic aerosols in China. These studies help to improve the understanding of the current state of air pollution in diverse Chinese cities. Analysis of chemical signatures of $PM_{2.5}$ could further assist as an input to model simulations and in policy making for emission control.

Keywords Particulate matters • Chemical composition • Spatial distribution • Sources

---

S. Guo (✉) • J. Zheng • M. Hu
State Key Joint Laboratory of Environmental Simulation and Pollution Control, College of Environmental Sciences and Engineering, Peking University, Beijing 100871, China
e-mail: songguo@pku.edu.cn; jingzheng@pku.edu.cn; minhu@pku.edu.cn

I. Bouarar et al. (eds.), *Air Pollution in Eastern Asia: An Integrated Perspective*,
ISSI Scientific Report Series 16, DOI 10.1007/978-3-319-59489-7_12

12.1 Introduction

China has experienced increasing urbanization, industrialization, and anthropogenic pollution over the last quarter of the twentieth century (Tie and Cao 2009). High concentrations of gaseous pollutants such as sulfur dioxide (SO_2), ozone (O_3) and particulate matter (PM) from diverse sources coexist in the atmosphere, which far exceed the atmospheric self-purification capacity, and the complicate interactions among them lead to the formation of a complex air pollution that is difficult to disentangle (Shao et al. 2006). Air pollution comes from a complex mixture of sources such as traditional coal combustion, vehicular emissions, and secondary pollution (Kumar et al. 2011; Huang et al. 2014). In 2014, the annual average concentration of $PM_{2.5}$ (PM with aerodynamic equivalent diameters less than 2.5 μm) in 161 cities reached 62 μg m^{-3}, while only 11.2% of these cities met the Grade II of China National Ambient Air Quality Standard (Protection 2015). In recent years, severe and long-lasting haze affected China on many occasions. The pollution episodes affected 1.3 million km^2 and 800 million people, and attracted worldwide attention during the first quarter of 2013(Huang et al. 2014).

Significant efforts have been dedicated to understanding the characteristics and formation of fine particle pollution in megacities, especially for city clusters and megacities with high population density such as the Beijing, Tianjin and Hebei (BTH) region (Guo et al. 2010; He et al. 2001), Yangtze River Delta (YRD) region (Wang et al. 2006; Du et al. 2011), Pearl River Delta (PRD) region (Hu et al. 2008, 2012), and Sichuan Basin (Tao et al. 2013; Chen and Xie 2014). According to the pollution density, geographic location (i.e. distance of the measurement station from large pollution source), and energy structure, the cities in those regions can be broadly classified into four different types, including urban sites, regional sites, and coastal site. The regional sites represent both the natural emission of air pollutants and contribution from the long range transports of urban pollution.

12.2 Chemical Composition of $PM_{2.5}$ in China

12.2.1 Spatial Distribution of Chemical Species in $PM_{2.5}$

12.2.1.1 $PM_{2.5}$ Concentration

In general, the mass concentrations of fine particles at different types of sites are 2–5 times higher than those of corresponding sites in European and American cities (Hidy 2009; Putaud et al. 2010). As shown in Fig. 12.1, by summarizing the field campaign data collected from 2008 to 2012 in major regions of China, it is found that high levels of $PM_{2.5}$ concentration (exceeding 100 μg m^{-3}) can be observed at

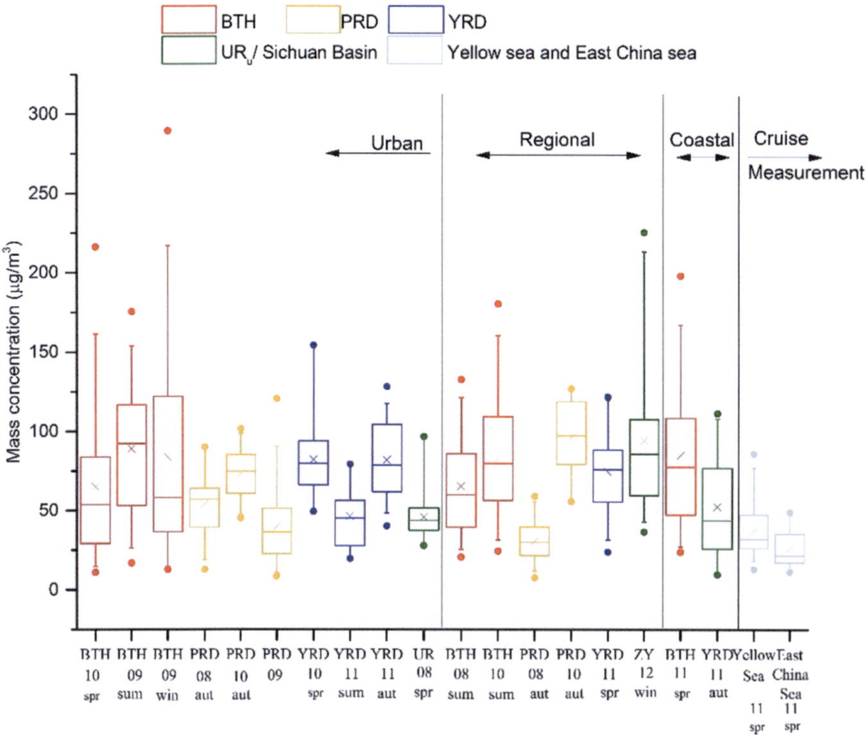

Fig. 12.1 Box-Whisker plot for $PM_{2.5}$ mass concentration of each filed campaign. The *box* denotes the 25, 75 percentiles. The whiskers denote the 5th and 95th percentiles. The *points* represent max/min value. *Crosses* and *horizontal lines* represent the mean value and median value of $PM_{2.5}$ concentration for entire study period, respectively

most of sites and covered from the urban to regional scale, indicating pollution episodes do not occur at or dependent on one particular season or site, but spread all over China. Standard deviations for each campaign are typically 40–70% of the averages, indicating that the $PM_{2.5}$ concentrations fluctuated and are influenced by extreme haze events. The average $PM_{2.5}$ concentrations are nearly identical at urban and regional pair sites of BTH, PRD and YRD regions, indicating the regional feature in the whole city cluster. The mean concentrations at coastal sites lie well above the median values due to the strong increases during injection of continental air mass with large amounts of air pollutants, which lead to skewed distributions of concentrations. The $PM_{2.5}$ concentrations measured at East China sea and Yellow sea are relatively higher than that of Indian seas (Quinn and Bates 2005), indicating that they are strongly influenced by the inland anthropogenic activity.

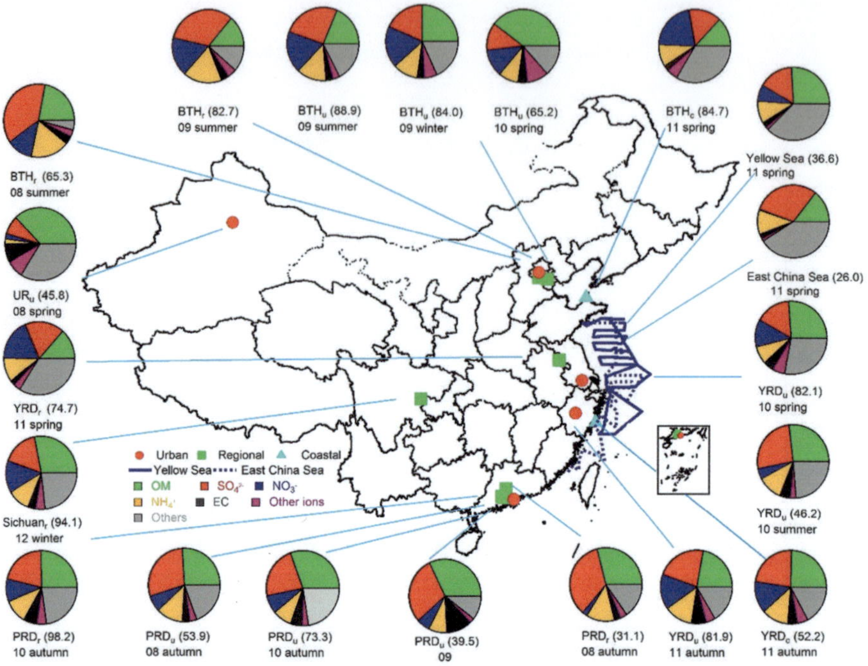

Fig. 12.2 The mass concentration of $PM_{2.5}$ and its major chemical speciation at different locations in China. (unit: μg m^{-3}). Please note that subscripts *u*, *r*, and *c* indicate urban, regional, coastal and cruise site types, respectively

12.2.1.2 Chemical Speciation

Relative contributions of each species to $PM_{2.5}$ at different sites are plotted in Fig. 12.2 Organic matter (OM) concentrations are reported by scaling OC by a constant factor of 1.8 (OM = 1.8× OC) for all sites (Turpin and Lim 2001). However it is possible that the actual scaling factor is inconstant and it might be higher at rural sites for organic aerosols due to a higher degree of aging (Zhang et al. 2011). The most abundant species in $PM_{2.5}$ in all campaigns are OM, sulfate, nitrate, ammonium, and EC. As shown in Fig. 12.2, despite the dominant components are same in China, the fractions of each compound vary between different sites and sampling periods. The chemical speciations measured in China are similar to the European studies in the sense, that OM accounts for the highest fraction of $PM_{2.5}$ at urban sites and that sulfate contributes more at regional sites (Putaud et al. 2004). The long lifetime of $PM_{2.5}$ combining with the wide spread formation of secondary aerosol results in the uniformity in the chemical composition of fine particles.

Carbonaceous Species

Carbonaceous aerosols in China are generated from the widespread use of coal, biomass burning, and petroleum products. Residential coal contributes significantly to black carbon in China (Bond et al. 2013). Carbonaceous species represent a large fraction of $PM_{2.5}$, accounting for around 30%. The field campaign average OC concentration ranged from 2.4 to 17.9 μg m^{-3}, while the EC levels showed a smaller range. This difference is probably attributed to the seasonal influence on secondary aerosol formation and residential heating and the different local source emission at each site (Sillanpää et al. 2006). The variations of OC/EC ratio may be used as an index reflecting emission sources and secondary organic aerosol (SOA) formation (Zeng and Wang 2011). In this study, OC/EC ratios ranged from 2.74 to 7.46, within the ranges reported by the literatures of China (Lin et al. 2009; Zhang et al. 2012). Comparatively higher OC/EC ratios are observed in East China Sea (Zheng et al. 2016), due to the SOA formation during long-range transportation from inland to the ocean and biogenically driven OC's contribution to marine aerosol (O'Dowd et al. 2004).

For urban sites, the results are comparable to the concentrations of carbonaceous aerosols at some Asia urban sites (Lin et al. 2009; Park et al. 2005; Kim et al. 2007), but are significantly higher than observations in Zografou, Greece (Remoundaki et al. 2013). The heating season with a greater consumption of coals in northern China is a plausible explanation for the high OC concentration in Beijing. The OC concentration in the wintertime in Ziyang (ZY_r) in Sichuan Basin ranked second after Beijing wintertime. The stagnant dispersion conditions caused by mountain-basin topography in this region enhance the accumulation of pollutants, leading to a high concentration of OC. In addition, biomass burning in this region also contributes to the high concentration of carbonaceous aerosol (Chen and Xie 2014). The regional OC and EC in BTH, YRD and PRD region are comparatively higher than those from other regional station in the world (Yu et al. 2009; Polidori et al. 2006; Viana et al. 2007). This may be due to the strong anthropogenic influences from urban region, respectively. The measurement at the coastal site is comparable to the previous observation during the spring of 2003 (Feng et al. 2007). Asian continental flow carried pollutants to this downstream site. Given the fact that diverse chemical analytical methods used to determine the OC and EC concentration can result in significant differences in the relative concentrations with consistent Total carbon (sum of OC plus EC) (Chow et al. 2004), here we only use the TOT concentration for comparison.

Inorganic Species

The anthropogenic emissions of precursors (SO_2, NOx, and NH_3) strongly influence the concentration and compositions of SNA aerosol (sum of the mass concentration of NH_4^+, SO_4^{2-} and NO_3^-) (Wang et al. 2013). The high concentration of precursor gaseous pollutants can partially explain the high SNA concentrations in China. The contributions of sulfate are almost equal to those of OM in urban field campaigns, and higher contributions of sulfate are observed in regional sites. The concentrations of sulfate are five to ten times the measured concentrations in Europe and United States (Hidy 2009; Putaud et al. 2004). The fractions of SO_4^{2-} showed clear seasonal variation in BTH region in 2009. In the summer, enhanced photochemistry provides a higher concentration of oxidants (OH, O_3, and H_2O_2) and favorable metrological condition (e.g. high temperatures and humidity) lead to increased sulfate (Ziemba et al. 2007; Yang et al. 2011b; Ianniello et al. 2011).

Nitric acid is mainly formed by the reaction of NO_2 with OH radical during daytime or through hydrolysis of N_2O_5 during nighttime. The former pathway dominates HNO_3 formation (Alexander et al. 2009). Although the PRD region has higher NOx emission intensity (Wang et al. 2013), the concentrations and the relative contributions of nitrate in this region were lower than those of other regions in 2010 (below 10% of $PM_{2.5}$). The reason is partly attributed to the difficulties in precisely measuring particulate nitrate due to its volatilization under high temperatures.

Total ammonia emission in 2006 of China was estimated to be 9.8 Tg, contributing to 15% of global emissions. Since China is a large agricultural country, the major sources of ammonia are identified as livestock manure management (53%) and the use of synthetic N fertilizers (33%) (Huang et al. 2012). It is found that NH_4^+ concentration ere higher at rural sites than those of urban sites due to higher emission of NH_3 (Shen et al. 2011).

In general, SNA contributed to more than 90% of the total inorganic ionic species mass and accounted for ~50% of $PM_{2.5}$. Regional sites present a comparatively higher percentage of SNA than urban sites, indicating high contribution of secondary aerosol formation. It is consistent with European results that the regional background aerosol of Spain exhibited similar secondary characteristics (Salvador et al. 2012).

For the sites strongly influenced by biomass burning, the concentrations of K^+ could be more than 1 μg m^{-3}, accounting for ~2% in $PM_{2.5}$. Higher chloride concentration at coastal site is mainly attributed to sea salt aerosols, for the coastal site is influenced by marine air masses (Wu et al. 2006). Relative higher concentrations of Ca^{2+} were observed at Urumchi (UR_u) in spring of 2008. It has been reported that in spring, soil dust is transported from Jungger Basin and Guebantongute desert in the north of Urumchi, resulting in a higher contribution of Ca^{2+} (Li et al. 2008).

The contribution of unidentified fraction (others) is estimated by taking the difference of the sum of OM, EC, and inorganic constituents from the $PM_{2.5}$, accounting for 2–40% among all the field campaigns. The unidentified matter mainly consists of crustal dust and sea salt. Compared to sulfate, nitrate and ammonium mentioned above, crustal species tend to play less important role in $PM_{2.5}$ (around 10%) (Zhou et al. 2016). However, given their importance in source apportionment, detailed analysis of elements will be done in future analysis.

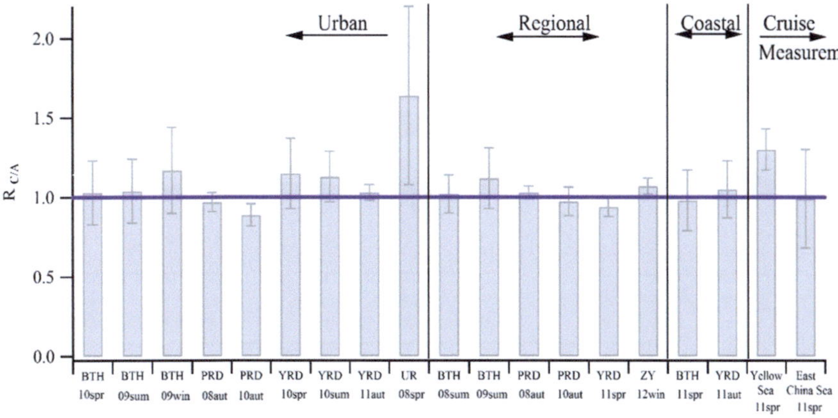

Fig. 12.3 $R_{C/A}$ of different sites

12.2.2 Chemical Characterizations of Fine Particles

12.2.2.1 Particle Acidity

Acidity is an important chemical property to measure because it governs the aerosol phase reactions in terms of catalyzation of the secondary organic aerosol formation reaction (Surratt et al. 2007). It is affected by the relative concentration of SNA, which is further controlled by factors such as the levels of precursors, oxidants, and micro-meteorological conditions (Squizzato et al. 2013; Pathak et al. 2009). It is reasonable to assume that SO_4^{2-}, NO_3^-, and NH_4^+ control fine particle acidity since SNA dominate the ion concentration (Ziemba et al. 2007; Yang et al. 2011b). The neutralizing levels of $PM_{2.5}$ are determined by the equivalent charge ratio of the acidic sulfate and nitrate to ammonia (He et al. 2012; Pathak et al. 2009). The neutralization ratio is defined as following:

$$R = \frac{[NH_4^+]}{[NO_3^-] + [SO_4^{2-}]} \tag{12.1}$$

In Eq. (12.1), the aerosol acidity is evaluated by possible neutralization of the two major inorganic acids with ammonium. A ratio less than one indicates a partial neutralization of acidic aerosols. Excess ammonium favors the gas-phase reaction to neutralized H_2SO_4 and HNO_3. In the BTH and YRD regions, this ratio is near 1, ranging from 0.89 to 1.10, indicating sulfate and nitrate can be almost fully neutralized by ammonia (See Fig. 12.3). For Urumchi, given the dominant role of Ca^{2+} in the chemical properties of $PM_{2.5}$ as mentioned above, the contribution of Ca^{2+} should be taken into the consideration. The property of aerosol would switch from the acidic to alkaline mode.

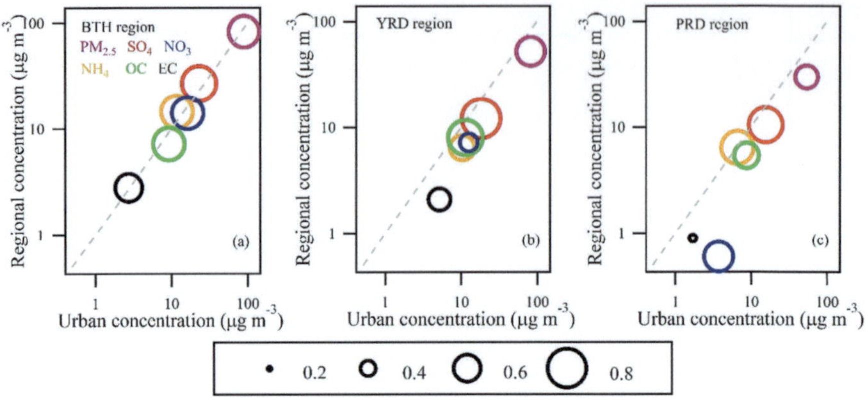

Fig. 12.4 Comparison of mass concentration of $PM_{2.5}$ and major chemical species for urban and regional sites in each region. Each *color* corresponds to a chemical species. The size of each cycle is proportional to the correlation values between the temporal trends of each species at urban and regional site

12.2.2.2 Characteristic of Regional Air Pollution

The role of regional air pollution due to the rapid urbanization has been studied extensively in the past few years (Guo et al. 2010; Li et al. 2014; Yang et al. 2011a). Regional pollution is regarded as the regional scale elevated concentrations of air pollutants, which mainly due to the secondary formation and the horizontal and vertical homogeneity of the meteorological conditions of the whole region (Jia et al. 2008). Previous results revealed that persistent regional stagnation conditions favored the elevation of PM mass or SO_4^{2-} on a regional scale from the mid-western to northeastern of United States (Blanchard et al. 2013). Several studies have been performed at urban/rural pair sites to investigate regional air pollution in BTH, PRD and YRD region (Yang et al. 2011a; Guo et al. 2010).

The comparison of chemical properties of fine particles at a pair sites supports the assessment of regional air pollution. For most chemical species, urban concentrations are similar to their regional counterparts, indicating the uniform distribution of regional pollution. For EC, the urban concentration can reach 222% higher than that of regional sites, showing a stronger local contribution than other components (see Fig. 12.4). The linear correlations between temporal trends of each species (e.g. concentration of $PM_{2.5}$, OC, SO_4^{2-}, NO_3^-, NH_4^+) of pair sites are analyzed here in terms of the Spearman correlation analysis, for example the correlation of $PM_{2.5}$ concentrations of urban and regional site in BTH region, to investigate the regional pollutions of fine particles during the observation periods. Table 12.1 summarizes the results of the correlation coefficient of each species. The temporal trends of $PM_{2.5}$ at the urban sites typically correlate well with those of corresponding regional sites. The regional wide elevated PM concentrations indicate the occurrence of regional air pollution. The analysis also show generally high correlation values between the daily data of pair sites for the secondary inorganic

Table 12.1 Spearman correlation analysis of daily data from pair sites in three regions for $PM_{2.5}$ and different chemical components

Spearman correlations	$PM_{2.5}$	OC	EC	SO_4^{2-}	NO_3^-	NH_4^+
BTH	0.71**	0.73**	0.64**	0.78**	0.72**	0.73**
PRD	0.55**	0.60**	0.26	0.78**	0.68**	0.76**
YRD	0.66**	0.77**	0.54*	0.74**	0.46*	0.59**

*Correlation is significant at the 0.05 level (2-tailed)
**Correlation is significant at the 0.01 level (2-tailed)

species. The SO_4^{2-} correlations for BTH, YRD, and PRD region ranged from 0.74 to 0.78. This is consistent with a regional formation mechanism with SO_4^{2-}. For the primary traffic-related components, EC, the correlation coefficient decreased to around 0.5 for the three regions, indicating the influence of local emissions (see Fig. 12.4). High correlations between the temporal trends of $PM_{2.5}$ and different chemical components of urban and regional sites highlight the uniformly distributed air pollutants within one region. The regional influences are greater for secondary species than primary components.

12.2.2.3 Formation of Heavy Pollution Episodes

Given the coupling interactions between complex meteorological conditions, pollution sources, and atmospheric transformation processes, characterization of formation mechanisms of severe haze with high $PM_{2.5}$ levels at different sites presents a major challenge (Guo et al. 2014). Figure 12.5 shows the relative contribution of OM, EC, SO_4^{2-}, NO_3^-, NH_4^+ and others to $PM_{2.5}$ as a function of $PM_{2.5}$ mass concentration. In Northern China (mainly BTH region), the fraction of OM in fine particle decreases with increasing $PM_{2.5}$ concentration, while the relative contributions of secondary inorganic species increases at the urban, regional, and coastal sites. When $PM_{2.5}$ concentration reached peak values, the fractions of SNA can reach around 60–80%, indicating the importance of secondary formation in the pollution episodes. Previous research with back trajectory analysis revealed that the clean air masses from the north had a larger contribution from primary aerosol emissions, characterized by a higher contribution of OM, while haze pollution occurred when stagnant air masses with secondary regional aerosols dominated (Huang et al. 2010; Sun et al. 2010). A broadly consistent pattern was also observed in Taiyuan, which is a heavily polluted city in Northern China (Meng et al. 2007). Lower fraction of OC was observed on heavy polluted days (24%) compared to relative clean days (50%). However in the Southern China (YRD and PRD region), the relative contributions of different chemical components remain relatively constant (the increase of secondary aerosols concentration is lower than 15%) as the $PM_{2.5}$ concentration increases. No significant increase of SNA fraction in $PM_{2.5}$ was also observed during a heavy polluted episode in the YRD region, with 42% on moderate days and 52% on the heavily polluted days (Fu et al. 2008). A similar

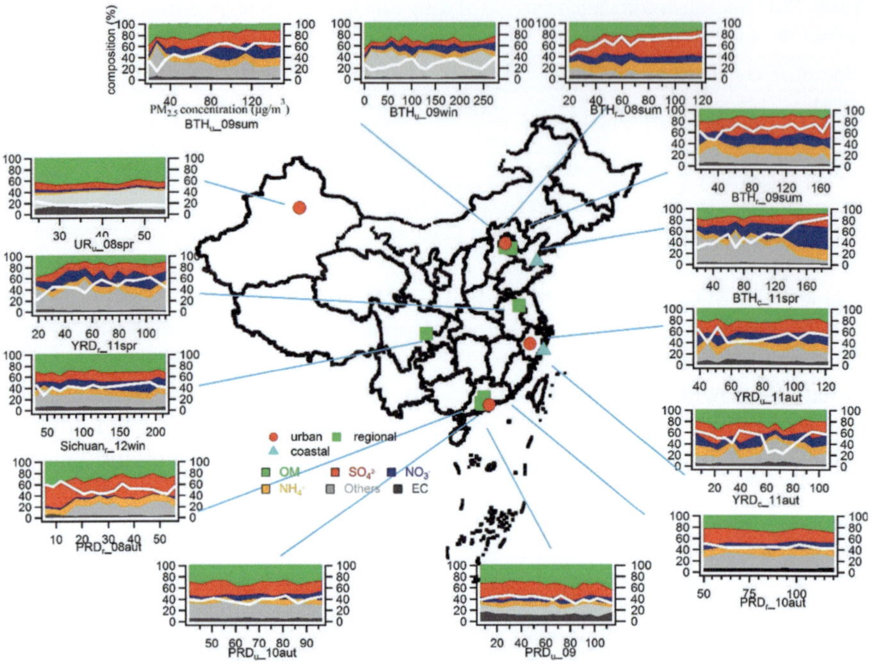

Fig. 12.5 Relative contribution of each species (OM, SO_4^{2-}, NO_3^-, NH_4^+, EC, and others) to $PM_{2.5}$ is shown as a function of total mass concentration of $PM_{2.5}$. The *grey line* denotes the trend of fraction of SNA as a function of $PM_{2.5}$ level

phenomenon was also observed in Xiamen, a city in Southern China (Zhang et al. 2013).

Although secondary formation is a major cause of particulate pollution in the whole China, the discrepancy of relative contribution from different chemical components to $PM_{2.5}$ level implies diverse formation mechanisms of heavy air pollution at different sites of China. Comparatively stronger emissions of gas precursors (e.g. SO_2, NO_x) in Northern China would have a potential effect on SNA formation (Wang et al. 2013). It may be a plausible explanation for the dominant contribution of SNA. It is of interest to perform detail analyses on the relationship of geographical features with formation mechanisms of heavy pollution at specific sites in China.

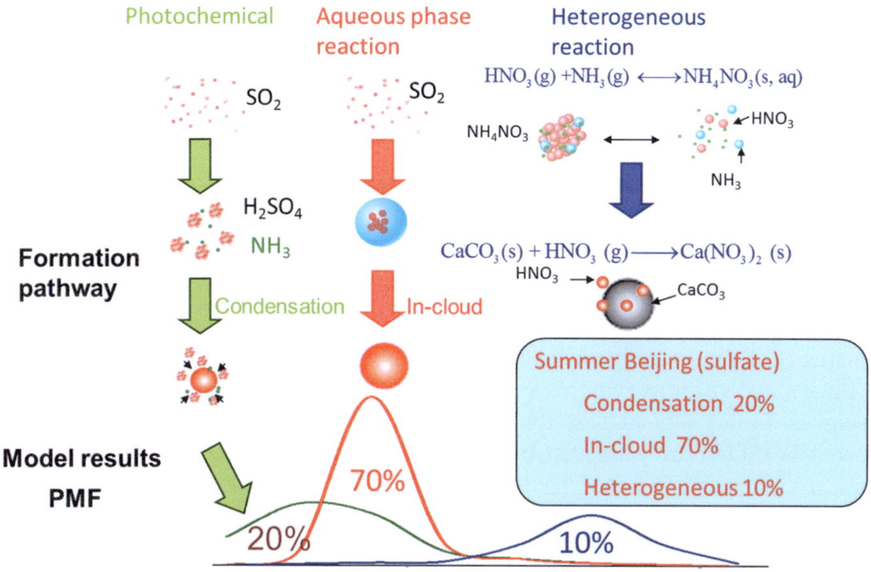

Fig. 12.6 Quantification of secondary formation pathways to secondary inorganic aerosols: a case of sulfate formation in summer of Beijing

12.2.3 Quantification of the Formation Pathways to Secondary Inorganic Aerosols

Formation pathways of secondary inorganic aerosols, i.e. sulfate, nitrate, and ammonium, are relatively well understood. Several possible formation processes have been proposed to explain the droplet mode sulfate, including condensation and coagulation of smaller particles, in-cloud processes and the growth of the condensation mode by addition of sulfate and water (aerosol droplet process), and heterogeneous reactions of inorganic gases on soil particles. However, the contributions of these formation pathways alter with temporal and spatial variation. It is important for models to quantify the contributions of different formation pathways. By using PMF model, different modes of secondary inorganic aerosols can be resolved from total particle mass, and the concentrations of resolved condensation mode, droplet mode and coarse mode can be regarded as the particles that are formed by gas-to-particle condensation process, in-cloud or aerosol droplet process and heterogeneous reaction process, respectively. Figure 12.6 shows a case of sulfate formation in summer of Beijing (Guo et al. 2010). As a result, in summer of Beijing, droplet mode is dominant, and an average fraction of 80% and 70% is attributed to in-cloud or aerosol droplet process at urban and rural site respectively. Correspondingly, 14 and 22% of the sulfate is attributed to gaseous H_2SO_4 condensation. The rest is contributed by heterogeneous reaction process.

12.3 Primary Sources and Secondary Formation of Organic Aerosols

Organic compounds constitute a significant fraction of atmospheric fine particulate matter (PM), and a comprehensive understanding of their impacts on human health, climate, and visibility requires a detailed characterization of their composition and sources. In this session, contributions of primary particles and secondary organic aerosols (SOA) are estimated by chemical mass balance (CMB) modeling and tracer-yield method based on the filter samples collected in a typical city of China, i.e. Beijing, in the year 2000, 2005, 2006, and 2008. The SOA results are also compared with those in other places in the world.

12.3.1 Primary and Secondary Sources of Ambient Organic Aerosols

Primary and secondary organic tracers are applied in CMB model and tracer-yield method to estimate the contributions of primary and secondary particulate organic sources to total OC. Five primary sources and four secondary organic carbons (SOCs) derived from isoprene, α-pinene, β-caryophyllene are apportioned. The results are listed in Table 12.2 (Wang et al. 2009; Zheng et al. 2005; Guo et al. 2013).

Table 12.2 Comparison of primary source contributions to fine particulate organic carbon in summer of Beijing (%)

	2000	2005	2006	2008	2008
Year/site	PKU	PKU	PKU	PKU	Yufa
Gasoline vehicles	19	10.6 ± 2.8	7.1 ± 1.5	10.3 ± 8.7	7.9 ± 6.2
Diesel vehicles		9.6 ± 6.2	5.7 ± 1.5	16.2 ± 5.9	14.5 ± 4.1
Vegetative detritus	1.5	0.3 ± 0.2	0.3 ± 0.2	1.4 ± 0.8	1.9 ± 1.5
Dust	25.6				
Wood burning		10.4 ± 3.5	8.2 ± 2.0	7.5 ± 5.5	8.9 ± 4.9
Straw burning		1.3 ± 0.7	1.7 ± 2.8		
Coal burning	5.8	2.1 ± 2.2	1.8 ± 0.8	5.8 ± 5.5	7.8 ± 4.6
Cigarette	2.2				
Cooking		24.5 ± 4.2	23.8 ± 4.3		
Other OC	45.9	41.1 ± 9.7	51.4 ± 5.6	60.7 ± 13.2	58.8 ± 11.0
Reference	Zheng et al. (2005)	Wang et al. (2009)		Guo et al. (2013)	

Vehicle emission is a very important source of fine organic particles in Beijing. It has been reported that the number of vehicles in Beijing increased 20–25% per year (China Statistical Yearbook 2011, http://www.stats.gov.cn/tjsj/ndsj/2011/indexeh.htm). The increasing vehicle amount has led to serious fine particle pollution in Beijing (Chan and Yao 2008). In 2008, diesel and gasoline vehicle emissions were the major contributors to ambient organic particles with the contributions of $16.2 \pm 5.9\%$ and $14.5 \pm 4.1\%$ by diesel engines and $10.3 \pm 8.7\%$ and $7.9 \pm 6.2\%$ by gasoline vehicles at Beijing urban and rural area, respectively. The higher contributions at urban area are reasonable because there are more vehicles in urban area. Vehicle emission contributed more in 2008 than previous years. However, considering the OC concentrations, the absolute concentrations attributing to vehicle emission are lowest in 2008 (20.9 $\mu g/m^3$, 16.1 $\mu g/m^3$, 22.1 $\mu g/m^3$, and 9.2 ± 3.3 $\mu g/m^3$, in 2000, 2005, 2006, and 2008, respectively). A diesel engine contribution is reduced more significant than gasoline engines.

Biomass burning is also an important source for ambient particles in Beijing. Although open biomass burning has been strictly forbidden in Beijing, and its contribution to organic particles could be considered to be zero (Cao et al. 2005), the open fire from nearby province could still influence the air quality of Beijing. Moreover, wood consumed as domestic fuel in rural area also greatly contributed to ambient fine particles. However, the biomass burning contribution keeps decreasing more recently.

Coal is the dominant fuel in Beijing. About 68.7% of the total energy consumption of China in 2008 was coal (China Statistical Yearbook 2008, http://www.stats.gov.cn/tjsj/ndsj/2008/indexeh.htm). In 2008, coal burning accounted for $5.8 \pm 5.5\%$ and $7.8 \pm 4.6\%$ of the measured OC. This result was comparable with the result in 2000, but higher than that reported in 2005 and 2006. For the total OC was much lower in 2008, the absolute coal burning OC was the lowest in recent years. The alleviation of coal burning pollution was mainly due to the more strict restriction. However, the rural area was still affected significantly by coal combustion. Although Coal cannot be used within the central urban area of Beijing, large amount of coal was still consumed in rural area of Beijing. In addition, the most common coal usage in rural area was residential coal, which has higher emission factors (Zhang et al. 2008).

There is no tracer-based SOA study in Beijing except the year of 2008, so the results in 2008 is exhibited here as an example. The contributions of the secondary aerosol sources were $34.5 \pm 13.9\%$ at the urban site in 2008, with $17.4 \pm 7.6\%$ from toluene, $9.7 \pm 5.4\%$ from isoprene, $5.1 \pm 2.0\%$ from α-pinene, and $2.3 \pm 1.7\%$ from β-caryophyllene. At the rural site, the secondary sources were responsible for $38.4 \pm 14.4\%$ of OC, among which toluene SOC accounted for $17.3 \pm 6.9\%$, isoprene for $13.9 \pm 9.1\%$, α-pinene for $5.6 \pm 1.9\%$ and β-caryophyllene for $1.7 \pm 1.0\%$. On average, $73.8 \pm 9.7\%$ and $79.6 \pm 10.1\%$ of the total OC could be apportioned by primary and secondary sources at the two sites, indicating major

Table 12.3 Comparison of the concentrations of secondary organic carbon (SOC) derived from isoprene, α-pinene, β-caryophyllene and toluene in Beijing and other regions (μg C/m^3)

	PKU	Yufa	RTP[a]	YL[b]	CL[c]
Isoprene SOC	0.92 ± 0.65	1.30 ± 1.01	1.13 ± 0.38	0.19 ± 0.21	0.21 ± 0.13
α-pinene SOC	0.49 ± 0.28	0.52 ± 0.21	0.73 ± 0.10	0.17 ± 0.11	0.21 ± 0.04
Toluene SOC	1.67 ± 0.98	1.53 ± 0.64	0.52 ± 0.24	0.21 ± 0.23	1.06 ± 0.36
β-caryophyllene SOC	0.21 ± 0.18	0.16 ± 0.10	1.10 ± 0.18	0.54 ± 0.73	0.06 ± 0.02

[a]RTP(US), Research Triangle Park, North California, US, suburban site located on a grass covered field surrounded
[b]YL(HK), Yuen Long, Hong Kong, urban site in densely populated residential area
[c]CL(US), Cleveland, OH, US, three sites in heavily industrialized area

primary and secondary sources of OC in Beijing. The unapportioned OC ranged from 0.3 to 5.5 μg m^{-3}, accounting for 3.5–49.6% of the total OC, likely being comprised of uncharacterized primary or secondary sources such as dust, cooking, and other SOC. Unapportioned OC of 20–25% also indicates that large gaps remained concerning the OC sources in Beijing.

12.3.2 Comparison of Secondary Sources Within Beijing and Other Regions

A comparison of the secondary sources in Beijing and other regions is summarized in Table 12.3. Within the Beijing region, SOCs from biogenic gas-phase precursors (e.g., isoprene and α-pinene) contribute more at rural area due to more biogenic emissions at the rural site. However, the toluene SOC at the rural site contributes as much as that at the urban site. Considering that no other toluene sources are near the two sites, such as petroleum refinery, the major source of toluene is probably from vehicles. A pair t-test result shows that there is no difference between toluene SOC concentrations at the two sites. This may be due to the comparable level of toluene at two sites (Wang et al. 2010). Moreover, a good correlation of toluene SOC between the two sites with the correlation coefficient of 0.74 also suggests that toluene SOC is regional. This result implies that some anthropogenic components are widespread in Beijing and represents a major factor in affecting the regional air quality.

The SOA results in Beijing are compared with other places in the world. Both biogenic and anthropogenic SOA concentrations are high in Beijing. Yufa (Beijing rural site) and RTP (Research Triangle Park, US suburban site) have higher biogenic SOA than the urban sites, i.e., PKU (Peking University, Beijing urban site), CL (Cleveland, OH, US, three sites in heavily industrialized area) and YL (Hong Kong, urban site in densely populated residential area). However, among the urban sites, PKU has much higher biogenic SOA, and especially the isoprene SOA concentration is as much as that at Yufa and RTP. Although Yufa is a rural site, the toluene SOA is very high, and the concentration is even higher than those at other urban regions, i.e., CL and YL. Compared with other places in the world, anthropogenic SOA presents a significant fraction in China, not only in urban area but also in rural area. This result is important regard to reducing organic aerosol pollution. It is considered extremely difficult to control SOA, except control the gaseous precursors. Biogenic SOA seems impossible to be controlled, since one does not want to cut off all the plants. However, things become easier in China, where anthropogenic SOA is dominant. Reducing anthropogenic VOCs is an efficient way to reduce organic aerosols in China.

12.4 Summary

The session presents comprehensive assessments of the chemical property of fine aerosol particles at different sites in China. Datasets of chemical characteristics of fine particles are analyzed to understand the sources and transformation of $PM_{2.5}$ in the atmosphere.

Heavy pollution episodes with $PM_{2.5}$ higher than 100 μg m^{-3} can be observed during most of the sites, independent of sites and seasons. High $PM_{2.5}$ mass concentrations of coastal sites and during cruise measurements indicate offshore regions of China are strongly influenced by continental air pollutants. Organics and sulfate are the most abundant species in $PM_{2.5}$. SNA account for about 50% of the $PM_{2.5}$ concentrations. Simultaneous measurements conducted at three urban-regional pairs of sites demonstrate the urban and regional sites have similar concentrations of pollutants. The similar variations of temporal trends of $PM_{2.5}$ and secondary inorganic species of pairs sites indicate a uniform distribution of pollution.

Secondary inorganic species are the dominant contributors to high $PM_{2.5}$ concentration in Northern China, indicating a high level of the secondary formation during haze days. The relative contributions of each species as a function of $PM_{2.5}$ mass concentration demonstrate that Northern and Southern China have different formation mechanisms of heavy pollution episode.

Vehicle emission, biomass burning, and coal combustion are major contributors of ambient organic particle in China. Secondary organic aerosol becomes a more and more important fraction. Anthropogenic SOA presents a dominant fraction compared with other places in the world, which make it easier to control the organic aerosol pollution of China. Reducing anthropogenic VOCs is an efficient way to reduce organic aerosols in China.

References

Alexander, B., Hastings, M. G., Allman, D. J., Dachs, J., Thornton, J. A., & Kunasek, S. A. (2009). Quantifying atmospheric nitrate formation pathways based on a global model of the oxygen isotopic composition ($\Delta^{17}O$) of atmospheric nitrate. *Atmospheric Chemistry and Physics, 9*, 5043–5056.

Blanchard, C. L., Hidy, G. M., Tanenbaum, S., Edgerton, E. S., & Hartsell, B. E. (2013). The Southeastern Aerosol Research and Characterization (SEARCH) study: Spatial variations and chemical climatology, 1999–2010. *Journal of the Air & Waste Management Association, 63*, 260–275. doi:10.1080/10962247.2012.749816.

Bond, T. C., Doherty, S. J., Fahey, D. W., Forster, P. M., Berntsen, T., DeAngelo, B. J., Flanner, M. G., Ghan, S., Kärcher, B., Koch, D., Kinne, S., Kondo, Y., Quinn, P. K., Sarofim, M. C., Schultz, M. G., Schulz, M., Venkataraman, C., Zhang, H., Zhang, S., Bellouin, N., Guttikunda, S. K., Hopke, P. K., Jacobson, M. Z., Kaiser, J. W., Klimont, Z., Lohmann, U., Schwarz, J. P., Shindell, D., Storelvmo, T., Warren, S. G., & Zender, C. S. (2013). Bounding the role of black carbon in the climate system: A scientific assessment. *Journal of Geophysical Research, 118*, 5380–5552. doi:10.1002/jgrd.50171.

Cao, G., Zhang, X., Wang, D., & Zheng, F. (2005). Inventory of emissions of pollutants from open burning crop residue. *Journal of Agriculture and Environmental Sciences, 24*, 800–804.

Chan, C. K., & Yao, X. (2008). Air pollution in mega cities in China. *Atmospheric Environment, 42*, 1–42. doi:10.1016/j.atmosenv.2007.09.003.

Chen, Y., & Xie, S.-d. (2014). Characteristics and formation mechanism of a heavy air pollution episode caused by biomass burning in Chengdu, Southwest China. *The Science of the Total Environment, 473–474*, 507–517. doi:10.1016/j.scitotenv.2013.12.069.

Chow, J. C., Watson, J. G., Chen, L.-W. A., Arnott, W. P., & Moosmuller, H. (2004). Equivalence of elemental carbon by thermal/optical reflectance and transmittance with different temperature protocols. *Environmental Science & Technology, 38*, 4414–4442.

Du, H., Kong, L., Cheng, T., Chen, J., Du, J., Li, L., Xia, X., Leng, C., & Huang, G. (2011). Insights into summertime haze pollution events over Shanghai based on online water-soluble ionic composition of aerosols. *Atmospheric Environment, 45*, 5131–5137. doi:10.1016/j.atmosenv.2011.06.027.

Feng, J., Guo, Z., Chan, C. K., & Fang, M. (2007). Properties of organic matter in PM2.5 at Changdao Island, China—a rural site in the transport path of the Asian continental outflow. *Atmospheric Environment, 41*, 1924–1935. doi:10.1016/j.atmosenv.2006.10.064.

Fu, Q., Zhuang, G., Wang, J., Xu, C., Huang, K., Li, J., Hou, B., Lu, T., & Streets, D. G. (2008). Mechanism of formation of the heaviest pollution episode ever recorded in the Yangtze River Delta, China. *Atmospheric Environment, 42*, 2023–2036. doi:10.1016/j.atmosenv.2007.12.002.

Guo, S., Hu, M., Wang, Z. B., Slanina, J., & Zhao, Y. L. (2010). Size-resolved aerosol water-soluble ionic compositions in the summer of Beijing: implication of regional secondary formation. *Atmospheric Chemistry and Physics, 10*, 947–959.

Guo, S., Hu, M., Guo, Q., Zhang, X., Schauer, J. J., & Zhang, R. (2013). Quantitative evaluation of emission controls on primary and secondary organic aerosol sources during Beijing 2008 Olympics. *Atmospheric Chemistry and Physics, 13*, 8303–8314. doi:10.5194/acp-13-8303-2013.

Guo, S., Hu, M., Zamora, M. L., Peng, J., Shang, D., Zheng, J., Du, Z., Wu, Z., Shao, M., Zeng, L., Molina, M. J., & Zhang, R. (2014). Elucidating severe urban haze formation in China. *Proceedings of the National Academy of Sciences of the United States of America*, 201419604. doi:10.1073/pnas.1419604111.

He, K. B., Yang, F. M., Ma, Y. L., Zhang, Q., Yao, X. H., Chan, C. K., Cadle, S., Chan, T., & Mulawa, P. (2001). The characteristics of PM2.5 in Beijing, China. *Atmospheric Environment, 35*, 4959–4970. doi:10.1016/S1352-2310(01)00301-6.

He, K., Zhao, Q., Ma, Y., Duan, F., Yang, F., Shi, Z., & Chen, G. (2012). Spatial and seasonal variability of PM2.5 acidity at two Chinese megacities: Insights into the formation of secondary inorganic aerosols. *Atmospheric Chemistry and Physics, 12*, 1377–1395. doi:10.5194/acp-12-1377-2012.

Hidy, G. M. (2009). Surface-level fine particle mass concentrations: From hemispheric distributions to megacity sources. *Journal of the Air & Waste Management Association, 59*, 770–789. doi:10.3155/1047-3289.59.7.770.

Hu, M., Wu, Z., Slanina, J., Lin, P., Liu, S., & Zeng, L. (2008). Acidic gases, ammonia and water-soluble ions in PM2.5 at a coastal site in the Pearl River Delta, China. *Atmospheric Environment, 42*, 6310–6320. doi:10.1016/j.atmosenv.2008.02.015.

Hu, W. W., Hu, M., Deng, Z. Q., Xiao, R., Kondo, Y., Takegawa, N., Zhao, Y. J., Guo, S., & Zhang, Y. H. (2012). The characteristics and origins of carbonaceous aerosol at a rural site of PRD in summer of 2006. *Atmospheric Chemistry and Physics, 12*, 1811–1822. doi:10.5194/acp-12-1811-2012.

Huang, X. F., He, L. Y., Hu, M., Canagaratna, M. R., Sun, Y., Zhang, Q., Zhu, T., Xue, L., Zeng, L. W., Liu, X. G., Zhang, Y. H., Jayne, J. T., Ng, N. L., & Worsnop, D. R. (2010). Highly time-resolved chemical characterization of atmospheric submicron particles during 2008 Beijing Olympic Games using an Aerodyne High-Resolution Aerosol Mass Spectrometer. *Atmospheric Chemistry and Physics, 10*, 8933–8945. doi:10.5194/acp-10-8933-2010.

Huang, X., Song, Y., Li, M., Li, J., Huo, Q., Cai, X., Zhu, T., Hu, M., & Zhang, H. (2012). A high-resolution ammonia emission inventory in China. *Global Biogeochemical Cycles, 26*, n/a–n/a, 10.1029/2011gb004161.

Huang, R. J., Zhang, Y., Bozzetti, C., Ho, K. F., Cao, J. J., Han, Y., Daellenbach, K. R., Slowik, J. G., Platt, S. M., Canonaco, F., Zotter, P., Wolf, R., Pieber, S. M., Bruns, E. A., Crippa, M., Ciarelli, G., Piazzalunga, A., Schwikowski, M., Abbaszade, G., Schnelle-Kreis, J., Zimmermann, R., An, Z., Szidat, S., Baltensperger, U., El Haddad, I., & Prevot, A. S. (2014). High secondary aerosol contribution to particulate pollution during haze events in China. *Nature, 514*, 218–222. doi:10.1038/nature13774.

Ianniello, A., Spataro, F., Esposito, G., Allegrini, I., Hu, M., & Zhu, T. (2011). Chemical characteristics of inorganic ammonium salts in PM2.5 in the atmosphere of Beijing (China). *Atmospheric Chemistry and Physics, 11*, 10803–10822. doi:10.5194/acp-11-10803-2011.

Jia, Y., Rahn, K. A., He, K., Wen, T., & Wang, Y. (2008). A novel technique for quantifying the regional component of urban aerosol solely from its sawtooth cycles. *Journal of Geophysical Research, 113*, 10.1029/2008jd010389.

Kim, H.-S., Huh, J.-B., Hopke, P. K., Holsen, T. M., & Yi, S.-M. (2007). Characteristics of the major chemical constituents of PM2.5 and smog events in Seoul, Korea in 2003 and 2004. *Atmospheric Environment, 41*, 6762–6770. doi:10.1016/j.atmosenv.2007.04.060.

Kumar, P., Gurjar, B. R., Nagpure, A. S., & Harrison, R. M. (2011). Preliminary estimates of nanoparticle number emissions from road vehicles in megacity Delhi and associated health impacts. *Environmental Science & Technology, 45*, 5514–5521. doi:10.1021/es2003183.

Li, J., Zhuang, G., Huang, K., Lin, Y., Xu, C., & Yu, S. (2008). Characteristics and sources of air-borne particulate in Urumqi, China, the upstream area of Asia dust. *Atmospheric Environment, 42*, 776–787. doi:10.1016/j.atmosenv.2007.09.062.

Li, W., Wang, C., Wang, H., Chen, J., Yuan, C., Li, T., Wang, W., Shen, H., Huang, Y., Wang, R., Wang, B., Zhang, Y., Chen, H., Chen, Y., Tang, J., Wang, X., Liu, J., Coveney Jr., R. M., & Tao, S. (2014). Distribution of atmospheric particulate matter (PM) in rural field, rural village and urban areas of northern China. *Environmental Pollution, 185*, 134–140. doi:10.1016/j.envpol.2013.10.042.

Lin, P., Hu, M., Deng, Z., Slanina, J., Han, S., Kondo, Y., Takegawa, N., Miyazaki, Y., Zhao, Y., & Sugimoto, N. (2009). Seasonal and diurnal variations of organic carbon in PM2.5in Beijing and the estimation of secondary organic carbon. *Journal of Geophysical Research, 114*, 10.1029/2008jd010902.

Meng, Z. Y., Jiang, X. M., Yan, P., Lin, W. L., Zhang, H. D., & Wang, Y. (2007). Characteristics and sources of PM2.5 and carbonaceous species during winter in Taiyuan, China. *Atmospheric Environment, 41*, 6901–6908. doi:10.1016/j.atmosenv.2007.07.049.

O'Dowd, C. D., Facchini, M. C., Cavalli, F., Ceburnis, D., Mircea, M., Decesari, S., Fuzzi, S., Yoon, Y. J., & Putaud, J.-P. (2004). Biogenically driven organic contribution to marine aerosol. *Nature, 437*, 676–680. doi:10.1038/nature02970.

Park, S. S., Bae, M. S., Schauer, J. J., Ryu, S. Y., Kim, Y. J., Yong Cho, S., & Kim, S. J. (2005). Evaluation of the TMO and TOT methods for OC and EC measurements and their characteristics in PM2.5 at an urban site of Korea during ACE-Asia. *Atmospheric Environment, 39*, 5101–5112. doi:10.1016/j.atmosenv.2005.05.016.

Pathak, R. K., Wu, W. S., & Wang, T. (2009). Summertime PM2.5 ionic species in four major cities of China: nitrate formation in an ammonia-deficient atmosphere. *Atmospheric Chemistry and Physics, 9*, 1711–1722.

Polidori, A., Turpin, B. J., Lim, H.-J., Cabada, J. C., Subramanian, R., Pandis, S. N., & Robinson, A. L. (2006). Local and regional secondary organic aerosol: Insights from a year of semi-continuous carbon measurements at pittsburgh. *Aerosol Science and Technology, 40*, 861–872. doi:10.1080/02786820600754649.

Protection, M. o. E (2015). Report on the state of the environment in China.

Putaud, J.-P., Raes, F., Van Dingenen, R., Brüggemann, E., Facchini, M. C., Decesari, S., Fuzzi, S., Gehrig, R., Hüglin, C., Laj, P., Lorbeer, G., Maenhaut, W., Mihalopoulos, N., Müller, K., Querol, X., Rodriguez, S., Schneider, J., Spindler, G., Brink, H. t., Tørseth, K., & Wiedensohler, A. (2004). A European aerosol phenomenology—2: Chemical characteristics of particulate matter at kerbside, urban, rural and background sites in Europe. *Atmospheric Environment, 38*, 2579–2595. doi:10.1016/j.atmosenv.2004.01.041.

Putaud, J. P., Van Dingenen, R., Alastuey, A., Bauer, H., Birmili, W., Cyrys, J., Flentje, H., Fuzzi, S., Gehrig, R., Hansson, H. C., Harrison, R. M., Herrmann, H., Hitzenberger, R., Hüglin, C., Jones, A. M., Kasper-Giebl, A., Kiss, G., Kousa, A., Kuhlbusch, T. A. J., Löschau, G., Maenhaut, W., Molnar, A., Moreno, T., Pekkanen, J., Perrino, C., Pitz, M., Puxbaum, H., Querol, X., Rodriguez, S., Salma, I., Schwarz, J., Smolik, J., Schneider, J., Spindler, G., ten Brink, H., Tursic, J., Viana, M., Wiedensohler, A., & Raes, F. (2010). A European aerosol phenomenology – 3: Physical and chemical characteristics of particulate matter from 60 rural, urban, and kerbside sites across Europe. *Atmospheric Environment, 44*, 1308–1320. doi:10.1016/j.atmosenv.2009.12.011.

Quinn, P. K., & Bates, T. S. (2005). Regional aerosol properties: Comparisons of boundary layer measurements from ACE 1, ACE 2, Aerosols99, INDOEX, ACE Asia, TARFOX, and NEAQS. *Journal of Geophysical Research, 110*. doi:10.1029/2004jd004755.

Remoundaki, E., Kassomenos, P., Mantas, E., Mihalopoulos, N., & Tsezos, M. (2013). Composition and mass closure of PM2.5 in urban environment (Athens, Greece). *Aerosol and Air Quality Research, 13*, 72–82. doi:10.4209/aaqr.2012.03.0054.

Salvador, P., Artíñano, B., Viana, M., Alastuey, A., & Querol, X. (2012). Evaluation of the changes in the Madrid metropolitan area influencing air quality: Analysis of 1999–2008 temporal trend of particulate matter. *Atmospheric Environment, 57*, 175–185. doi:10.1016/j.atmosenv.2012.04.026.

Shao, M., Tang, X., Zhang, Y., & Li, W. (2006). City clusters in China: Air and surface water pollution. *Frontiers in Ecology and the Environment, 4*, 353–361.

Shen, J., Liu, X., Zhang, Y., Fangmeier, A., Goulding, K., & Zhang, F. (2011). Atmospheric ammonia and particulate ammonium from agricultural sources in the North China Plain. *Atmospheric Environment, 45*, 5033–5041. doi:10.1016/j.atmosenv.2011.02.031.

Sillanpää, M., Hillamo, R., Saarikoski, S., Frey, A., Pennanen, A., Makkonen, U., Spolnik, Z., Van Grieken, R., Braniš, M., & Brunekreef, B. (2006). Chemical composition and mass closure of particulate matter at six urban sites in Europe. *Atmospheric Environment, 40*, 212–223. doi:10.1016/j.atmosenv.2006.01.063.

Squizzato, S., Masiol, M., Brunelli, A., Pistollato, S., Tarabotti, E., Rampazzo, G., & Pavoni, B. (2013). Factors determining the formation of secondary inorganic aerosol: A case study in the Po Valley (Italy). *Atmospheric Chemistry and Physics, 13*, 1927–1939. doi:10.5194/acp-13-1927-2013.

Sun, J., Zhang, Q., Canagaratna, M. R., Zhang, Y., Ng, N. L., Sun, Y., Jayne, J. T., Zhang, X., Zhang, X., & Worsnop, D. R. (2010). Highly time- and size-resolved characterization of submicron aerosol particles in Beijing using an Aerodyne Aerosol Mass Spectrometer. *Atmospheric Environment, 44*, 131–140. doi:10.1016/j.atmosenv.2009.03.020.

Surratt, J. D., Lewandowski, M., Offenberg, J. H., Jaoui, M., Kleindienst, T. E., Edney, E. O., & Seinfeld, J. H. (2007). Effect of acidity on secondary organic aerosol formation from isoprene. *Environmental Science & Technology, 41*, 5363–5369. doi:10.1021/es0704176.

Tao, J., Chen, T., Zhang, R., Cao, J., Zhu, L., Wang, Q., Luo, L., & Zhang, L. (2013). Chemical composition of PM2.5 at an urban site of Chengdu in Southwestern China. *Advances in Atmospheric Sciences, 30*, 1070–1084. doi:10.1007/s00376-012-2168-7.

Tie, X., & Cao, J. (2009). Aerosol pollution in China: Present and future impact on environment. *Particuology, 7*, 426–431. doi:10.1016/j.partic.2009.09.003.

Turpin, B. J., & Lim, H.-J. (2001). Species contributions to PM2.5 mass concentrations: Revisiting common assumptions for estimating organic mass. *Aerosol Science and Technology, 35*, 602–610. doi:10.1080/02786820119445.

Viana, M., Maenhaut, W., ten Brink, H. M., Chi, X., Weijers, E., Querol, X., Alastuey, A., Mikuška, P., & Večeřa, Z. (2007). Comparative analysis of organic and elemental carbon concentrations in carbonaceous aerosols in three European cities. *Atmospheric Environment, 41*, 5972–5983. doi:10.1016/j.atmosenv.2007.03.035.

Wang, Y., Zhuang, G., Zhang, X., Huang, K., Xu, C., Tang, A., Chen, J., & An, Z. (2006). The ion chemistry, seasonal cycle, and sources of PM2.5 and TSP aerosol in Shanghai. *Atmospheric Environment, 40*, 2935–2952. doi:10.1016/j.atmosenv.2005.12.051.

Wang, Q., Shao, M., Zhang, Y., Wei, Y., Hu, M., & Guo, S. (2009). Source apportionment of fine organic aerosols in Beijing. *Atmospheric Chemistry and Physics, 9*, 8573–8585.

Wang, B., Shao, M., Lu, S. H., Yuan, B., Zhao, Y., Wang, M., Zhang, S. Q., & Wu, D. (2010). Variation of ambient non-methane hydrocarbons in Beijing city in summer 2008. *Atmospheric Chemistry and Physics, 10*, 5911–5923. doi:10.5194/acp-10-5911-2010.

Wang, Y., Zhang, Q. Q., He, K., Zhang, Q., & Chai, L. (2013). Sulfate-nitrate-ammonium aerosols over China: Response to 2000–2015 emission changes of sulfur dioxide, nitrogen oxides, and ammonia. *Atmospheric Chemistry and Physics, 13*, 2635–2652. doi:10.5194/acp-13-2635-2013.

Wu, D., Tie, X., & Deng, X. (2006). Chemical characterizations of soluble aerosols in southern China. *Chemosphere, 64*, 749–757. doi:10.1016/j.chemosphere.2005.11.066.

Yang, F., Huang, L., Duan, F., Zhang, W., He, K., Ma, Y., Brook, J. R., Tan, J., Zhao, Q., & Cheng, Y. (2011a). Carbonaceous species in $PM_{2.5}$ at a pair of rural/urban sites in Beijing, 2005–2008. *Atmospheric Chemistry and Physics, 11*, 7893–7903. doi:10.5194/acp-11-7893-2011.

Yang, F., Tan, J., Zhao, Q., Du, Z., He, K., Ma, Y., Duan, F., Chen, G., & Zhao, Q. (2011b). Characteristics of PM2.5 speciation in representative megacities and across China. *Atmospheric Chemistry and Physics, 11*, 5207–5219. doi:10.5194/acp-11-5207-2011.

Yu, X. Y., Cary, R. A., & Laulainen, N. S. (2009). Primary and secondary organic carbon downwind of Mexico City. *Atmospheric Chemistry and Physics, 9*, 6793–6814.

Zeng, T., & Wang, Y. (2011). Nationwide summer peaks of OC/EC ratios in the contiguous United States. *Atmospheric Environment, 45*, 578–586. doi:10.1016/j.atmosenv.2010.10.038.

Zhang, Y. X., Schauer, J. J., Zhang, Y. H., Zeng, L. M., Wei, Y. J., Liu, Y., & Shao, M. (2008). Characteristics of particulate carbon emissions from real-world Chinese coal combustion. *Environmental Science & Technology, 42*, 5068–5073. doi:10.1021/Es7022576.

Zhang, Q., Jimenez, J. L., Canagaratna, M. R., Ulbrich, I. M., Ng, N. L., Worsnop, D. R., & Sun, Y. (2011). Understanding atmospheric organic aerosols via factor analysis of aerosol mass spectrometry: A review. *Analytical and Bioanalytical Chemistry, 401*, 3045–3067. doi:10.1007/s00216-011-5355-y.

Zhang, F., Xu, L., Chen, J., Yu, Y., Niu, Z., & Yin, L. (2012). Chemical compositions and extinction coefficients of PM2.5 in peri-urban of Xiamen, China, during June 2009–May 2010. *Atmospheric Research, 106*, 150–158. doi:10.1016/j.atmosres.2011.12.005.

Zhang, F., Chen, J., Qiu, T., Yin, L., Chen, X., & Yu, J. (2013). Pollution characteristics of PM2.5 during a typical haze episode in Xiamen, China. *Atmospheric and Climate Sciences, 03*, 427–439. doi:10.4236/acs.2013.34044.

Zheng, M., Salmon, L. G., Schauer, J. J., Zeng, L. M., Kiang, C. S., Zhang, Y. H., & Cass, G. R. (2005). Seasonal trends in PM2.5 source contributions in Beijing, China. *Atmospheric Environment, 39*, 3967–3976. doi:10.1016/j.atmosenv.2005.03.036.

Zheng, J., Hu, M., Peng, J., Wu, Z., Kumar, P., Li, M., Wang, Y., & Guo, S. (2016). Spatial distributions and chemical properties of PM2.5 based on 21 field campaigns at 17 sites in China. *Chemosphere, 159*, 480–487. doi:10.1016/j.chemosphere.2016.06.032.

Zhou, X., Cao, Z., Ma, Y., Wang, L., Wu, R., & Wang, W. (2016). Concentrations, correlations and chemical species of PM2.5/PM10 based on published data in China: Potential implications for the revised particulate standard. *Chemosphere, 144*, 518–526. doi:10.1016/j.chemosphere.2015.09.003.

Ziemba, L.D., Fischer, E., Griffin, R.J., & Talbot, R.W. (2007). Aerosol acidity in rural New England: Temporal trends and source region analysis. *Journal of Geophysical Research, 112*, 10.1029/2006JD007605.

Part IV
Space Observations

Chapter 13
Observation of Air Pollution in Asia Using UV/Visible Space Sensors

Andreas Richter, Andreas Hilboll, and Thomas Wagner

Abstract In this chapter, the basic principles of ultraviolet (UV)/visible satellite remote sensing of tropospheric pollutants are reviewed. Results from satellite observations over Asia are discussed, focusing on NO_2 and SO_2 retrievals. The discussion covers measurements of the spatial distribution of pollutants, their temporal evolution on regional and local scales, and the inversion of emission estimates. Finally, the link between satellite pollution measurements and haze events in Asia is briefly discussed.

Keywords Remote sensing • Satellite • Pollution • Asia • DOAS

13.1 Introduction

Over the last 100 years, air pollution has transformed from being a local phenomenon, mainly affecting the population of industrial centres, to a global problem, impacting on the life of a majority of people living today. The main reasons of this transformation are the rapid economic and technological development with massive increase in fossil fuel use for transportation and energy generation, and fast population growth leading to urbanisation and increased demands on resources. While originally mainly North America and Europe were affected, rapid economic development in Asia over the last decades has made the continent a hot-spot of air pollution. In this text, we will focus on the development in China and India,

A. Richter (✉)
Institute of Environmental Physics, University of Bremen, Bremen, Germany
e-mail: richter@iup.physik.uni-bremen.de

A. Hilboll
Institute of Environmental Physics, University of Bremen, Bremen, Germany

MARUM, University of Bremen, Bremen, Germany
e-mail: hilboll@uni-bremen.de

T. Wagner
Max-Planck-Institute für Chemie, Mainz, Germany
e-mail: thomas.wagner@mpic.de

I. Bouarar et al. (eds.), *Air Pollution in Eastern Asia: An Integrated Perspective*,
ISSI Scientific Report Series 16, DOI 10.1007/978-3-319-59489-7_13

although the Middle East has seen similar increases in air pollution, while Japan has benefitted from strict emission controls (see Hilboll et al. 2013).

As the effects of air pollution on human health, agriculture, and the environment became apparent, measures have been taken to control emissions of air pollutants and to develop networks of instruments monitoring air quality and emissions. At the same time, research on air quality and the chemical and physical mechanisms involved in the formation of smog and other forms of air pollution intensified, increasing the need for reliable data on atmospheric composition.

Most atmospheric composition measurements are performed using in-situ methods which provide accurate data on an air volume probed. However, such measurements are localized and the number of measurements that can be taken is limited by the number of instruments which are available and where they can be placed. They are also often strongly influenced by local emission sources and topography (for example a street canyon), and not necessarily representative of a larger area.

Remote sensing observations from the ground or from airborne platforms have originally been mainly used to take measurements in the stratosphere and upper atmosphere which are difficult to access with in-situ methods. As satellite platforms became available, remote sensing of atmospheric composition from space became possible in the second half of the twentieth century, for the first time enabling global measurements by single instruments. The capabilities of these instruments have continuously been extended and improved, and today, a range of satellite instruments provide data on atmospheric composition using different techniques and spectral ranges. Compared to other data sources, such systems provide global coverage, a very large number of consistent observations, usually over several years of operation, and with rapid access to the data. However, the uncertainties on the individual measurements are comparatively large and measurements average over a large air volume, integrating both horizontally and in the vertical. For a detailed discussion of satellite remote sensing of atmospheric composition and many references see (Burrows et al. 2011). The application of the IASI sensor to air pollution measurements over China is presented in Chap. 14.

13.2 UV/Visible Remote Sensing of Atmospheric Composition

Remote sensing in the UV and visible spectral range was first developed for stratospheric ozone observations, but a number of other trace gases relevant for air pollution can also be retrieved at these wavelengths, including nitrogen dioxide (NO_2), sulphur dioxide (SO_2), formaldehyde (HCHO), glyoxal (CHOCHO), and aerosols.

The basic principle of remote sensing of atmospheric composition in the UV and visible spectral domain is absorption spectroscopy (see Richter and Wagner 2011 for a more detailed discussion). Light passing through the atmosphere interacts with air molecules, clouds, and particles and thereby changes its intensity and spectral composition.

13.2.1 Basic Principle

Rayleigh scattering on air molecules and Mie scattering on cloud droplets and aerosols changes the direction of the photons, thereby reducing the intensity of transmitted radiation while increasing the proportion of scattered light. As scattering efficiency changes slowly with wavelength, no narrow band spectral features are introduced by these processes.

Absorption of light by molecules reduces the total intensity. As the wavelength dependence of the absorption efficiency is characteristic for each molecule and often has narrow spectral structures, individual absorbers can be identified in the spectra of transmitted light using their "fingerprint" absorption features. The overall reduction in intensity follows an exponential law (Lambert Beer's law) depending on the number of absorbing molecules along the light path and thus is a function of absorber density and light path length.

All current satellite-borne UV/visible sensors measuring atmospheric composition use solar light, either in transmission for profile measurements in the upper atmosphere or by observing the light scattered and reflected from the surface and the atmosphere, extending the range of measurements to the troposphere. The basic geometry of nadir viewing observations is sketched in Fig. 13.1. Light from the sun enters the atmosphere, is scattered and absorbed in the atmosphere and reflected from the surface. The satellite receives a combination of light scattered at the surface and in the atmosphere. By comparing the light received from the atmosphere with that obtained from a direct sun measurement without interfering atmosphere, the optical depth of the atmosphere along the light path can be determined. From the optical depth the slant column (SC) is derived using the Differential Optical Spectroscopy (DOAS) method (Platt and Stutz 2008). The SC is defined as the absorber density, integrated along the light path. This quantity depends on the measurement geometry, the vertical profile of the absorber and

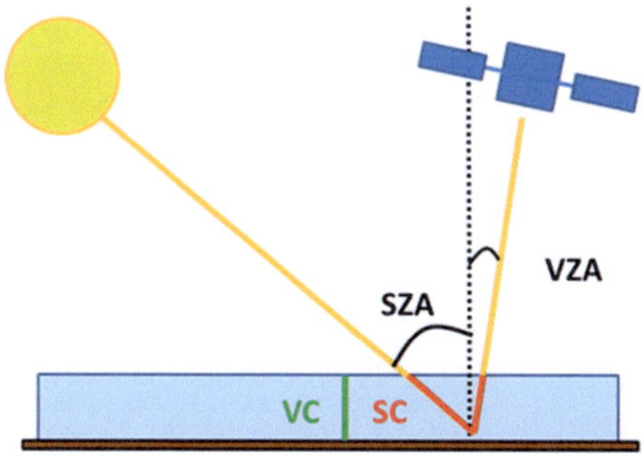

Fig. 13.1 Schematic of the observation geometry for a UV/vis nadir measurement of tropospheric composition. The slant column (*SC*) depends on the solar zenith angle (*SZA*), the viewing zenith angle (*VZA*), the vertical distribution of the absorber and the surface reflectance. The vertical column (*VC*) is independent of the observation geometry. Note that only the ray reflected at the surface is shown while light scattered in the atmosphere will also contribute to the signal observed at satellite

interfering aerosols and clouds and on surface reflectance. It therefore has to be converted to the vertical column (VC) which is defined as the integral of absorber concentration along the vertical axis and does not depend on measurement conditions.

13.2.2 *Vertical Sensitivity*

As only the integrated column amount can be determined from UV/vis nadir observations, the vertical distribution of an absorber can usually not be derived. In particular this method cannot directly measure the surface concentration of air pollutants such as NO_2 which would be of direct relevance for air quality studies. In fact, for UV/visible retrievals from nadir satellite observations, an a priori assumption has to be made on the vertical distribution of the trace gas of interest. This profile usually comes from atmospheric chemical transport models and strongly impacts on the results. The reason for this is the rapid decrease of measurement sensitivity towards the ground in particular over dark surfaces as illustrated in Fig. 13.2. This change in vertical sensitivity can be modelled with radiative transfer models (RTM) and a correction with a so called air mass factor (AMF) is applied to the measurements by weighting the vertical sensitivity with the relative a priori vertical profile of the absorber.

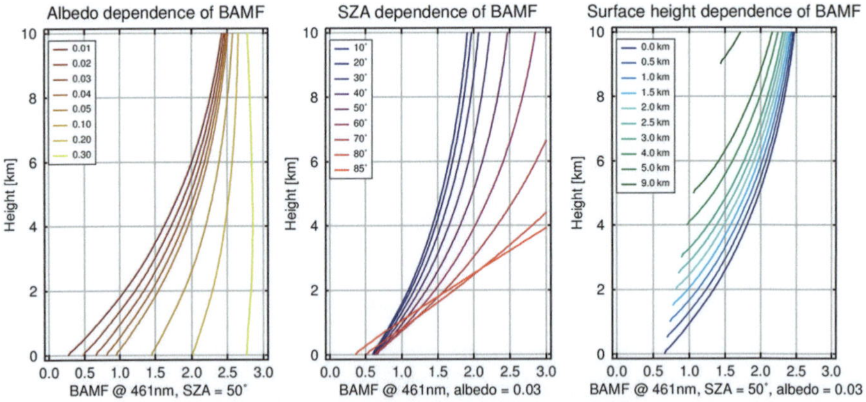

Fig. 13.2 Examples of the vertical sensitivity of nadir NO_2 observations as function of albedo (*left*), solar zenith angle (*middle*), and surface altitude (*right*). Shown is the box air mass factor (BAMF) which describes the sensitivity of the measurement to the presence of NO_2 in a given layer as a function of altitude. All scenarios are at 50° solar zenith angle, 461 nm, 0 m surface height, no aerosols and no clouds unless stated otherwise in the figure caption

13.2.3 Effect of Clouds and Aerosol

Additional complication may arise from the presence of aerosols and/or clouds. While the effect of Rayleigh scattering on air molecules can easily be modelled and accounted for in the retrievals, clouds and aerosols are highly variable in the atmosphere and therefore difficult to model. Qualitatively, the impact of both can easily be understood as they both behave as strongly scattering layers (see Fig. 13.3). If a cloud or aerosol layer is thick, most of the radiation will be scattered back to the satellite instrument from this layer and sensitivity to absorption below the layer will be reduced (shielding). As a result an UV/vis satellite measurement is "blind" to what lies below a thick cloud. However, the large number of scattering events within the cloud or aerosol layer increases the effective path length and thus the sensitivity of the measurements is increased within the layer (light path enhancement). Above the cloud or aerosol layer, the number of reflected photons is increased leading to enhanced sensitivity just as over bright surfaces (albedo effect, see Fig. 13.2). As a result, for pollution layers there often are counteracting effects of scattering from the top of the layer and multiple scattering within the layer, making the overall effect relatively small.

The details of the aerosol effect depend on the vertical distribution and the single scattering albedo which is the ratio of scattered to the sum of scattered and absorbed light. Strongly absorbing aerosol will reduce the light path length and thus lead to smaller tropospheric slant columns.

In current retrieval algorithms, cloud information is usually coming from measurements of the instrument itself by evaluating the measured intensity and the signal of absorbers with well-known vertical distribution (O_2, O_4) (Acarreta et al. 2004; Koelemeijer et al. 2001). From this a cloud fraction and an effective cloud top height can be derived which are then used for an explicit cloud correction. Situations where more than 50% of the radiation originates from a cloud (corresponding

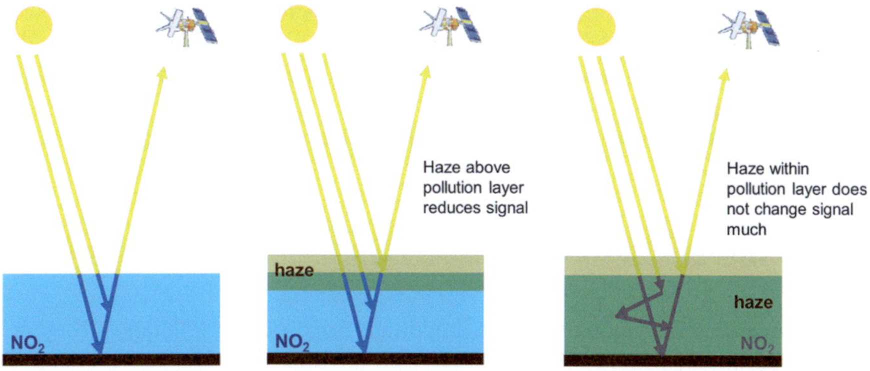

Fig. 13.3 Sketch of the effect of an aerosol layer on the light path of UV/vis nadir observations depending on its position relative to the absorber layer. Elevated aerosol layers mostly shield the signal (*centre*) while well mixed pollution layers have little impact on the measurements (*right*)

to ~20% geometrical cloud fraction), are usually excluded from tropospheric retrievals, as the information content on the layers close to the surface is deemed to be insufficient. In contrast to the cloud correction, an explicit aerosol correction is rarely applied in UV/visible retrievals, mainly because the information needed on the aerosols is not available. However, the cloud correction scheme will implicitly correct for aerosols to some degree (Boersma et al. 2011), at least if they are elevated relative to the absorbers. On the other hand, the presence of aerosols can also lead to erroneous cloud retrievals if for example a high cloud is situated over a low aerosol layer.

13.2.4 *Instruments*

The first satellite instrument to provide tropospheric trace gas observations beyond ozone was the Global Ozone Monitoring Experiment (GOME), launched into a sun-synchronous orbit on ERS-2 in 1995 (Burrows et al. 1999). It was followed by the SCIAMACHY instrument on ENVISAT, which was launched in 2002 and provided more species, better spatial resolution, and additional measurement modes (Bovensmann et al. 1999). The Ozone Monitoring Instrument (OMI) on Aura offers further improved spatial resolution and daily global coverage since 2004 (Levelt et al. 2006). The series of GOME-2 instruments operated on the MetOp satellites continues the GOME time series with better spatial resolution and temporal coverage since 2007 (Munro et al. 2016). The OMPS instruments can also provide some information on tropospheric pollutants (Yang et al. 2014) but are optimized for ozone column observations.

For the future, the TROPOMI instrument on Sentinel-5P will take measurements at unprecedented spatial resolution and coverage (Veefkind et al. 2012), followed by the UVN instrument on MetOp-SG/Sentinel-5.

While all the above listed instruments are in low earth orbits (LEO) resulting in global coverage but at best daily resolution, the upcoming geostationary instruments GEMS (Kim 2012), TEMPO (Zoogman et al. 2016), and Sentinel-4 (Ingmann et al. 2012) are aiming at several measurements per day albeit for a limited geographic area only.

An overview of the sensors is given in Table 13.1.

13.3 Nitrogen Dioxide

Nitrogen oxides are among the most important tropospheric pollutants. They are formed during all combustion processes, both anthropogenic (energy production, transport, cement production, biomass burning) and natural (wild fires). Additional sources are bacterial soil emissions, especially in regions of extensive agriculture, and lightning. Emissions of nitrogen oxides are mostly in the form of NO which

Table 13.1 Overview of UV/visible satellite instruments

Instrument	Operation	Orbit	Coverage	Spatial resolution
GOME	09.1995–04.2003	LEO 10:30	3 days	40×320 km^2
SCIAMACHY	08.2002–04.2012	LEO 10:00	6 days	30×60 km^2
OMI	10.2004 – today	LEO 13:45	1 day	13×24 km^2 at nadir
GOME2-A	01.2007 – today	LEO 09:30	~1 day	40×80 km^2
GOME2-B	01.2012 – today	LEO 09:30	~1 day	40×80 km^2
GOME2-C		LEO 09:30	~1 day	40×80 km^2
OMPS	10.2011 – today	LEO 13:20	1 day	50×50 km^2
TROPOMI		LEO 13:30	1 day	3.5×7 km^2
UVNS (S-5)		LEO 09:30	1 day	7.5×7.5 km^2
GEMS		GEO Asia	Asia hourly	8×7 km^2
TEMPO		GEO North-America	North-America hourly	9×5 km^2
UVN (S-4)		GEO Europe	Europe hourly	8×8 km^2

rapidly reacts to form NO_2 in the atmosphere. It is therefore often discussed as NOx ($NO + NO_2$). The main loss of NO_2 in the atmosphere is through reaction with OH during the day and O_3 at night. NOx is a driver of ozone chemistry in the troposphere and together with volatile organic compounds (VOCs) it leads to photochemical smog. Through formation of HNO_3 it is also linked to acid rain and aerosol formation. Nitrogen dioxide itself adversely affects the human respiratory system and therefore is a regulated pollutant.

The atmospheric life time of NO_2 is short (only a few hours) in the lower troposphere and as a result it is mainly found close to its emissions. This together with its combustion sources makes NO_2 a good tracer of anthropogenic pollution.

Owing to its large and structured absorption bands around 450 nm, NO_2 is a good target for UV/visible satellite observations. Under polluted conditions, uncertainties in the spectral retrieval of NO_2 are only a few percent but errors from light path estimation can be of the order of 30%. As NO_2 columns in the stratosphere cannot be neglected, the satellite derived slant columns (which contain both stratospheric and tropospheric signals) need first to be corrected for stratospheric NO_2 before the tropospheric column can be determined. This correction is usually based on model data or spatial smoothing approaches and introduces additional uncertainties. Over pollution hotspots in Asia with their large tropospheric signals these uncertainties are however small.

Maps of tropospheric NO_2 columns from satellite observations as in Fig. 13.4 have been widely used to visualize the spatial distribution and hot-spots of NO_2 pollution. As already pointed out, the short atmospheric lifetime of NO_2 results in higher NO_2 columns over regions with large emissions. It is therefore reasonable to assume that regions with large NO_2 columns are also areas where air quality at ground level is poor, and in this sense the satellite maps give an impression of

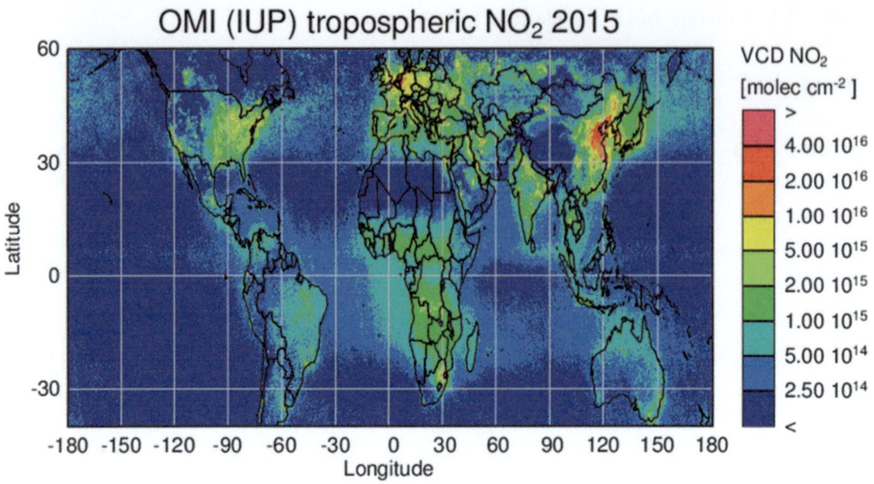

Fig. 13.4 Global map of NO_2 tropospheric columns as derived from OMI data for the year 2015. The OMI columns shown are from the IUP-UB analysis

regional air quality. However, it is important to keep in mind that the NO_2 column is a vertically and horizontally integrated quantity, and NO_2 mixing ratios at street level which are of relevance for human health might behave differently than the satellite observed columns.

13.3.1 Observations of NO_2 Changes

One of the most successful applications of satellite NO_2 data is the analysis of temporal changes. As the space-borne instruments are relatively stable over time and measurement conditions do usually not change, the results can be compared between different years and the temporal evolution of the NO_2 loading be evaluated. The first studies of this type were based on GOME and SCIAMACHY data (van der A et al. 2006; Richter et al. 2005) and showed decreases of NO_2 columns over the US, Europe, and Japan, and strong increases over eastern central China. At that time, this was in contradiction to traditional bottom-up inventories which predicted a stabilization of NOx emissions in China. Re-evaluation of the input into these inventories could then close the gap, confirming the satellite derived trend at least qualitatively (Zhang et al. 2007). The reasons for the changes observed are the rapid economic development in Asia and in China in particular which lead to larger energy demand and therefore increased fossil fuel combustion. The decreasing NO_2 values in other parts of the world were linked to improved emission reduction technology in combination with a switch to cleaner fuels. Following the early studies, the satellite series were extended and data from additional instruments (OMI, GOME-2) were included, confirming the original findings but providing more robust trends and also more spatial details (van der A et al. 2006; Duncan et al. 2015; Hilboll et al. 2013; Schneider et al. 2015). In particular, the OMI data with their high spatial resolution enable analysis of more local changes, and highlight the increase in NO_2 levels in many large cities worldwide. For China, the increasing trend in NO_2 continued over many years but since about 2012, NO_2 levels over the central eastern part of China started to stabilize and then to decrease (see Fig. 13.5). The reasons for this change are an active environmental policy in China which is leading to the implementation of improved emission reduction technologies and the close down of strongly emitting industrial and energy producing facilities. A recent decrease in economic growth might also contribute.

Other studies have investigated more local or short term variations in NO_2 levels in China. Zhang et al. (2009) evaluated OMI NO_2 data over a region where new power plants came into operation and could show that the NO_2 signal of these individual plants could clearly be detected in the OMI data (see Fig. 13.6). Other studies investigated the effect of short term reductions in NOx emissions which are sometimes imposed during large international events in China to improve air quality. As shown in Ding et al. (2015) and Mijling et al. (2009), local emission reductions lead to locally decreased NO_2 columns in satellite data if the effects of meteorology are accounted for. This is in agreement with surface observations also indicating the benefit of emission reductions. Economic crisis can also reduce NOx emissions or at least slow down their upward trend, and this has been evaluated in Lin and McElroy (2011) for the 2008 crisis in China.

Similar to China, India has also seen rapid economic growth over the last decades, resulting in severe pollution problems in many parts of the country. Analysis of NO_2 column changes in satellite data revealed large upward trends

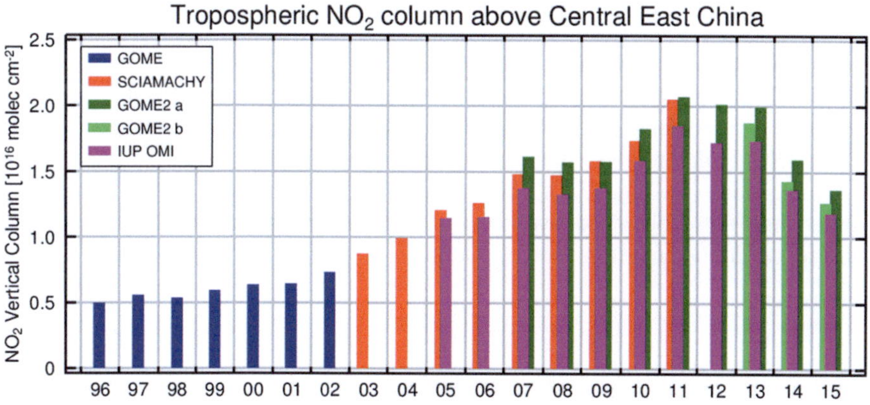

Fig. 13.5 Development of satellite observed NO_2 columns over eastern central China. All data are from the IUP-UB data analysis using consistent a priori assumptions and a 20% cloud screening. Lower values for OMI are expected because of the noon overpass time; differences between GOME, SCIAMACHY, and the GOME-2 instruments are partly due to overpass time but also reflect instrumental biases which are not corrected here (Hilboll et al. 2013)

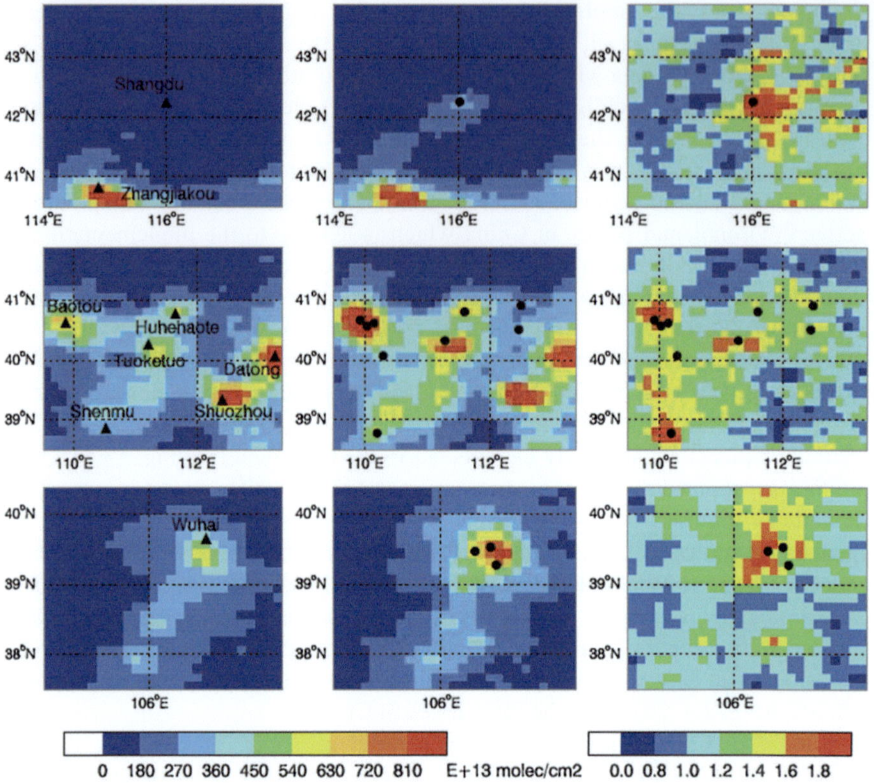

Fig. 13.6 NO_2 columns retrieved from OMI data over Inner Mongolia in summer 2005 (*left*) and 2007 (*middle*) and their ratio (*right*). Also indicated are the positions of newly build and extended coal-fired power plants (*black dots*) (From Zhang et al. 2009)

depending on region (Bucsela et al. 2013; Ghude et al. 2008, 2011, 2013; Hilboll et al. 2017a; Krotkov et al. 2016), and large point emission sources such as steel factories can be identified in the satellite maps. Similarly to the situation in East China, NO_2 columns over the North Indian Plain have peaked in 2012 and are declining since (see Fig. 13.7). The absolute NO_2 columns over India are however lower than those observed over central eastern China, presumably because of differences in the emission sources leading to lower NOx levels in India.

13.3.2 *Inversion of NOx Emissions*

For NO_2 trends, satellite data are usually treated in a relative way by using a specific year as the reference and evaluating changes relative to that year only (see Fig. 13.8). This has the advantage that many relative errors will cancel and

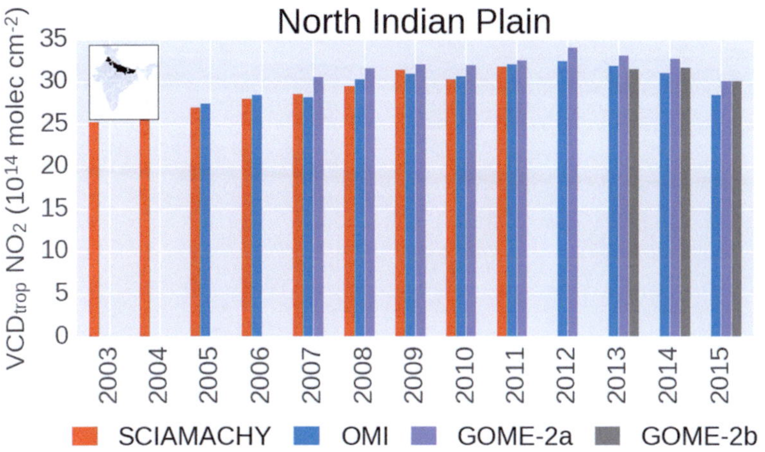

Fig. 13.7 Annual mean NO_2 columns over the North Indian Plain, as observed by the instruments SCIAMACHY, OMI, and GOME-2 (From Hilboll et al. 2017b)

therefore uncertainties of the relative changes are much lower than uncertainties of the absolute columns.

Changes in satellite NO_2 columns are linked to NOx emissions as enhancements in NOx emissions will lead to an increase in NO_2 columns. However, other reasons for the observed increase in NO_2 columns also have to be considered.

As discussed in Sect. 13.2.2, the sensitivity of UV/visible satellite observations depends on a number of parameters such as surface reflectance, the presence of aerosols and clouds and also the vertical distribution of the NO_2. Any change in one of these parameters which is not accounted for in the analysis will lead to a change in the tropospheric columns and may be misinterpreted as change in NO_2 levels. In particular a change in aerosol loading is of concern over China, as particle concentrations from construction, combustion, and secondary aerosol formation are high in polluted regions. These effects have been analysed in several studies (Leitão et al. 2010; Lin et al. 2014) and while aerosols impact on the NO_2 results, the overall effect on satellite trends appears to be limited.

Even if the NO_2 columns are assumed to be correct, the link between changes in columns and in emissions is not necessarily linear as it depends on the atmospheric life time of NO_2 which in turns depends on the photochemical regime in the atmosphere. This problem has been addressed by many studies using increasingly more sophisticated methods. A simple and fast approach is to assume that locally, a linear relationship exists between NOx emissions and NO_2 columns, and that this can be provided by an atmospheric model (van der A et al. 2006; Han et al. 2009; Martin et al. 2002, 2003). However, this does not account for the fact that the model might not be using the right emissions and not fully represent atmospheric chemistry, and therefore the atmospheric lifetime of NO_2 from the model might not be correct.

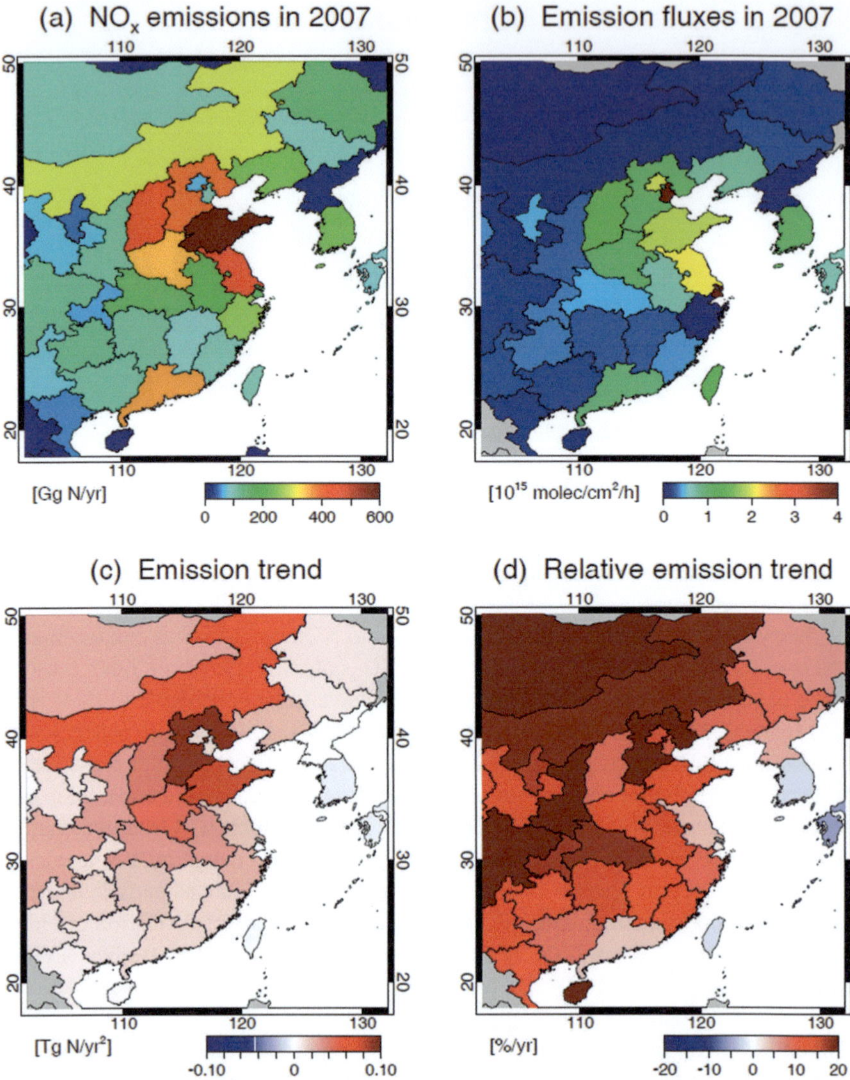

Fig. 13.8 NOx emissions as estimated from a trend model on GOME2 data (**a**) per region, (**b**) per area, (**c**) the annual trend per region for 2007–2001 and (**d**) the relative regional emission changes (From Mijling and van der A 2012)

There are approaches to determine the NO_2 lifetime from the measurements themselves by utilising the wind speed and wind direction dependence of the plume shape (Beirle et al. 2011; Fioletov et al. 2013). Such approaches work best for localised point sources. However, a recent extension of the method (Liu et al. 2016) demonstrated good agreement with bottom-up inventories also for large cities (see Fig. 13.9). A more general solution to the problem is the use of the model adjoint,

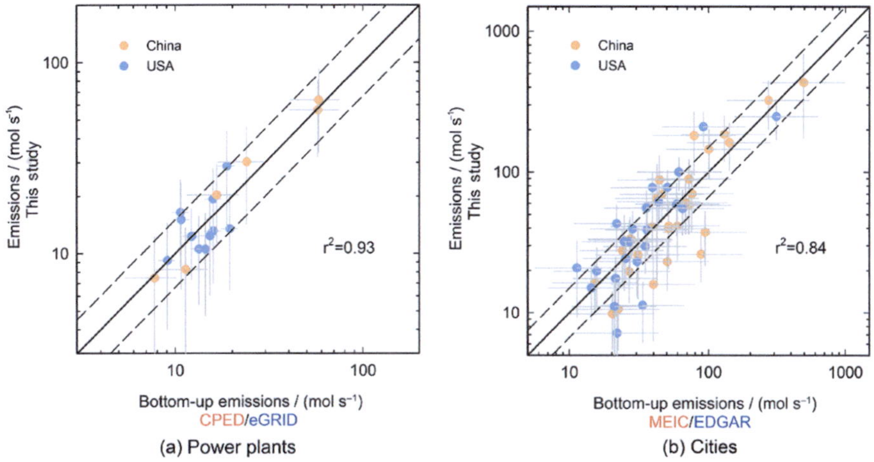

Fig. 13.9 NOx emission estimates from OMI satellite data estimated by deriving the atmospheric lifetime of NO_2 directly from the measurements. The results are compared to bottom-up emission estimates and good agreement is found in particular for power plants (From Liu et al. 2016)

which provides full consistency between emissions and modelled NO_2 columns (Stavrakou et al. 2008, 2013).

13.4 Sulphur Dioxide

Sulphur dioxide is an important pollutant at ground level, leads to the formation of sulphate aerosols and to acid rain. Anthropogenic sources are mainly the combustion of sulphur rich coal for energy production or heating, but industrial sources and emissions from oil and gas mining also contribute. As China and India are both using coal to cover a large fraction of their energy needs, they are the most prominent SO_2 sources worldwide today. Removal of SO_2 from exhaust gases is technologically possible and has been implemented in most industrialised countries. In China, flue gas desulphurization has been introduced rapidly since 2006 at least for large power plants.

In spite of the large absorption cross-section of SO_2, the retrieval from space borne instruments is less sensitive than for NO_2 for several reasons. The main difficulty is the spectral range of the absorption which is in the UV where solar intensity is reduced, transmission through the ozone layer limited and Rayleigh scattering in the atmosphere strong. As a result, only large SO_2 amounts can be detected from spacc and the sensitivity to the lowest layers in the atmosphere is limited. This also implies that SO_2 can be detected more easily in satellite data when it is elevated, for example during pollution export events.

In contrast to NO_2, there is usually no stratospheric SO_2 layer which has to be accounted for in the analysis. However, sporadic volcanic eruptions can lead to local and in some cases global SO_2 plumes which interfere with the retrieval of pollution SO_2 and need to be excluded.

13.4.1 Observations of SO_2 Changes

As for NO_2, satellite data of pollution SO_2 are mainly used to map their spatial distribution, to estimate SO_2 emissions, and to evaluate changes of atmospheric SO_2 levels over time. In Fig. 13.10, OMI SO_2 columns over China are shown for three time periods (Krotkov et al. 2016). Large signals are found over the industrial parts in the eastern part of the country, reflecting the main emission regions. Comparison of the data taken in different time periods shows a rapid decrease in SO_2 levels since 2005 in most regions, much earlier and faster than the reduction observed for NO_2. This is the result of introduction of flue gas desulphurisation (FGD) in coal fired power plants.

This can also be shown in data taken over individual large power plants as shown by (Li et al. 2010) and in Fig. 13.11 (Wang et al. 2015). FGD was introduced in

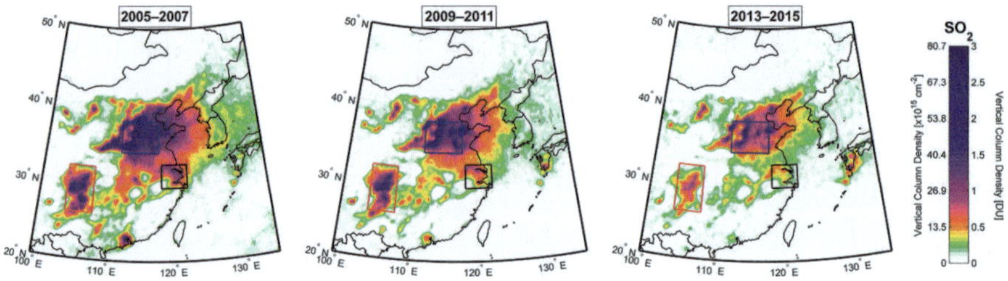

Fig. 13.10 Change in OMI SO_2 columns over China over the time period 2005–2015 (From Krotkov et al. 2016)

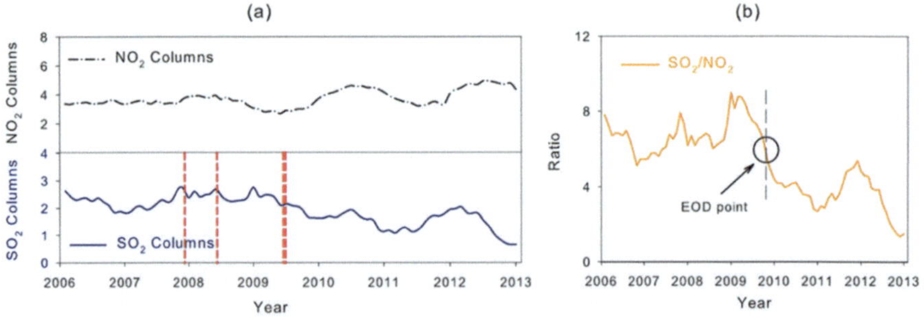

Fig. 13.11 *Left*: OMI NO_2 and SO_2 columns over a power plant in China. The data are a 12 month running average to reduce noise. The *red lines* indicate the dates when flue gas desulphurisation came into operation. *Right*: Ratio of OMI SO_2 to NO_2 columns indicating the reduction in SO_2 emissions once FGD was implemented. The EOD (effective operation date) point is the time at which the FGD became efficient (From Wang et al. 2015)

several steps to this particular power plant, and as analysis of the measurements over a large number of power plants showed, the full effect of FGD became apparent only in 2008/2009, 2 years after the installation of the first devices, showing the duration of implementing the new regulations.

Inversion of SO_2 columns to emission estimates of SO_2 is possible as in the case of NO_2 (Lee et al. 2011; Wang et al. 2011), but uncertainties are larger due to the larger measurement error. The use of measurement derived life-times for an emission estimate which does not rely on model simulations as demonstrated for NO_2 also works for SO_2 point sources (Fioletov et al. 2011), and was used to identify a large number of SO_2 sources not listed in current emission inventories (McLinden et al. 2016).

13.5 Connection to Haze Events

In addition to NO_2 and SO_2, also two VOCs can be retrieved from UV/visible satellite observations: formaldehyde (HCHO) and glyoxal (CHOCHO) (Lerot et al. 2010; De Smedt et al. 2008, 2015; Vrekoussis et al. 2009, 2010; Wittrock et al. 2006). While main sources of these VOCs and their precursors are biogenic emissions and wild fires, there also are anthropogenic components and in fact, large columns are observed over central east China with some indication for upward trends.

All of the above mentioned pollutants are present in large concentrations during haze events in Asia, contributing to their adverse health effects. In addition, they are involved in the formation of aerosols; sulphate aerosols are formed from SO_2 emissions, nitrate aerosols from NO_2, ammonium sulphate from a combination of both and secondary organic aerosols from the oxidation of VOCs. They are thus key ingredients in the formation of the dense aerosol which is characteristic of the haze events.

Over the last 20 years, upward trends were observed in the satellite time series for NO_2, SO_2, and the VOCs. It is thus not surprising to see an increased frequency and intensity of haze events. On the other hand, the significant reductions of pollutant levels in China in the last years and in particular the decrease in SO_2 emissions should reduce the occurrence of haze events which is not yet observed. However, many other parameters, in particular meteorology, play important roles in the haze formation and might obscure the beneficial effects of reduced pollutant levels.

The satellite measurements and their uncertainty are also affected by haze events. As discussed in Sect. 13.2.2, aerosols can both increase and decrease the sensitivity of satellite observations to boundary layer pollution, and this is difficult to correct in the analysis. In the absence of detailed data on aerosol composition and vertical distributions, most of today's algorithms do not explicitly correct for the

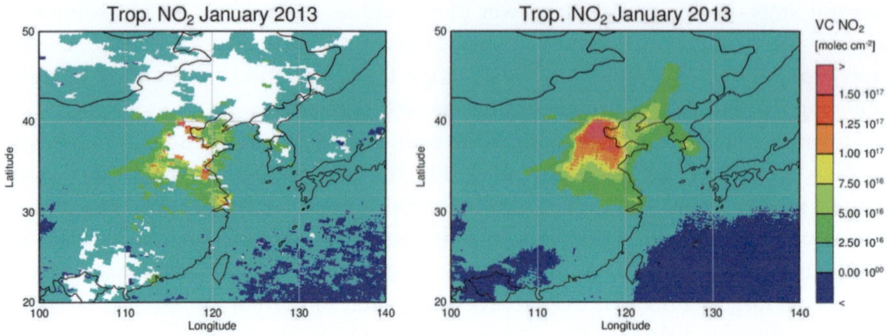

Fig. 13.12 Example for the sampling effect introduced by the miss-classification of haze events as cloudy pixels which are excluded from the analysis. *Left*: GOME-2/MetOp-A NO_2 using the standard IUP-UB data analysis for January 2013 which applies a 20% cloud screening based on the FRESCO+ algorithm (Wang et al. 2008). *Right*: The same data but without cloud screening. As can be seen, all the high values during the haze event are classified as cloudy and removed from the standard data. Vertical columns in the *right figure* are only indicative as the effect of haze and clouds is not accounted for

effects, increasing the uncertainties of satellite products during haze events (Leitão et al. 2010; Lin et al. 2014).

If haze events are strong enough they will be classified as low clouds in the retrieval algorithms. This leads to a sampling bias over Asia when during the most intense haze periods satellite measurements are flagged as cloudy and not included in the analysis. As a result, the very large pollution levels during such events are not included in averages of satellite data leading to a low bias (see Fig. 13.12 for an example).

13.6 Summary and Conclusions

UV/visible remote sensing data from satellite platforms provide important long-term records of several key pollutants on a global scale. The data indicate large upward trends in all pollutant amounts over Asia in general and China in particular until about the year 2012. This is the direct effect of rapid economic growth and increased use of fossil fuels, and has severe impacts on air quality and human health. For the most recent years, a stagnation and even decrease can be observed in the satellite data of NO_2 and SO_2 over central eastern China and the North Indian Plain, showing the effect of ambitious environmental legislation, technological progress, and slow down of growth rates and the resulting emission reductions.

Satellite derived column amounts of pollutants are directly linked to surface emissions but complex mathematical inversion is needed to fully account for the nonlinear relationship introduced by atmospheric chemistry and meteorology (see Chap. 8). Several methods have been developed for simplified estimates, including the promising approach of deriving atmospheric lifetime of pollutants directly from the measurements. The observed changes of NO_2, SO_2, and VOC amounts over China have direct relevance for the occurrence of haze events in Asia as all of them are aerosol precursors. However, interpretation of satellite measurements during haze events is complicated by the aerosol induced changes in light path, and more work is needed to fully exploit the potential of satellite observations under these conditions.

Acknowledgments This work has received research funding from the European Community's Seventh Framework Programme (FP7) under the PANDA project grant agreement n° 606719.

References

Acarreta, J. R., De Haan, J. F., & Stammes, P. (2004). Cloud pressure retrieval using the O2-O2 absorption band at 477 nm. *Journal of Geophysical Research, 109*(D5), D05204. doi:10.1029/2003JD003915.

Beirle, S., Boersma, K. F., Platt, U., Lawrence, M. G., & Wagner, T. (2011). Megacity emissions and lifetimes of nitrogen oxides probed from space. *Science (80-), 333*, 1737–1739. doi:10.1126/science.1207824.

Boersma, K. F., Eskes, H. J., Dirksen, R. J., van der A, R. J., Veefkind, J. P., Stammes, P., Huijnen, V., Kleipool, Q. L., Sneep, M., Claas, J., Leitão, J., Richter, A., Zhou, Y., & Brunner, D. (2011). An improved tropospheric NO2 column retrieval algorithm for the ozone monitoring instrument. *Atmospheric Measurement Techniques, 4*(9), 1905–1928. doi:10.5194/amt-4-1905-2011.

Bovensmann, H., Burrows, J. P., Buchwitz, M., Frerick, J., Noël, S., Rozanov, V. V., Chance, K. V., & Goede, A. P. H. (1999). SCIAMACHY: Mission objectives and measurement modes. *Journal of the Atmospheric Sciences, 56*(2), 127–150. doi:10.1175/1520-0469(1999)056<0127:SMOAMM>2.0.CO;2.

Bucsela, E. J., Krotkov, N., Celarier, E., Lamsal, L. N., Swartz, W. H., Bhartia, P. K., Boersma, K. F., Veefkind, J. P., Gleason, J. F., & Pickering, K. E. (2013). A new stratospheric and tropospheric NO2 retrieval algorithm for nadir-viewing satellite instruments: Applications to OMI. *Atmospheric Measurement Techniques, 6*(10), 2607–2626. doi:10.5194/amt-6-2607-2013.

Burrows, J. P., Weber, M., Buchwitz, M., Rozaniv, V., Ladstätter-Weißenmayer, A., Richter, A., DeBeek, R., Hoogen, R., Bramstedt, K., Eichmann, K.-U., Eisinger, M., & Al, E. (1999). The global ozone monitoring experiment (GOME): Mission concept and first scientific results. *Journal of the Atmospheric Sciences, 56*, 151–175. doi:10.1175/1520-0469(1999)0562.0.CO;2.

Burrows, J. P., Platt, U., Borrell, P. (2011). The remote sensing of tropospheric composition from space. In *The physics of earth and space environments*. Springer.

De Smedt, I., Müller, J.-F., Stavrakou, T., van der A, R., Eskes, H., & Van Roozendael, M. (2008). Twelve years of global observations of formaldehyde in the troposphere using GOME and SCIAMACHY sensors. *Atmospheric Chemistry and Physics, 8*(16), 4947–4963. doi:10.5194/acp-8-4947-2008.

De Smedt, I., Stavrakou, T., Hendrick, F., Danckaert, T., Vlemmix, T., Pinardi, G., Theys, N., Lerot, C., Gielen, C., Vigouroux, C., Hermans, C., Fayt, C., Veefkind, P., Müller, J.-F., & Van Roozendael, M. (2015). Diurnal, seasonal and long-term variations of global formaldehyde columns inferred from combined OMI and GOME-2 observations. *Atmospheric Chemistry and Physics, 15*(8), 12519–12545. doi:10.5194/acpd-15-12241-2015.

Ding, J., van der A, R. J., Mijling, B., Levelt, P. F., & Hao, N. (2015). NOx emission estimates during the 2014 Youth Olympic games in Nanjing. *Atmospheric Chemistry and Physics, 15* (16), 9399–9412. doi:10.5194/acp-15-9399-2015.

Duncan, B. N., Lamsal, L. N., Thompson, A. M., Yoshida, Y., Lu, Z., Streets, D. G., Hurwitz, M. M., & Pickering, K. E. (2015). A space-based, high-resolution view of notable changes in urban NO x pollution around the world (2005–2014). *Journal of Geophysical Research – Atmospheres*. doi:10.1002/2015JD024121.

Fioletov, V. E., McLinden, C. A., Krotkov, N., Moran, M. D., & Yang, K. (2011). Estimation of SO2 emissions using OMI retrievals. *Geophysical Research Letters, 38*(21), 1–5. doi:10.1029/2011GL049402.

Fioletov, V. E., McLinden, C. A., Krotkov, N., Yang, K., Loyola, D. G., Valks, P., Theys, N., Van Roozendael, M., Nowlan, C. R., Chance, K., Liu, X., Lee, C., & Martin, R. V. (2013). Application of OMI, SCIAMACHY, and GOME-2 satellite SO_2 retrievals for detection of large emission sources. *Journal of Geophysical Research – Atmospheres, 118*(19), 11399–11418. doi:10.1002/jgrd.50826.

Ghude, S. D., Fadnavis, S., Beig, G., Polade, S. D., & van der A, R. J. (2008). Detection of surface emission hot spots, trends, and seasonal cycle from satellite-retrieved NO2 over India. *Journal of Geophysical Research – Atmospheres, 113*(2), 1–13. doi:10.1029/2007JD009615.

Ghude, S. D., Kulkarni, P. S., Kulkarni, S. H., Fadnavis, S., & van der A, R. J. (2011). Temporal variation of urban NOx concentration in India during the past decade as observed from space. *International Journal of Remote Sensing, 32*(3), 849–861. doi:10.1080/01431161.2010.517797.

Ghude, S. D., Pfister, G. G., Jena, C., van der A, R. J., Emmons, L. K., & Kumar, R. (2013). Satellite constraints of nitrogen oxide (NOx) emissions from India based on OMI observations and WRF-Chem simulations. *Geophysical Research Letters, 40*(x), 423–428. doi:10.1029/2012GL053926.

Han, K. M., Song, C. H., Ahn, H. J., Park, R. S., Woo, J. H., Lee, C. K., Richter, A., Burrows, J. P., Kim, J. Y., & Hong, J. H. (2009). Investigation of NOx emissions and NOx-related chemistry in East Asia using CMAQ-predicted and GOME-derived NO2 columns. *Atmospheric Chemistry and Physics, 9*(3), 1017–1036. doi:10.5194/acp-9-1017-2009.

Hilboll, A., Richter, A., & Burrows, J. P. (2013). Long-term changes of tropospheric NO2 over megacities derived from multiple satellite instruments. *Atmospheric Chemistry and Physics, 13* (8), 4145–4169. doi:10.5194/acp-13-4145-2013.

Hilboll, A., Richter, A., & Burrows, J. P. (2017a). On the impact of rapid economic growth on NO2 pollution over India, as seen from space. (2), 1–12. Submitted.

Hilboll, A., Richter, A., & Burrows, J. P. (2017b). NO2 pollution over India observed from space – The impact of rapid economic growth, and a recent decline. *Atmospheric Chemistry and Physics Discussions*. doi:10.5194/acp-2017-101, in review.

Ingmann, P., Veihelmann, B., Langen, J., Lamarre, D., Stark, H., & Courrèges-Lacoste, G. B. (2012). Requirements for the GMES atmosphere service and ESA's implementation concept: Sentinels-4/-5 and -5p. *Remote Sensing of Environment, 120*, 58–69. doi:10.1016/j.rse.2012.01.023.

Kim, J. (2012). GEMS (Geostationary Enviroment Monitoring Spectrometer) onboard the GeoKOMPSAT to monitor air quality in high temporal and spatial resolution over Asia-Pacific region. In *EGU general assembly*. Vienna. 22–27 April 2012, EGU2012-4051.

Koelemeijer, R. B. A., Stammes, P., Hovenier, J. W., & de Haan, J. F. (2001). A fast method for retrieval of cloud parameters using oxygen A band measurements from the Global Ozone Monitoring Experiment. *Journal of Geophysical Research, 106*(D4), 3475. doi:10.1029/2000JD900657.

Krotkov, N. A., McLinden, C. A., Li, C., Lamsal, L. N., Celarier, E. A., Marchenko, S. V., Swartz, W. H., Bucsela, E. J., Joiner, J., Duncan, B. N., Boersma, K. F., Veefkind, J. P., Levelt, P. F., Fioletov, V. E., Dickerson, R. R., He, H., Lu, Z., & Streets, D. G. (2016). Aura OMI observations of regional SO2 and NO2 pollution changes from 2005 to 2015. *Atmospheric Chemistry and Physics, 16*(7), 4605–4629. doi:10.5194/acp-16-4605-2016.

Lee, C., Martin, R. V., van Donkelaar, A., Lee, H., Dickerson, R. R., Hains, J. C., Krotkov, N., Richter, A., Vinnikov, K., & Schwab, J. J. (2011). SO2 emissions and lifetimes: Estimates from inverse modeling using in situ and global, space-based (SCIAMACHY and OMI) observations. *Journal of Geophysical Research, 116*(D6), D06304. doi:10.1029/2010JD014758.

Leitão, J., Richter, A., Vrekoussis, M., Kokhanovsky, A., Zhang, Q. J., Beekmann, M., & Burrows, J. P. (2010). On the improvement of NO2 satellite retrievals – aerosol impact on the airmass factors. *Atmospheric Measurement Techniques, 3*(2), 475–493. doi:10.5194/amt-3-475-2010.

Lerot, C., Stavrakou, T., De Smedt, I., Müller, J. F., & Van Roozendael, M. (2010). Glyoxal vertical columns from GOME-2 backscattered light measurements and comparisons with a global model. *Atmospheric Chemistry and Physics, 10*(24), 12059–12072. doi:10.5194/acp-10-12059-2010.

Levelt, P. F., Van den Oord, G. H. J., Dobber, M. R., Malkki, A., Visser, H., de Vries, J., Stammes, P., Lundell, J. O. V., & Saari, H. (2006). The ozone monitoring instrument. *IEEE Transactions Geoscience Remote Sensing, 44*(5), 1093–1101. doi:Urn:nbn:nl:ui:25-648485.

Li, C., Zhang, Q., Krotkov, N. A., Streets, D. G., He, K., Tsay, S.-C., & Gleason, J. F. (2010). Recent large reduction in sulfur dioxide emissions from Chinese power plants observed by the ozone monitoring instrument. *Geophysical Research Letters, 37*(8), L08807. n/a-n/a. doi:10.1029/2010GL042594.

Lin, J. T., & McElroy, M. B. (2011). Detection from space of a reduction in anthropogenic emissions of nitrogen oxides during the Chinese economic downturn. *Atmospheric Chemistry and Physics, 11*(15), 8171–8188. doi:10.5194/acp-11-8171-2011.

Lin, J. T., Martin, R. V., Boersma, K. F., Sneep, M., Stammes, P., Spurr, R., Wang, P., Van Roozendael, M., Clémer, K., & Irie, H. (2014). Retrieving tropospheric nitrogen dioxide from the ozone monitoring instrument: Effects of aerosols, surface reflectance anisotropy, and vertical profile of nitrogen dioxide. *Atmospheric Chemistry and Physics, 14*(3), 1441–1461. doi:10.5194/acp-14-1441-2014.

Liu, F., Beirle, S., Zhang, Q., Dörner, S., He, K., & Wagner, T. (2016). NOx lifetimes and emissions of cities and power plants in polluted background estimated by satellite observations. *Atmospheric Chemistry and Physics, 16*(8), 5283–5298. doi:10.5194/acp-16-5283-2016.

Martin, R. V., Chance, K., Jacob, D. J., Kurosu, T. P., Spurr, R. J. D., Bucsela, E., Gleason, J. F., Palmer, P. I., Bey, I., Fiore, A. M., Li, Q., Yantosca, R. M., & Koelemeijer, R. B. (2002). An improved retrieval of tropospheric nitrogen dioxide from GOME. *Journal Geophysics Research D Atmospheric, 107*(2), 4611. doi:10.1029/2001JD001027.

Martin, R. V., Jacob, D. J., Chance, K., Kurosu, T. P., Palmer, P. I., & Evans, M. J. (2003). Global inventory of nitrogen oxide emissions constrained by space-based observations of NO2 columns. *Journal of Geophysical Research, 108*(D17), 4537. doi:10.1029/2003JD003453.

McLinden, C. A., Fioletov, V., Shephard, M. W., Krotkov, N., Li, C., Martin, R. V., Moran, M. D., & Joiner, J. (2016). Space-based detection of missing sulfur dioxide sources of global air pollution. *Natural Geoscience* (May), 1–7. doi:10.1038/ngeo2724.

Mijling, B., & van der A, R. J. (2012). Using daily satellite observations to estimate emissions of short-lived air pollutants on a mesoscopic scale. *Journal of Geophysical Research – Atmospheres, 117*(x), 1–20. doi:10.1029/2012JD017817.

Mijling, B., van der A, R. J., Boersma, K. F., Van Roozendael, M., De Smedt, I., & Kelder, H. M. (2009). Reductions of NO2 detected from space during the 2008 Beijing Olympic Games. *Geophysical Research Letters, 36*(2), 1–6. doi:10.1029/2009GL038943.

Munro, R., Lang, R., Klaes, D., Poli, G., Retscher, C., Lindstrot, R., Huckle, R., Lacan, A., Grzegorski, M., Holdak, A., Kokhanovsky, A., Livschitz, J., & Eisinger, M. (2016). The GOME-2 instrument on the Metop series of satellites: Instrument design, calibration, and level 1 data processing – an overview. *Atmospheric Measurement Techniques, 9*(3), 1279–1301. doi:10.5194/amt-9-1279-2016.

Platt, U., & Stutz, J. (2008). *Differential Optical Absorption Spectroscopy (DOAS), principles and applications*. Berlin: Springer.

Richter, A., & Wagner, T. (2011). Solar backscattered radiation: UV, visible and near IR – trace gases. In J. P. Burrows, U. Platt, & P. Borrell (Eds.), *The remote sensing of tropospheric composition from space* (pp. 67–122). Heidelberg: Springer.

Richter, A., Burrows, J. P., Nüss, H., Granier, C., & Niemeier, U. (2005). Increase in tropospheric nitrogen dioxide over China observed from space. *Nature, 437*(7055), 129–132. doi:10.1038/nature04092.

Schneider, P., Lahoz, W. A., & van der A, R. (2015). Recent satellite-based trends of tropospheric nitrogen dioxide over large urban agglomerations worldwide. *Atmospheric Chemistry and Physics, 15*(17), 1205–1220. doi:10.5194/acp-15-1205-2015.

Stavrakou, T., Müller, J. F., Boersma, K. F., De Smedt, I., & van der A, R. J. (2008). Assessing the distribution and growth rates of NOx emission sources by inverting a 10-year record of NO2 satellite columns. *Geophysical Research Letters, 35*(2), 1–5. doi:10.1029/2008GL033521.

Stavrakou, T., Müller, J. F., Boersma, K. F., van der A, R. J., Kurokawa, J., Ohara, T., & Zhang, Q. (2013). Key chemical NOx sink uncertainties and how they influence top-down emissions of nitrogen oxides. *Atmospheric Chemistry and Physics, 13*, 9057–9082. doi:10.5194/acp-13-9057-2013.

van der A, R. J., Peters, D. H. M. U., Eskes, H., Boersma, K. F., Van Roozendael, M., De Smedt, I., & Kelder, H. M. (2006). Detection of the trend and seasonal variation in tropospheric NO2 over China. *Journal of Geophysical Research – Atmospheres, 111*(x), 1–10. doi:10.1029/2005JD006594.

Veefkind, J. P., Aben, I., McMullan, K., Förster, H., de Vries, J., Otter, G., Claas, J., Eskes, H. J., de Haan, J. F., Kleipool, Q., van Weele, M., Hasekamp, O., Hoogeveen, R., Landgraf, J., Snel, R., Tol, P., Ingmann, P., Voors, R., Kruizinga, B., Vink, R., Visser, H., & Levelt, P. F. (2012). TROPOMI on the ESA sentinel-5 precursor: A GMES mission for global observations of the atmospheric composition for climate, air quality and ozone layer applications. *Remote Sensing of Environment, 120*(2012), 70–83. doi:10.1016/j.rse.2011.09.027.

Vrekoussis, M., Wittrock, F., Richter, A., & Burrows, J. P. (2009). Temporal and spatial variability of glyoxal as observed from space. *Atmospheric Chemistry and Physics, 9*(13), 4485–4504. doi:10.5194/acp-9-4485-2009.

Vrekoussis, M., Wittrock, F., Richter, A., & Burrows, J. P. (2010). GOME-2 observations of oxygenated VOCs: What can we learn from the ratio glyoxal to formaldehyde on a global scale? *Atmospheric Chemistry and Physics, 10*(21), 10145–10160. doi:10.5194/acp-10-10145-2010.

Wang, P., Stammes, P., van der A, R., Pinardi, G., & van Roozendael, M. (2008). FRESCO+: An improved O_2 A-band cloud retrieval algorithm for tropospheric trace gas retrievals. *Atmospheric Chemistry and Physics Discussions, 8*(3), 9697–9729. doi:10.5194/acpd-8-9697-2008.

Wang, S., Xing, J., Chatani, S., Hao, J., Klimont, Z., Cofala, J., & Amann, M. (2011). Verification of anthropogenic emissions of China by satellite and ground observations. *Atmospheric Environment, 45*(35), 6347–6358. doi:10.1016/j.atmosenv.2011.08.054.

Wang, S., Zhang, Q., Martin, R. V., Philip, S., Liu, F., Li, M., Jiang, X., & He, K. (2015). Satellite measurements oversee China's sulfur dioxide emission reductions from coal-fired power plants. *Environmental Research Letters, 10*(11), 114015. doi:10.1088/1748-9326/10/11/114015.

Wittrock, F., Richter, A., Oetjen, H., Burrows, J. P., Kanakidou, M., Myriokefalitakis, S., Volkamer, R., Beirle, S., Platt, U., & Wagner, T. (2006). Simultaneous global observations of glyoxal and formaldehyde from space. *Geophysical Research Letters, 33*(16), L16804. doi:10.1029/2006GL026310.

Yang, K., Carn, S. A., Ge, C., Wang, J., & Dickerson, R. R. (2014). Advancing measurements of tropospheric NO 2 from space: New algorithm and first global results from OMPS. *Geophysics Research Letters, 41*(2), 4777–4786., n/a-n/a. doi:10.1002/2014GL060136.

Zhang, Q., Streets, D. G., He, K., Wang, Y., Richter, A., Burrows, J. P., Uno, I., Jang, C. J., Chen, D., Yao, Z., & Lei, Y. (2007). NOx emission trends for China, 1995–2004: The view from the ground and the view from space. *Journal of Geophysical Research, 112*(D22), D22306. doi:10.1029/2007JD008684.

Zhang, Q., Streets, D. G., & He, K. (2009). Satellite observations of recent power plant construction in Inner Mongolia, China. *Geophysical Research Letters, 36*(15), 1–5. doi:10.1029/2009GL038984.

Zoogman, P., Liu, X., Suleiman, R. M., Pennington, W. F., Flittner, D. E., Al-Saadi, J. A., Hilton, B. B., Nicks, D. K., Newchurch, M. J., Carr, J. L., Janz, S. J., Andraschko, M. R., Arola, A., Baker, B. D., Canova, B. P., Chan Miller, C., Cohen, R. C., Davis, J. E., Dussault, M. E., Edwards, D. P., Fishman, J., Ghulam, A., González Abad, G., Grutter, M., Herman, J. R., Houck, J., Jacob, D. J., Joiner, J., Kerridge, B. J., Kim, J., Krotkov, N. A., Lamsal, L., Li, C., Lindfors, A., Martin, R. V., McElroy, C. T., McLinden, C., Natraj, V., Neil, D. O., Nowlan, C. R., O'Sullivan, E. J., Palmer, P. I., Pierce, R. B., Pippin, M. R., Saiz-Lopez, A., Spurr, R. J. D., Szykman, J. J., Torres, O., Veefkind, J. P., Veihelmann, B., Wang, H., Wang, J., & Chance, K. (2016). Tropospheric emissions: Monitoring of pollution (TEMPO). *Journal of Quantitative Spectroscopy and Radiative Transfer*, 1–23. doi:10.1016/j.jqsrt.2016.05.008.

Chapter 14
Observation of Air Pollution over China Using the IASI Thermal Infrared Space Sensor

Cathy Clerbaux, Sophie Bauduin, Anne Boynard, Lieven Clarisse, Pierre Coheur, Maya George, Juliette Hadji-Lazaro, Daniel Hurtmans, Sarah Safieddine, Martin Van Damme, and Simon Whitburn

Abstract In this chapter we describe what is achievable in terms of pollutant tracking from space using observations provided by thermal infrared remote sensors. After a general introduction on infrared remote sensing, we exploit the data provided by the Infrared Atmospheric Sounding Interferometer (IASI) missions onboard the Metop series of satellite to illustrate pollution detection at various spatial and temporal scales. Then, we focus on air pollution over China and discuss three case studies involving different pollutants. The first example discusses the geophysical conditions for detection of ammonia (NH_3) and sulfur dioxide (SO_2), both precursors of particulate matter (PM). The second case illustrates the seasonal variation of ozone (O_3), in particular during the monsoon period. The third case shows the local accumulation of enhanced levels of carbon monoxide (CO) when pollution episodes occur.

Keywords IASI • Infrared remote sensing • Spectroscopy • Air pollution

---

C. Clerbaux (✉)
LATMOS/IPSL, UPMC Sorbonne Universités, UVSQ, CNRS, Paris, France

Spectroscopie de l'Atmosphère, Chimie Quantique et Photophysique, Université libre de Bruxelles (ULB), Brussels, Belgium
e-mail: cathy.clerbaux@latmos.ipsl.fr

S. Bauduin • L. Clarisse • P. Coheur • D. Hurtmans • M. Van Damme • S. Whitburn
Spectroscopie de l'Atmosphère, Chimie Quantique et Photophysique, Université libre de Bruxelles (ULB), Brussels, Belgium
e-mail: sbauduin@ulb.ac.be; lclariss@ulb.ac.be; pfcoheur@ulb.ac.be; dhurtma@ulb.ac.be; martin.van.damme@ulb.ac.be; swhitbur@ulb.ac.be

A. Boynard • M. George • J. Hadji-Lazaro • S. Safieddine
LATMOS/IPSL, UPMC Sorbonne Universités, UVSQ, CNRS, Paris, France
e-mail: anne.boynard@latmos.ipsl.fr; maya.george@latmos.ipsl.fr; juliette.hadji-lazaro@latmos.ipsl.fr; sarah.safieddine@latmos.ipsl.fr

I. Bouarar et al. (eds.), *Air Pollution in Eastern Asia: An Integrated Perspective*, ISSI Scientific Report Series 16, DOI 10.1007/978-3-319-59489-7_14

14.1 What Do Thermal Infrared Space Sensors See?

Remote sensing observation from satellite is increasingly viewed as a way to measure pollution, to monitor atmospheric changes, and to control the implementation of environmental policies. Space-borne sensors using the thermal infrared (IR) spectral range measure the atmospheric composition by using the spectroscopic properties of molecules and particles to derive their atmospheric concentrations (see Fig. 14.1, representing an atmospheric spectrum).

Most atmospheric constituents absorb and re-emit infrared photons, and the strongest spectral features are due to rotational-vibrational transitions allowed by the quantum mechanical selection rules. In particular, when transitions that lead to a change of the molecular dipole moment upon absorption or emission of a photon are allowed. The emission spectrum of Earth's atmosphere contains many spectral features associated with the different molecules that absorb and emit radiation. Furthermore, due to the variable vertical temperature profile in the atmosphere, the observed spectrum is a superposition of all these processes at different altitudes. A simulation of what a thermal infrared instrument onboard a satellite would observe can be obtained by radiative transfer calculations taking into account all geophysical variables and molecular properties affecting the emission/absorption of radiation through the successive atmospheric layers. Figure 14.2 provides an example for such a spectrum as measured in the nadir (downward looking) observation mode. Aside from water vapor (H_2O) and the long-lived greenhouse gases (carbon dioxide (CO_2), methane (CH_4), nitrous oxide (N_2O)) that provide the main absorption features observable in this spectral range, atmospheric abundances can be obtained for more reactive species such as carbon monoxide (CO), ozone (O_3), ammonia (NH_3) and to some extent sulfur dioxide (SO_2).

Nadir infrared space-borne instruments have the advantage of providing observations both during day and night. In contrast to in-situ measurements, the satellite

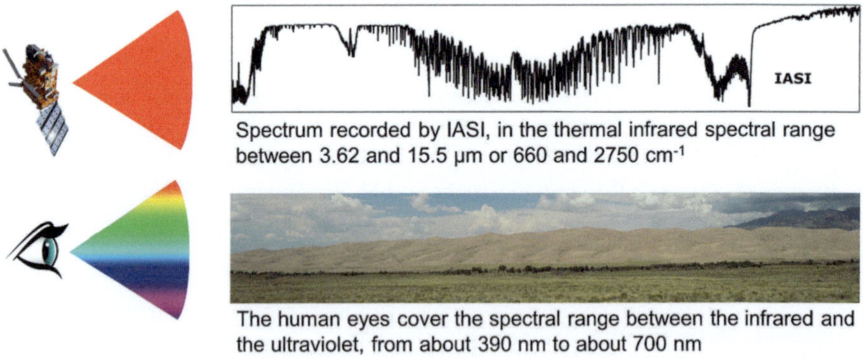

Fig. 14.1 The satellite-borne remote sensor records a spectrum in the infrared, from which trace gas concentrations can be derived (*top*). The visible spectrum is the only part the human eye can see (*bottom*)

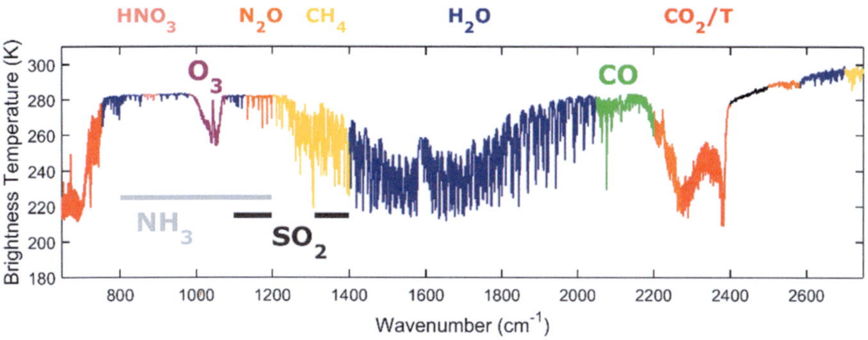

Fig. 14.2 Atmospheric spectrum (in brightness temperature units) as seen by a nadir looking space borne IR sensor with the main absorbing species identified: CO and O_3 have strong absorption signatures, whereas NH_3 and SO_2 absorption lines are weak and of the same order of magnitude than the instrumental noise

observation is less accurate, and provides averaged information over larger areas determined by the instrument's footprint, and integrated information on the vertical. The more challenging step is the conversion of the measured spectra into atmospheric abundance for the absorbing gases. This step is known as the inverse problem or "retrieval". Different methods exist to retrieve gas concentrations from the atmospheric spectra (optimal estimation, regularization techniques, neural networks, etc.). They rely on minimization techniques, e.g., to adjust simulated spectra (generated using radiative transfer simulations) with actual measured radiance data, together with a constraint upon the solution (called the a priori), penalizing estimates that deviate greatly from the profile shape provided in the prior estimate (Rodgers 2000).

For cloud free conditions, nadir-looking thermal infrared instruments can measure the atmospheric radiation down to the ground and some vertical information can be derived from the shape of the absorption lines, provided that the spectral resolution is sufficient. The thermal contrast (temperature difference between the surface and the boundary layer) determines to what extent the detection of species close to the surface is possible. When the Earth's surface heats up or cools down faster than the atmosphere, the thermal contrast increases. Thermal contrasts are generally higher over land during the day and lower over water and over land at night. Large thermal contrasts are associated with a better sensitivity towards the surface. The capability of an infrared nadir sounder to probe the lower atmospheric layers, where local pollution occurs, therefore strongly depends on location, temperature, type of surface (emissivity) and time of the day. Along with the instrumental characteristics (spectral resolution and radiometric noise) thermal contrast determines the amount of vertical information that can be retrieved for a given species in the nadir geometry, and the uneven sensitivity to the different atmospheric layers.

14.2 The IASI Mission Onboard the Metop Satellite

After this introduction, we will illustrate what an advanced thermal infrared sensor can observe, in particular over China. For this purpose we use the data provided by the two Infrared Atmospheric Sounding Interferometer (IASI) (Clerbaux et al. 2009), which are flying onboard the Metop-A and -B satellites, which have provided an uninterrupted record of atmospheric spectra since 2007 from which several key atmospheric species can be monitored (Coheur et al. 2009; Clarisse et al. 2011; Clerbaux et al. 2015). The IASI Fourier transform spectrometers measure the upwelling radiances in the thermal infrared spectral range extending from 645 to 2760 cm^{-1} (3.6–15.5 μm), with high radiometric quality, using a 0.5 cm^{-1} spectral resolution (Cayla 1993). Each observation represents 8461 radiance channels, from which information on the surface properties, temperature, and the atmospheric composition can be derived using ad hoc retrieval algorithms (Hilton et al. 2012). In 8 s a total of 120 views is collected over a swath of ~2200 km using a stare-and-stay mode of 30 arrays of four individual elliptical pixels, each of which has a 12 km diameter at nadir (increasing at larger viewing angles) with two calibrations views (dark space and black body reference).

The Metop satellites are flying at about 817 km on a polar Sun-synchronous orbit, with around 14 orbits per day, Metop-A and Metop-B being half an orbit apart (Fig. 14.3). The local overpass time for the morning and evening orbits is around 9:30 local time. More than 1.2 million spectra per day are collected by each IASI instrument, from which a significant part is affected by the presence of clouds in the instrument's field of view and therefore discarded for the atmospheric composition retrievals. Every orbit's batch of data is made available, so that a near real-time processing of the data is possible. A third IASI will be launched in the fall of 2018 (on Metop-C), and the program has been secured in the longer term with the three IASI-new generation (IASI-NG) instruments (Clerbaux and Crevoisier 2013; Crevoisier et al. 2014) to be launched on the Metop second generation satellites in 2021 and onwards.

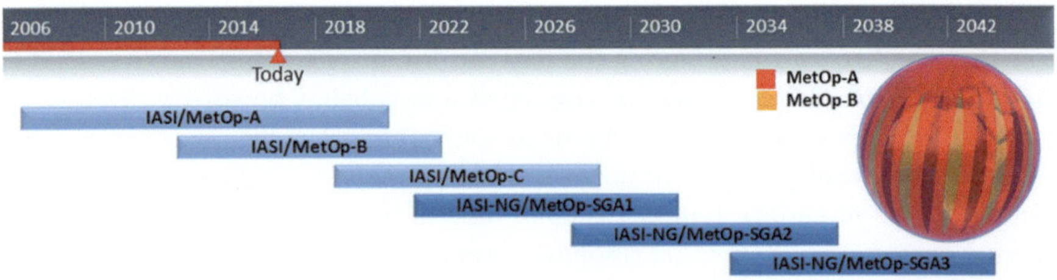

Fig. 14.3 Spatial and temporal coverages provided by the IASI missions and its follows-on: at the end of 2016 two IASIs were operating onboard Metop-A (*red*) and Metop-B (*yellow*), covering each the whole globe twice a day. After the launch of MetOp-C, an improved IASI, IASI-NG, with a spectral resolution and signal to noise performance improved by a factor of 2, will fly on the next MetOp generation

14.3 Global Scale Distribution of Pollutants

14.3.1 Carbon Monoxide Maps

Carbon monoxide (CO) is a combustion-related pollutant that is mainly emitted by motor vehicle traffic, heating/cooking systems, industrial activities, and biomass burning. With a lifetime of a few weeks, it can be transported over long distances. It is an excellent tracer of pollution, indicative of anthropogenic emissions, biomass burning and atmospheric circulation patterns. It is also a key atmospheric species as it regulates the oxidizing capacity of the troposphere through its destruction cycle involving the hydroxyl radical (OH), and as it is an ozone and carbon dioxide precursor.

The top left panel of Fig. 14.4 shows the monthly average for CO total column distribution measured by IASI for April 2013 (daytime data, Metop-A observations) (Hurtmans et al. 2012; George et al. 2015). As expected, it can be seen that large concentrations of CO are found near emission sources, and plumes are transported downwind. In the Northern Hemisphere (NH) elevated levels of CO are found above the West and East USA coasts, over Europe, and over East Asia.

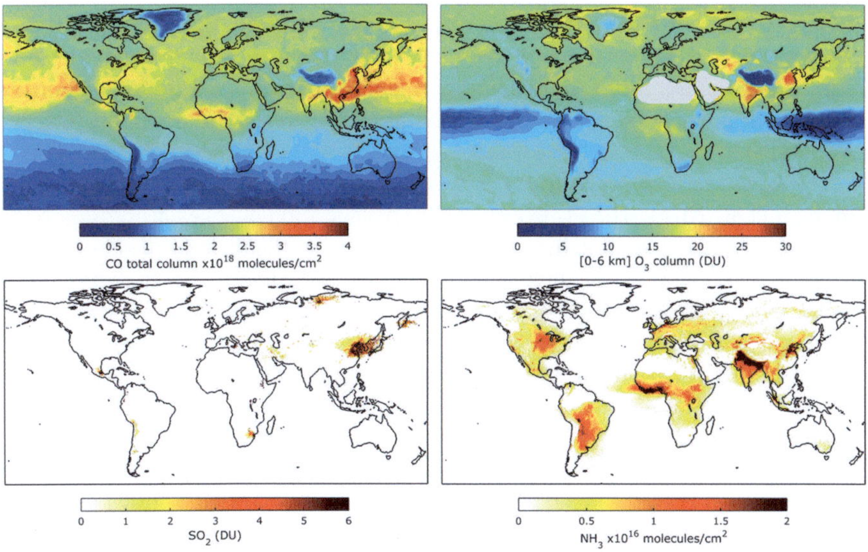

Fig. 14.4 Global distribution of column concentrations for different pollutants as measured by IASI. For CO (*top left*) and O_3 (*top right*) the observation is always possible, provided there is no cloud in the field of view. For SO_2 (*bottom left*) and NH_3 (*bottom right*), as the absorption signature is weaker, the detection is only possible over a given detection limit (which varies with the location and the geophysical conditions). Note that high-altitude SO_2 due to explosive degassing has been removed, as it is of sporadic nature, and that desert areas are colored in *grey* in the ozone plot, as dust absorption interfere with that of O_3. The periods plotted are April 2013 (CO), 15 May to 15 July 2014 (O_3) and 2008–2015 (SO_2 and NH_3)

Due to long range transport, high CO concentrations are also observed over the Northern Pacific and Atlantic oceans. In the tropics the CO maximum is mainly associated with fires occurring in the Amazon basin, in central and southern Africa and sometimes over Australia, with maximum in August-November. NH concentrations peak in April, after accumulating during winter, and drop off gradually until late summer as the increasing solar insolation activates tropospheric chemistry (except over Siberia's and Alaska's fire regions where CO concentrations increase in summer e.g., R'Honi et al. 2013). The whole IASI CO record (2007–present) has the highest concentrations associated with major fires that occurred in Russia (July 2010), Siberia (July 2012) and Indonesia (fall 2014 and 2015) (Whitburn et al. 2016a).

14.3.2 Tropospheric Ozone Distribution Maps

Breathing ozone (O_3) can cause a variety of health problems. The sources of O_3 in the troposphere are either downward transport from the stratosphere or local chemical production within the troposphere. In urban areas, for example, ozone can be formed through sunlight-initiated oxidation of volatile organic compounds (VOCs) in the presence of nitrogen oxides (NO_x) and CO. During summer time urban population is often exposed to O_3 levels exceeding air quality standard values (e.g., Honoré et al. 2008). Depending on the location and local limitations of the emission of precursors, decreasing/increasing ozone trends are observed (Cooper et al. 2014).

The signature of tropospheric O_3 is well captured by IASI (Hurtmans et al. 2012; Boynard et al. 2016). The O_3 tropospheric concentration changes according to the latitude and season. High values are especially detected in densely populated areas in midlatitudes due to the high emission of O_3 precursors and the subsequent photochemical production of O_3. Figure 14.4 (right panel) shows the ozone column, integrated from the surface to 6 km, and averaged over 2 months (from 15 May to 15 July 2014). The [0–6 km] was chosen as it is representative of lowest layer with the vertical resolution achievable by IASI. High concentrations are found in the latitudes between 0° and 45°N, where the main anthropogenic sources of tropospheric O_3 are located (e.g., Dufour et al. 2010; Safieddine et al. 2013). The high values detected between 0° and 45°S correspond to the biomass burning source of O_3 precursors in the lower troposphere and to O_3 source from lighting NO_x in the mid-troposphere. The high values detected in the northern high latitudes are explained by residual stratospheric intrusions, which inject O_3-rich air into the 0–6 km column. More generally, the highest detected O_3 values in the two hemispheres are in spring (March-April-May for NH and September-October-November for SH), which coincides to the period where the intrusions from the stratosphere are the highest. The smallest values for the O_3 tropospheric column are observed in the high and middle latitudes of the SH as well as around the Tropics, all year long except when fires occur. The figure also shows the transport of O_3, over the oceans as well as the transport of pollution from China to North America.

14.3.3 Ammonia and Sulfur Dioxide Maps

Ammonia (NH_3) and sulfur dioxide (SO_2) are both pollutants, and also contribute to the formation of inorganic particles (e.g., ammonium sulfate and ammonium nitrate) in the atmosphere. Fine particulate matter (PM) affects the human health, as it triggers respiratory and cardiovascular diseases and induces a decrease in life expectancy. A study by Lelieveld et al. (2015) has shown that PM is responsible for 3.3 million premature deaths per year worldwide, predominantly in Asia.

Both NH_3 and SO_2 have weak absorption features in the thermal infrared spectral range, and hence it proved to be difficult to measure them from the IASI spectra. By averaging over longer time periods to increase the signal over the background noise, emission hotspots could be identified across the globe (Clarisse et al. 2009; Bauduin et al. 2014). A major effort to develop specific processing tools adapted to weak absorbers was undertaken following the ideas developed in Walker et al. (2011) and Carboni et al. (2012), leading to more systematic day and night retrieved observations (Van Damme et al. 2014; Whitburn et al. 2016b; Bauduin et al. 2016). Emission sources can be better assessed when averaging the daily data over long time periods. Figure 14.4 (bottom) illustrates the spatial distribution of SO_2 and NH_3 after accumulating the data over the lifetime of the IASI mission. For SO_2, hot spots are linked with industrial activity, coal consumption and to degassing volcanoes. NH_3 high concentrations are associated with cattle and agriculture activities, but large fires also contribute to the high abundancies observed locally.

14.4 Zoom over China and Three Cases Studies

Figure 14.5 is similar to Fig. 14.4, but zoomed over East Asia. It can be seen that all pollutants listed here show concentration peaks in the North China Plain, mainly over Beijing and its surroundings. In the following we will report information extracted from three studies: The first study (Bauduin et al. 2016) describes the atmospheric conditions needed to detect SO_2, the second one (Safieddine et al. 2016) analyses the impact of strong dynamic phenomena such as the monsoon event on the ozone concentrations, and finally the last case study (Boynard et al. 2014) illustrates the local increase of CO as seen during the night when a high pollution event occurs.

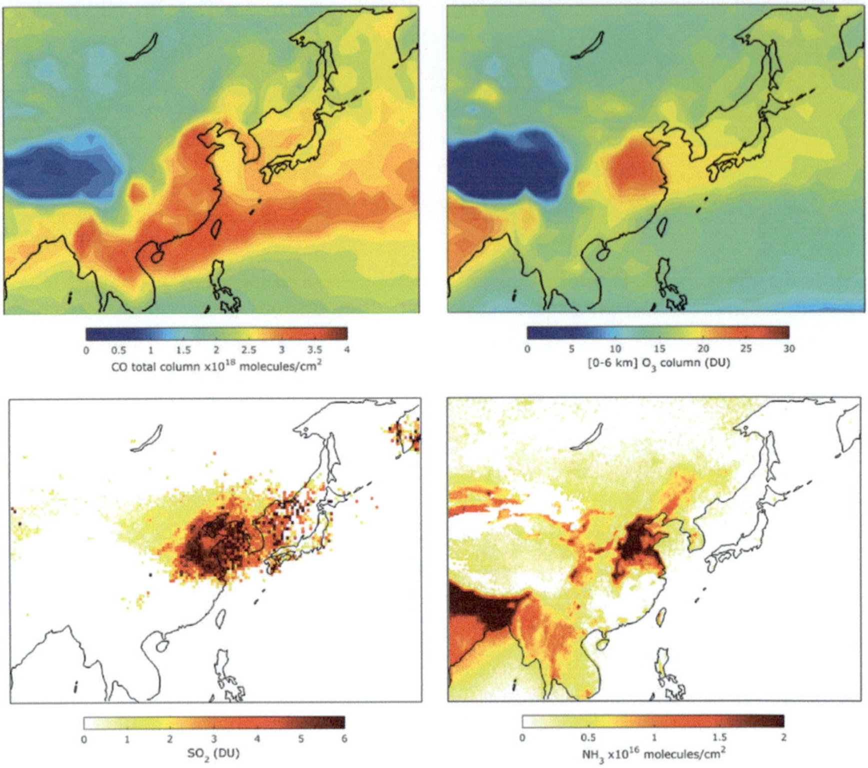

Fig. 14.5 Maps over China with column concentrations for different pollutants as measured by IASI (zoom of Fig. 14.4). The periods plotted are April 2013 (CO), 15 May to 15 July 2014 (O_3) and 2008–2015 (SO_2 and NH_3)

14.4.1 Case 1: Geophysical Conditions for Detection of Sulfur Dioxide and Ammonia

In China the SO_2 budget is dominated by anthropogenic sources mainly due to energy supply through coal/oil combustion producing large amounts of SO_2, whereas NH_3 is associated with livestock and agricultural activities through the use of nitrogen fertilizer. China is one of the world's largest emission sources of SO_2, and enhanced levels can be seen over the industrial areas, with emissions peaking close to Beijing and concentrations decreasing westwards. For NH_3 high levels are found in the Indo-Gangetic plain, and again near Beijing.

Figure 14.6 reports the IASI SO_2 observations over China and details the key parameters driving the possibility to measure it. The plot shows a sensitivity analysis using 7 years of IASI observations (1 January 2008–30 September 2014)

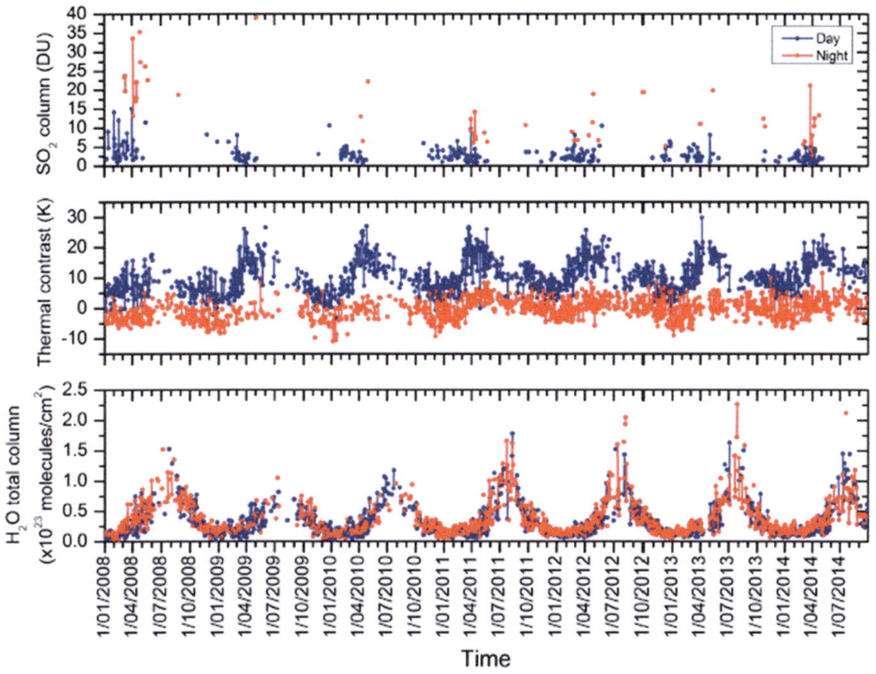

Fig. 14.6 Daily averages of near-surface SO_2 (DU), thermal contrast (K) and H_2O total column (molecules/cm^2) around Beijing. Averages have been calculated within a radius of 125 km of Beijing. Daytime measurements are indicated in *blue* and nighttime measurements are indicated in *red* (Bauduin et al. 2016)

above Beijing (Bauduin et al. 2016). Daily averages of near-surface SO_2 columns, thermal contrast and H_2O total column are shown, separately for the morning (in blue) and evening (in red) overpasses of IASI. The daily-averaged SO_2 columns from the morning overpass vary around 3 Dobson Units (DU), with maxima that can reach 15 DU. The time series is incomplete, with successful SO_2 retrievals from October-December to May-June associated with fairly high thermal contrast (10 K on average but up to 20 K – see middle panels) and low humidity. Favorable thermal contrast conditions persist mostly year-round in Beijing but the humidity is too high during the other months (May-June to October-November) to allow the near-surface detection. The intense water vapor lines then interfere with the SO_2 weak lines. It is inferred from Fig. 14.6 that IASI is mostly not sensitive to surface SO_2 in the evening due to the drop of thermal contrast to close to 0.

Ammonia is less affected by the presence of water vapor, but similarly to SO_2, thermal contrast is also a key factor for detection and derivation of reliable concentrations. For a fixed NH_3 column, a larger thermal contrast will give rise to larger spectral signatures and vice versa (Whitburn et al. 2016b).

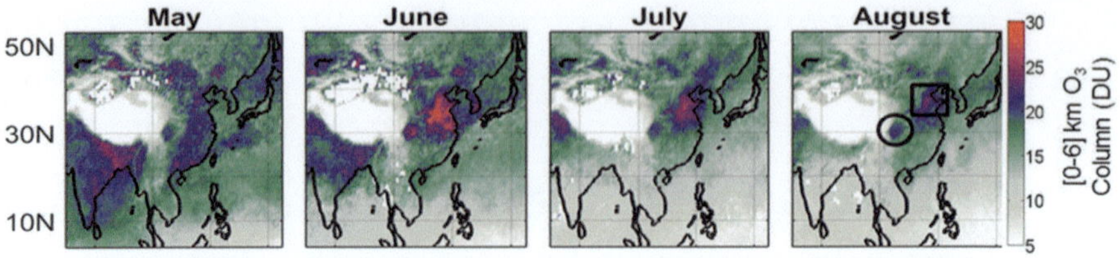

Fig. 14.7 Monthly averaged tropospheric [0–6] km O_3 column from IASI, for each of the months of May to August 2011. The *black square* and *circle* are the regions the least affected by the monsoon and discussed more in the text (Safieddine et al. 2016)

14.4.2 Case 2: Tropospheric Ozone Measurements During the Summer Monsoon Period

During the East Asian summer monsoon the anticyclonic flow dominates the seasonal circulation pattern, and ozone is a good tracer of this weather pattern. Since most of the O_3 resides in the stratosphere, the remote observation of tropospheric ozone with IASI remains tricky as absorption/emission contributions from both stratospheric and tropospheric ozone are mixed in the spectra. During the East Asian summer monsoon, the southwesterly monsoon flow brings warm, wet and clean air masses from the Indian Ocean to South, South-East and East Asia. In June, in particular, the monsoon becomes stronger as the wind force and the cloud cover increase over the regions below 20°N and a decrease in tropospheric O_3 is recorded over India, Bay of Bengal and South East China (Safieddine et al. 2016 and Fig. 14.7). During July and August, the monsoon reaches its maximal strength. Due to high cloud cover and strong winds, the tropospheric O_3 columns show a large decrease over Asia in general, although the area of Beijing, Tianjin and the North China Plain (see the black square in Fig. 14.7) in general are less affected by the monsoon and they show much weaker O_3 decrease. The other interesting region in China that shows little or no change is the area between the Chongqing and Sichuan province (black circle in Fig. 14.7). This region does not exhibit any monsoon characteristics with low cloud cover and weak winds. This region is also between two mountains, favoring the persistence of O_3 during summer.

14.4.3 Case 3: Carbon Monoxide During a Local Pollution Event

From the numerous IASI observations, CO maps over China can be derived each morning and each evening, both over land and sea. The altitude of peak sensitivity for the bottom layer of retrieved CO generally ranges between around 2 and around 10 km, with most of the information coming from the mid troposphere. This implies that the instrument is mostly sensitive to uplifted and horizontally transported CO,

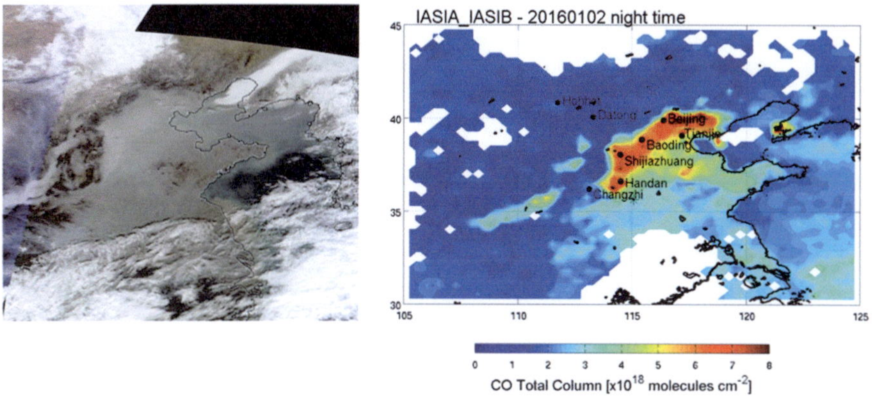

Fig. 14.8 MODIS/Terra (visible image) and IASI CO observation during the pollution event that occurred on January 2, 2016 (following the methodology used in Boynard et al. 2014)

whereas it can miss local pollution events if these are confined near the ground. Interestingly, as pollution peaks are often associated with a temperature inversion in the atmospheric boundary layer (and hence thermal contrast), detection at surface level is possible in these cases. This type of events occurs each winter in China, leading to large pollution exceedances and accumulation of pollutants near the surface. Figure 14.8 illustrates such an event for January 2, 2016, using the methodology described in Boynard et al. (2014). The left plot represents the visible view as provided by the MODIS remote sensor whereas the right plot shows the CO concentration peaking over the Beijing area during nighttime, when temperature inversion is the strongest.

14.5 Conclusion

Infrared spectroscopy from satellite is a powerful technique to observe distributions of pollutants from local to global scales. This chapter describes the basic principles used to derive trace gas concentrations from upwelling radiance spectra recorded in the thermal infrared spectral range. Using the numerous data provided by the IASI mission flying onboard the Metop series of satellites, examples of retrieved concentrations are provided for carbon monoxide, ozone, sulfur dioxide, and ammonia. Large abundancies are reported over East Asia for these four pollutants and three case studies allow to provide a more in-depth description of their spatial/temporal evolution.

Acknowledgments The IASI mission is a joint mission of Eumetsat and the Centre National d'Etudes Spatiales (CNES, France). The IASI L1 data are distributed in near real time by Eumetsat through the Eumetcast system distribution. The authors acknowledge the AERIS French atmospheric database for providing the IASI data. The French scientists are grateful to CNES for scientific collaboration and financial support. The Belgian scientists are grateful to F.R.S.-FNRS and the Belgian State Federal Office for Scientific, Technical and Cultural Affairs and the European Space Agency (ESA-Prodex arrangements) for supporting their research.

References

Bauduin, S., Clarisse, L., Clerbaux, C., Hurtmans, D., & Coheur, P.-F. (2014). IASI observations of sulfur dioxide (SO_2) in the boundary layer of Norilsk. *Journal of Geophysical Research – Atmospheres, 119*, 4253–4263. doi:10.1002/2013JD021405.

Bauduin, S., Clarisse, L., Hadji-Lazaro, J., Theys, N., Clerbaux, C., & Coheur, P.-F. (2016). Retrieval of near-surface sulfur dioxide (SO2) concentrations at a global scale using IASI satellite observations. *Atmospheric Measurement Techniques, 9*, 721–740. doi:10.5194/amt-9-721-2016.

Boynard, A., Clerbaux, C., Clarisse, L., Safieddine, S., Pommier, M., Van Damme, M., Bauduin, S., Oudot, C., Hadji-Lazaro, J., Hurtmans, D., & Coheur, P.-F. (2014). First simultaneous space measurements of atmospheric pollutants in the boundary layer from IASI: A case study in the North China Plain. *Geophysical Research Letters, 41*, 645–651. doi:10.1002/2013GL058333.

Boynard, A., Hurtmans, D., Koukouli, M. E., Goutail, F., Bureau, J., Safieddine, S., Lerot, C., Hadji-Lazaro, J., Wespes, C., Pommereau, J.-P., Pazmino, A., Zyrichidou, I., Balis, D., Barbe, A., Mikhailenko, S. N., Loyola, D., Valks, P., Van Roozendael, M., Coheur, P.-F., & Clerbaux, C. (2016). Seven years of IASI ozone retrievals from FORLI: Validation with independent total column and vertical profile measurements. *Atmospheric Measurement Techniques, 9*, 4327–4353. doi:10.5194/amt-9-4327-2016.

Carboni, E., Grainger, R., Walker, J., Dudhia, A., & Siddans, R. (2012). A new scheme for sulphur dioxide retrieval from IASI measurements: Application to the Eyjafjallajökull eruption of April and May 2010. *Atmospheric Chemistry and Physics, 12*, 11417–11434. doi:10.5194/acp-12-11417-2012.

Cayla, F. R. (1993). IASI infrared interferometer for operations and research. In A. Chedin, M. T. Chahine, & N. A. Scott (Eds.), *High spectral resolution infrared remote sensing for earth's weather and climate studies, NATO ASI Series* (Vol. I 9). Berlin: Springer.

Clarisse, L., Clerbaux, C., Dentener, F., Hurtmans, D., & Coheur, P.-F. (2009). Global ammonia distribution derived from infrared satellite observations. *Nature Geoscience, 2*, 479–483. doi:10.1038/ngeo551.

Clarisse, L., R'Honi, Y., Coheur, P.-F., Hurtmans, D., & Clerbaux, C. (2011). Thermal infrared nadir observations of 24 atmospheric gases. *Geophysical Research Letters, 38*, L10802. doi:10.1029/2011GL047271.

Clerbaux, C., & Crevoisier, C. (2013). New directions: Infrared remote sensing of the troposphere from satellite: Less, but better. *Atmospheric Environment, 72*, 24–26. doi:10.1016/j.atmosenv.2013.01.057.

Clerbaux, C., Boynard, A., Clarisse, L., George, M., Hadji-Lazaro, J., Herbin, H., Hurtmans, D., Pommier, M., Razavi, A., Turquety, S., Wespes, C., & Coheur, P.-F. (2009). Monitoring of atmospheric composition using the thermal infrared IASI/MetOp sounder. *Atmospheric Chemistry and Physics, 9*, 6041–6054. doi:10.5194/acp-9-6041-2009.

Clerbaux, C., Hadji-Lazaro, J., Turquety, S., George, M., Boynard, A., Pommier, M., Safieddine, S., Coheur, P.-F., Hurtmans, D., Clarisse, L., & Van Damme, M. (2015). Tracking pollutants from space: Eight years of IASI satellite observation. *Comptes Rendus Geoscience, 347*, 134–144. doi:10.1016/j.crte.2015.06.001.

Coheur, P.-F., Clarisse, L., Turquety, S., Hurtmans, D., & Clerbaux, C. (2009). IASI measurements of reactive trace species in biomass burning plumes. *Atmospheric Chemistry and Physics, 9*, 5655–5667. doi:10.5194/acp-9-5655-2009.

Cooper, O. R., Parrish, D. D., Ziemke, J., Balashov, N. V., Cupeiro, M., Galbally, I. E., Gilge, S., Horowitz, L., Jensen, N. R., Lamarque, J.-F., Naik, V., Oltmans, S. J., Schwab, J., Shindell, D. T., Thompson, A. M., Thouret, V., Wang, Y., & Zbinden, R. M. (2014). Global distribution and trends of tropospheric ozone: An observation-based review. *Element Science Anthropology, 2*, 000029. doi:10.12952/journal.elementa.000029.

Crevoisier, C., Clerbaux, C., Guidard, V., Phulpin, T., Armante, R., Barret, B., Camy-Peyret, C., Chaboureau, J.-P., Coheur, P.-F., Crépeau, L., Dufour, G., Labonnote, L., Lavanant, L., Hadji-Lazaro, J., Herbin, H., Jacquinet-Husson, N., Payan, S., Péquignot, E., Pierangelo, C., Sellitto, P., & Stubenrauch, C. (2014). Towards IASI-New Generation (IASI-NG): Impact of improved spectral resolution and radiometric noise on the retrieval of thermodynamic, chemistry and climate variables. *Atmospheric Measurement Techniques, 7*, 4367–4385. doi:10.5194/amt-7-4367-2014.

Dufour, G., Eremenko, M., Orphal, J., & Flaud, J.-M. (2010). IASI observations of seasonal and day-to-day variations of tropospheric ozone over three highly populated areas of China: Beijing, Shanghai, and Hong Kong. *Atmospheric Chemistry and Physics, 10*, 3787–3801. doi:10.5194/acp-10-3787-2010.

George, M., Clerbaux, C., Bouarar, I., Coheur, P.-F., Deeter, M. N., Edwards, D. P., Francis, G., Gille, J. C., Hadji-Lazaro, J., Hurtmans, D., Inness, A., Mao, D., & Worden, H. M. (2015). An examination of the long-term CO records from MOPITT and IASI: Comparison of retrieval methodology. *Atmospheric Measurement Techniques, 8*, 4313–4328. doi:10.5194/amt-8-4313-2015.

Hilton, F., Armante, R., August, T., Barnet, C., Bouchard, A., Camy-Peyret, C., Capelle, V., Clarisse, L., Clerbaux, C., Coheur, P.-F., Collard, A., Crevoisier, C., Dufour, G., Edwards, D., Faijan, F., Fourrié, N., Gambacorta, A., Goldberg, M., Guidard, V., Hurtmans, D., Illingworth, S., Jacquinet-Husson, N., Kerzenmacher, T., Klaes, D., Lavanant, L., Masiello, G., Matricardi, M., McNally, A., Newman, S., Pavelin, E., Payan, S., Péquignot, E., Peyridieu, S., Phulpin, T., Remedios, J., Schlüssel, P., Serio, C., Strow, L., Stubenrauch, C., Taylor, J., Tobin, D., Walter, W., & Zhou, D. (2012). Hyperspectral earth observation from IASI: Five years of accomplishments. *Bulletin of the American Meteorological Society, 93*(3), 347–370. doi:10.1175/BAMS-D-11-00027.1.

Honoré, C., Rouïl, L., Vautard, R., Beekmann, M., Bessagnet, B., Dufour, A., Elichegaray, C., Flaud, J.-M., Malherbe, L., Meleux, F., Menut, L., Martin, D., Peuch, A., Peuch, V.-H., & Poisson, N. (2008). Predictability of European air quality: Assessment of 3 years of operational forecasts and analyses by the PREV'AIR system. *Journal of Geophysical Research, 113*, D04301. doi:10.1029/2007JD008761.

Hurtmans, D., Coheur, P.-F., Wespes, C., Clarisse, L., Scharf, O., Clerbaux, C., Hadji-Lazaro, J., George, M., & Turquety, S. (2012). FORLI radiative transfer and retrieval code for IASI. *Journal of Quantitative Spectroscopy and Radiative Transfer, 113*(11), 1391–1408. doi:10.1016/j.jqsrt.2012.02.036.

Lelieveld, J., Evans, J. S., Fnais, M., Giannadaki, D., & Pozzer, A. (2015). The contribution of outdoor air pollution sources to premature mortality on a global scale. *Nature, 525*, 367–371. doi:10.1038/nature15371.

R'Honi, Y., Clarisse, L., Clerbaux, C., Hurtmans, D., Duflot, V., Turquety, S., Ngadi, Y., & Coheur, P.-F. (2013). Exceptional emissions of NH_3 and HCOOH in the 2010 Russian wildfires. *Atmospheric Chemistry and Physics, 13*, 4171–4181. doi:10.5194/acp-13-4171-2013.

Rodgers, C. D. (2000). *Inverse methods for atmospheric sounding: Theory and practice*. World Scientific Series on Atmospheric, Oceanic and Planetary Physics, 2. Hackensack.

Safieddine, S., Clerbaux, C., George, M., Hadji-Lazaro, J., Hurtmans, D., Coheur, P.-F., Wespes, C., Loyola, D., Valks, P., & Hao, N. (2013). Tropospheric ozone and nitrogen dioxide measurements in urban and rural regions as seen by IASI and GOME-2. *Journal of Geophysical Research – Atmospheres, 118*, 10555–10566. doi:10.1002/jgrd.50669.

Safieddine, S., Boynard, A., Hao, N., Huang, F., Wang, L., Ji, D., Barret, B., Ghude, S. D., Coheur, P.-F., Hurtmans, D., & Clerbaux, C. (2016). Tropospheric ozone variability during the East Asian summer monsoon as observed by satellite (IASI), aircraft (MOZAIC) and ground stations. *Atmospheric Chemistry and Physics, 16*, 10489–10500. doi:10.5194/acp-16-10489-2016.

Van Damme, M., Clarisse, L., Heald, C. L., Hurtmans, D., Ngadi, Y., Clerbaux, C., Dolman, A. J., Erisman, J. W., & Coheur, P.-F. (2014). Global distributions, time series and error characterization of atmospheric ammonia (NH_3) from IASI satellite observations. *Atmospheric Chemistry and Physics, 14*, 2905–2922. doi:10.5194/acp-14-2905-2014.

Walker, J. C., Dudhia, A., & Carboni, E. (2011). An effective method for the detection of trace species demonstrated using the MetOp infrared atmospheric sounding interferometer. *Atmospheric Measurement Techniques, 4*, 1567–1580. doi:10.5194/amt-4-1567-2011.

Whitburn, S., Van Damme, M., Clarisse, L., Turquety, S., Clerbaux, C., Coheur, P.-F. (2016a). Peat fires doubled annual ammonia emissions in Indonesia during the 2015 El Niño. *Geophysical Research Letters, 43*. doi:10.1002/2016GL070620.

Whitburn, S., Van Damme, M., Clarisse, L., Bauduin, S., Heald, C. L., Hadji-Lazaro, J., Hurtmans, D., Zondlo, M. A., Clerbaux, C., & Coheur, P.-F. (2016b). A flexible and robust neural network IASI-NH_3 retrieval algorithm. *Journal of Geophysical Research-Atmospheres, 121*(11), 6581–6599. doi:10.1002/2016JD024828.

Chapter 15
Monitoring Aerosol Properties in East Asia from Geostationary Orbit: GOCI, MI and GEMS

Jhoon Kim, Mijin Kim, and Myungje Choi

Abstract With the launch of Geostationary Ocean Color Imager (GOCI) and Meteorological Imager (MI) onboard the Communication, Oceanography, and Meteorology Satellite (COMS) over Asia in 2010, hourly monitoring of various aerosol properties has been realized. Algorithms and aerosol data products are presented for the period of 5 years since its launch. Seasonal cycle of aerosol optical depth (AOD) and its decreasing trend were observed. Together with the plan to launch Geostationary Environment Monitoring Spectrometer (GEMS) in 2019, monitoring of trace gas concentration for ozone, aerosol and their precursors will be possible in high temporal and spatial resolution. In this study, results and plan to monitor aerosol and trace gas concentration from geostationary earth orbit (GEO) are presented.

Keywords GOCI • MI • GEMS • Aerosol • Trace gas • Remote sensing • Geostationary earth orbit

15.1 Introduction

There have been extensive satellite missions to measure atmospheric composition including aerosols and trace gas concentrations. Moderate Resolution Imaging Spectroradiometer (MODIS) has provided long-term global measurements of aerosol properties since 2000 for Terra and 2002 for Aqua. Figure 15.1 shows average AOD over Asia from MODIS aerosol products for the period from 2004 to 2014 (Kim 2016; Remer et al. 2005). High AODs have been observed over Asia

J. Kim (✉)
Department of Atmospheric Sciences, Yonsei University, Seoul, Republic of Korea

On leave at Harvard Smithonian Center for Astrophysics, Cambridge, MA, USA
e-mail: jkim2@yonsei.ac.kr

M. Kim • M. Choi
Department of Atmospheric Sciences, Yonsei University, Seoul, Republic of Korea
e-mail: mysky0110@yonsei.ac.kr; choi816@yonsei.ac.kr

I. Bouarar et al. (eds.), *Air Pollution in Eastern Asia: An Integrated Perspective*,
ISSI Scientific Report Series 16, DOI 10.1007/978-3-319-59489-7_15

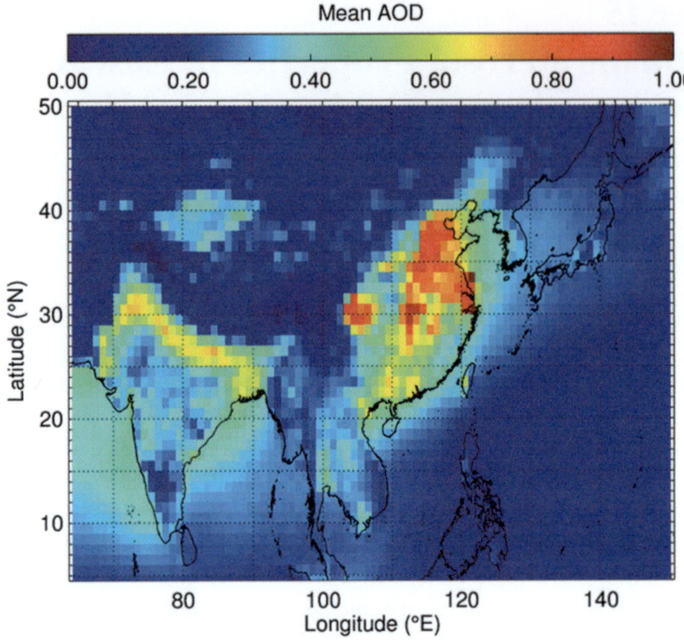

Fig. 15.1 Average MODIS AOD at 550 nm over Asia region for 11 years from 2004 to 2014 (From Kim 2016)

compared to other parts of the globe with its rapid industrialization and increasing population.

To date, most of these space-borne measurements have been from low earth orbit (LEO) allowing one or two measurements per day. Although conventional geostationary meteorological satellite missions have also provided AOD retrievals from a broad visible channel, their accuracy has been rather limited. Recently, with the launch of geostationary satellites in meteorology and ocean color such as Himawari 8 in 2014 and Geostationary Ocean Color Imager (GOCI) in 2010, aerosol properties have been retrieved in high temporal and spatial resolution (Choi et al. 2016). Their data qualities have been shown to be comparable to those from LEO, such as MODIS. The observations of trace gas concentrations are expected to be realized from geostationary earth orbit (GEO) with the launch of dedicated payloads in 2019–2021 time frame, following up the Ozone monitoring Instrument (OMI) (Levelt et al. 2006), Global Ozone Monitoring Experiment (GOME) in LEO. Geostationary Environment Monitoring Spectrometer (GEMS) is planned to be in GEO over Asia by 2019 to provide measurements of aerosol and ozone with their precursors, together with NASA's Tropospheric Emission: Monitoring Pollution (TEMPO) over North America (Zoogman et al. 2016) and ESA's Sentinel 4 over Europe.

15.2 Aerosol Property Retrieval from GOCI and MI

GEO-KOMPSAT is a geostationary program of the Korea Aerospace Research Institute (KARI) to observe atmosphere and ocean. GEO-KOMPSAT-1, also known as Communication, Oceanography and Meteorology Satellite (COMS) was launched in June 27, 2010, with a conventional five-channel Meteorological Imager (MI) and eight-channel GOCI. GOCI is the first ocean color imager in GEO, with the spatial resolution of 500 m covering spectral bands in visible and near infrared (NIR). These two instruments have provided information on aerosol in high temporal and spatial resolution over East Asia in terms of optical depth, size and type classification. The channels in visible and IR for the GOCI and MI allow us to retrieve aerosol property information from GEO both in high temporal and spatial resolution.

GEO-KOMPSAT-2 is planned for launch in 2018–2019 as twin satellites, 2A as weather and 2B as atmospheric environment and ocean color satellite, with a 16-channel Advanced Meteorological Imager (AMI), a UV-Visible scanning spectrometer, GEMS, and GOCI-2. Major specification of these instruments onboard the GEO-KOMPSAT series is summarized in Table 15.1.

Table 15.1 Major specification of instruments onboard GEO-KOMPSAT series

	GEO-KOMPSAT-1 (COMS) (2010-present)		GEO-KOMPSAT 2 (2018–2028)		
			2A	2B	
Payload	MI	GOCI	AMI	GOCI-2	GEMS
Lifetime	7 years		10 years		
No. of channels	5	8	16	13	1000
Spectral range	0.6–12 μm	0.41–0.86 μm	0.4–13 μm	0.375–0.86 μm	300–500 nm
Spatial resolution	1 km (Vis) 4 km (IR)	500 m	0.5/1 km (Vis) 2 km (IR)	250 m @ eq 1 km (FD)	7×8 km^2 @ Seoul 3.5×8 km^2 (aerosol)
Temporal resolution	1 h	1 h	10 min (FD)	1 h	1 h
Major products	CTP CTT CF AOD SST ... (16 prods)	Ocn current, chloryphyl, ... AOD FMF	CTP, CTT, CF, AOD, FMF, OLR, SI, CSR, SST, LST, AMV, ... (56 prods)	Ocn. current, chloryphyl, DOM, Phyto-plankton, ...	AOD, AI, SSA, ACH, CCH, CRF, NO_2, O_3, SO_2, UVI, HCHO, CHOCHO

15.2.1 GOCI

GOCI, the first and unique ocean color imager onboard the GEO has eight spectral bands in visible to NIR including 412 ± 20, 443 ± 20, 490 ± 20, 555 ± 20, 660 ± 20, 680 ± 10, 745 ± 20, and 865 ± 40 nm. The spatial resolution is 500 m and the temporal resolution is 1 h. The flow chart of GOCI aerosol algorithm is shown in Fig. 15.2. Look up table (LUT) approaches were used in the algorithm with the consideration of nonsphercity and the surface reflectance from 30-day clear sky composite. GOCI aerosol algorithm provides AOD, fine mode fraction (FMF), single scattering albedo (SSA), and aerosol type (Choi et al. 2016; Lee et al.

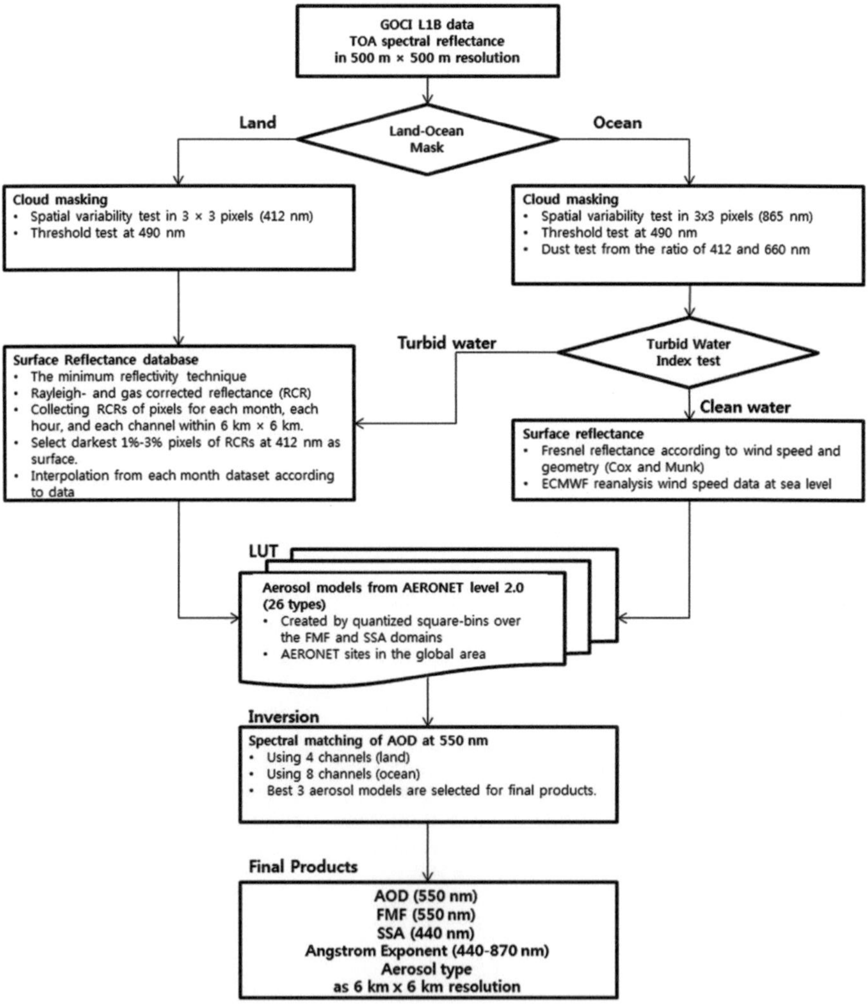

Fig. 15.2 Flow chart of GOCI Aerosol algorithm

2010; Kim et al. 2007). Due to the lack of IR channels in GOCI, standard deviation test was applied for cloud masking over 3 × 3 pixels, in addition to the reflectance threshold tests. For the surface reflectance, 30-day minimum technique is adopted. The algorithm retrieves AOD at 550 nm using all channels by selecting minimized standard deviation of AODs at different wavelengths. Examples of retrieved aerosol properties on May 6, 2012 during the DRAGON-NE Asia Campaign are shown in Fig. 15.3. There was thick haze flowing across the coastal line of China from Qingdao to Shanghai through the south of the Korean Peninsula reaching to Japan. The white area represents cloud-masked pixels, where the aerosol products are not retrieved.

The retrieved values showed AOD up to 2.0 (Fig. 15.3b), FMF of about 0.8 (Fig. 15.3c), Angstrom exponent (AE) of ~1.2, and SSA between 0.95 and 0.99 (Fig. 15.3e), thus indicated highly reflecting fine-mode aerosol, such as sulfate and/or nitrate (Fig. 15.3f). From the series of hourly image of the retrieved values from GOCI, the FMF, AE, SSA and thus their aerosol type detection results tend to remain the same throughout the day.

15.2.2 MI

A visible channel in MI covering wavelengths from 0.55 to 0.80 μm was used to retrieve AOD in the spatial resolution of 4 km (Kim et al. 2016). IR channels were used for cloud masking. The AOD algorithm also adopts 30-day minimum reflectance technique for surface reflectance, background AOD for bias correction, and critical reflectance for SSA as necessary. The details of the algorithm are shown in Fig. 15.4 as a flow chart (Kim et al. 2016). Retrieved AOD at selected hours on April 27, 2012 from MI are shown in Fig. 15.5. As shown in the true color images, there were dust events from China to the northern Korean Peninsula. The AODs were up to 2.0. Two scenes of retrieved AODs from MODIS Terra and Aqua for morning and afternoon, respectively, are also shown for comparison. Also note that the white area represents cloud-masked pixels.

15.3 Regional Trends of Retrieved Aerosol Optical Depth over Asia

Both GOCI and MI have provided valuable information on aerosol properties over Asia. With their high temporal resolution of up to 15 min for MI and 1 h for GOCI, many interesting events have been captured, including Asian dust events on April 27th, 2012 as shown in Fig. 15.5 with selected true color images from GOCI and AODs from MI on the day. MI captured the thick dust plume moving from China to North Korea with AOD up to 2, which matched well with dust layer features in the

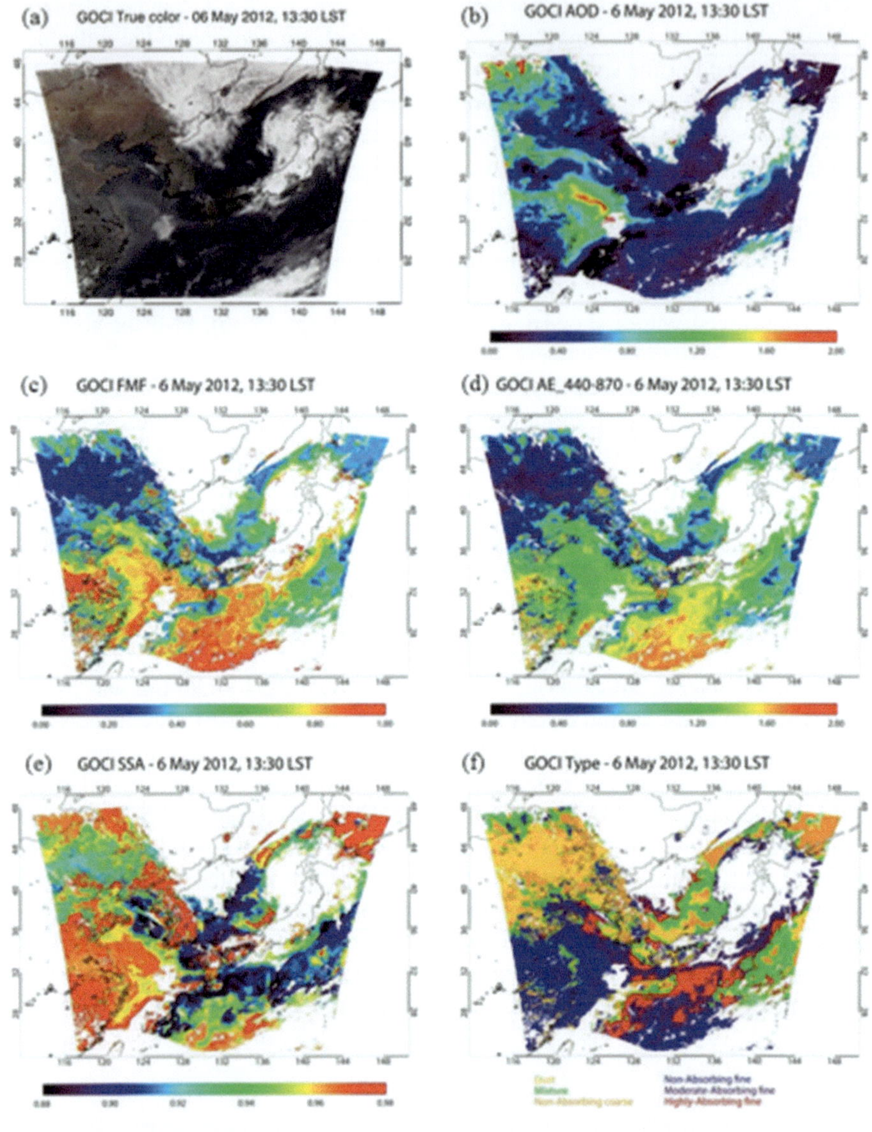

Fig. 15.3 Images of (**a**) GOCI true color, (**b**) AOD at 550 nm, (**c**) FMF at 550 nm, (**d**) AE between 440 and 870 nm, (**e**) SSA at 440 nm, and (**f**) type for 6 May 2012, 13:30 LST. Aerosol types are colored *yellow* (dust), *green* (mixture), *orange* (non-absorbing coarse type), *blue* (non-absorbing fine type), *purple* (moderately absorbing fine type), and *red* (highly absorbing fine type)

true color images. For the comparison, two AOD maps are included from MODIS Terra and Aqua for morning and afternoon overpass, respectively. Retrieval of AODs from MI is robust, capturing the aerosol plume very well without discontinuity across the coastal line. Dataset of aerosol properties have been accumulated at

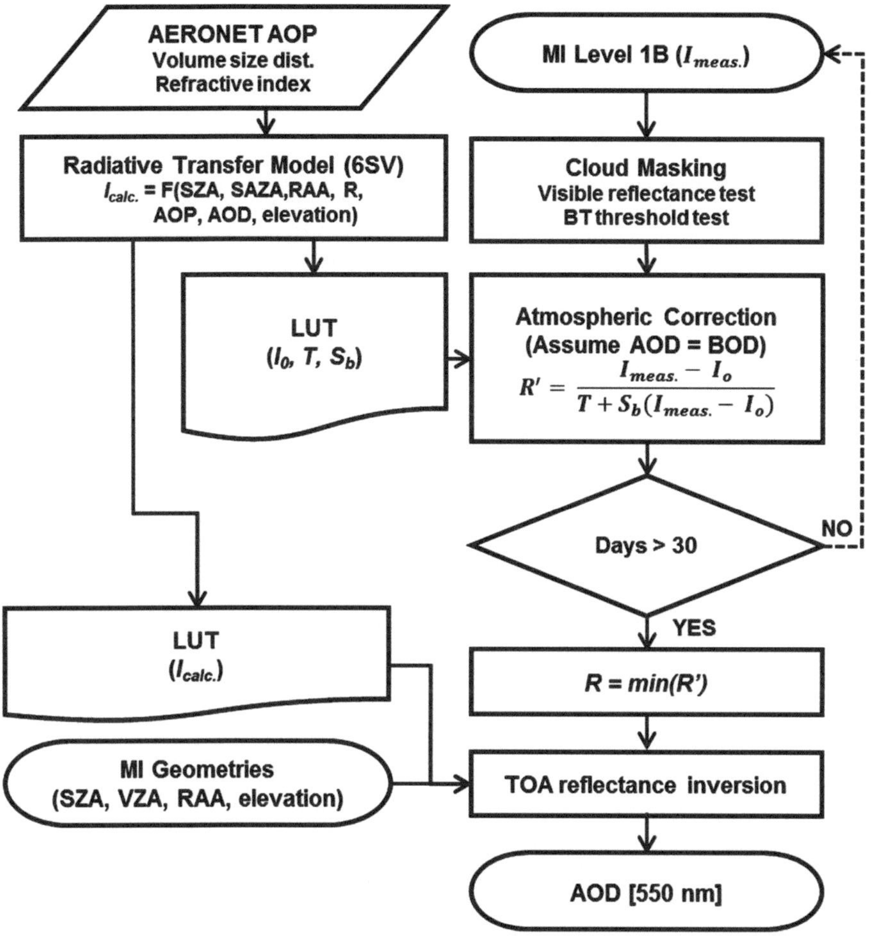

Fig. 15.4 Flow chart of MI aerosol algorithm

every hour since March, 2011. The qualities of the retrieval by GOCI and MI were shown to be comparable to those of MODIS (Choi et al. 2016; Kim et al. 2016).

Trends of monthly average AODs for different sub-region (S1–S12) and cities of Beijing (S13), Seoul (S14) and Tokyo (S15) in Asia are shown in Fig. 15.6. Sub-region numbers are given from northwest to southeast in north to south direction, in general. For most of the region, AODs show clear seasonal cycle with maxima right before summer monsoon and minima in late fall where additional peak appears occasionally. For the trend analysis for 5 years from March 2011 to February 2016, only the deseasonalized AOD values are shown here. It is interesting to note that the overall linear trend of the deseasonalized AOD is decreasing in East Asian region for both GOCI and MODIS, despite some systematic difference in AOD values in north-east (S1), middle-east (S2) and south-east

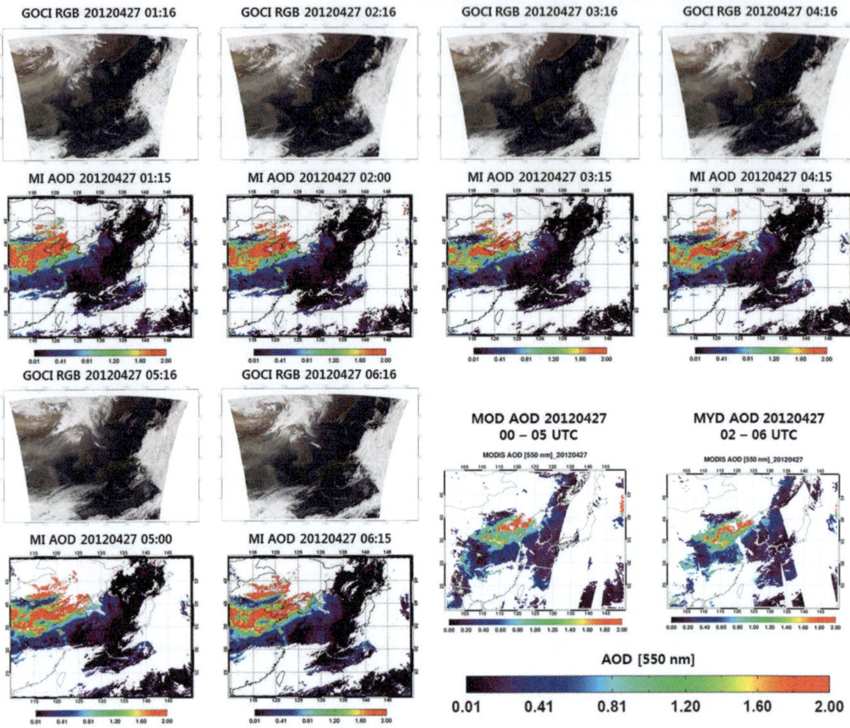

Fig. 15.5 Retrieved AOD from MI at selected hours on April 27th, 2012 with RGB images from GOCI. *Lower right panel* shows AOD from MODIS Terra and Aqua for the same day

(S3) China, as noted by comparing their average values. It is also noteworthy for the high variability of AOD anomalies in Beijing. In general, the amplitudes of AOD anomalies tend to decrease as one goes from west of the domain to east. Similar characteristics were observed from the time series of MI AODs. The average AODs for the 5 year period are 0.776 over Beijing, 0.466 over Seoul, and 0.240 over Tokyo, reflecting local emission, long-range transport from neighboring region and air quality regulation, where the respective trend is -0.002 year^{-1}, -0.008 year^{-1}, and -0.006 year^{-1}.

Although long term dataset of SSA and FMF are available, their trends cannot be assessed due to their lack of required accuracies.

15.4 Atmospheric Composition Observation from GEMS

With the recent development of spectrometers in UV-visible with sub-nm spectral resolution and development of retrieval algorithms, we now can generate estimates of the column amounts of aerosols, O_3, and their precursors including NO_2, SO_2,

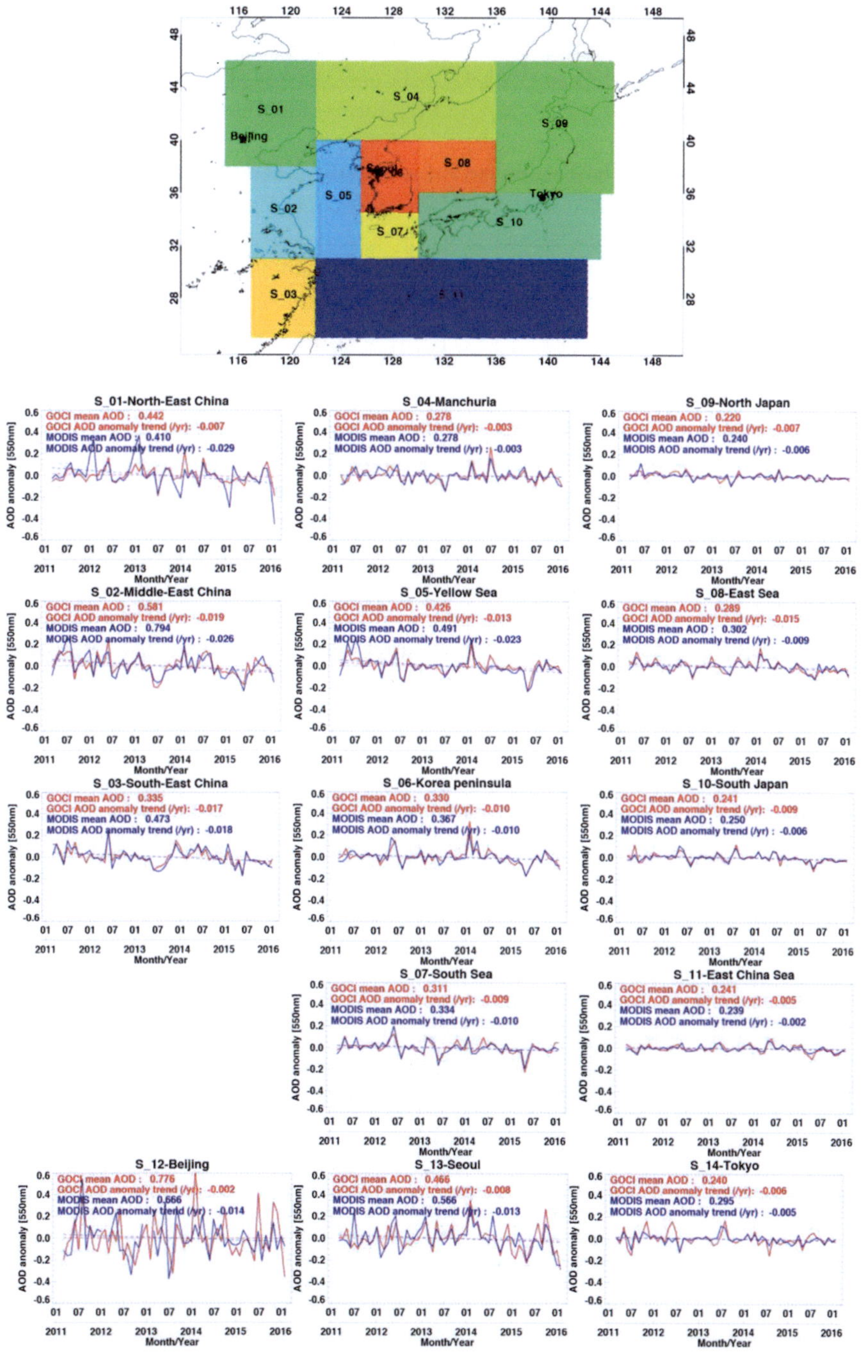

Fig. 15.6 Time series of monthly average AOD from GOCI for different sub-regions in Asia for 5 years from 2011 to 2015. Sub-regions of S1–S11 are shown in the *top panel* of the figure. Time series at Beijing (S12), Seoul (S13) and Tokyo (S14) are also shown at the *bottom panel*

HCHO in the troposphere. To date, all the UV-visible satellite missions to monitor trace gas concentrations in the atmosphere have been in LEO, usually allowing one observation per day, including OMI, GOME, and SCIAMACHY. With the advent of new UV-visible instruments on GEO platforms, the diurnal variation of these components now can be captured.

The GEMS is to be launched into orbit in 2019 over Asia. The instrument is basically a step-and-stare scanning UV-visible imaging spectrometer, with scanning Schmidt telescope and Offner spectrometer. A UV-enhanced 2D CCD takes images, with one axis spectral and the other north-south spatial, with east-west scanning over time. On orbit calibrations are planned, making daily solar measurements, and weekly LED light source linearity checks. For the solar calibration, there are two transmissive diffusers, a daily working one and a reference diffuser used twice a year to check the degradation of the working one. Dark current measurements are planned twice a day, before and after the daytime imaging. In order to avoid dark current issues and random telegraph signal (RTS), the CCD is cooled to temperatures well below 0 °C. Spectral stability is required to be better than 0.02 nm over a daily measurement, straylight less than 2%, polarization sensitivity less than 2% at the instrument level, and the instrument system level MTF better than 0.3. Preflight characterization of the instrument and subsystem is underway. Scientific algorithm has been under the development based on direct radiance fitting, differential optical absorption spectroscopy, optimal estimation, TOMS-type ozone algorithm, and multi-spectral algorithm for aerosols.

By early 2020s, the geostationary orbits are expected to be filled with the three UV visible spectrometers (Fig. 15.7): the NASA's TEMPO over North America, the ESA's Sentinel-4 Ultraviolet Visible Near infrared (UVN) spectrometer over Europe, the GEMS over Asia, with the KNMI's Tropospheric Monitoring

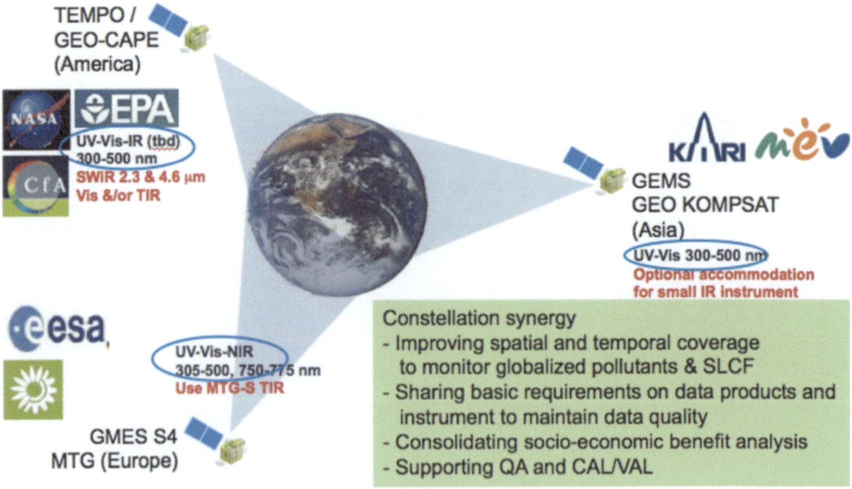

Fig. 15.7 Geostationary constellation of atmospheric composition observation

Instrument (TROPOMI) flying underneath in LEO. Recognized by the Committee on Earth Observing Satellite (CEOS) Atmospheric Composition Constellation (ACC) (CEOS 2011), the geostationary constellation of UV-visible spectrometers will enlighten us on the global distribution of ozone, aerosol, and their precursors.

Acknowledgement Authors acknowledge support from Development of the integrated data processing system for GOCI-II, by the Ministry of Oceans and Fisheries, Korea, KMA Research and Development Program under Grand KMIPA (KMIPA 2015-5010), and the Public Technology Program based on Environmental Policy(2017000160001). Authors also acknowledge Korea Aerospace Research Institute (KARI) for the support of GEO-KOMPSAT Mission.

References

CEOS Atmospheric Composition Constellation. (2011). *A geostationary satellite constellation for observing global air quality: An international path forward.* http://ceos.org/document_management/Virtual_Constellations/ACC/Documents/ACC_White-Paper-A-Geostationary-Satellite-Cx-for-Observing-Global-AQ-v4_Apr2011.pdf. Accessed on 02.01.2016.

Choi, M., Kim, J., Lee, J., Kim, M., Je Park, Y., Jeong, U., Kim, W., Holben, B., Eck, T. F., Lim, J. H., & Song, C. K. (2016). GOCI Yonsei Aerosol Retrieval (YAER) algorithm and validation during DRAGON-NE Asia 2012 campaign. *Atmospheric Measurement Techniques, 9*, 1377–1398. doi:10.5194/amt-9-1377-2016.

Kim, J. Y. (2016). Long-term trend of aerosol optical depth (AOD) derived from MODIS aqua using linear regression and ensemble empirical mode decomposition over Asia. M.S. Thesis, Yonsei University, Seoul, Korea.

Kim, J., Lee, J., Lee, H. C., Higurashi, A., Takemura, T., & Song, C. H. (2007). Consistency of the aerosol type classification from satellite remote sensing during the Atmospheric Brown Cloud-East Asia Regional Experiment campaign. *Journal of Geophysical Research-Atmospheres, 112*, D22S33. doi:10.1029/2006JD008201.

Kim, M., Kim, J., Jeong, U., Kim, W., Hong, H., Holben, B., Eck, T.F., Lim, J.H., Song, C.K., Lee, S., & Chung, C.Y. (2016). Aerosol optical properties derived from the DRAGON-NE Asia campaign, and implications for a single-channel algorithm to retrieve aerosol optical depth in spring from Meteorological Imager (MI) on-board the Communication, Ocean, and Meteorological Satellite (COMS). *Atmospheric Chemistry and Physics, 16*, 1789–1808.

Lee, J., Kim, J., Song, C. H., Ryu, J.-H., Ahn, Y.-H., & Song, C. K. (2010). Algorithm for retrieval of aerosol optical properties over the ocean from the geostationary ocean color imager. *Remote Sensing of Environment, 114*, 1077–1088. doi:10.1016/j.rse.2009.12.021.

Levelt, P. F., Hilsenrath, E., Leppelmeier, G. W., van den Oord, G. H. J., Bhartia, P. K., Tamminen, J., de Haan, J. F., & Veefkind, J. P. (2006). The ozone monitoring instrument. *IEEE Transactions on Geoscience and Remote Sensing, 44*(5), 1093–1101.

Remer, L. A., Kaufman, Y. J., Tanré, D., Mattoo, S., Chu, D. A., Martins, J. V., Li, R.-R., Ichoku, C., Levy, R. C., Kleidman, R. G., Eck, T. F., Vermote, E., & Holben, B. N. (2005). The MODIS aerosol algorithm, products, and validation. *Journal of the Atmospheric Sciences, 62*(4), 947–973.

Zoogman, et al. (2017). Tropospheric emissions: Monitoring of pollution (TEMPO). *Journal Quantitative Spectroscopy and Radiative Transfer, 186*, 17–39.

Chapter 16
Space Observation of Aerosols from Satellite Over China During Pollution Episodes: Status and Perspectives

Jacques Pelon, David M. Winker, Gérard Ancellet, Mark A. Vaughan, Damien Josset, Arianne Bazureau, and Nicolas Pascal

Abstract High pollution events frequently occurred in China during the last years, as industrial activity, traffic, and urban heating has been increasing sources of pollutants (gases and particles). High particle concentrations, well above the recommended threshold values have been measured in large urban areas of eastern and southern China. A major contributor to summertime pollution episodes is dust. Main dust sources are located in the western part of China but mineral surface active in dry periods may contribute to aerosol sources over eastern urban areas in winter. To allow a comprehensive regional survey, ground-based measurements have been developed and satellite radiometric observations (as for example MODIS, MISR or OMI) have been extensively used to derive pollution index and particulate matter (PM) concentration estimates from optical properties over the column (aerosol optical depth -AOD- and size information). Satellites have revealed that pollution was extending over large areas at the regional scale, and that they could be used as first estimates of pollution with good accuracy on average. However, the dispersion due to the nature of the particles in space and time, and the sensitivity of their size to relative humidity has been a limitation in this observational approach. To better characterize and predict pollution events, chemical transport models are now used as integration tools to account for aerosol

J. Pelon (✉) • G. Ancellet • A. Bazureau
LATMOS/IPSL, UPMC Sorbonne Universités, UVSQ, CNRS, Paris, France
e-mail: Jacques.Pelon@latmos.ipsl.fr; Gerard.Ancellet@latmos.ipsl.fr; Ariane.Bazureau@latmos.ipsl.fr

D.M. Winker • M.A. Vaughan
NASA Langley Research Center, MS/475, Hampton, VA, USA
e-mail: David.M.Winker@nasa.gov; Mark.A.Vaughan@nasa.gov

D. Josset
Naval Research Laboratory, Stennis Space Center, Hancock County, MS, USA
e-mail: Damien.Josset@nrlssc.navy.mil

N. Pascal
ICARE/AERIS Data and Services Center, 59 658, Villeneuve d'Ascq, France
e-mail: Nicolas.Pascal@icare.univ-lille1.fr

I. Bouarar et al. (eds.), *Air Pollution in Eastern Asia: An Integrated Perspective*, ISSI Scientific Report Series 16, DOI 10.1007/978-3-319-59489-7_16

formation and composition, meteorological and physical constraints on links between microphysical and optical particle properties. Matching observed and modeled parameters and understanding biases is however a prerequisite for that. Radiometric satellite AODs have been first used in models, but it has further been shown that the integration of active sensor profiles allowed significant information to be added on the vertical to account for moisture impact, using data from ground-based lidar (acronym of light detection and ranging, also identified as an optical radar) networks and more widely from the space lidar CALIOP (Cloud and Aerosol Lidar with Orthogonal Polarization) onboard CALIPSO (Cloud and Aerosol Lidar and Infrared Pathfinder Space Observations). CALIOP is providing data on aerosol type and distribution which are further expected to bridge the gap between in situ and radiometric remote sensing space observations, better constrain models, and help deriving more accurate retrieval of particle extinction near the surface. Although CALIPSO satellite track and repeat cycle do not allow to revisit the same place within days, the duration and extent of pollution episodes over eastern China is such that CALIOP measurements offer a way to constrain regional models to better analyze such episodes, accounting for a better categorization of aerosols. After discussing the context of observations and analyses, a presentation of satellite analyses is presented in the light of examples on severe winter regional pollution episodes.

Keywords Space observation • CALIPSO • CALIOP • MODIS • Particulate Matter, PM • Aerosol Optical Depth, AOD

16.1 Introduction

Haze pollution events are frequently occurring in China and affect large areas. In most of the studies recently conducted, it has been shown that episodes with optical depths of aerosols much larger than one could be rather frequent over the vertical column. Large concentrations of particles (PM_{10}) reaching 150 μg.m^{-3} at the surface and frequently exceeding values of 300 μg.m^{-3} have been reported, as for example over Hebei and Liaoning regions. In Beijing, values exceeding admitted safe particle contents have been measured in $PM_{2.5}$ (excluding a potential contribution of desert dust particles as compared to PM_{10}) more than half the year in 2014 and 2015, according to in situ measurements performed. Extreme $PM_{2.5}$ values exceeding 1000 μg.m^{-3} have even been measured in Shenyang (east of Beijing) end of 2015. Studies have also been conducted over other regions such as eastern and southern China, where large pollution events are also occurring (Zheng et al. 2013).

These pollution episodes are usually long duration episodes (several days), especially in the densely populated, highly industrialized regions of eastern China (Ma et al. 2010; Wang et al. 2012). Such regional severe air pollution episodes can be divided into dust and non-dust types and are depending on the season. The dust type air pollution mainly occurs in spring, whereas non-dust type severely polluted processes correspond to haze and mostly occur in fall and winter. Stable meteorological conditions have been evidenced to favor the accumulation of species that catalyze the formation of pollutants transported from anthropogenic sources surrounding Beijing city (Gao et al. 2015). The composition of $PM_{2.5}$ has been analyzed during severe pollution haze in various seasons, and was found to be linked to secondary aerosol formation (Zhao et al. 2013; Guo et al. 2014; Cheng et al. 2015). Winter episodes have been however more intense, and characteristics appeared to change with impact of moisture (Cheng et al. 2015). Although the secondary ion species (sulfate, nitrate and ammonium) in particles were increasingly generated during all haze episodes and linked to increased relative humidity, the abundance of Fe in such haze episodes was identified to be rather large, as for dusty episodes. The levels of January 2013 and more generally of recent years were very critical, and it was suggested that heterogeneous reactions could play a significant role (Cheng et al. 2015; Zheng et al. 2015).

The importance of pollution events has led to the rapid development of a dense network of in situ measurements to monitor the occurrence of such events at the local and regional scales and better understand processes leading to such pollution peaks. Before the development of this network, estimates of PMs have been looked at the regional scale from satellite remote sensing measurements on a day-to-day basis. Analyses have been first done using satellite AOD measurements using polar orbiting satellites (SeaWiFS, MODIS, MISR, OMI, GOME2, IASI, etc.) and also geostationary satellites (GOES, MSG/SEVIRI, COMS/GOCI etc.) to benefit from their good spatial coverage and high time resolution (see chapter on "Monitoring Atmospheric Composition" in this book, by Kim et al., and recent paper of Choi et al. 2016).

Studies were started over the US in the 1990s, and validations were taking advantage of the well-developed $PM_{2.5}$ ground-based observations from the air quality in situ network (Wang and Christopher 2003; Liu et al. 2004). In such studies, and those further conducted over China, comparisons of AODs to $PM_{2.5}$ and PM_{10} usually show large dispersions especially in wintertime (Wang and Christopher 2003; Zha et al. 2010), mostly because of the spatial heterogeneity, varying composition of particles, and particle hygroscopicity which control the mass-to-extinction relationship. Humidity and aerosol speciation indeed play an important role in the relationship between AOD (measured in real atmospheric conditions) and $PM_{2.5}$ (measured in heated conditions). Particles will have different growth factors with humidity (roughly varying by a factor 2 between 70% and 95% relative humidity) depending on their speciation (see for example Snider et al. 2015). Introduction of a global correction of the humidity effect has shown some improvements in the AOD to PM relationship (Zheng et al. 2013). Multi-parameter regression models have been used to account for multiple physical parameters such as relative humidity, boundary layer height or surface type (Paciorek et al. 2008; Zheng et al. 2013). Ma et al. (2014) showed that although the correlation was improved using such models, the dispersion between modeled and observed $PM_{2.5}$ remained high. Further statistical techniques have been explored, such as the maximum covariance and spectral decomposition (Li et al. 2014), allowing to identify modes to optimize satellite AOD and in situ $PM_{2.5}$ correspondence, which showed variable success. Recent analyses use Chemical transport Models as a more comprehensive frame of analysis and integration.

Vertical distribution is crucial for analysis of transport and impact. These studies pointed out a few discrepancies depending on the vertical structure and composition of aerosol particles (Wind-blown dust can be an important source of particles (Gao et al. 2015), which may be trapped in stagnant air-masses. This emphasizes the need of refinements in the knowledge of aerosol vertical distribution (Wang and Christopher 2003; Liu et al. 2011; Li et al. 2015), that could be obtained from in situ aircraft or balloon measurements. Lidar observations can also provide valuable information on the vertical, but as it provides more complex optical measurements, a specific signal analysis needs to be done.

First studies using lidar observations were conducted in the US in the early 1980s, later more developed combining in situ and remote sensing in northern America and Europe (Hoff et al. 1996; Raut and Chazette 2009). First extensive studies of pollution in China combining ground-based lidar to get information on vertical and in situ chemical composition have been done in Beiging at the end of the 1990s (Bergin et al. 2001). Further analyses were conducted with Raman lidar (Tesche et al. 2007; Hänel et al. 2012; Shin et al. 2015), allowing to characterize optical parameters such as the lidar ratio (ratio of the extinction to backscattering coefficients), depending on the aerosol type. Typical values corresponding to urban pollution aerosol (60 sr) were found in the boundary layer. Lidar depolarization measurements were shown to be very helpful to identify the presence of transported mineral dust (Shin et al. 2015). As mineral particles are not spherical a large depolarization is occurring in the scattered signals and can be identified as such to separate dust from pollution (Tesche et al. 2009: Sugimoto et al. 2015). Characterization of the presence of submicronic particles, more likely linked to non-dust pollution can also be identified from the analysis of the signal backscattered at two wavelengths (O'Neill et al. 2003). Another advantage of lidar measurements is to allow the derivation of the boundary layer height from the gradient of particulate backscattering at the top of the mixed layer (Menut et al. 1999). Since then, spaceborne lidar observations from CALIPSO offered a new way to look at the vertical and regional extent of aerosols over Chinese regions (Tao et al. 2014; Wang et al. 2014; Shin et al. 2015) and PBL height (Liu et al. 2015).

However, the haze episodes occurring over China, especially in winter, appear to be more complex. Indeed, in the analysis of severe pollution episodes over Beijing, it was emphasized that the low wind and high humidity meteorological conditions as well as the decrease of solar flux linked to the increasing AOD may explain pollution increase through heterogeneous reactions (Zheng et al. 2015). This makes observations and the links with model simulations necessary to better understand undergoing processes and forecast pollution episodes. The use of chemical transport models was explored as a possible integration tool to better account for the link between AOD and $PM_{2.5}$ over the US (Liu et al. 2004; van Donkelaar et al. 2006). Global analyses using MODIS and MISR showed the validity of the approach (van Donkelaar et al. 2010). These studies were satisfying on average, but concentrations and extinctions observed over eastern China are at the extreme range of usually measured values, so that a specific analysis needs to be conducted. Besides the need of a better identification of aerosol composition, the combination of such an approach with the use of vertical profiles of the particulate extinction coefficient as derived from CALIPSO appeared to improve the analysis (van Donkelaar et al. 2013; Geng et al. 2015).

The combination of advanced space observations combined to in situ and remote sensing network should however lead in the future to improved observation-derived estimations, and a better use of models. As part of it, the new CALIPSO version 4 operational products to be issued in the autumn 2016 timeframe, is expected to provide a better monitoring and understanding of these episodes. It will benefit from significant advances as compared to version 3 data. Additional research products are being developed to further take advantage of the vertical resolution and combination between observations within the platform and the A-Train.

After a brief description of aerosol characteristics observed from in situ measurements, we will identify critical parameters useful to the analysis of pollution from optical active remote sensing and briefly discuss some case studies made possible using CALIPSO before presenting the evolution of CALIPSO products including both operational and research products.

16.2 In Situ Characterization of Pollution Events

Haze and fog events over big cities in China have been increasing quite strongly during the last decade. In wet cases droplets are formed leading to the apparition of fog and clouds. Haze cases can cause visibility to decrease below 10 km, but fog may then more severely reduce it to a km or so.

Averaged over China, haze occurrence has been experienced on more than 20 days per year between 1961 and 2013 (Su et al. 2015). The number of annual haze days has been steadily growing until 2000, and its growth rate was multiplied by 4 since then (Fig. 16.1). In 2013, the number of haze days reached about 90 days. In comparison, the annual fog days in China average at around 22 day per year. It is to be noted that humidity increase with height due to temperature decrease, and haze at the surface may be part of a foggy (or cloudy) situation when looking to upper levels.

As large concentrations of small particles may cause severe health problems, the development of measurements in operational networks and campaigns to monitor and characterize such events has been strengthened during the last decade.

16.2.1 In Situ Measurements

A new PM monitoring network was very rapidly developed over China and offers now very dense measurements in the eastern part, clustering in the more polluted cities as seen in Fig. 16.2. There are more than 800 measurements sites.

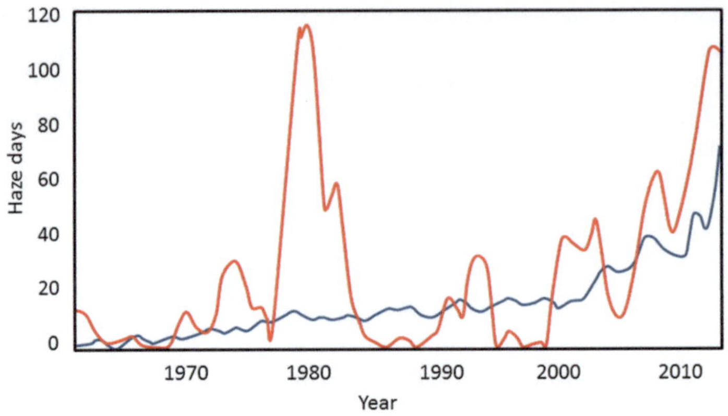

Fig. 16.1 Average haze days over China (in *blue*) and in Beijing (in *red*) (After Su et al. 2015)

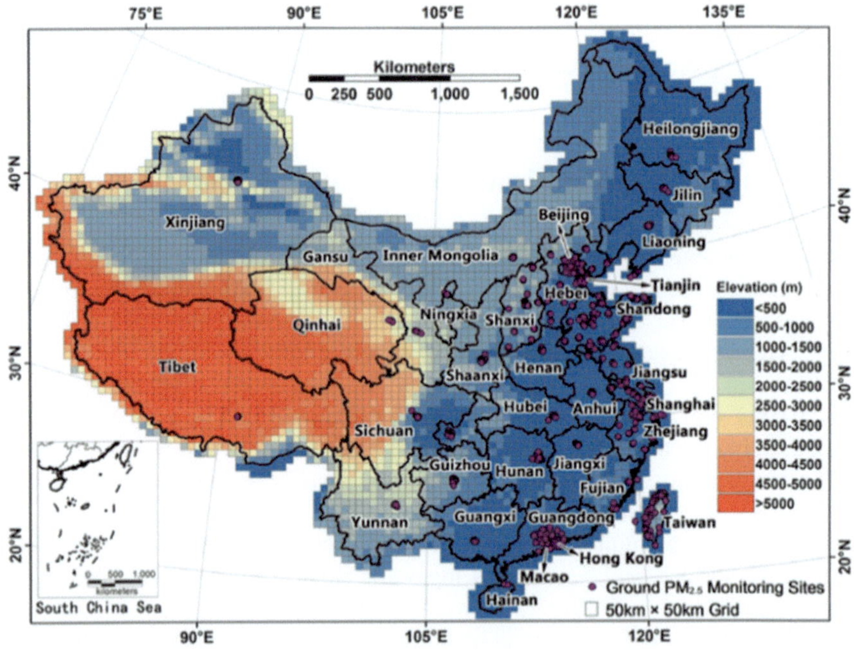

Fig. 16.2 Spatial distribution of $PM_{2.5}$ monitoring sites over China and elevation map (Adapted with permission from Ma et al. 2014)

Measurements are made using the Tapered Element Oscillating Microbalance (TEOM) method. Data are centralized at the China Environmental Monitoring Center (CEMC) (http://113.108.142.147:20035/emcpublish/). Some regions and cities have established additional monitoring sites that are not included in the CEMC's Web site.

In addition to operational PM measurements, chemical analysis of aerosols are performed regularly and during campaigns (Wang et al. 2015; Zhang et al. 2013). In order to perform more observations, cost effective $PM_{2.5}$ measurements combined with impact filter to sample black carbon, water soluble ions and nephelometer and sunphotometer measurements to improve links between in situ and satellite AODS have also been proposed as a new network at the global scale (Snider et al. 2015).

Composition of $PM_{2.5}$ is an important parameter to determine to understand relationships between PM and optical parameters. It has been more and more analyzed over various sites during field campaigns. Over the area of Xi'an (capital of the province of Shaanxi, south-west of Beijing) for example, measurements of chemical species were performed over six sites that allowed to clearly identify significant variability in time and space. It has been identified that $PM_{2.5}$ composition was dominated by organic carbon (OC) in winter, soil dust in spring, and sulfate, nitrate, and ammonium in summer and autumn, as shown in Fig. 16.3. Emissions from fossil fuel combustion (traffic and coal burning) were identified to be the main source responsible for the fine particle pollution. Seasonal distributions of cumulated sulfate, nitrate and ammonium relative to total $PM_{2.5}$ showed increase in autumn and decrease in winter (Wang et al. 2015). In studies conducted over this region to estimate ground-based $PM_{2.5}$ from satellite AODs, correlations between satellite and in situ measurements were shown to be improved using a non-linear model for predictions considering several parameters controlling the composition, dilution and relative humidity induced growth in size of particles in the boundary layer (You et al. 2015).

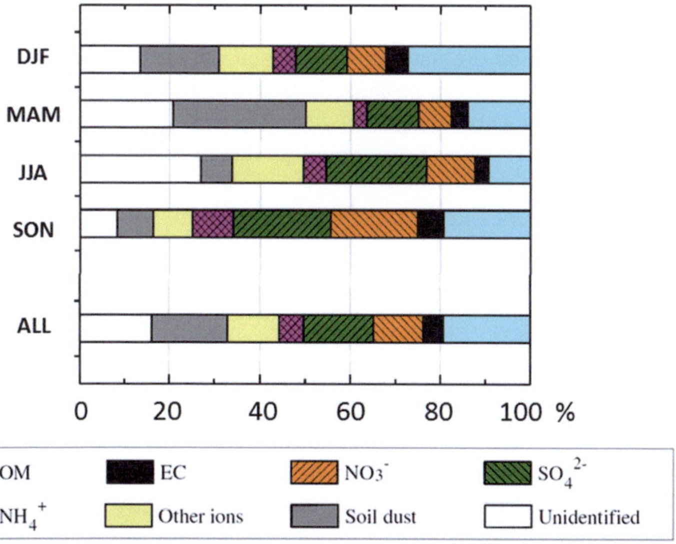

Fig 16.3 Seasonal variation of $PM_{2.5}$ composition measured over six sites in Xi'an area. OM is estimated as 1.6×OC. Soil dust is taken equal to 2.5×Fe (Adapted from Wang et al. 2015)

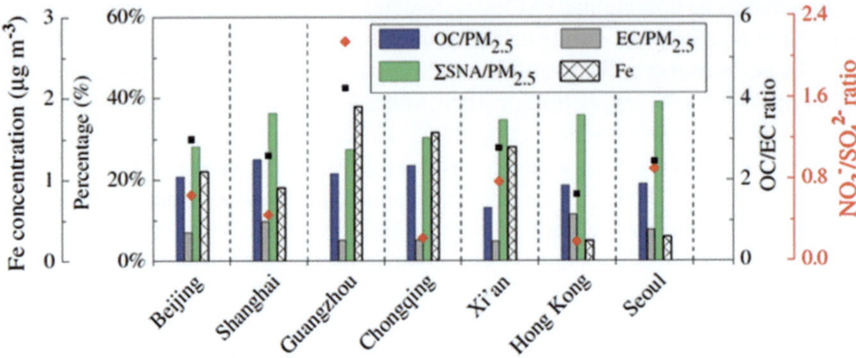

Fig. 16.4 Comparison of the concentration of major chemical species (*OC* organic carbon, *EC* Elemental carbon, *SNA* sulfate, nitrate and ammonium) contributing to $PM_{2.5}$ relative to the total $PM_{2.5}$ concentration for various Chinese and Korean cities (Adapted from Wang et al. 2015)

In Beijing, in the beginning of 2000s organic compounds were measured to be about 30% of the $PM_{2.5}$ mass, being unusually high as compared to elemental carbon. Sulfate, nitrate and ammonium (SNA) compounds were at about 30%.

Mineral aerosols have been found to be about 16% of the total mass (Bergin et al. 2001). As seen in Fig. 16.4 the ratio of OC to EC is still high, and if the ratio of SNA to $PM_{2.5}$ is comparable, the ratio of OC to $PM_{2.5}$ is now closer to 20%. Comparing these results to measurements in other big Chinese and Korean cities derived from several studies (Wang et al. 2015, see Fig. 16.4), similar ratio of OC to total $PM_{2.5}$ were measured on average (about 20%). Although the ratio of nitrate to sulfate was largely varying by a factor 4–5, the cumulated sulfate, nitrate and ammonium relative to total $PM_{2.5}$ was observed to be rather constant staying about 30%. Identified Fe could be originating from crustal dust in autumn. In winter it might be primarily emitted from anthropogenic sources (iron and steel smelting), but may also be due to previously transported dust. It is also seen that iron concentration is much smaller in Hong-Kong and Seoul as due to large scale circulation pattern bringing less dust.

It has been further evidenced that dust particles are good sites for heterogeneous chemistry reactions producing additional water soluble secondary aerosols. Dust particles may act as sinks for acidic pollution gases, and it has been evidenced that after dust days, calcium carbonate and sulfate concentrations were increased (Wang et al. 2005). The determination of dust (or anthropogenic dust defined as dust produced in urban areas and/or mixed with pollution particles), in the atmosphere is thus important to $PM_{2.5}$ analysis and characterization of pollution events.

The role of humidity in the pollution increase was further emphasized as the concentration of soluble particles (inorganic ions) in $PM_{2.5}$ was observed to increase with relative humidity over Beijing (Cheng et al. 2015). It is a key parameter in the relationship between microphysical and optical parameters of aerosols.

The nitrate to sulfate ratio has increased as compared to measurements in the year 2000s, possibly linked to the increase in nitrogen oxides emissions due to traffic. According to Cheng et al. (2015), further changes were noticed such as the production of non-soluble secondary aerosols and moist winters also appear to me more frequent in the last decade.

The moistening of the PBL, and the large number of condensation nuclei produced by the increased pollution are expected to lead to non-precipitating clouds composed of small cloud droplets which may further contribute to the pollution development cycle.

16.2.2 Characterization of Particles from Remote Sensing Measurements

Besides in situ observations, allowing to regularly characterize aerosol properties close to the surface or at identified levels above the surface from masts, or during specific campaigns also involving aircraft, regular total column and profiling measurements can be done from ground-based sun-photometers and lidar, respectively.

Links between particle microphysical concentration such as $PM_{2.5}$ and an optical parameter such as extinction or scattering coefficients has been studied during a large number of field experiments using in situ particle counters and nephelometers, able to make collocated in situ measurements.

The total aerosol extinction coefficient of a mixture of particles can be written as

$$\alpha_\lambda(z) = \sum_i \alpha_{\lambda,i}(z) = \pi \sum_i \left(\int_0^\infty Q_{e,i}(r,\lambda) r^2 N_i(r) \right) dr \tag{16.1}$$

where $Q_{e,i}$ (r, l) is the scattering efficiency of the particles from species i, depending on size r and wavelength λ, and Ni(r) is their size distribution. $Q_{e,i}$ is usually determined from Mie calculations for spherical particles.

To further determine links between AOD and PMs, relationships between integrated optical and local microphysical properties of urban pollution need to account for variation over the vertical column, which introduces an additional degree of complexity. The AOD is the integral of the extinction over the vertical as

$$AOD_\lambda = \int_0^\infty \alpha_\lambda(z) dz \tag{16.2}$$

The total $PM_{2.5}$ at a given altitude z is given by the integrated number of particles over the particle size distribution (PSD) for a given species, then summed over all species. Let the aerosol mass density to be ρ_i for each species, the $PM_{2.5}$ derived as

$$PM2.5(z) = \frac{4}{3}\pi \sum_i \int_0^{2.5} \rho_i(r) r^3 N_i(r) dr \tag{16.3}$$

Relationship between AOD and $PM_{2.5}$ is thus not so direct. It can be simplified using several hypotheses, as assuming the mass density is constant with particle size (growth with relative humidity independent of size) and the efficiency is constant with size. We then obtain an estimate of total mass as

$$PM(z) = \frac{4}{3} \sum_i \rho_i r_{eff,i} \frac{\alpha_{\lambda,i}(z)}{Q_{e,i,\lambda}} \tag{16.4}$$

So ideally this relationship stands for vertically resolved measurements or simulations. Particle size distribution is reflecting their composition. Soluble pollution and OC particles are small size particles (about 0.1 μm) whereas the average sizes of dust and sea-salt particles are about 20 times larger. Small particles of the order of 0.05–0.2 μm belong to the accumulation mode (also called fine mode), and particles which size distribution is peaking at about 1–3 μm, belong to the coarse mode.

An important information that can be derived for optical measurements is the contribution of small size particles as compared to those in coarse mode. The Angström exponent in extinction A(z), is defined from the ratio of extinction coefficients α at two wavelengths, λ_1 and λ_2, as

$$A(z) = \left[Ln\left(\frac{\alpha_{\lambda_2}(z)}{\alpha_{\lambda_1}(z)}\right)\right] / \left[Ln\left(\frac{\lambda_1}{\lambda_2}\right)\right] \tag{16.5}$$

The value of A is indicative of the ratio of the importance of scattering by small particles (in the accumulation mode about 0.1 μm) to large ones (in the coarse mode above 1 μm). This has been emphasized in several studies and proposed as a mean to determine the ratio between the two modes (Cattrall et al. 2005; Kleidman et al. 2005).

$PM_{2.5}$ includes the contribution of the accumulation mode of the PSD, and part of the coarse mode, whereas PM_{10} is closer to the total PM, including the contribution of the coarse mode. So optical parameters determined for $PM_{2.5}$ are expected to be different from those related to PM_{10}. Models are however able to deal with a more representative calculation accounting for particles classes in the size distributions of contributing species and related varying properties.

Another important parameter to retrieve extinction coefficient profile from lidar measurement is the lidar ratio defined as the ratio of the extinction and backscattering coefficients for a given species. The lidar ratio for aerosols Sa is dependent on the particle type and microphysical properties, and may differ in time and space.

Table 16.1 Aerosol properties in Asian cities. Ratios are referring to ratio of values obtained at 550 and 1020 nm

City	Country	Period	Nb obs	Lidar ratio (sr)	Ratio of lidar ratio	Back ratio	Ext. ratio	Angst. Coeff.
Beijing	China	2001–2004	281	58	1.5	1.7	2.5	1.3
Anmyon	S. Korea	1999–2004	259	57	1.4	1.6	2.3	1.2
Chen-Kung	Taiwan	2001–2004	202	61	1.5	1.7	2.5	1.3

Adapted from Cattrall et al. (2005)

Results derived from AERONET measurements over Asian cities are reported in Table 16.1. This is a critical parameter in the determination of the extinction coefficient, and in polluted areas, the AOD is usually large, and a small uncertainty in the lidar ratio may induce important error on the extinction coefficient near the surface (about a factor 2 increase in the near surface extinction coefficient for a 15% increase in Sa – from 65 to 75 sr at an AOD = 0.35 -see for example Pelon et al. 2002). Due to the fact that small particles characterizing pollution air masses have large Sa values, and that particle size grows with relative humidity modifying their indices, relationships between Sa and A can be used to infer lidar ratio from Angtsröm coefficient value for a given aerosol type. However, lidar ratio is not only depending on the size of particles but also on their absorption properties as it is inversely proportional to single scattering albedo.

It is furthermore necessary to have consistent properties between CTM and lidar analyses to accurately use extinction information. It is to be noticed that Sa can be directly derived (without a priori hypothesis on particle type) from vertically resolved high resolution spectral lidar (HSRL) measurements (Burton et al. 2012) or to a lesser extent integrated lidar measurements from CALIOP space observations using SODA analysis (see Sect. 16.4).

In order to analyze range-resolved CALIOP observations, aerosol properties were further quantified from extended AERONET sun-photometer measurements (covering the period 1993–2002) with identification of aerosol types (Omar et al. 2005, 2009). Six clusters were first identified from a global scale classification to determine representative average properties (Omar et al. 2005). Results obtained from this cluster analysis are reported in Table 16.2. Likely Composition of Aerosol Types corresponding to the cluster Analysis has been estimated from their imaginary indices. It was noticed that the large imaginary part of the refractive index of dirty pollution suggests that this type of aerosol contains a significantly larger (than polluted continental) fraction of soot. Accumulation mode radii varies from 0.12 to 0.16 μm, whereas coarse mode are ranging from 2.8 (desert dust) to 3.7 μm (BB).

From previous analyses and remarks, the question related to aerosol composition and size growth with humidity then needs to be addressed to perform radiative closure of aerosol microphysical and optical properties on the vertical. Studies on this topic started in the 1980s in the US, and estimates of the pollution aerosol mass efficiency in urban environment were found to be close to 3.2 $g.m^{-2}$ in low relative

Table 16.2 Cluster analyses results by category. The values are the center of each cluster. The membership coefficient is an indication of the dispersion in each cluster

Cluster type	Desertdust	Pollutedmarine	Pollutedcontinental	Background/rural	Biomass burning	Heavypollution
Likely composition	Mineraldust	NaCl + OC + soot	SO4 2–, NO3-,OC, NH4+ soot	SO4 2–, NO3–,NH4 +, OC	Soot + OC	SO4 2–, NO3–,NH4+ + OC + soot
Number of records	22,202	6527	55,667	20,307	26,662	12,548
Single scattering albedo (673 nm)	0.93	0.93	0.92	0.88	0.80	0.72
Optical depth (673 nm)	0.327	0.140	0.191	0.036	0.190	0.100
Real refractive index (673 nm)	1.4520	1.3943	1.4098	1.4494	1.5202	1.4104
Imaginary ref. index (673 nm)	0.0036	0.0044	0.0063	0.0092	0.0245	0.0337
Angstrom coefficient (673 nm/873 nm)	0.486	0.678	1.536	1.381	1.332	1.232
Fine fraction by volume	0.22	0.26	0.53	0.38	0.33	0.49
Membership	0.92	0.91	0.93	0.94	0.93	0.89
Sa at 532 nm (sr)	40		70		70	
Sa at 1064 nm (sr)	55		30		40	

Adapted from Omar et al. (2005)

humidity values (Waggoner et al. 1981). Studies conducted in Canada using in situ and remote sensing (lidar) gave similar results for Ontario (Hoff et al. 1996). This value is consistent with a dry sulphate aerosol, and a factor 2 smaller than for sulphate aerosol at 60–70% humidity that is currently used to estimate radiative impact on climate (Box and Trautmann 1994). Further analyses have been performed in the frame of previous Aerosol Characterization Experiments (ACE). During ACE-1, Quinn and Coffman (1998) concluded that the remaining uncertainty was large due to the poor knowledge of particle growth with relative humidity, and strong disagreements were also observed during ACE-Asia including ground-based lidar derived extinctions and airborne measurements (Huebert et al. 2003; Murayama et al. 2003). In the frame of ACE-2, comparisons were made in a pollution outflow using airborne lidar and in situ observations and agreement with satellite AODs were rather satisfactory although the optical depth (<0.2) remained rather low (Flamant et al. 2000). In all these cases measurements were taken over the sea, and sea salt was one of the major contributor in the boundary layer.

Mass scattering efficiencies derived from mass distributions and nephelometer measurement on the different species analyzed, showed that the contribution of submicron particles to light scattering was on average about 75%, mostly related to combustion-related products and products of chemical reactions. Values of the mass scattering efficiencies were ranging from 2.5 to 6 $g.m^{-2}$. This large range of values is explaining the large variability in the retrieved $PM_{2.5}$.

16.2.3 Results from Radiometric Satellite Observations in CT-Models

The comparison of PMs to satellite AODs has concentrated a lot of efforts, using empirical models or chemical transport models to better estimate near surface PM (see for example Liu et al. 2004, 2009; van Donkelaar et al. 2006). MODIS and MISR AODs have been largely used in these analyses. Testing AOD and non-AOD models also showed that AOD models have a better prediction performance, although more biased (Liu et al. 2009).

An example of PMs retrieved from GEOS-Chem simulations using MODIS and MISR satellite observations over USA is given in Fig. 16.5 on a day-to-day basis (van Donkelaar et al. 2010). Comparison with in-situ measurements from US Air Quality network show a good agreement with a correlation coefficient of 0.77 and a slope of 1.07. Results obtained over Asia are given in Fig. 16.6.

MODIS has been extensively used for characterizing AODs, in particular looking at their climatology over China over the last 10 years (Luo et al. 2014). Dust particles are however not well retrieved by standard MODIS algorithms but MODIS Deep Blue aerosol products (Hsu et al. 2001) are providing this information taking advantage of absorption in the blue. Comparisons between standard

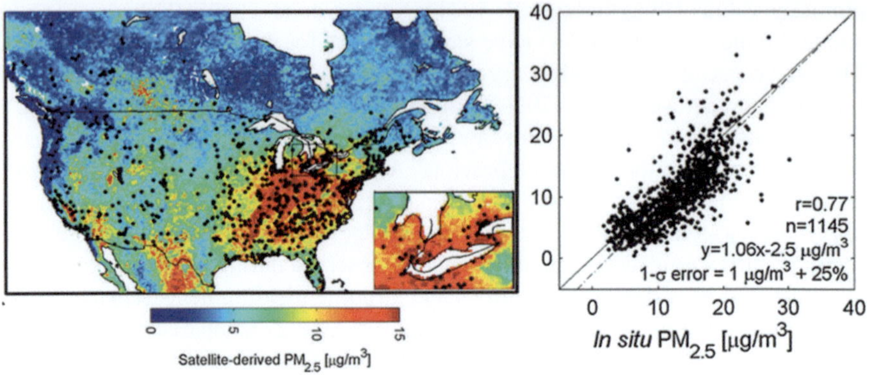

Fig. 16.5 Estimated $PM_{2.5}$ using MODIS and MISR AODs and comparisons to ground-based observations of USair quality network (Adapted from van Donkelaar et al. 2010)

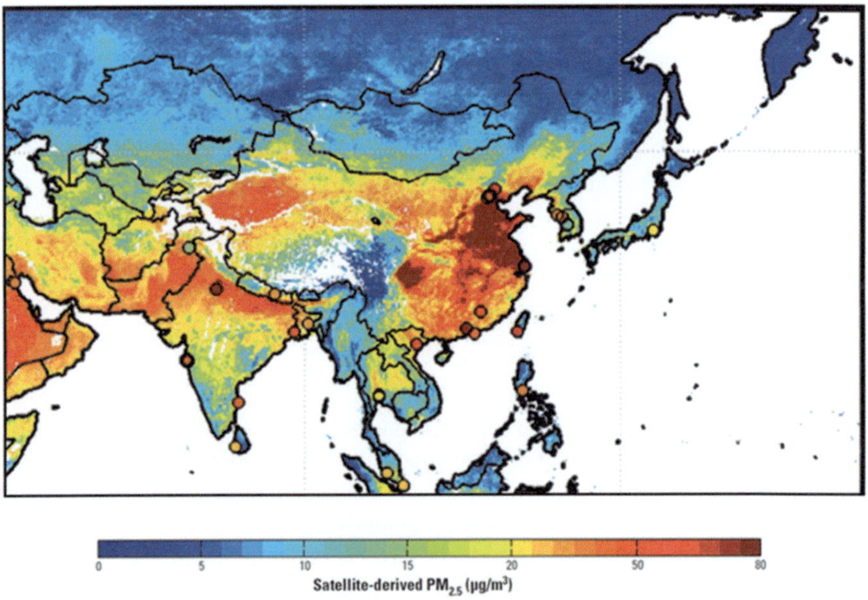

Fig. 16.6 Estimated $PM_{2.5}$ from MODIS and MISR AODs and GEOS-chem (Adapted from van Donkelaar et al. 2010)

(550 nm) and Deep Blue MODIS retrievals (collection 4) and MISR for monthly AODs can thus show significant differences (Hsu et al. 2001).

Besides the multispectral MODIS or multispectral and multiangular MISR spectro-radiometers, or multichannel radiometers on geostationary satellites other instruments such as the ozone monitoring Instrument (OMI) on AURA, have been used to infer aerosol content from ultraviolet (UV) observations. The contrast in two UV bands of the Total Ozone Mapping Spectrometer (TOMS) measurements was first used onboard Nimbus-7, in 1979, to identify the aerosol content of absorbing particles (mineral dust and carbonaceous aerosols) (Herman et al. 1997; Torres et al. 1998). OMI, implemented onboard AURA (Levelt et al. 2006) and more recently the Total Ozone Unit (TOU) Chinese instrument allow the determination of Absorbing Aerosol Optical Depth (AAOD). As such measurements rely on absorption, results are sensitive to the altitude of the aerosol layers, and the inversion relies on the vertical distribution of mass (an atmospheric mass factor (AMF) is defined to account for altitude dependence).

The Absorbing Aerosol Index (AAI) derived from the measured ratio of the UV absorption at 340 and 380 nm corrected from clear atmosphere contribution (see Herman et al. 1997; Torres et al. 1998) can be considered as a qualitative proxy of the AAOD, avoiding the use of AMF. The large horizontal resolution of OMI (~20 km) as compared to MODIS and MISR ones (~1 km), may lead to a larger cloud contamination. AAOD and AAI derivations are less sensitive to cloud contamination, induced by the larger pixel size of these observations.

The analysis the trend and spatiotemporal variations of haze in China using OMI data on Aura satellite showed that the occurrence of large AAI values in the period from 2005 to 2013 (Fig. 16.7 (right)) shows that contribution of absorbing particles has strongly increased in the north east region as compared to the period from 1979 to 2004 derived from TOMS (AAI are comparable as OMI and TOMS have similar channels, Zhang et al. 2015). This is in particular the case in the region of Hebei. Note in both figures the SW-NE axis of the spatial distribution of elevated AAI linked to polluted areas. In their analysis Wang et al. (2015) noticed that hazy periods were characterized by a weak high-pressure center northeast of Beijing. This configuration is favoring south-westerly flows over Beijing.

Recently, new analyses have been performed looking to dust particle 3D distribution (Cuesta et al. 2015) which showed transport of dust plumes over the boundary layer under cyclonic conditions. Results from this analysis combining

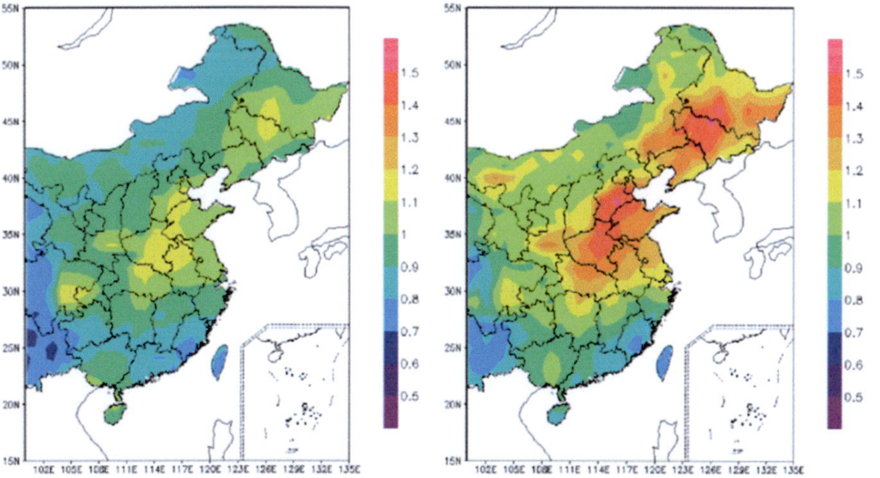

Fig. 16.7 Pollution derived from TOMS and OMI AAI values derived over eastern China for the period 1979–2004 (*left*) and (2005–2013) for wintertime (After Zhang et al. 2015)

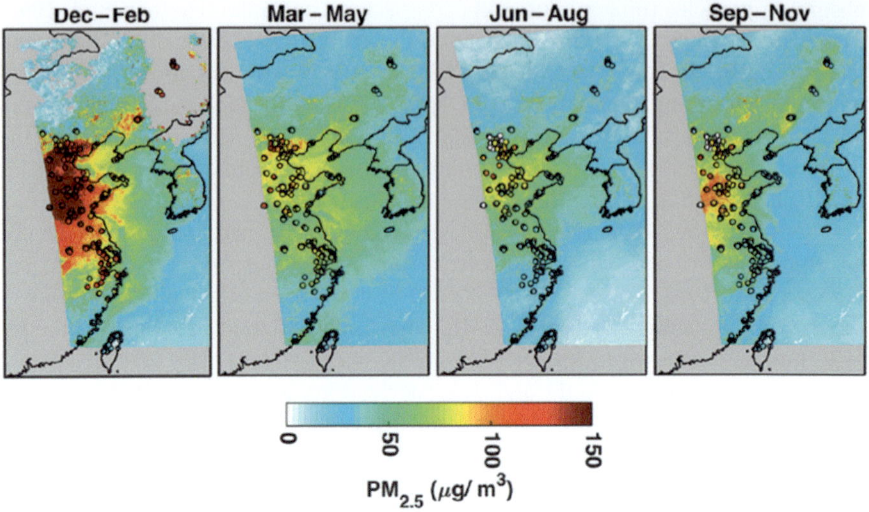

Fig. 16.8 Maps of seasonal $PM_{2.5}$ concentrations simulated from the GEOS-Chem model for 2013, as based on in situ and satellite (GOCI) measurements (Adapted from Xu et al. 2015)

measurements in several channels with different weighting functions are showing good agreement with CALIOP vertical cross-sections. Further analyses combining observations from IASI and GOME 2 (operating in the UV as OMI and TOMS) onboard METOP are considered to extend retrievals to other aerosol particles as well.

Considering the combination of spaceborne AOD measurements and CTMs, the use of the Korean Geostationnary Ocean Color Imager (GOCI) showed that the use of geostationary AOD retrievals at scales close to those of MODIS could be very efficient for looking at pollution. As reported in Figs. 16.8 and 16.9, results obtained with GOCI data used in GEOS-Chem model showed a very good agreement with ground-based measured PMs in detecting peak pollution areas and period (Xu et al. 2015). As shown in Fig. 16.9, the monthly averages of ground-based and GOCI-derived $PM_{2.5}$ were strongly correlated. A regression coefficient of about 0.66 with a slope near unity was found, looking to daily surface measurements, which is comparable to previous results.

16.3 Use of CALIPSO Observations

CALIPSO Version3 data are available since 2009 and have been extensively used, in particular for studying dust transport form eastern desert areas and pollution events. Various products have been developed (Winker et al. 2009, 2010). In the operational algorithm, level 1 data are analyzed and inverted to derive vertical profiles of extinction and provide a classification (Winker et al. 2009). A specific cloud-cleared attenuated backscatter product based on level 1 data is also proposed

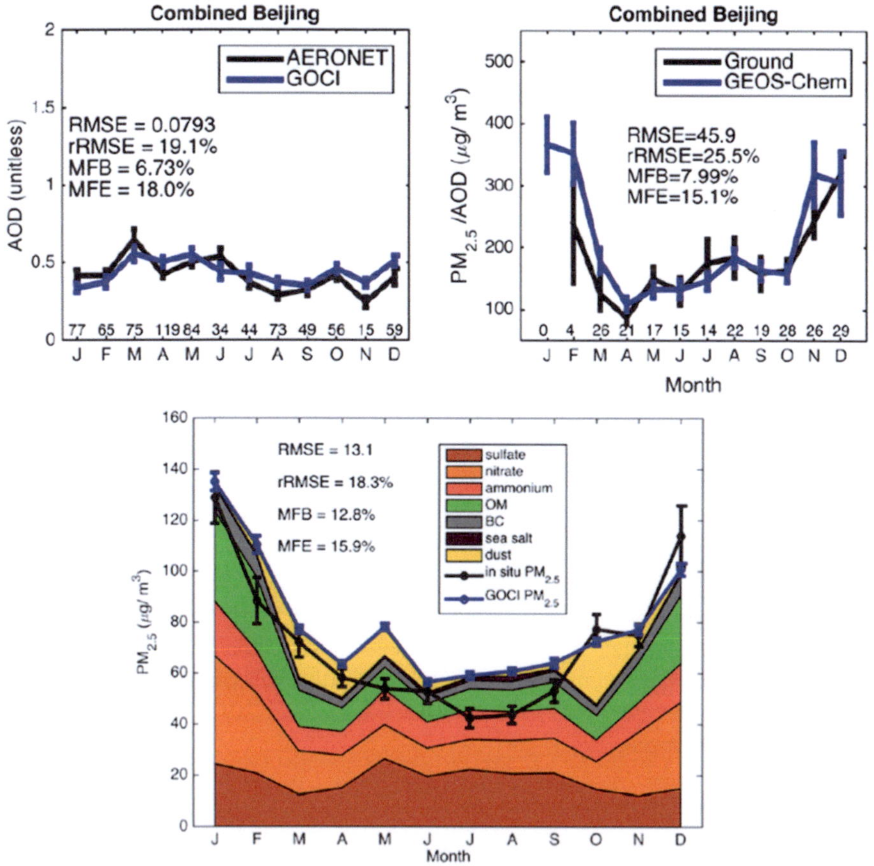

Fig. 16.9 Monthly variation of the AOD and ratio of $PM_{2.5}$ to AOD (*upper figures*) and $PM_{2.5}$ composition derived from observations and model simulations for Beijing and surrounding region (adapted from Xu et al. 2015)

in particular to operational agencies for assimilation. For climatological analyses, a quasi 3D product has been elaborated (Winker et al. 2013). In this product CALIOP Version 3 Level 2532 nm aerosol extinction data have been aggregated onto a global $2 \times 5°$ latitude-longitude grid. Vertical resolution is 60 m, from −0.5 to 12 km. This will be further detailed in Sect. 16.4.

As previously discussed, dust may be an important catalyst of chemical reactions. In industrial and urban areas, during dry winter season, soil erosion may lead to dust that can be mixed with pollution. Polluted dust is a direct product of CALIOP using information on depolarization (Omar et al. 2009; Winker et al. 2009). Recently a study combining MODIS and CALIOP observations has been conducted that allowed to identify the polluted dust burden in the planetary boundary layer over several regions over the globe including (Huang et al. 2015). Results reported in Fig. 16.10 show that contents are more important in eastern China, south of Beijing.

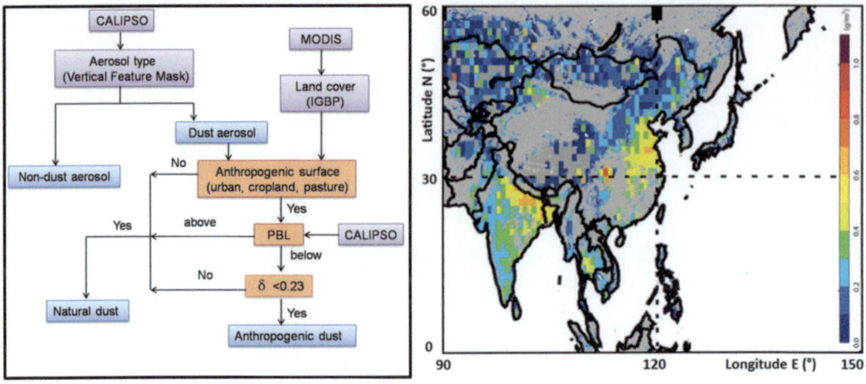

Fig. 16.10 Schematic detection flow chart combing CALIPSO and land cover data set from MODIS, for the detection of anthropogenic dust and anthropogenic dust burden in winter season (Adapted from Huang et al. 2015)

16.3.1 Examples from Case Studies

Taking advantage of A-Train co-located observations from MODIS and CALIPSO, studies can be conducted from a 2D horizontal observation of the AODs and a vertical distribution derived from CALIOP. We further discuss in this section two winter observation cases from January 2013 and October 2015 using MODIS and CALIPSO observations.

16.3.1.1 January 2013

Wang et al. (2014) thus report observations related to a severe pollution haze episode lasting almost 10 days during January 2013over eastern China. This haze episode was one of the more intense pollution episodes in mainland China in recent years in terms of the influenced region, duration and pollution intensity. A large area was concerned and particle concentrations PM_{10} were exceeding 650 μg/m^3 on 11 January in Zhengzhou, 979 μg/m^3 on 12 and 974 μg/m^3 13 January in Shijiazhuang (capital of Hebei province, see Fig. 16.1), 524 μg/m^3 on 12 January in Tianjin, closer to the coast south-east of Beijing. 518 μg/m^3 were reported in Beijing measurements of PM_{10} on January 13, 2013. MODIS AODs were exceeding values of 1 during several days from 10 to 13 January, reaching values close to 2.5, as shown in Fig. 16.11.

Using CALIPSO lidar observations, Wang et al. (2014) concluded that most of the pollution was maintained in the boundary layer below 1500 m (see Fig. 16.12).

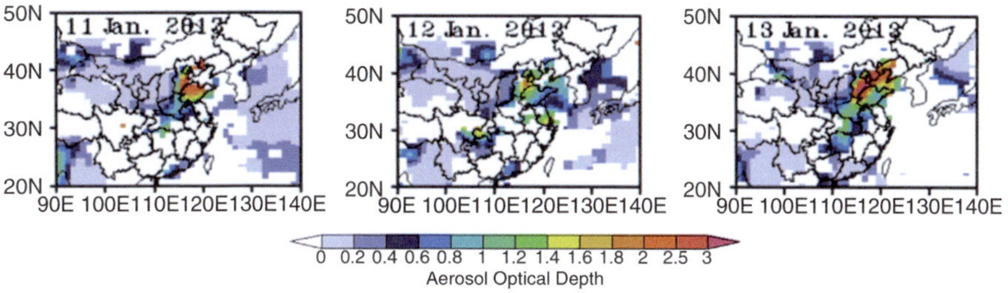

Fig. 16.11 MODIS AOD observations on 11–13 January 2013 showing a large increase in AOD linked to pollution over Beijing area (Adapted from Wang et al. 2014)

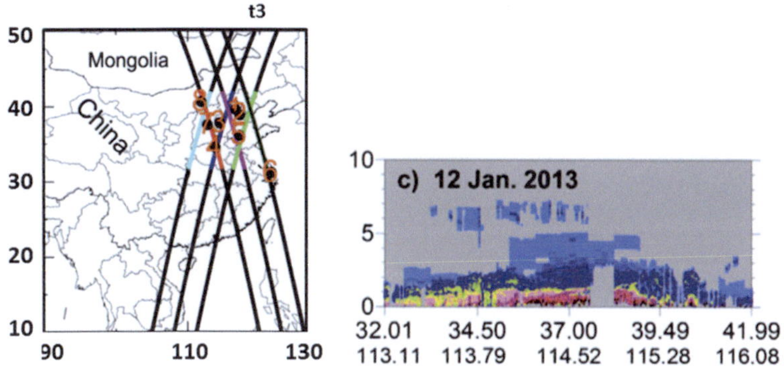

Fig. 16.12 CALIPSO tracks of CALIPSO overpasses during the haze episode in January 2013 over eastern China, and 2D of the extinction coefficient and profile derived on January 12 (track t3) (From Wang et al. 2014)

16.3.1.2 October 2015 Case

CALIOP and MODIS Observations made during a strong pollution case that occurred end of October 2015 are reported in Fig. 16.13. A large area, where the optical depth identified by MODIS is larger than 1, is extending over north-eastern China between 30 and 40 N, and 110 and 130 E (see Fig. 16.13 left part). Three CALIPSO overpasses for daytime and nighttime are reported in the plot of MODIS AODs, showing corresponding vertical structure of aerosol backscattering coefficient between 25 and 45N in the right part of Fig. 16.13. During nighttime (blue overpass in the middle), the AOD is so large at 38N that the surface is hardly detected by the lidar at 532 nm (see corresponding vertical cross-section on the right). This is not the case at 1064 nm (not shown), indicating that absorption is much stronger in the visible corresponding to a probable signature of a larger part of carbonated aerosols. Most of the particles are located below 3 km in altitude, except over the north-eastern part where it appears to be lifted by the mountains, leading to cloud formation.

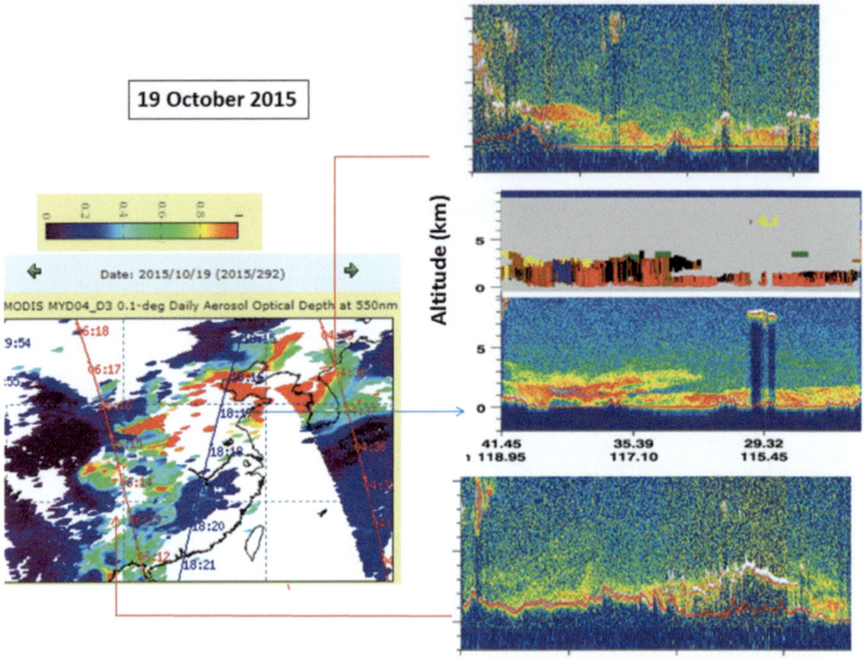

Fig. 16.13 MODIS and CALIPSO observations on 19 October 2015 over eastern China: MODIS daily AOD and corresponding CALIOP backscattering cross-sections with morning and evening tracks reported on MODIS image

Classification of aerosols from operational V3 algorithm shown in Fig. 16.13 for the nighttime orbit mostly corresponds to polluted continental (red) and smoke (black). The average lidar ratio in this area for this nighttime orbit (18:20 UTC) is thus about 60 sr. When looking at the daytime orbit over the sea (red orbit at 04:40 UTC close to Korea in Fig. 16.10), the aerosol typing looks slightly different with a more pronounced occurrence of continental and sea-salt aerosols leading to smaller lidar ratios even in the northerly part.

Looking to the retrieval of the aerosol optical depths from CALIOP and from MODIS as reported in Fig. 16.14, one can see that CALIOP appears to underestimate the AOD over this orbit. Using a constant lidar ratio equal to the average value of nighttime orbit over land (60 sr) gives a better agreement between the two observations. This is confirmation of the importance of the aerosol type identification and of the need to use appropriate values to minimize biases in extinction coefficients. These biases may also induce distortion in the vertical extinction profiles which are used in the scaling of model outputs when combining active and passive remote sensing observations and models analyses.

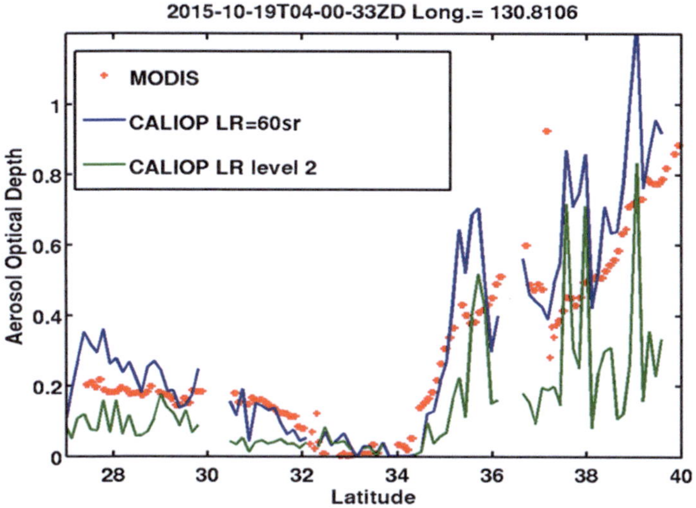

Fig. 16.14 Aerosol optical depths deduced from CALIOP V3 (*green solid line*) and MODIS (Coll. 6, *red dots*) operational retrievals compared to CALIOP AODs retrieved using a constant value of the lidar ratio (*blue solid line*)

16.3.2 CALIPSO and Model Analyses

Several studies have been conducted using ground-based and remote sensing (either from ground or from space) and chemical transport models to analyze pollution events over China. More recent ones are involving CALIPSO observations either directly or for comparisons (Ma et al. 2014; Hara et al. 2011; Geng et al. 2015; Xu et al. 2015; van Donkelaar et al. 2015).

In their study using GEOS-Chem, van Donkelaar et al. (2006) however evidenced that the use of simulated profiles have successfully improved the AOD–$PM_{2.5}$ correlation at regional to global scales. Analyses made using the shape of CALIOP extinction profiles (V3) and MODIS AODs in an optimal estimation approach (van Donkelaar et al. 2013). Significant differences between CALIOP and initial model profiles were evidenced in major biomass burning regions, but agreement was otherwise of the order of 25%. Overall agreement using CALIOP profiles over USA Canada and Europe were better in terms of dispersion when looking to surface $PM_{2.5}$ but degraded in terms of slope, showing the existence of residual biases in the determination of the extinction profiles.

16.4 Perspective of CALIPSO V4: New Inputs for Going Further

Three products are or will be soon available to characterize aerosol properties as given in Table 16.3. The operational CALIPSO product, now evolving to the newly available version 4, and two other research products, developed in France. The first one (LATMOS) is based on L1 to derive L2 extinction profiles is introducing additional tests and retrievals, whereas the second (SODA), based on ocean surface reflectance analysis (Josset et al. 2008, 2011), allows to derives AODs and lidar ratios above the ocean or clouds. Table 16.3 summarizes these products available from the CALIPSO team, both at NASA/ASDC on the US side and ICARE /AERIS on the French side, and a short description is given hereafter.

Table 16.3 CALIPSO products available for the characterization of aerosol properties

Data level	Operational V4 products (global) NASA/ASDC: http://www-calipso.larc.nasa.gov/products/ ICARE/AERIS:http://www.icare.univ-lille1.fr/	Research products	
		LATMOS (aerosol global) to be distributed by ICARE	SODA (ICARE) http://www.icare.univ-lille1.fr/projects/soda/
1	Lidar attenuated backscatter profiles at 532 and 1064 nm, depolarization ratio at 532 nm, 3 band IIR and WFC radiances	Lidar attenuated backscatter profiles at 532 and 1064 nm, depolarization ratio at 532 nm, 3 band IIR	Lidar integrated attenuated surface backscatter at 532 and 1064 nm, normalized to surface radar reflectivity
	Meteorological profiles		
	Lidar aerosol and cloud browse images		
1.5	Lidar backscatter profiles cleaned from clouds using L2-CALIOP	Lidar backscatter profiles cleaned from clouds using L2-CALIOP and IIR	
2a	Lidar aerosol and cloud layer heights/thicknesses	Lidar aerosol profiles (extinction, lidar ratio, depolarization, color ratio)	Optical depth of aerosols and ST clouds over the ocean
	Layer properties (extinction, depolarization, color ratio, …)		Optical depth of aerosols and ST clouds over low water clouds
	Cloud ice/water phase		Integrated lidar ratio
	Cloud emissivity and particle size (IIR)		
3	Aerosol extinction profiles based on L2 product (Winker et al. 2013)		Aerosol optical depths based on L2 SODA product

16.4.1 Operational Products

Expedited products using climatological ancillary data are available 1 day after data collection. Final data are produced following receipt of all required ancillary data (e.g., meteorological profiles from MERRA-2 re-analyses). They are archived and publicly available at NASA/LaRC and AERIS/ICARE data Centers. The calibration of the level 1 data in V4 has been improved to deliver greater detection sensitivity to spatially diffuse scattering layers and reduce uncertainties in the retrieval of optical properties, as compared to V3(available since 2010). Calibration biases at 532 nm will be reduced to less than 1.0%, and numerous artifacts eliminated from the 1064 nm calibration coefficients. More accurate color ratios on aerosols will be available within version 4.

The probability density functions that drive the CALIOP CAD algorithm (Liu et al. 2009) have been completely recalculated in order to accommodate the substantial changes to color ratio introduced by the new V4 calibrations. Additionally, the CAD algorithm is now applied at single shot resolution to minimize false identification of strongly scattering aerosol layers as clouds. The extinction coefficient determination in the operational product requires several steps. The first one is the identification of the altitude of the scattering layer using the Selective Iterated Boundary Locator (SIBYL algorithm, Vaughan et al. 2009). The SIBYL algorithm vertically scans lidar profiles and identifies regions of enhanced scattering, and records the location and characteristics of these atmospheric features. Then the Scene Classification Algorithm (SCA), which is actually a set of algorithms, is used to classify these layers by type (cloud or aerosol) and then by sub-type. It relies on a statistical analysis of observed parameters (Liu et al. 2009), and assigns a lidar ratio to the feature correspondingly for the extinction processing in HERA (Young and Vaughan 2009). If the feature is non-lofted, a lidar ratio is selected based on the model corresponding to the identified feature type. Aerosol models were developed using data from AERONET (Omar et al. 2009), and new lidar ratios will be used for extinction calculations. A new aerosol type (polluted marine aerosols) has been introduced in V, and stratospheric aerosols are now identified as part of the unified aerosol analysis scheme.

16.4.2 LATMOS Product

Products elaborated with LATMOS algorithm are based on level 1 CALIPSO data product revisited to check for L2 operational aerosol plus cloud discrimination using IIR cloud signature. It includes more physical information on the aerosol than the operational CALIOP products as aerosol properties in elevated layers (such as lidar ratio and Angström coefficient) that are directly retrieved using a constrained analysis similar to the one used for clouds in the operational cloud product.

The algorithm performs an analysis on lidar ratio using averaged profiles combined with layer information, and extinction coefficient profiles and optical depths

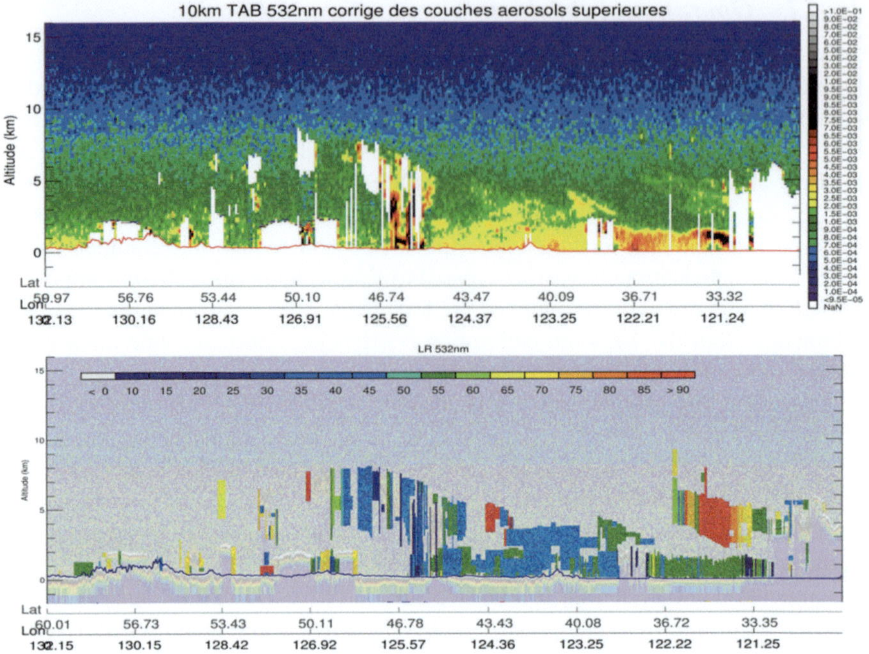

Fig. 16.15 Example of retrievals using the LATMOS algorithm of total backscattering coefficient and lidar ratio over eastern China for a low pollution case (AOD <0.5) in December 2008. *White* parts in *upper* figure are linked to the presence of clouds. Lidar ratio values in the *right* figure are ranging from 25 sr (sea salt particles) to 80 sr (heavy polluted)

are calculated using calculated lidar ratios. Aerosol types are presently classified using operational identification possibly modified according to depolarization, color and lidar ratio values. Classification may thus be different from the operational one. An example of retrieval is given in Fig. 16.15. It is expected that the analysis on elevated layers can be included in a further version of the operational analysis.

16.4.3 SODA

The SODA algorithm allows to retrieve the optical properties of optically thin atmospheric features over the ocean and dense water clouds from transmission analysis. The new SODA product, produced and archived at ICARE, now provides the optical depth, the color ratio, and lidar ratio of aerosols and semi-transparent clouds over the ocean and over dense liquid water clouds. Over the ocean, radar and lidar surface echoes are used to give a surface reference used in the lidar transmission analysis (Josset et al. 2008). Over dense water clouds the cloud reflectance is determined from cloud lidar ratio and multiple scattering correction using depolarization (Hu et al. 2007).

The aerosol and cloud optical depth are retrieved at the two CALIOP wavelengths through three steps. Only profiles over the ocean are kept in the analysis of semi-transparent aerosol and cloud layers only. In this case and in order to facilitate classification, data profiles are analyzed only if a maximum of the lidar signal is located at the ocean surface level. This allows to identify and separate profiles with low level dense clouds which totally removes attenuated signal near the surface. Over land the altitude given by the digital elevation model is used to determine the intensity of the surface echo as given by the operational CALIOP algorithm.

A first step is to calibrate the data over the ocean using lidar-radar cross-calibration, to correct any monthly calibration bias coming from the lidar or radar. It is based only on the clear-sky observations and does not rely on any underlying assumption of surface roughness behavior. The algorithm then calculates the fourth order polynomial coefficients linking radar normalized surface scattering cross section and the lidar ocean integrated attenuated surface backscatter coefficient. AODs are then calculated from the ratio of the measured lidar surface reflectance to the theoretical one derived from normalization to radar.

The determination of AODs using SODA does not require the knowledge of a lidar ratio, avoiding biases linked to the knowledge of particle properties, but on the reverse the average lidar ratio on the vertical can be derived from the ratio of the aerosol optical depth to the integrated particle backscatter coefficient (Josset et al. 2011).

As an example, average values of the lidar ratios retrieved from previous analyses for December-January-February season (Dawson et al. 2015), are given in Fig. 16.16. As shown in the eastern Asia region, values are exceeding 50 sr indicative of high pollution export over the ocean, as emphasized in previous sections.

SODA and MODIS AODs above the ocean have been found in good agreement (Josset et al. 2008, 2011; Dawson et al. 2015). The SODA retrieved lidar ratios have been compared to the ones retrieved from High Spectral Resolution Lidar measurements and those used in CALIOP inversion algorithm for single layer (Josset et al. 2011), and is was shown that some significant difference may be observed leading to uncertainties in the extinction coefficient near the surface.

A dual layer approach tested during INDOEX (Pelon et al. 2002) may further extend SODA capabilities, as well as a combination of MODIS and CALIOP analyses (Kaufman et al. 2003), but it is expected that merging additional surface information in the operational CALIOP analysis will further improve the retrieval of the vertical profile, allowing improved performance not only over the ocean but also over land surface.

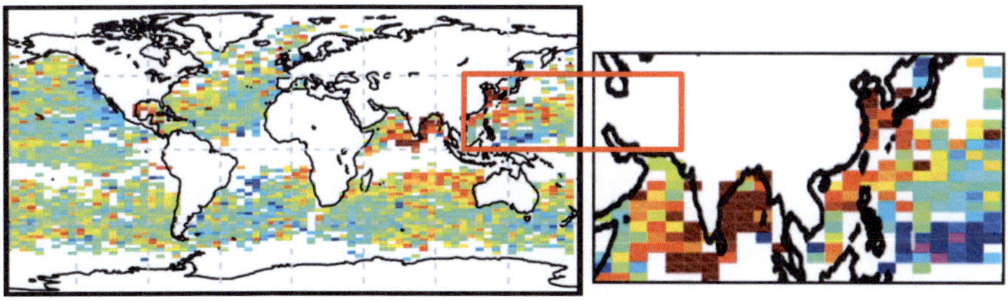

Fig. 16.16 *Upper* figure: Global map of the vertically averaged lidar ratio retrieved from SODA and *lower* figure: focus in the area delimited by the *red-rectangle* in the *upper* figure (After Dawson et al. 2015)

16.5 Conclusion

Encouraging results have been obtained using satellite observations for air quality analysis, but several challenges remain to improve the use of remote sensing products provided by space missions in the analysis and prevision of severe pollution events. Results already obtained using satellite observations show that PM can be rather well estimated from statistical models, but best results are obtained with chemical transport models. In this context, the vertical information on aerosol extinction profiles as given by CALIOP allows to provide additional constraints and potential improvement. Comparisons between simulated and observed PMs are very encouraging but dispersion remains large. Severe pollution events remain a challenge to better understand interacting processes. It is the objective of satellite observations and more particularly the NASA-CNES CALIPSO and then the ESA EarthCare mission (planned launch 2019) embarking a high spectral resolution lidar to provide up-to-date high quality products over land and ocean. New products will be available, and their use in an integrative model approach should enable to better characterize and quantify the impact of pollution events.

References

Aaron van Donkelaar, Randall V. Martin, Michael Brauer, Brian L. Boys. (2014). Use of satellite observations for long-term exposure assessment of global concentrations of fine particulate matter. *Environmental Health Perspectives*.

Bergin, M. H., Cass, G. R., Xu, J., Fang, C., Zeng, L. M., Yu, T., Salmon, L. G., Kiang, C. S., Tang, X. Y., Zhang, Y. H., & Chameides, W. L. (2001). Aerosol radiative, physical, and chemical properties in Beijing during June 1999. *Journal of Geophysical Research: Atmospheres, 106* (D16), 17969–17980.

Box, M. A., & Trautmann, T. (1994). Computation of anthropogenic sulphate aerosol forcing using radiative perturbation theory. *Tellus, 46B*, 33–39.

Burton, S. P., Ferrare, R. A., Hostetler, C. A., Hair, J. W., Rogers, R. R., Obland, M. D., Butler, C. F., Cook, A. L., Harper, D. B., & Froyd, K. D. (2012). Aerosol classification using airborne high spectral resolution lidar measurements – Methodology and examples. *Atmospheric Measurement Techniques, 5*, 73–98. doi:10.5194/amt-5-73-2012.

Cattrall, C., Reagan, J., Thome, K., & Dubovik, O. (2005). Variability of aerosol and spectral lidar and backscatter and extinction ratios of key aerosol types derived from selected Aerosol Robotic Network locations. *Journal of Geophysical Research, 110*, D10S11. doi:10.1029/2004JD005124.

Cheng, Y., He, K.-B., Du, Z.-Y., Zheng, M., & Duan, F. K. (2015). Humidity plays an important role in the PM pollution in Beijing. *Environmental Pollution, 196*, 68–75. doi:10.1036/j.envpol.2014.11.028.

Choi, M., Kim, J., Lee, J., Kim, M., Je Park, Y., Jeong, U., Kim, W., Holben, B., Eck, T. F., Lim, J. H., & Song, C. K. (2016). GOCI Yonsei Aerosol retrieval algorithm and validation during DRAGON-NE Asia 2012 campaign. *Atmospheric Measurement Techniques, 9*, 1377–1398. doi:10.5194/amt-9-1377-2016.

Cuesta, J., Eremenko, M., Flamant, C., Dufour, G., Laurent, B., Bergametti, G., Höpfner, M., Orphal, J., & Zhou, D. (2015). Three-dimensional distribution of a major desert dust outbreak over East Asia in March 2008 derived from IASI satellite observations. *Journal of Geophysical Research – Atmospheres, 120*, 7099–7127. doi:10.1002/2014JD022406.

Dawson, K. W., Meskhidze, N., Josset, D., & Gassó, S. (2015). Spaceborne observations of the lidar ratio of marine aerosols. *Atmospheric Chemistry and Physics, 15*, 3241–3255. doi:10.5194/acp-15-3241-2015.

Flamant, C., Pelon, J., Chazette, P., Trouillet, V., Quinn, P., Frouin, R., Bruneau, D., Leon, J.-F., Bates, T., Johnson, J., & Livingston, J. (2000). Airborne lidar measurements of aerosol spatial distribution and optical properties over the Atlantic Ocean during an European pollution outbreak of ACE-2. *Tellus, 52B*, 662–677.

Gao, J., Tian, H., Cheng, K., Lu, L., Zheng, M., Wang, S., Hao, J., Wang, K., Hua, S., Zhu, C., & Wang, Y. (2015). The variation of chemical characteristics of PM2.5 and PM10 and formation causes during two haze pollution events in urban Beijing, China. *Atmospheric Environment, 107*, 1–8.

Geng, G., Zhang, Q., Martin, R. V., van Donkelaar, A., Huo, H., Che, H., Lin, J., & He, K. (2015). Estimating long-term PM concentrations in China using satellite-based aerosol optical depth and a chemical transport model. *Remote Sensing of Environment, 166*, 262–270. doi:10.1016/j.rse.2015.05.016.

Guo, S., Hu, M., Zamora, M. L., Peng, J., Shang, D., Zheng, J., Du, Z., Wu, Z., Shao, M., Zeng, L., Molina, M. J., & Zhang, R. (2014). Elucidating severe urban haze formation in China. *PNAS, 111*(49), 17373–17378. doi:10.1073/pnas.1419604111.

Hänel, A., Baars, H., Althausen, D., Ansmann, A., Engelmann, R., & Sun, J. Y. (2012). One-year aerosol profiling with EUCAARI Raman Lidar at Shangdianzi GAW station: Beijing plume and seasonal variations. *Journal of Geophysical Research, 117*, D13201. doi:10.1029/2012JD017577.

Hara, Y., Uno, I., Shimizu, A., Sugimoto, N., Matsui, I., Yumimoto, K., Kurokawa, J., Ohara, T., & Liu, Z. (2011). Seasonal characteristics of spherical aerosol distribution in Eastern Asia: Integrated analysis using ground/space-based lidars and a chemical transport model. *SOLA, 7*, 121–124. doi:10.2151/sola.2011-031.

Herman, J. R., Bhartia, P. K., Torres, O., Hsu, C., Seftor, C., & Celarier, E. (1997). Global distribution of UV-absorbing aerosols from Nimbus-7/TOMS data. *Journal of Geophysical Research, 102*, 16,911–16,922.

Hoff, R. M., Guise-Bagley, L., Staebler, R. M., Wiebe, H. A., Brook, J., Georgi, B., & Düsterdiek, T. (1996). Lidar, nephelometer, and in situ aerosol experiments in southern Ontario. *Journal of Geophysical Research, 101*(D14), 19199–19209.

Hsu, N. C., Tsay, S.-C., King, M. D., & Herman, J. R. (2001). Deep blue retrievals of Asian aerosol properties during ACE-Asia. *IEEE Transactions on Geoscience and Remote Sensing, 44*(11), 3180–3195.

Hu, Y., Vaughan, M., Liu, Z., Powell, K., & Rodier, S. (2007). Retrieving optical depths and lidar ratios for transparent layers above opaque water clouds from CALIPSO lidar measurements. *IEEE Translation Geoscience and Remote Sensing Letters, 4*, 523–526.

Huang, J. P., Liu, J. J., Chen, B., & Nasiri, S. L. (2015). Detection of anthropogenic dust using CALIPSO lidar measurements. *Atmospheric Chemistry and Physics, 15*, 11653–11665. doi:10.5194/acp-15-11653-2015.

Huebert, B. J., Bates, T., Russell, P. B., Shi, G., Kim, Y. J., Kawamura, K., Carmichael, G., & Nakajima, T. (2003). An overview of ACE-Asia: Strategies for quantifying the relationships between Asian aerosols and their climatic impacts. *Journal of Geophysical Research, 108* (D23), 8633. doi:10.1029/2003JD003550.

Josset, D., Pelon, J., Protat, A., & Flamant, C. (2008). New approach to determine aerosol optical depth from combined CALIPSO and CloudSat ocean surface echoes. *Geophysical Research Letters, 35*, L10805. doi:10.1029/2008GL033442.

Josset, D., Rogers, R., Pelon, J., Hu, Y., Liu, Z., Omar, A., & Zhai, P. (2011). CALIPSO lidar ratio retrieval over the ocean. *Optics Express, 19*, 18696–18706. doi:10.1364/OE.19.018696.

Kaufman, Y. J., Tanre, D., Leon, J.-F., & Pelon, J. (2003). Retrievals of profiles of fine and coarse aerosols using lidar and radiometric space measurements. *IEEE Transactions on Geoscience and Remote Sensing, 41*(8), 1743–1754.

Kleidman, R. G., O'Neill, N. T., Remer, L. A., Kaufman, Y. J. Eck, T. F., Didier Tanré, Oleg Dubovik, & Holben, B. N. (2005). Comparison of Moderate Resolution Imaging Spectroradiometer (MODIS) and Aerosol Robotic Network (AERONET) remote-sensing retrievals of aerosol fine mode fraction over ocean. *Journal of Geophysical Research, 110*(D22).

Levelt, P., et al. (2006). Science objectives of the ozone monitoring instrument. *IEEE Transactions on Geoscience and Remote Sensing, 44*(5), 1199–1208. doi:10.1109/TGRS.2006.87233.

Li, J., Carlson, B. E., & Lacis, A. A. (2014). Application of spectral analysis techniques to the intercomparison of aerosol data – Part 4: Synthesized analysis of multi-sensor satellite and ground-based AOD measurements using combined maximum covariance analysis. *Atmospheric Measurement Techniques, 7*, 2531–2549. doi:10.5194/amt-7-2531-2014.

Li, J., Carlson, B. E., & Lacis, A. A. (2015). How well do satellite AOD observations represent the spatial and temporal variability of $PM_{2.5}$ concentration for the United States ? *Atmospheric Environment, 102*, 260–273.

Liu, Y., Park, R. J., Jacob, D. J., Li, Q., Kilaru, V., & Sarnat, J. A. (2004). Mapping annual mean ground-level PM concentrations using multiangle imaging spectroradiometer aerosol optical thickness over the contiguous United States. *Journal of Geophysical Research, 109*, D22206. doi:10.1029/2004JD005025.

Liu, Z., et al. (2009). The CALIPSO lidar cloud and aerosol discrimination: Version 2 algorithm and initial assessment of performance. *Journal of Atmospheric and Oceanic Technology, 26*, 1198–1213.

Liu, Y., Paciorek, C. J., & Koutrakis, P. (2009b). Estimating regional spatial and temporal variability of PM concentrations using satellite data, meteorology, and land use information. *Environmental Health Perspectives, 117*(6), 886–892.

Liu, J., Wang, Z., Wang, J., Welton, E. J., Ferrare, R. A., & Newson, R. K. (2011). The effect of aerosol vertical profiles on satellite-estimated surface particle sulfate concentrations. *Remote Sensing of Environment, 15*, 508–513.

Liu, J., Huang, J., Chen, B., Zhou, T., Yan, H., Jin, H., Huang, Z., & Zhang, B. (2015). Comparisons of PBL heights derived from CALIPSO and ECMWF reanalysis data over China. *Journal of Quantitative Spectroscopy and Radiative Transfer, 153*, 102–112.

Luo, Y., Zheng, X., Zhao, T., & Chen, J. (2014). A climatology of aerosol optical depth over China from recent 10 years of MODIS remote sensing data. *International Journal of Climatology, 34*, 863–870. doi:10.1002/joc.3728.

Ma, J., Chen, Y., Wang, W., Yan, P., Liu, H., Yang, S., Hu, Z., & Lelieveld, J. (2010). Strong air pollution causes widespread haze – Clouds over China. *Journal of Geophysical Research, 115*, D18204. doi:10.1029/2009JD013065.

Ma, Z., Hu, X., Huang, L., Bi, J., & Liu, Y. (2014). Estimating ground level $PM_{2.5}$ in China using satellite remote sensing. *Environmental Science & Technology, 48*, 7436–7444.

Menut, L., Flamant, C., Pelon, J., & Flamant, P. H. (1999). Urban boundary layer height determination from lidar measurements over the Paris area. *Applied Optics, 38*, 945–954.

Murayama, T., et al. (2003). An intercomparison of lidar-derived aerosol optical properties with airborne measurements near Tokyo during ACEAsia. *Journal of Geophysical Research, 108* (D23), 8651. doi:10.1029/2002JD003259.

O'Neill, N. T., Eck, T. F., Smirnov, A., Holben, B. N., & Thulasiraman, S. (2003). Spectral discrimination of coarse and fine mode optical depth. *Journal of Geophysical Research, 108* (D17), 4559. doi:10.1029/2002JD002975.

Omar, A., Winker, D., Kittaka, C., Vaughan, M., Liu, Z., Hu, Y., Trepte, C., Rogers, R., Ferrare, R., Kuehn, R., & Hostetler, C. (2009). The CALIPSO automated aerosol classification and lidar ratio selection algorithm. *Journal of Atmospheric and Oceanic Technology, 26*. doi:10.1175/2009JTECHA1231.1.

Omar, A. H., Won, J.-G., Winker, D. M., Yoon, S.-C., Dubovik, O., & McCormick, M. P. (2005). Development of global aerosol models using cluster analysis of Aerosol Robotic Network (AERONET) measurements. *Journal of Geophysical Research, 110*, D10S14. doi:10.1029/2004JD004874.

Paciorek C. J., Liu, Y., Moreno-Macias, H., & Kondragunta, S. (2008). *Spatio-temporal associations between GOES aerosol optical depth retrievals and ground-level PM2.5*. Harvard University Biostatistics Working Paper Series, http://www.bepress.com/harvardbiostat/paper74.

Pelon, J., Flamant, C., Chazette, P., Leon, J.-F., Tanre, D., Sicard, M., & Satheesh, S. K. (2002). Characterization of aerosol spatial distribution and optical properties over the Indian Ocean from airborne LIDAR and radiometry during INDOEX'99. *Journal of Geophysical Research, 107*(D19), 8029. doi:10.1029/2001JD000402.

Quinn, P. K., & Coffman, D. J. (1998). Local closure during the first aerosol characterization experiment (ACE 1)' Aerosol mass concentration and scattering and backscattering coefficients. *Journal of Geophysical Research, 103*(D13), 16,575–16,596.

Raut, J.-C., & Chazette, P. (2009). Assessment of vertically-resolved PM_{10} from mobile lidar observations. *Atmospheric Chemistry and Physics, 9*, 8617–8638.

Shin, S.-K., Müller, D., Lee, C., Lee, K. H., Shin, D., Kim, Y. J., & Noh, Y. M. (2015). Vertical variation of optical properties of mixed Asian dust/pollution plumes according to pathway of air mass transport over East Asia. *Atmospheric Chemistry and Physics, 15*, 6707–6720. doi:10.5194/acp-15-6707-2015.

Snider, G., Weagle, C. L., Martin, R. V., van Donkelaar, A., Conrad, K., Cunningham, D., Gordon, C., Zwicker, M., Akoshile, C., Artaxo, P., Anh, N. X., Brook, J., Dong, J., Garland, R. M., Greenwald, R., Griffith, D., He, K., Holben, B. N., Kahn, R., Koren, I., Lagrosas, N., Lestari, P., Ma, Z., Vanderlei Martins, J., Quel, E. J., Rudich, Y., Salam, A., Tripathi, S. N., Yu, C., Zhang, Q., Zhang, Y., Brauer, M., Cohen, A., Gibson, M. D., & Liu, Y. (2015). SPARTAN: A global network to evaluate and enhance satellite-based estimates of ground-level particulate matter for global health applications. *Atmospheric Measurement Techniques, 8*, 505–521. doi:10.5194/amt-8-505-2015.

Su, B., Zhan, M., Zhai, J., Wang, Y., & Fisher, T. (2015). Spatio-temporal variation of haze days and atmospheric circulation pattern in China (1961–2013). *Atmospheric Environment, 380–381*, 14–21. doi:10.1016/j.quaint.2014.11.044.

Sugimoto, N., Nishizawa, T., Shimizu, A., Matsui, I., & Kobayashi, H. (2015). Detection of internally mixed Asian dust with air pollution aerosols using a polarization optical particle counter and a polarization-sensitive two-wavelength lidar. *Journal of Quantitative Spectroscopy and Radiative Transfer, 150*, 107–113.

Tao, M., et al. (2014). A study of urban pollution and haze clouds over northern China during the dusty season based on satellite and surface observations. *Atmospheric Environment, 82*, 183–192.

Tesche, M., Ansmann, A., Müller, D., Althausen, D., Engelmann, R., Hu, M., & Zhang, Y. (2007). Particle backscatter, extinction, and lidar ratio profiling with Raman lidar in south and north China. *Applied Optics, 46*, 6302–6308.

Tesche, M., Ansmann, A., Muller, D., Althausen, D., Engelmann, R., Freudenthaler, V., & Groß, S. (2009). Vertically resolved separation of dust and smoke over Cape Verde using multiwavelength Raman and polarization lidars during Saharan Mineral Dust Experiment 2008. *Journal of Geophysical Research, 114*, D13202. doi:10.1029/2009JD011862.

Torres, O., Barthia, P. K., Herman, J. R., Ahmad, Z., & Gleason, J. (1998). Derivation of aerosol properties from satellite measurements of backscattered ultraviolet radiation: Theoretical basis. *Journal of Geophysical Research, 103*, 17 099–17 110.

van Donkelaar, A., Martin, R. V., & Park, R. J. (2006). Estimating ground-level PM using aerosol optical depth determined from satellite remote sensing. *Journal of Geophysical Research, 111*, D21201. doi:10.1029/2005JD006996.

van Donkelaar, A., Martin, R. V., Brauer, M., Kahn, R., Levy, R., Verduzco, C., & Villeneuve, P. J. (2010). Global estimates of ambient fine particulate matter concentrations from satellite based aerosol optical depth: Development and application. *Environmental Health Perspectives, 118*, 847–855. doi:10.1289/ehp.0901623.

van Donkelaar, A., Martin, R. V., Spurr, R. J. D., Drury, E., Remer, L. A., Levy, R. C., & Wang, J. (2013). Optimal estimation for global groundlevel fine particulate matter concentrations. *Journal of Geophysical Research – Atmospheres, 118*, 5621–5636. doi:10.1002/jgrd.50479.

van Donkelaar, R., Martin, V., Brauer, M., & Boys, B. L. (2015) .Use of satellite observations for long-term exposure assessment of global concentrations of fine particulate matter. *Environ Health Perspect, 123*, 135–143. http://dx.doi.org/10.1289/ehp.1408646.

Vaughan, M. A., Powell, K. A., Kuehn, R. E., Young, S. A., Winker, D. M., Hostetler, C. A., Hunt, W. H., Liu, Z., McGill, M. J., & Getzewich, B. J. (2009). Fully automated detection of cloud and aerosol layers in the CALIPSO lidar measurements. *Journal of Atmospheric and Oceanic Technology, 26*, 2034–2050.

Waggoner, A. P., Weiss, R. E., Ahlquist, N. C., Covert, D. S., Will, S., & Charlson, R. J. (1981). Optical characteristics of atmospheric aerosols. *Atmospheric Environment, 15*, 1891–1909.

Wang, J., & Christopher, S. A. (2003). Intercomparison between satellite-derived aerosol optical thickness and PM mass: Implications for air quality studies. *Geophysical Research Letters, 30* (21), 2095. doi:10.1029/2003GL018174.

Wang, Y., Zhuang, G. S., Sun, Y. L., & An, Z. S. (2005). Water-soluble part of the aerosol in the dust storm season –evidence of the mixing between mineral and pollution aerosols. *Atmospheric Environment, 39*, 7020–7029.

Wang, L., Xu, J., Yang, J., Zhao, X., Wei, W., Cheng, D., Pan, X., & Su, J. (2012). Understanding haze pollution over the southern Hebei area of China using the CMAQ model. *Atmospheric Environment, 56*, 69–79.

Wang, H., et al. (2014). A multisource observation study of the severe prolonged regional haze episode over eastern China in January 2013. *Atmospheric Environment, 89*, 807–815.

Wang, P., Cao, J.-J., Shen, Z.-X., Han, Y.-M., Lee, S.-C., Huang, Y., Zhu, C.-S., Wang, Q.-Y., Xu, H.-M., & Huang, R.-J. (2015). Spatial and seasonal variations of PM2.5 mass and species during 2010 in Xi'an, China. *Science of the Total Environment, 508*., 1 March 2015, 477–487.

Winker, D. M., Vaughan, M. A., Omar, A., Hu, Y., Powell, K. A., Liu, Z., Hunt, W. H., & Young, S. A. (2009). Overview of the CALIPSO mission and CALIOP data processing algorithms. *Journal of Atmospheric and Oceanic Technology, 26*, 2310–2323.

Winker, D. M., Pelon, J., Coakley, J. A., Jr., Ackerman, S. A., Charlson, R. J., Colarco, P. R., Flamant, P., Fu, Q., Hoff, R., Kittaka, C., Kubar, T. L., LeTreut, H., McCormick, M. P., Mégie, G., Poole, L., Powell, K., Trepte, C., Vaughan, M. A., & Wielicki, B. A. (2010). The CALIPSO mission: A global 3D view of aerosols and clouds. *Bulletin of the American Meteorological Society, 91*, 1211–1229. doi:10.1175/2010BAMS3009.1.

Winker, D. M., Tackett, J. L., Getzewich, B. J., Liu, Z., Vaughan, M. A., & Rogers, R. R. (2013). The global 3-D distribution of tropospheric aerosols as characterized by CALIOP. *Atmospheric Chemistry and Physics, 13*, 3345–3361. doi:10.5194/acp-13-3345.

Xu, J.-W., et al. (2015). Estimating ground-level PM in Eastern China using aerosol optical depth determined from the GOCI satellite instrument. *Atmospheric Chemistry and Physics, 15*, 13133–13144. doi:10.5194/acp-15-13133-2015.

You, W., Zang, Z., Pan, X., Zhang, L., & Chen, D. (2015). Estimating PM2.5 in Xi'an, China using aerosol optical depth: A comparison between the MODIS and MISR retrieval models. *Science of the Total Environment, 505*, 1156–1165.

Young, S. A., & Vaughan, M. A. (2009). The retrieval of profiles of particulate extinction from Cloud-Aerosol Lidar Infrared Pathfinder Satellite Observations (CALIPSO) data: Algorithm description. *Journal of Atmospheric and Oceanic Technology, 26*, 1105–1111.

Zha, Y., Gao, J., Jiang, J., Lu, H., & Huang, J. (2010). Monitoring of urban air pollution from MODIS aerosol data: Effect of meteorological parameters. *Tellus, 62B*, 109–116. doi:10.1111/j.1600-0889.2010.00451.x.

Zhang, R., Jing, J., Tao, J., Hsu, S.-C., Wang, G., Cao, J., Lee, C. S. L., Zhu, L., Chen, Z., Zhao, Y., & Shen, Z. (2013). Chemical characterization and source apportionment of PM in Beijing: Seasonal perspective. *Atmospheric Chemistry and Physics, 13*, 7053–7074. doi:10.5194/acp-13-7053-2013.

Zhang, X., Wang, L., Wang, W., Cao, D., & Xe, D. (2015). Long-term trend and spatiotemporal variations of haze over China by satellite observations from 1979 to 2013. *Atmospheric Environment, 119*, 362–373. doi:10.1016/j.atmosenv.2015.08.053.

Zhao, X. J., Zhao, P. S., Xu, J., Meng, W., Pu, W. W., Dong, F., He, D., & Shi, Q. F. (2013). Analysis of a winter regional haze event and its formation mechanism in the North China Plain. *Atmospheric Chemistry and Physics, 13*(11), 5685–5696.

Zheng, J., Che, W., Zheng, Z., Chen, L., & Zhong, L. (2013). Analysis of spatial and temporal variability of pm10 concentrations using MODIS aerosol optical thickness in the Pearl River Delta region, China. *Aerosol and Air Quality Research, 13*, 862–876. doi:10.4209/aaqr.2012.09.0234.

Zheng, G. J., Duan, F. K., Su, H., Ma, Y. L., Cheng, Y., Zheng, B., Zhang, Q., Huang, T., Kimoto, T., Chang, D., Pöschl, U., Cheng, Y. F., & He, K. B. (2015). Exploring the severe winter haze in Beijing: The impact of synoptic weather, regional transport and heterogeneous reactions. *Atmospheric Chemistry and Physics, 15*, 2969–2983. doi:10.5194/acp-15-2969-2015.

Chapter 17
Space Observations of Dust in East Asia

Paul Ginoux and Adrien Deroubaix

Abstract Using 13-year satellite based data, dust events in East Asia are analyzed in term of their frequency and surface properties. Three distinct regions are defined based on their geomorphology: the western basins surrounded by high elevation mountains, the central Mongolian and Loess Plateaus, and the northeastern low level plains. Dust activity peaks in spring but persists during the other seasons. The western basins are active all year long with a minimum in winter. The central plateaus have low dust activity in fall, while dust events in the northeastern plains are rare in winter and summer. We find that the sources in the northeast and along the southern margin of the Loess Plateau are associated with land use. Analysis of surface conditions indicates very strong interannual variations of snow cover which may affect the central and northeastern sources, while vegetation cover modulates seasonal variation of the northeastern sources. Strong surface winds peak in spring for all sources, but are more frequent over central and northeastern sources. Dust emission is initiated when surface wind speed is greater than a threshold of wind erosion which has been retrieved for the first time over the entire East Asia dust sources. The mean values for the western, central, and northeastern source regions are, respectively, 6.5, 9.1 and 8.6 m/s, in excellent agreement with previous studies. This work shows the power of using satellite retrievals to derive high resolution properties over large areas, although some products have large uncertainties. This drawback has been partially alleviated by comparing our results with previous studies based on in-situ data.

Keywords Dust • Emission • MODIS • China

P. Ginoux (✉)
NOAA Geophysical Fluid Dynamics Laboratory, Princeton University, Forrestal Campus, 201 Forrestal Road, Princeton, NJ 08540, USA
e-mail: paul.ginoux@noaa.gov

A. Deroubaix
Princeton University, Atmospheric and Oceanic Sciences, 300 Forrestal Road, Princeton, NJ 08544, USA

Laboratoire de Meteorologie Dynamique, Ecole Polytechnique, Route de Saclay, Palaiseau 91128, France
e-mail: adrien.deroubaix@lmd.polytechnique.fr

I. Bouarar et al. (eds.), *Air Pollution in Eastern Asia: An Integrated Perspective*, ISSI Scientific Report Series 16, DOI 10.1007/978-3-319-59489-7_17

17.1 Introduction

Dust is one of the most abundant aerosols in the atmosphere, and through its interactions with radiation, cloud microphysics, and gas phase chemistry, it affects air quality and climate. Dust emission from East Asia is around 150 Mtons/year, which corresponds to 10% globally (Huneeus et al. 2011). This value is highly uncertain with model estimations varying between 27 and 727 Mtons/year (Huneeus et al. 2011). Most of the suspended dust will deposit over land. The remaining will be transported over long distance (its lifetime is about 5 days). Dust will eventually reach North America, where its amount is as large as locally produced dust (Yu et al. 2012). In some occasions dust will leapfrog North America to end up in Europe (Grousset et al. 2003). Major dust sources have been associated with topographic depression, which have been intermittently flooded during the Quaternary while others have been recently activated by human activities (Prospero et al. 2002). However, some researchers have highlighted the role of anthropogenic processes in increasing dust emission in China (Wang et al. 2006). Some studies claimed that desertification in northern China has significantly enhanced dust emission. On the other hand, it was shown that these anthropogenic sources were secondary, and meteorology and climate are more important factors in controlling long term dust variability in East Asia (Zhang et al. 2003). Some model studies suggest that dust from agriculture is contributing less than 10% globally (Tegen et al. 2004), while satellites based estimates indicate a contribution of the order of 20–25%, with strong regional variation (Ginoux et al. 2012a). Although there appears to be a convergence in global estimates, anthropogenic dust is still a contentious subject with large uncertainties at regional scale.

Our main objective is to determine the main controlling factors of dust activity in East Asia. To untangle the spatiotemporal dependency of these factors, we first group dust sources into regions with similar geomorphologic characteristics. Then the seasonal and interannual variability of dust events in each selected regions are analyzed in term of the underlying variability of snow cover, vegetation cover, soil moisture and land use. To reduce uncertainties, our study relies on high-resolution satellite data as well as meteorological reanalysis. Finally, the threshold of wind erosion, the most uncertain although most important parameter of dust emission, is determined at 0.1° resolution over East Asian dust sources.

17.2 Detection of Dust from Space

Detection of dust events from satellite observations relies on specific optical properties of dust. Satellite retrieval algorithms test different combination of aerosol components, and select the aerosol composition that provides the best fit between calculated and measured radiances. The most important optical characteristics of dust are:

1. Increasing absorption from visible wavelength to the ultraviolet. The near-ultraviolet backscattered radiances measured by the Total Ozone Mapping Spectrometer (TOMS) and Ozone Monitoring Instrument (OMI) were used to calculate an absorbing Aerosol Index (Herman et al. 1997) and Aerosol Optical Depth (Torres et al. 2002; Torres et al. 2013). Although the TOMS Aerosol Index is a qualitative measure of the presence of dust, it has been widely used to study the transport (Ginoux et al. 2004) and sources (Goudie and Middleton 2001; Prospero et al. 2002) of dust, as well as its refractive index (Colarco et al. 2002);
2. Increasing absorption from red to deep blue. Selection of three backscattered radiances from the deep blue to red has been used to retrieve aerosol properties from the Moderate Resolution Imaging Spectroradiometer (MODIS), and the Sea-Viewing Wide Field-of-View Sensor (SeaWiFS) (Hsu et al. 2004, 2006);
3. Strong absorption band at 10 m or 1000 cm^{-1} was used to calculate an Infrared Difference Dust Index (IDDI) from Meteosat (Legrand et al. 2001), a dust index from MSG-Spinning Enhanced Visible and Infrared Imager (SEVIRI) (Schepanski et al. 2007), Dust Optical Depth and height from the Atmospheric Infrared Sounder (AIRS) (DeSouza-Machado et al. 2010), and Aerosol Optical Depth, mean altitude and effective radius of aerosols from IASI (Peyridieu et al. 2013)
4. Non-sphericity of dust particles affect the scattering phase function which is used to retrieve dust sources from multi-viewing angle cameras such as on the Multi-angle Imaging Spectro-Radiometer (MISR) satellite (Lee et al. 2016). Non-sphericity also depolarizes light which is used operationally by the Cloud-Aerosol Lidar with Orthogonal Polarization (CALIOP) retrieval algorithm to determine dust feature mask and calculate aerosol extinction profile (Winker et al. 2009). A unique aspect of the CALIOP retrieval is it provides not only the horizontal but also the vertical distribution of dust events (Liu et al. 2008).

The retrieval of physical quantities for dust is complicated by their dependency on mineralogy, aerosol size distribution, and mixing with other aerosol types. Uncertainties in satellite retrievals also depend on the presence of undetected clouds within a pixel (sub-pixel cloud contamination), and, in the near ultraviolet, the interactions with molecular scattering (Torres et al. 2002). The larger the pixel size the larger the bias. In the infrared, uncertainties are also associated with variability in surface temperature and total column water (Peyridieu et al. 2013). Among these different satellite aerosol retrievals, it was demonstrated that MODIS Deep Blue is able to distinguish dust plumes from fine-mode pollution particles even in complex aerosol environments such as the one over Beijing (Hsu et al. 2006). The other advantages of MODIS are the twice daily measurements (10:30

AM for MODIS on Terra platform, and 1:30 PM for MODIS on Aqua platform), high resolution (10 km at Nadir), long-term records (since 2000 for Terra and 2002 for Aqua), and cross-validation with other satellite products (Sayer et al. 2013). For this study, we use Level 2 MODIS Deep Blue Collection 6 aerosol products over land (Hsu et al. 2013; Sayer et al. 2013). The products include aerosol optical depth (AOD) and single scattering albedo (SSA) at 412 nm (deep blue), 470 nm (blue), and 670 nm (red). The single scattering albedo measures the amount of scattering relative to the total extinction (sum of scattering and absorption). Its value varies from 1 for purely scattering aerosol to 0.9 or less for absorbing aerosols. Another useful property to detect dust is the Angstrom exponent (α), which can be easily calculated from two measures of AOD using the following formula,

$$\alpha = -\log(\tau_1/\tau_2)/\log(\lambda_1/\lambda_2) \tag{17.1}$$

where τ_1 and $\tau_2$are the measured AOD at wavelengths $\lambda_1 = 470$ nm and $\lambda_2 = 670$ nm, respectively. AOD at 550 nm (τ_{550}) is calculated from retrieved AOD values at 470 nm and 670 nm using Eq. 17.1. The Angstrom exponent is often used as a qualitative indicator of aerosol particle size, with values greater than 2 indicating small particles associated with combustion byproducts, and values less than 1 indicating large particles like dust (Schuster et al. 2006). As dust is absorbing, the SSA is always less than 1. We have adapted these criteria in this work for MODIS Deep Blue Collection 6 quality assured aerosol products, by assuming that the aerosol optical depth is entirely due to dust (DOD = 550) when <0.3 and SSA <1. As the Level 2 products are provided at the pixel level with a varying shape and dimension, it is more convenient to map, the data on a regular fixed geographical grid. Therefore, the daily MODIS Aqua Level 2 products from January 2003 to December 2015 are first mapped on a regular 0.1° latitude by 0.1° longitude grid, weighting pixel quantities by its surface area. Then, using the two criteria described above, DOD is extracted from AOD retrieved with the highest MODIS quality flag (flag = 3). Finally, the frequency of occurrence (FoO) of DOD greater than 0.2 is calculated seasonally or yearly, then average over 13 years (2003–2015). The maxima of FoO are indicative of the location of dust sources. The higher the selection of minimum DOD the more intense the sources. The validity of this method was demonstrated over the Chihuahuan desert, where FoO maxima are clearly located over erodible geomorphological surfaces (Baddock et al. 2016).

17.3 Major Dust Regions

Figure 17.1 shows the distribution of annual mean FoO (2003–2015) over East Asia with the topography in the background. Three dusty regions can be distinguished relatively to their surrounding topography. In each of these regions the FoO occupies low-level ground relative to the surroundings, but with different elevation gradients. The sharpest gradient covers the Xianjing province in northwestern

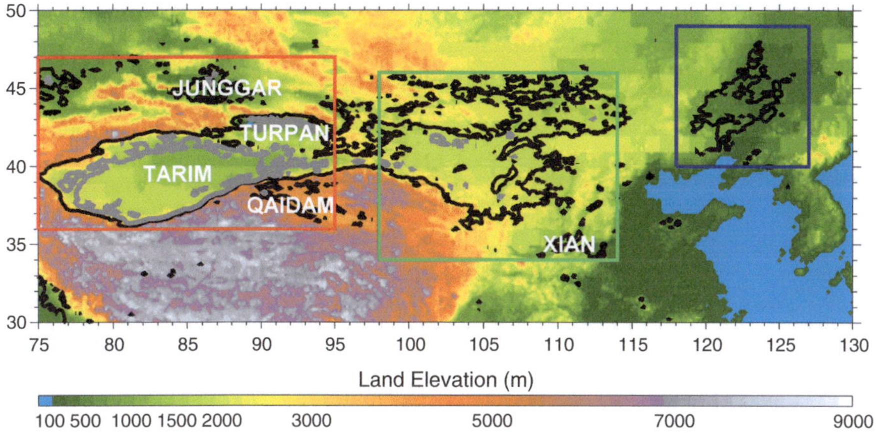

Fig. 17.1 Annual mean (2003–2015) Frequencies of Occurrence (FoO) of MODIS Dust Optical Depth (DOD) greater than 0.2 are shown in *black* (5% or 18 days/year) and *grey* (25% or 91 days/year) contours. The *colored boxes* delineate the three major dust sources of China: *red* for the western basins, *green* for the Mongolian and Loess plateau, and *blue* for the northeastern plains. The Earth's surface elevation is *color shaded*

China, and includes four large basins: Junggar (northernmost), Turpan (north-easternmost), Tarim (westernmost), and Qaidam (south easternmost). The Tien Shan Mountains to the north, the Pamirs to the west, the Kunlun Mountains to the south, and the Altun Mountains to the East enclose the Turpan, Tarim and Qaidam basins. The Tien Shan and the Mongolian Altai Mountains delimit the Junggar basin. These basins are all active dust sources with on average 18 days with dust events per year. The largest and most active source is the Taklamakan desert within the Tarim basin where FoO reaches 90 days of dust events per year. The Turpan depression is a fault trough descending ultimately to 155 m below sea level on its western part. The second region (green box in Fig. 17.1) includes the semi-deserts and Gobi deserts of the Mongolian and Loess plateau. The sources are mostly located in Inner Mongolia. A detached source on the southeast of the green box is located at the confluence of the Yellow, Wei and Fen rivers forming the Guangzhong Plain along the southern margin of the Loess plateau and the Qinling Range. The alluvial soil of lower the parts of the Wei and Fen rivers are used to cultivate wheat. Chemical analysis of PM10 in the city of Xi'an indicates that 29% is composed of dust (Shen et al. 2011). Analysis of the origin of dust in Xi'An indicates a primary contribution from northern sources, but also contribution from Loess plateau and local anthropogenic sources (Wang et al. 2006). Between the western basins (red box) and the Mongolian plateau (green box in Fig. 17.1), some high FoO spots are aligned along the Hexi corridor. Finally the eastern region (blue box) includes the Horqin sandy Plain in the northern farming-pastoral transition zone, which has seen rapid land degradation and rangeland modification between 1970 and 2000 (Nie and Clarke 2011). Although this eastern dust source seems secondary compare to the other sources, an increase of dust activity over the past 20 years has been observed (Zhang et al. 2003).

17.4 Time Series of AOD and DOD

Satellite data and model simulations indicate an increase of AOD over eastern China over the first decade of the twenty-first century (Chin et al. 2014). The same study shows a decrease of DOD in western China, but no change of AOD. This model result suggests a compensation of dust decrease by increase of fine mode aerosols. Figure 17.2 shows the time series of AOD and DOD over the three regions. Over the western region, DOD contributes 40% of AOD, while it drops to 20% and 10% for the central plateaus and northeastern plains, respectively. The western basins are also characterized with the highest annual mean DOD around 0.1, while values between 0.05 and 0.02 are observed over the two other regions. The time series do not show any significant trends either in DOD or AOD over the three dusty regions.

17.5 Seasonal Variation

17.5.1 Seasonal Variation of DOD

In this section, we analyze the seasonal variation of dust events and their controlling factors (i.e. surface winds, vegetation, soil moisture and snow cover). Figure 17.3 shows the seasonal mean (2003–2015) FoO and DOD. As it is well documented, dust events are more frequent, widespread, and intense during spring (March–April–May) than any other season (Prospero et al. 2002). The driving force of dust outbreaks in spring is the surface wind associated with cold frontal systems

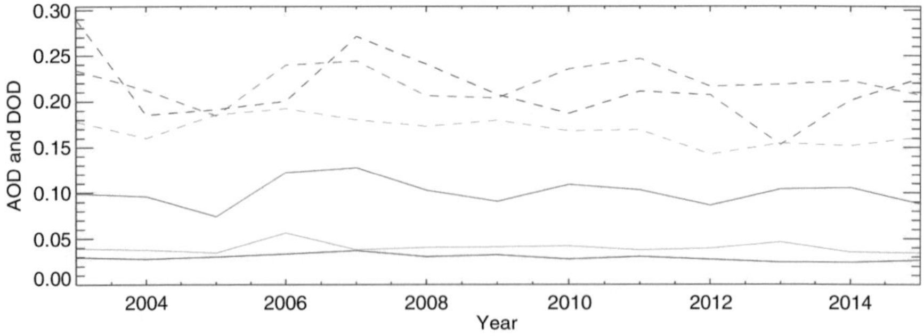

Fig. 17.2 Annual mean time series of AOD (*dash lines*) and DOD (*solid lines*) averaged over the western basins (*red lines*), central plateaus (*green lines*), and northeastern plains (*blue lines*)

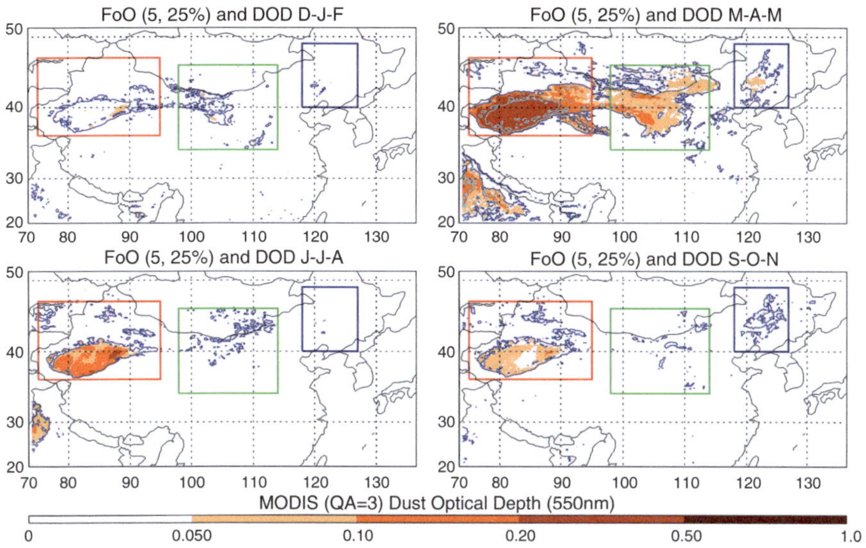

Fig. 17.3 Seasonal mean (2003–2015) distribution of MODIS Dust Optical Depth (DOD) are shaded in *red* hues, and frequencies of occurrence of DOD greater than 0.2 are shown in *dark blue* (5%) and *light blue* (25%) isocontours for winter (*D-J-F*), spring (*M-A-M*), summer (*J-J-A*) and fall (*S-O-N*). The three major dust regions are delineated with colored boxes (see Fig. 17.1 for details)

passing over dust sources. The highest seasonal mean DOD, with values greater than 1, is observed along the edges of the Taklamakan desert. In the other western basins, spring DOD is lower with values around 0.2 and 0.5. In the eastern plains, spring is also the peak of dust activity but with DOD values less than 0.2. The western basin sources are active all year long but with much more reduced DOD in winter. Except for spring, sources in the Mongolian plateau are mostly confined to Inner Mongolia in China and in winter, with very little activity in Mongolia. Also, dust events in the Guangzhong Plain are observed in winter and spring. In winter, the northern sources are weakly emitting dust, which suggest that this would be the period of local contribution. Dust activity in the Horqin sandy Plains is limited to spring and fall.

Figure 17.3 shows considerable seasonal variability and strength between our three regions, which can be associated with surface factors as discussed in the next section.

17.5.2 *Seasonal Variation of Surface Properties*

Soil particles can be mobilized once the surface wind velocity reaches a certain threshold (threshold of wind erosion), which increases rapidly with soil moisture, snow, and vegetation cover. A simple formulation of dust emission is given by the following parametrization (Ginoux et al. 2001):

$$Fp = CSu_{10m}{}^{2}(u_{10m} - ut) \tag{17.2}$$

where Fp is the vertical flux of dust (kg m^{-1} s^{-1}), $C = 1$ g s^2 m^{-1} is a dimensional constant, S is the fraction of erodible soil, u_{10m} is the horizontal wind speed at 10 m, ut is the threshold velocity of wind erosion. The fraction of erodible material will depend on vegetation cover, snow cover, and litter on the ground. The threshold of wind erosion is a function of soil texture, soil moisture, and surface roughness (Gillette and Passi 1988). The surface roughness is a measure of surface texture, which depends on spatial resolution. At 1 m resolution pebbles or stones will define the surface texture, but at 100 km resolution it will be the orography. For smooth surfaces, such as ephemeral lakes, typical values of ut are around 6.5 m s^{-1}. In West Africa, the observed minimum and maximum were 5 and 12.5 m s^{-1}, respectively (Helgren and Prospero 1987). In this section, we analyze the potential dependency of FoO with surface properties affecting dust emission, in particular the vegetation cover, snow cover, and soil moisture. The seasonal mean (2003–2015) values of the different parameters are given for each of the three source regions in Tables 17.1, 17.2, and 17.3, respectively. To focus on values relevant to dust emission, we average parameter values only where seasonal mean FoO is greater than 5%, as shown in Fig. 17.3. Consequently, the number of values used for averaging will vary between season and source regions, but will be similar (except for differences in missing satellite data) for all parameters within a season and dust region.

17.5.2.1 Snow Cover

Dust sources in East Asia are located in relatively high latitudes, which may be covered by snow for long periods of time. To study the effect of snow cover on dust emission we use the 8-day retrieved data from MODIS Collection 5 (MOD10C2) on Terra. The data are available on a $0.05 \times 0.05°$ grid (Hall et al. 2006). Figure 17.4 shows the seasonal mean (2003–2015) snow cover for March–April–May when dust events are most intense and frequent. Over the Tibetan plateau and the northern part of the domain, surface is covered entirely by snow, while only limited area are covered by snow within the 5% isocoutour of FoO. Tables 17.1, 17.2, and 17.3 show that snow cover peaks for all regions in spring with the highest value 18% on the northeastern plains. For the other seasons, snow cover is too small to affect dust emission. It is important to note that interannual variability is very high with standard deviation equivalent to the mean values. Some year snow cover in spring may be a limiting factor for dust emission, while others it is negligible. Interannual variability of snow cover in East Asia was used previously to study the effect of snow on dust outbreaks in spring (Kurosaki and Mikami 2004). In this study, they did find an effect in Mongolia but not in eastern China. However, their analysis did not include other factors such soil moisture or vegetation cover.

Table 17.1 Seasonal mean ± standard deviation of properties over the western basins (36° N-47°N, 75°E–95°E)

Property	Dec–Jan–Feb	Mar–Apr–May	Jun–Jul–Aug	Sep–Oct–Nov
DOD	0.025 ± 0.017	0.26 ± 0.07	0.08 ± 0.02	0.04 ± 0.02
AOD	0.07 ± 0.03	0.47 ± 0.08	0.23 ± 0.04	0.1 ± 0.03
Snow (%)	3 ± 1	11 ± 11	6 ± 6	1 ± 0.7
Soil moisture ($m^3\ m^{-3}$)	0.01 ± 0.01	0.03 ± 0.02	0.03 ± 0.02	0.02 ± 0.02
LAI ($m^2\ m^{-2}$)	0.03 ± 0.004	0.05 ± 0.01	0.13 ± 0.02	0.083 ± 0.013
W*max* ($m.s^{-1}$)	3.1 ± 0.14	4.6 ± 0.15	4.5 ± 0.18	4.6 ± 0.12
Freq W*max* ≥4 $m.s^{-1}$(%)	22 ± 4	51 ± 3	58 ± 4	32 ± 3
Freq W*max* ≥6 $m.s^{-1}$(%)	5 ± 2	21 ± 3	21 ± 3	10 ± 1
Freq W*max* ≥8 $m.s^{-1}$(%)	1 ± 0.7	8 ± 1.4	6 ± 1.6	3 ± 0.8
Freq W*max* ≥10 $m.s^{-1}$(%)	0.3 ± 0.2	2 ± 0.6	1 ± 0.6	0.7 ± 0.4

Table 17.2 Seasonal mean ± standard deviation of properties over the central plateau (34° N-46°N, 98°E–114°E)

Property	Dec–Jan–Feb	Mar–Apr–May	Jun–Jul–Aug	Sep–Oct–Nov
DOD	0.022 ± 0.007	0.09 ± 0.03	0.016 ± 0.006	0.013 ± 0.007
AOD	0.09 ± 0.015	0.26 ± 0.04	0.2 ± 0.035	0.12 ± 0.025
Snow (%)	3 ± 3	16 ± 13	7 ± 8	1 ± 2
Soil moisture ($m^3\ m^{-3}$)	0.01 ± 0.008	0.03 ± 0.02	0.036 ± 0.025	0.03 ± 0.02
LAI ($m^2\ m^{-2}$)	0.04 ± 0.01	0.07 ± 0.02	0.15 ± 0.026	0.09 ± 0.02
W*max* ($m.s^{-1}$)	4.4 ± 0.4	6 ± 0.35	5 ± 0.23	4.7 ± 0.26
Freq W*max* ≥4 $m.s^{-1}$(%)	40 ± 7	73 ± 5	63 ± 3	56 ± 5
Freq W*max* ≥6 $m.s^{-1}$(%)	17 ± 5	43 ± 5	26 ± 4	27 ± 5
Freq W*max* ≥8 $m.s^{-1}$(%)	7 ± 3	21 ± 4	8 ± 2	10 ± 3
Freq W*max* ≥10 $m.s^{-1}$(%)	2 ± 1	7 ± 2	2 ± 1	2 ± 1

Table 17.3 Seasonal mean ± standard deviation of properties over the northeastern plains (40°N-49°N, 118°E–127°E)

Property	Dec–Jan–Feb	Mar–Apr–May	Jun–Jul–Aug	Sep–Oct–Nov
DOD	0.01 ± 0.01	0.06 ± 0.02	0.01 ± 0.005	0.03 ± 0.01
AOD	0.09 ± 0.03	0.3 ± 0.06	0.28 ± 0.06	0.17 ± 0.04
Snow (%)	9 ± 14	18 ± 22	8 ± 9	3 ± 5
Soil moisture ($m^3\ m^{-3}$)	0.004 ± 0.004	0.04 ± 0.03	0.06 ± 0.04	0.03 ± 0.026
LAI ($m^2\ m^{-2}$)	0.12 ± 0.007	0.24 ± 0.014	1.1 ± 0.22	0.44 ± 0.05
W*max* ($m.s^{-1}$)	4.6 ± 0.3	5.7 ± 0.4	4.2 ± 0.3	4.8 ± 0.3
Freq W*max* ≥4 $m.s^{-1}$(%)	64 ± 6	74 ± 6	50 ± 7	59 ± 5
Freq W*max* ≥6 $m.s^{-1}$(%)	31 ± 4	40 ± 6	18 ± 5	23 ± 6
Freq W*max* ≥8 $m.s^{-1}$(%)	9 ± 3	16 ± 4	3 ± 2	6 ± 2
Freq W*max* ≥10 $m.s^{-1}$(%)	1 ± 0.8	3 ± 1	0.2 ± 0.3	0.5 ± 0.4

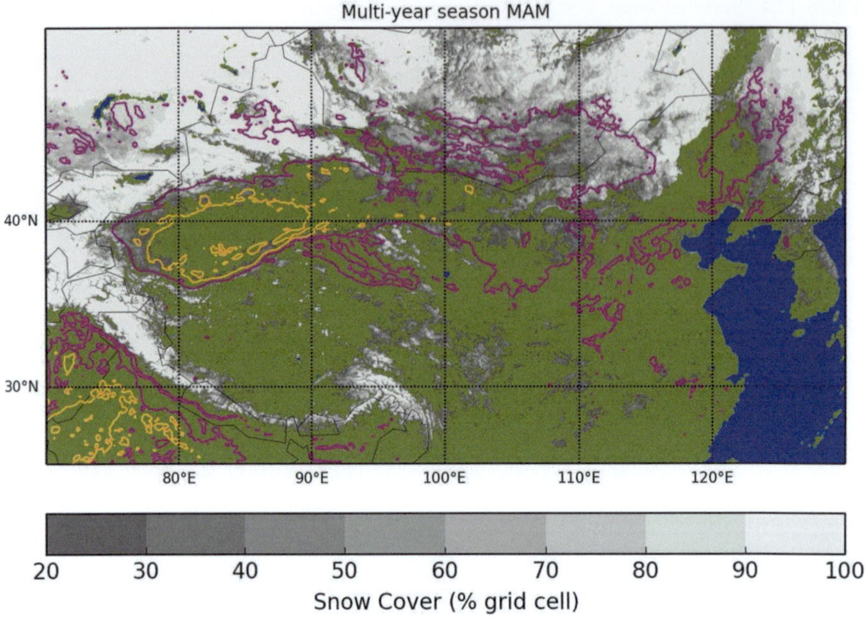

Fig. 17.4 Multi-year mean distribution of MODIS snow fraction (%) for March–April–May (*grey* shading), and 5 (*magenta* isocontour) and 25% (*orange* isocontour) FoO of DOD >0.2. Land is colored in *green*, and ocean in *blue*

17.5.2.2 Soil Moisture

For soil moisture, there are several products available from the Advanced Microwave Scanning Radiometer for the Earth Observing System (AMSR-E) measurements. The AMSR-E has a penetration of 1 cm, and retrieved values correspond to volumetric soil moisture. Inter-comparison of the different products indicates that the University of Montana soil moisture (Du et al. 2016) excels in flatter and less vegetated areas (Tuttle and Salvucci 2014). In addition, University of Montana data has been extended to 2015 from calibrated overlapping satellite passive microwave sensor observations from AMSR-E, FY-3B and AMSR2 (Du et al. 2016). For these reasons we choose soil moisture from the University of Montana. The data are provided twice (ascending and descending orbits) daily on a 0.05° EASE-grid cylindrical projection. We combine both orbits, and average overlapping values to derive a near global daily soil moisture from 2003 to 2015. Figure 17.5 shows the multi-year mean distribution of volumetric soil moisture for spring. A difficulty in relating soil moisture with *ut* is the strong dependency on soil texture (Fécan et al. 1999). This dependency delays the effect of soil moisture until it reaches a certain threshold value. The larger the clay (sand) content the larger (smaller) the threshold value. Once this threshold has been reached *ut* will double with less than a 5% increase of soil moisture. Referring to (Fécan et al. 1999), soil sensitive to gravimetric soil moisture less than 2.5% is sandy soil with very little clay content. By

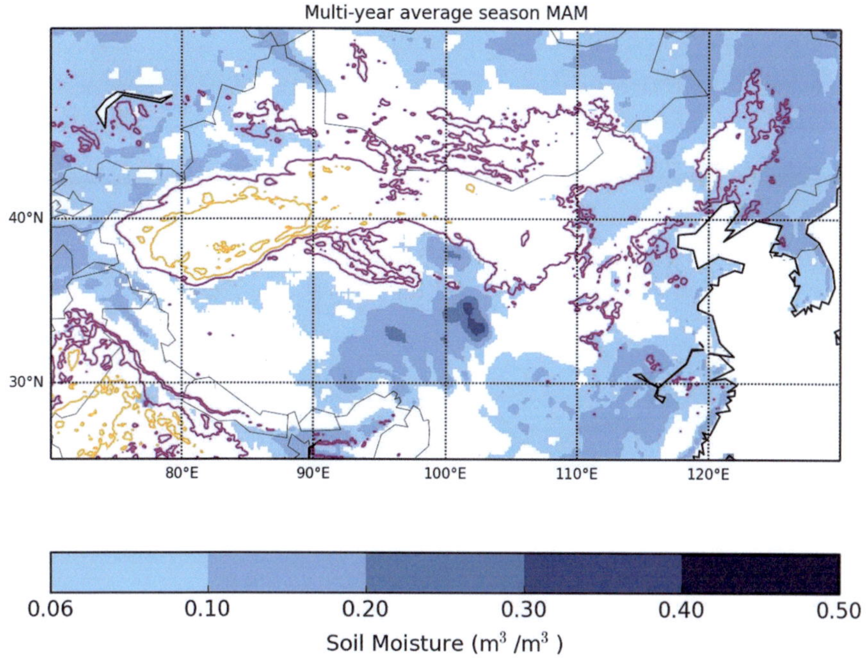

Fig. 17.5 Multi-year mean distribution of AMSR-E volumetric soil moisture (m^3 m^{-3}) for March–April–May

converting gravimetric to volumetric soil moisture by assuming a soil density of 2500 kg m^{-3}, values below 6% (first isovalue in Fig. 17.5) will have no effect on dust emission whatever the soil texture. We see in Fig. 17.5 that only the easternmost part of the 5% FoO is overlapping volumetric soil moisture greater than 6%. Seasonal mean values of soil moisture and their standard deviation, over the three source regions (Tables 17.1, 17.2, and 17.3), show that soil moisture is too low to affect dust emission in the western basins. It may occasionally be a factor in summer in the central plateaus and in spring in the eastern plains. But it will most likely affect dust emission in summer in the eastern plains.

17.5.2.3 Vegetation Cover

For vegetation cover, we use the Leaf Area Index (LAI) as a proxy. LAI is defined as the one-sided green leaf area per unit ground surface area in units of m^2 m^{-2}. Globally, values ($\pm$ standard deviation) range from 1.31 $\pm$ 0.85 for deserts to 8.72 $\pm$ 4.32 for tree plantation, with evergreen forests displaying the highest LAI. For dust emission, some modeling studies assume that dust emission shuts off for LAI >0.3 (Tegen et al. 2002; Mahowald et al. 2010). For this study we use the monthly gridded (0.05° $\times$ 0.05°) LAI retrieved from MODIS data by Boston

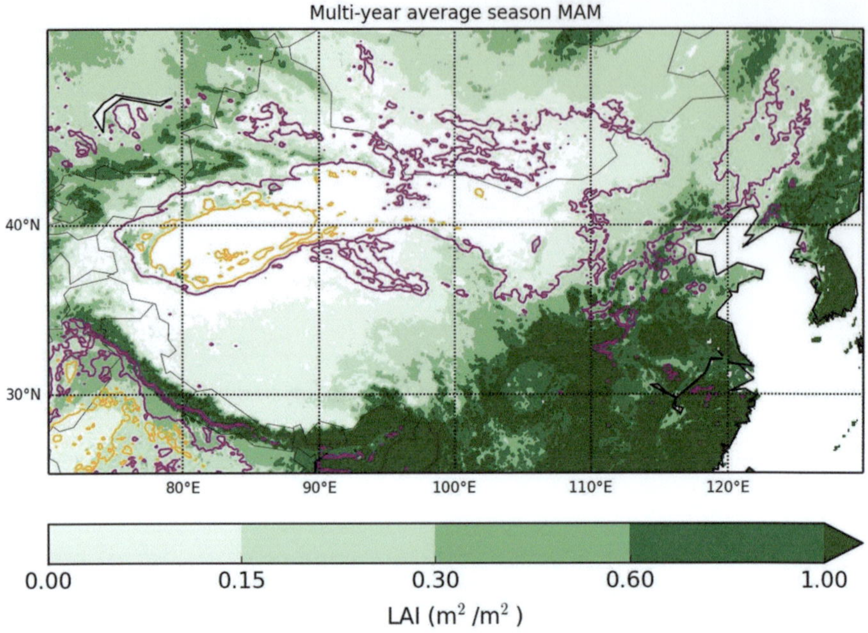

Fig. 17.6 Multi-year mean distribution of MODIS Leaf Area Index ($m^2\ m^{-2}$) for March–April–May

University (Yang et al. 2006; Bory et al. 2003). Figure 17.6 shows the multi-year mean distribution of MODIS LAI for spring, with the 5 and 25% FoO of DOD >0.2. Except for the eastern FoO and along the edges of the Tarim Basin, LAI is lower than 0.3, and will have no overall effect on dust emission. This seems to be true for other seasons, as shown in Tables 17.1, 17.2, and 17.3. The standard deviation is relatively small compared to the seasonal mean values, which indicates little interannual changes. The highest LAI value is observed during the wet season (June–July–August) in the northeastern plains, and should considerably limit dust emission.

17.5.2.4 Surface Wind

Equation 17.2 shows the strong dependency of dust emission on surface winds. Good correspondence was found between surface wind and dust outbreaks in year-to-year variations and in spatial distributions at meteorological stations over East Asia (Kurosaki and Mikami 2003). Here, we use instantaneous 10-m wind speed from ERA-Interim reanalysis (Dee et al. 2011). The data are provided on a regular 0.75° grid every 6 h.

Due to the cubic dependency of dust emission on surface wind, we use the daily maximum (*wmax*) rather than the daily mean. Tables 17.1, 17.2, and 17.3 provide the seasonal mean (2003–2015) w_{max}, and the frequencies of occurrence that *wmax* is greater than 4, 6, 8 and 10 m s^{-1}. Distinct differences appear between the three source regions but the standard deviation is quite small for all seasons and all regions. In other words, for the 2003–2015 period there seems to be little inter-annual variation of surface winds. This means that yearly variations of DOD for this period are controlled by surface conditions rather than surface winds. One common aspect between all regions is that mean *wmax* and FoO peak in spring, which explains the spring maximum in dust emission. The most frequent strong winds are observed over the Central plateaus, followed closely over northeastern plains. There is a 7% FoO to have surface winds greater than 10 m s^{-1} in the central plateaus but only 2% in the western basins. These results clearly show the key role played by the surface conditions in differentiating the frequency and intensity of dust events between dust sources. Otherwise, most dust events would originate from central and northeastern China and not from the western basin. This may eventually change depending on the effects of climate change on the hydrological cycle and surface winds. During the Last Glacial Maximum (LGM), isotopic analysis of dust in Greenland ice cores indicates an origin of dust from the Tarim Basin and not from Mongolia (Bory et al. 2003). Thus, in case of glacial period with reduced precipitation and strengthening of surface winds, dust activities in the western sources of China were further amplified.

17.6 Dust and Land Use

In the previous section, we showed that vegetation cover in the arid or semi-arid regions of dust sources is too small to affect dust emission, except along the edges of the Tarim Basin and in northeastern China. These regions are also affected by land use. There are also small scale sources related to land use. An example is along the Fen river valley down to historical Xi'An city. This valley was famous for its rich agriculture, but is now more oriented towards industry. Back-trajectory analysis in Xi'An indicated possible contributions from local agriculture (Wang et al. 2006). In addition, collocation of plumes of dust and NH3 (a byproduct of fertilizer used for agriculture) was observed from satellite data over that region (Ginoux et al. 2012b). It was found that 25% of dust over cultivated areas is mixed with NH3 in East Asia. With modernization of agriculture, harvest yields have increased by as much as 240% for wheat over the last four decades (Piao et al. 2010). In order to estimate percentage changes of landcover type for the last decade (2003–2012), we calculate the surface area of cropland and pasture over China retrieved from AVHRR satellite data by the University of Maryland (Hansen and Reed 2000). For croplands, we found only 2% change between 2003 and 2012 with most of it in eastern China. If landuse change may have increase dust sources in the past, it is not the case for the last decade. On the other hand, soil disturbed by agriculture may be more susceptible to wind erosion, as illustrated by the infamous Dust Bowl of the mid 1930s in the Midwest of the United States.

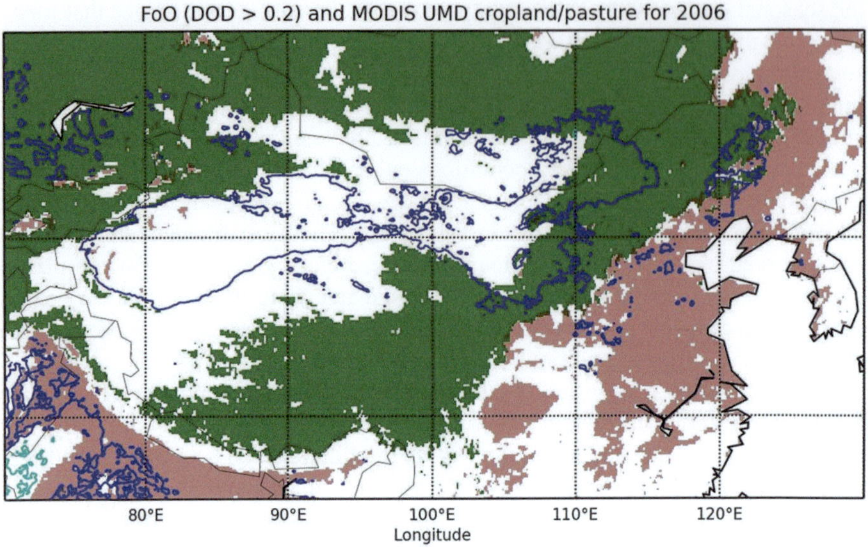

Fig. 17.7 Distribution of 5% FoO of DOD >0.2 in spring 2006 (*blue contours*) and annual distribution of pasture (*green shading*) and cropland (*tan shading*) for 2006

In order to estimate the contribution of dust from agriculture, we calculate the surface of 5% FoO of DOD >0.2 where there is landuse. Figure 17.7 shows the distribution of FoO in relation to grassland and cultivation for 2006, which was the year of the largest extent of FoO. If we compare the lowest spatial coverage of FoO in 2003 (82,104 km^2) with its largest in 2006 (179,104 km^2) we found that its covers a land use fraction of 8% and 30%, respectively. In other words, the maximum contribution of dust from agriculture is 30%. However, this depends also on the land use dataset. Using another land use dataset by Klein Goldewijk (2001), the fractions over landuse are 22% and 45% for 2003 and 2006, respectively.

Our results indicate that land use changes over the last decade did not modulate dust variability in China, but 8–45% of dust events originate from cropland and pasture depending on meteorological conditions, with a 50% uncertainty associated with the choice of land use dataset.

17.7 Threshold of Wind Erosion

In this section, we estimate the threshold of wind erosion (*ut*) used in Eq. 17.2 for every 0.1° grid cell where the annual mean (2003–2015) FoO of DOD >0.2 is at least 5%. The method consists of calculating the cumulative distribution of surface wind reaching a certain velocity with a step of 0.5 m s^{-1}. Knowing the FoO of DOD >0.2, *ut* can be retrieved from the cumulative distribution. For this calculation, we use the 10-meter wind speed from the ECMWF ERA-interim re-analysis. The data

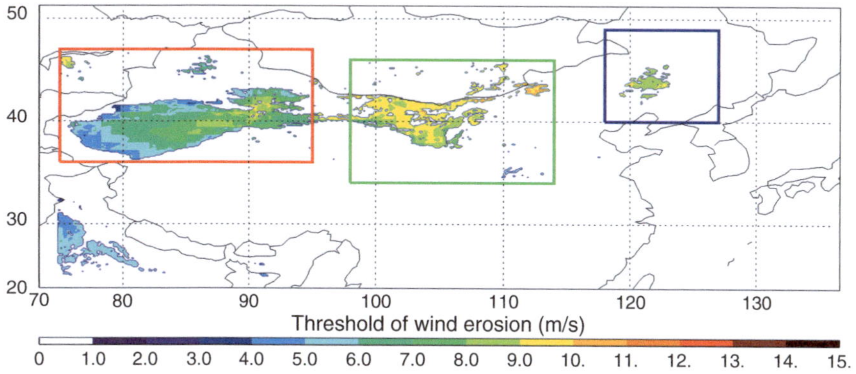

Fig. 17.8 Annual mean distribution of the threshold of wind erosion (m s^{-1}). The *three boxes* indicated the domain of the three major dust regions

are provided every 6 h on a 0.75° × 0.75° grid. As shown in Eq. 17.2 dust emission depends at the cubic power of surface winds. It is thus more realistic to use the maximum value rather than the daily mean. Thus, for each day we use the maximum six-hourly surface winds.

Figure 17.8 shows the spatial distribution of the threshold of wind erosion. This may be the most detailed satellite based distribution of *ut*. The spatial mean thresholds of wind erosion are 6.5, 9.1, and 8.6 m s^{-1} for the western, central, and northeastern sources, respectively. These values compare remarkably well with the values derived from meteorological stations in East Asia (Kurosaki and Mikami 2007). In that study, *ut* values are 6.7 and 9.4 m s^{-1} for the Tarim Basin and Loess Plateau, respectively. For the north- eastern plains they found higher value around 11 m s^{-1}. However, they calculate *ut* based on spring data only, while there are also dust events in fall (cf. Fig. 17.3). The mean value over for the three regions is 7.1 m s^{-1}. Within each region, there is considerable variability. In the Tarim Basin, *ut* varies between 5 and 9 m s^{-1}, with the highest value in the eastern part of the Basin. In Inner Mongolia the values reach 11 m s^{-1} along the Mongolian border. Within the Fen Valley, *ut* is as low as 5 m s^{-1}. On the other hand, the spatial resolution of the re-analysis may be too coarse to capture mesoscale flow within the valley.

17.8 Conclusions

Using 13-year (2003–2015) satellite based data, we have investigated the seasonal and annual variability of dust events in East Asia and its relation with surface conditions. As shown in previous studies, we found that dust sources are located in topographic depressions in the western basins, the Mongolian and Loess plateaus in Central China, and in the northeastern plains. The sources located in northeastern plains and along the southern margin of the Loess Plateau are associated with land use. Dust emit from such sources are associated with human activities, and is considered as anthropogenic. Sometimes attribution is ambiguous such as dust found in the lower part of the Wei and Fen river valleys. It is unclear if this dust is from local agricultural sources or has been transported from the northwest sources. We found that each of the three source regions have distinct seasonal variation. Dust activity from the largest sources peak in spring, which makes it the dusty season in China. For the secondary sources in the East, dust activity continues all season, except summer during the wet season when vegetation grows. On an annual basis, the spatial extent of dust activity varies by a factor of 2, and is related to changes in winds rather than soil properties.

Our results indicate that land use changes over the last decade did not modulate dust variability in China, but 8–45% of dust events originate from cropland and pasture depending on meteorological conditions, with a 50% uncertainty associated with the choice of land use dataset. Dust frequency is correlated with the frequency of surface winds above some threshold values, which vary between source regions. The mean values for the western, central, and northeastern source regions are, respectively, 6.5, 9.1 and 8.6 m s^{-1}, in excellent agreement with previous study. Within each source there is considerable variability with values as low as 5 m s^{-1} in the Tarim Basin and as high as 12 m s^{-1} in Inner Mongolia. The sources in the eastern plains are more sensitive to erodibility factors (snow cover, soil moisture, and vegetation) than the western basins. On the other hand, strong winds are more frequent in the central and northeastern regions. This implies that these sources are more sensitive to the effects of climate changes on surface properties.

This work shows the power of using satellite retrievals to derive high resolution properties over continents. However, they are subject to uncertainties which are difficult to quantify without careful comparisons with in-situ data. To alleviate in part these drawbacks, we have compared our results with previous studies based on in-situ data. Overall, we did not find major discrepancies which suggest that our methodology could be extended to dust sources on other continents.

Acknowledgements We are grateful to Prof. Ranga Myneni and Dr. Taejin Park (Boston University) for providing the Leaf Area Index product that they retrieved from MODIS, as well as Prof. John S. Kimball and Dr. Jinyang Du (University of Montana) for providing the extended soil moisture dataset. The MODIS Deep Blue aerosol products were acquired from the Level-1 and Atmosphere Archive and Distribution System (LAADS) Distributed Active Archive Center (DAAC), located in the Goddard Space Flight Center in Greenbelt, Maryland (https://ladsweb.nascom.nasa.gov/). The MODIS 8-day snow product MYD10C2 was obtained from the NASA National Snow and Ice Data Center Distributed Active Archive Center (https://nsidc.org/data/myd10c2). This research is partially funded by NASA under grants NNG14HH42I-MAP, NNH14ZDA001N-ACMAP, and by NOAA/NWS Next Generation Global Prediction System (NGGPS) project.

References

Baddock, M. C., Ginoux, P., Bullard, J. E., & Gill, T. E. (2016). Do MODIS-defined dust sources have a geomorphological signature? *Geophysical Research Letters, 43*, 2606–2613. doi:10.1002/2015GL067327.

Bory, A. J.-M., Biscaye, P. E., Piotrowski, A. M., & Steffensen, J. P. (2003). Regional variability of ice core dust composition and provenance in Greenland. *Geochemistry, Geophysics, Geosystems, 4*, 1107. doi:10.1029/2003GC000627.

Chin, M., Diehl, T., Tan, Q., Prospero, J. M., Kahn, R. A., Remer, L. A., Yu, H., Sayer, A. M., Bian, H., Geogdzhayev, I. V., Holben, B. N., Howell, S. G., Huebert, B. J., Hsu, N. C., Kim, D., Kucsera, T. L., Levy, R. C., Mishchenko, M. I., Pan, X., Quinn, P. K., Schuster, G. L., Streets, D. G., Strode, S. A., Torres, O., & Zhao, X.-P. (2014). Multi-decadal aerosol variations from 1980 to 2009: A perspective from observations and a global model. *Atmospheric Chemistry and Physics, 14*, 3657–3690. doi:10.5194/acp-14-3657-2014.

Colarco, P., Toon, O. B., Torres, O., & Rasch, P. J. (2002). Determining the UV imaginary index of refraction of Saharan dust particles from total ozone mapping spectrometer data using a three-dimensional model of dust transport. *Journal of Geophysical Research, 107*(D16), 4289. doi:10.1029/2001JD000903.

Dee, D. P., Uppala, S. M., Simmons, A. J., Berrisford, P., Poli, P., Kobayashi, S., Andrae, U., Balmaseda, M. A., Balsamo, G., Bauer, P., Bechtold, P., Beljaars, A. C. M., van de Berg, L., Bidlot, J., Bormann, N., Delsol, C., Dragani, R., Fuentes, M., Geer, A. J., Haimberger, L., Healy, S. B., Hersbach, H., Hlm, E. V., Isaksen, L., Kllberg, P., Khler, M., Matricardi, M., McNally, A. P., Monge-Sanz, B. M., Morcrette, J.-J., Park, B.-K., Peubey, C., de Rosnay, P., Tavolato, C., Thpaut, J.-N., & Vitart, F. (2011). The ERA-interim reanalysis: Configuration and performance of the data assimilation system. *Quarterly Journal of the Royal Meteorological Society, 137*, 553–597. doi:10.1002/qj.828.

DeSouza-Machado, S. G., et al. (2010). Infrared retrievals of dust using AIRS: Comparisons of optical depths and heights derived for a North African dust storm to other collocated EOS A-Train and surface observations. *Journal of Geophysical Research, 115*, D15201. doi:10.1029/2009JD012842.

Du, J., Kimball, J. S., & Jones, L. A. (2016). Passive microwave remote sensing of soil moisture based on dynamic vegetation scattering properties for AMSR-E. *IEEE Transactions on Geoscience and Remote Sensing, 54*(1), 597–608.

Fécan, F., Marticorena, B., & Bergametti, G. (1999). Parameterization of the increase of the aeolian erosion threshold wind friction due to soil moisture for semi-arid areas. *Annales de Geophysique, 17*, 149–157.

Gillette, D. A., & Passi, R. (1988). Modeling dust emission caused by wind erosion. *Journal of Geophysical Research, 93*(D11), 14233–14242. doi:10.1029/JD093iD11p14233.

Ginoux, P., Chin, M., Tegen, I., Prospero, J. M., Holben, B., Dubovik, O., & Lin, S.-J. (2001). Sources and distributions of dust aerosols simulated with the GOCART model. *Journal of Geophysical Research, 106*, 20255–20274.

Ginoux, P., Prospero, J. M., Torres, O., & Chin, M. (2004). Long-term simulation of global dust distribution with the GOCART model: Correlation with the North Atlantic Oscillation. *Environmental Modelling & Software, 19*, 113–128.

Ginoux, P., Clarisse, L., Clerbaux, C., Coheur, P.-F., Dubovik, O., Hsu, N. C., & Van Damme, M. (2012a). Mixing of dust and NH3 observed globally over anthropogenic dust sources. *Atmospheric Chemistry and Physics, 12*, 7351–7363. doi:10.5194/acp-12-7351-2012.

Ginoux, P., Prospero, J. M., Gill, T. E., Hsu, N. C., & Zhao, M. (2012b). Global-scale attribution of anthropogenic and natural dust sources and their emission rates based on MODIS Deep Blue aerosol products. *Reviews of Geophysics, 50*, RG3005. doi:10.1029/2012RG000388.

Goudie, A. S., & Middleton, N. J. (2001). Saharan dust storms: Nature and consequences. *Earth-Science Reviews, 56*, 179–204.

Grousset, F., Ginoux, P., Bory, A., & Biscaye, P. (2003). Case study of a Chinese dust plume reaching the French Alps. *Geophysical Research Letters, 30*, 1277. doi:10.1029/2002GL016833.

Hall, D. K., Salomonson, V. V., & Riggs, G. A. (2006). *MODIS/aqua snow cover 8- day L3 global 0.05Deg CMG, version 5*. Boulder: NASA National Snow and Ice Data Center Distributed Active Archive Center. doi:http://dx.doi.org/10.5067/Y39HQA66VXAZ.

Hansen, M. C., & Reed, B. (2000). A comparison of the IGBP DISCover and University of Maryland 1 km global land cover products. *International Journal of Remote Sensing, 21*(6–7), 1365–1373.

Helgren, D. M., & Prospero, J. M. (1987). Wind velocities associated with dust deflation events in the Western Sahara. *Journal of Climate and Applied Meteorology, 26*, 1147–1151.

Herman, J., Bhartia, P. K., Torres, O., Hsu, N. C., Seftor, C., & Celarier, E. (1997). Global distribution of UV-absorbing aerosols from Nimbus 7/TOMS data. *Journal of Geophysical Research, 102*(D14), 16911–16922.

Hsu, N. C., Tsay, S.-C., King, M., & Herman, J. R. (2004). Aerosol properties over bright-reflecting source regions. *IEEE Transactions on Geoscience and Remote Sensing, 42*, 557–569.

Hsu, N. C., Tsay, S.-C., King, M., & Herman, J. R. (2006). Deep blue retrievals of Asian aerosol properties during ACE-Asia. *IEEE Transactions on Geoscience and Remote Sensing, 44*, 3180–3195.

Hsu, N. C., Jeong, M.-J., Bettenhausen, C., Sayer, A. M., Hansell, R., Seftor, C. S., Huang, J., & Tsay, S.-C. (2013). Enhanced deep blue aerosol retrieval algorithm: The second generation. *Journal of Geophysical Research – Atmospheres, 118*, 9296–9315. doi:10.1002/jgrd.50712.

Huneeus, et al. (2011). Global dust model intercomparison in AeroCom phase I. *Atmospheric Chemistry and Physics, 11*, 7781–7816. doi:10.5194/acp-11-7781-2011.

Klein Goldewijk, K. (2001). Estimating global land use change over the past 300 years: The HYDE database. *Global Biogeochemical Cycles, 15*, 417–433.

Kurosaki, Y., & Mikami, M. (2003). Recent frequent dust events and their relation to surface wind in East Asia. *Geophysical Research Letters, 30*, 1736. doi:10.1029/2003GL017261.

Kurosaki, Y., & Mikami, M. (2004). Effect of snow cover on threshold wind velocity of dust outbreak. *Geophysical Research Letters, 31*, L03106. doi:10.1029/2003GL018632.

Kurosaki, Y., & Mikami, M. (2007). Threshold wind speed for dust emission in east Asia and its seasonal variations. *Journal of Geophysical Research, 112*, D17202. doi:10.1029/2006JD007988.

Lee, H., Kalashnikova, O. V., Suzuki, K., Braverman, A., Garay, M. J., & Kahn, R. A. (2016). Climatology of the aerosol optical depth by components from the Multi-angle Imaging SpectroRadiometer (MISR) and chemistry transport models. *Atmospheric Chemistry and Physics, 16*, 6627–6640. doi:10.5194/acp-16-6627-2016.

Legrand, M., Plana-Fattori, A., & N'doume, C. (2001). Satellite detection of dust using the IR imagery of Meteosat, 1, infrared difference dust index. *Journal of Geophysical Research, 106* (D16), 18251–18274.

Liu, Z., Liu, D., Huang, J., Vaughan, M., Uno, I., Sugimoto, N., Kittaka, C., Trepte, C., Wang, Z., Hostetler, C., & Winker, D. (2008). Airborne dust distributions over the Tibetan Plateau and surrounding areas derived from the first year of CALIPSO lidar observations. *Atmospheric Chemistry and Physics, 8*, 5045–5060. doi:10.5194/acp-8-5045-2008.

Mahowald, N. M., Kloster, S., Engelstaedter, S., Moore, J. K., Mukhopadhyay, S., Mc-Connell, J. R., Albani, S., Doney, S. C., Bhattacharya, A., Curran, M. A. J., Flanner, M. G., Hoffman, F. M., Lawrence, D. M., Lindsay, K., Mayewski, P. A., Neff, J., Rothenberg, D., Thomas, E., Thornton, P. E., & Zender, C. S. (2010). Observed 20th century desert dust variability: Impact on climate and biogeochemistry. *Atmospheric Chemistry and Physics, 10*, 10875–10893. doi:10.5194/acp-10-10875-2010.

Nie, Q., & Clarke, K. C. (2011). Desertification in China's Horquin area: A multi-temporal land use change analysis. *Journal of Land Use Science, 6*(1), 53–73.

Peyridieu, S., Che'din, A., Capelle, V., Tsamalis, C., Pierangelo, C., Armante, R., Crevoisier, C., Crépeau, L., Sime'on, M., Ducos, F., & Scott, N. A. (2013). Characterisation of dust aerosols in the infrared from IASI and comparison with PARASOL, MODIS, MISR, CALIOP, and AERONET observations. *Atmospheric Chemistry and Physics, 13*, 6065–6082. doi:10.5194/acp-13-6065-2013.

Piao, S., Ciais, P., Huang, Y., Shen, Z., Peng, S., Li, J., Zhou, L., Liu, H., Ma, Y., Ding, Y., & Friedlingstein, P. (2010). The impacts of climate change on water resources and agriculture in China. *Nature, 467*(7311), 43–51.

Prospero, J. M., Ginoux, P., Torres, O., Nicholson, S. E., & Gill, T. E. (2002). Environmental characterization of global sources of atmospheric soil dust identified with the Nimbus 7 Total Ozone Mapping Spectrometer (TOMS) absorbing aerosol product. *Reviews of Geophysics, 40* (1), 1002. doi:10.1029/2000RG000095.

Sayer, A. M., Hsu, N. C., Bettenhausen, C., & Jeong, M. J. (2013). Validation and uncertainty estimates for MODIS collection 6 "Deep Blue" aerosol data. *Journal of Geophysical Research, 118*(14), 7864–7872.

Schepanski, K., Tegen, I., Laurent, B., Heinold, B., & Macke, A. (2007). A new Saharan dust source activation frequency map derived from MSG-SEVIRI IR-channels. *Geophysical Research Letters, 34*, L18803. doi:10.1029/2007GL030168.

Schuster, G. L., Dubovik, O., & Holben, B. N. (2006). Angstrom exponent and bimodal aerosol size distributions. *Journal of Geophysical Research, 111*, Do727. doi:10.1029/2005JD006328.

Shen, Z., Cao, J., Liu, S., Zhu, C., Wang, X., Zhang, T., Xu, H., & Hu, T. (2011). Chemical composition of PM10 and PM2. 5 collected at ground level and 100 meters during a strong wintertime pollution episode in Xi'an, China. *Journal of the Air & Waste Management Association, 61*(11), 1150–1159.

Tegen, I., Harrison, S. P., Kohfeld, K., Prentice, I. C., Coe, M., & Heimann, M. (2002). Impact of vegetation and preferential source areas on global dust aerosol: Results from a model study. *Journal of Geophysical Research, 107*(D21), 4576. doi:10.1029/2001JD000963.

Tegen, I., Werner, M., Harrison, S. P., & Kohfeld, K. E. (2004). Relative importance of climate and land use in determining present and future global soil dust emission. *Geophysical Research Letters, 31*, L05105. doi:10.1029/2003GL019216.

Torres, O., Bhartia, P. K., Herman, J. R., Syniuk, A., Ginoux, P., & Holben, B. (2002). A long term record of aerosol optical depth from TOMS observations and comparison to AERONET measurements. *Journal of the Atmospheric Sciences, 59*, 398–413.

Torres, O., Ahn, C., & Chen, Z. (2013). Improvements to the OMI near-UV aerosol algorithm using A-train CALIOP and AIRS observations. *Atmospheric Measurement Techniques, 6*, 3257–3270. doi:10.5194/amt-6-3257-2013.

Tuttle, S. E., & Salvucci, G. D. (2014). A new approach for validating satellite estimates of soil moisture using large-scale precipitation: Comparing AMSR-E products. *Remote Sensing of Environment, 142*, 207–222.

Wang, Y. Q., Zhang, X. Y., & Arimoto, R. (2006). The contribution from distant dust sources to the atmospheric particulate matter loadings at XiAn, China during spring. *Science of the Total Environment, 368*(2), 875–883.

Winker, D. M., Vaughan, M. A., Omar, A., Hu, Y., Powell, K. A., Liu, Z., Hunt, W. H., & Young, S. A. (2009). Overview of the CALIPSO mission and CALIOP data processing algorithms. *Journal of Atmospheric and Oceanic Technology, 26*(11), 2310–2323.

Yang, W., Shabanov, N. V., Huang, D., Wang, W., Dickinson, R. E., Nemani, R. R., Knyazikhin, Y., & Myneni, R. B. (2006). Analysis of leaf area index products from combination of MODIS Terra and Aqua data. *Remote Sensing of Environment, 104*(3), 297–312.

Yu, H., Remer, L. A., Chin, M., Bian, H., Tan, Q., Yuan, T., & Zhang, Y. (2012). Aerosols from overseas rival domestic emissions over North America. *Science, 337*, 566–569.

Zhang, X. Y., Gong, S. L., Zhao, T. L., Arimoto, R., Wang, Y. Q., & Zhou, Z. J. (2003). Sources of Asian dust and role of climate change versus desertification in Asian dust emission. *Geophysical Research Letters, 30*, 2272. doi:10.1029/2003GL018206.

Part V
Modeling

Chapter 18
Predicting Air Pollution in East Asia

Idir Bouarar, Katinka Petersen, Claire Granier, Ying Xie, Bas Mijling, A. van der Ronald, Michael Gauss, Matthieu Pommier, Mikhail Sofiev, Rostislav Kouznetsov, Natalia Sudarchikova, Lili Wang, Guangqiang Zhou, and Guy P. Brasseur

Abstract Air quality forecast systems are being developed and used in many regions of the world to provide alerts on unhealthy situations. We present in this chapter an ensemble modeling system for air quality prediction in East Asia. The system includes several chemistry transport models that provide prediction of levels of ozone (O_3), particulate matter ($PM_{2.5}$) and other pollutants in East Asia for 3–5 days ahead. We discuss here the performance of this multi-model prediction system and demonstrate that the ensemble approach leads to more accurate forecasts with improved statistical skills compared to predictions that are based on individual models. We also discuss the different uncertainties affecting the performance of air quality prediction systems and the different approaches that can be used to reduce such uncertainties.

I. Bouarar (✉) • K. Petersen • N. Sudarchikova
Max Planck Institute for Meteorology, Bundesstr. 53, Hamburg 20146, Germany
e-mail: idir.bouarar@mpimet.mpg.de; katinka.petersen@mpimet.mpg.de; natalia.sudarchikova@mpimet.mpg.de

C. Granier
Laboratoire d'Aérologie, Toulouse, France

NOAA Earth System Research Laboratory, Boulder, CO, USA

LATMOS/IPSL, UPMC, University Paris 06, Sorbonne Universities, Paris, France

Cooperative Institute for Research in Environmental Sciences, University of Colorado, Boulder, CO, USA
e-mail: claire.granier@latmos.ipsl.fr

Y. Xie • G. Zhou
Yangtze River Delta Center for Environmental Meteorology Prediction and Warning, Shanghai 200030, People's Republic of China

Shanghai Key Laboratory of Meteorology and Health, Shanghai Meteorological Service, Shanghai 200030, People's Republic of China
e-mail: yxie33@outlook.com; zhougq21@163.com

I. Bouarar et al. (eds.), *Air Pollution in Eastern Asia: An Integrated Perspective*, ISSI Scientific Report Series 16, DOI 10.1007/978-3-319-59489-7_18

Keywords Air quality • East Asia • Multi-model forecast • Ozone • Fine particulate matter

18.1 Introduction

Using state-of-the-art models to provide real-time forecasting of ozone, particulate matter and other chemical and aerosol species relevant for air quality allows for advance notice to alert the population on unhealthy air associated with elevated levels of pollution and to reduce its adverse effects. In addition to health alerts, a system for forecasting air quality can also be useful for local environmental regulators in charge of emission control. In addition to long-term mitigation programs, short-term actions can be taken to avoid acute air pollution episodes, including compulsory traffic reduction or reduction in industrial operations. Reliable forecasts are therefore required to develop efficient measures at the appropriate time and for an appropriate period.

In this chapter, we present an ensemble model prediction system implemented to forecast air quality in East Asia. This system has been conceived and developed in the framework of two EU-funded projects: PANDA and MarcoPolo (http://www.marcopolo-panda.eu). The ensemble model system includes global and regional chemistry-transport models and allows daily prediction of the distribution of main air pollutants in East Asia. We discuss also the uncertainties in current air quality forecasting systems and the different methods that can be used to improve the accuracy of the forecasts, including the use of data assimilation methods and of other post-processing statistical techniques.

B. Mijling • A. van der Ronald
The Royal Netherlands Meteorological Institute (KNMI), Utrechtsweg 297, 3731 GA De Bilt, The Netherlands
e-mail: mijling@knmi.nl; avander@knmi.nl

M. Gauss • M. Pommier
The Norwegian Meteorological Institute, Postboks 43, Blindern, 0313 Oslo, Norway
e-mail: michael.gauss@met.no; matthieup@met.no

M. Sofiev • R. Kouznetsov
Finnish Meteorological Institute, P.O. Box 503, FI-00101 Helsinki, Finland
e-mail: Mikhail.Sofiev@fmi.fi; rostislav.kouznetsov@fmi.fi

L. Wang
State Key Laboratory of Atmospheric Boundary Layer Physics and Atmospheric Chemistry (LAPC), Institute of Atmospheric Physics, Chinese Academy of Sciences, Beijing 100029, China
e-mail: wll@mail.iap.ac.cn

G.P. Brasseur
National Center for Atmospheric Research, Boulder, CO, USA

Max Planck Institute for Meteorology, Bundesstr. 53, Hamburg 20146, Germany
e-mail: guy.brasseur@mpimet.mpg.de

18.2 Air Quality Prediction Systems

Although real-time air quality forecasting is a relatively new discipline in atmospheric science, it has evolved rapidly in recent years with the use of sophisticated chemical transport models. The first forecasting systems were based on statistical models (Cobourn 2007; Vlachogianni et al. 2011) and empirical methods (e.g. Wolff and Lioy 1978). Among the statistical techniques, the regression analysis approach uses historical data to identify linear relationships between specific meteorological parameters and ambient air pollution concentrations. The resultant linear regression equation is used to predict the expected levels of air pollution. Another approach, also based on the analysis of historical data, is provided by artificial neural networks (e.g. Perez and Reyes 2006) which, by pattern recognition, identify weather patterns that are associated with elevated levels of pollutants. These statistical and empirical approaches have the advantage of being reasonably simple to develop and to implement. They require only modest computing resources. Their weakness is their inability to fully capture the complexity and the non-linearity of pollution-weather relationships.

Since the 1990s, air quality forecasting systems based on 3-D chemical transport models (CTMs) have benefited from the rapid development of models and of computing technology. As a result, daily air quality forecasts can be produced at the global, regional and local scales (Schaap et al. 2008; Baklanov et al. 2008; Flemming et al. 2009). CTMs have the advantage of representing explicitly physical and chemical processes that lead to the formation, transformation and distribution of air pollutants. They are currently used in many countries as operative tools to provide air quality forecasts up to the city scales.

18.3 Multi-model Air Quality Prediction System for East Asia

Current air quality prediction systems in East Asia and particularly in China are based on individual forecast models. The performance of such individual systems depends on the adopted model inputs (e.g. emissions, meteorological data, boundary conditions) and the parameterizations used (e.g. boundary layer mixing, chemistry and aerosol schemes) as will be discussed in Sect. 18.4. The comparison of the simulations from such individual systems shows substantial differences that are not always easily understood. The combination of several prediction systems to form a single multi-model ensemble is a pragmatic approach that is believed to provide more accurate forecasts and reduces the underlying uncertainties. The ensemble approach is already widely used in weather forecasting systems. It has been adopted in the CAMS air quality forecasting system for Europe (Marécal et al. 2015; Sofiev et al. 2015). Experience shows that this methodology leads to an increase in the prediction skill, especially when combined with data-assimilation methods and bias correction techniques.

Within the EU-funded projects PANDA and MarcoPolo, an ensemble forecasting system for air quality in East Asia has been developed using seven individual models (see Table 18.1). The WRF-Chem model is used by two institutes (MPI-M and SCUEM) but with different chemistry and aerosol schemes and different physics options and model inputs. The seven individual systems have different horizontal and vertical resolutions, with horizontal resolutions varying between 06 and 50 km. The range of the individual forecasts is 72 h starting at 00:00 UTC. The individual systems are not constrained to use the same emissions, meteorological data or initial and boundary conditions. The forecast species include $PM_{2.5}$, PM_{10}, NO_2 and O_3 for which observations are available in near-real time. In addition to the individual forecasts, the mean and median values for the seven models are calculated for 34 Chinese cities of more than three million inhabitants.

Figure 18.1 shows a comparison between observed $PM_{2.5}$ concentrations in Beijing for particular time periods and the forecast from different models. Among the five models used in the comparison, only two models (CIFS and MPI) captured the increase of $PM_{2.5}$ in Beijing around noon on 13 April 2016. The two models predicted accurately the pollution increase with, however, an overestimation of the peak value. In contrast, other models (e.g. CHIM, SCUEM) showed a rapid decrease in $PM_{2.5}$. In another example corresponding to May, 06 2016, $PM_{2.5}$ concentrations increase from around 50 to 300 μg/m^3 within a short period of time. Only CIFS was able to capture that large increase in the abundance of $PM_{2.5}$, but the model overestimates the observed concentrations, especially during the morning hours. The other models showed a decrease in the concentrations to less than 50 μg/m^3.

The analysis of several similar cases highlights the fact that individual models do not necessarily provide accurate estimates of air quality situations. The many uncertainties and sources of forecast errors, which explain such discrepancies, are

Table 18.1 Chemical transport models used for operational air quality forecasts in East Asia

Model	Institute	Acronym	Reference
CHIMERE	The Royal Netherlands Meteorological Institute (KNMI)	CHIM	Menut et al. (2013)
EMEP/ MSC-W	MET Norway	EMEP	Simpson et al. (2012)
WRF-Chem	Max Planck Institute for Meteorology (MPI-M)	MPI	Grell et al. (2005)
SILAM	Finnish Meteorological Institute (FMI)	SILAM	Sofiev et al. (2008)
CIFS	European Centre for Medium-range Weather Forecasts (ECMWF)	CIFS	Flemming et al. (2015)
WRF-Chem	Shanghai Center for Urban Environmental Meteorology (SCUEM)[a]	SCUEM	Grell et al. (2005)
LOTOS-EUROS	The Netherlands Organisation for applied scientific research (TNO)	LOTOS	Schaap et al. (2008)

[a]Now as Yangtze River Delta center for environmental meteorology prediction and warning

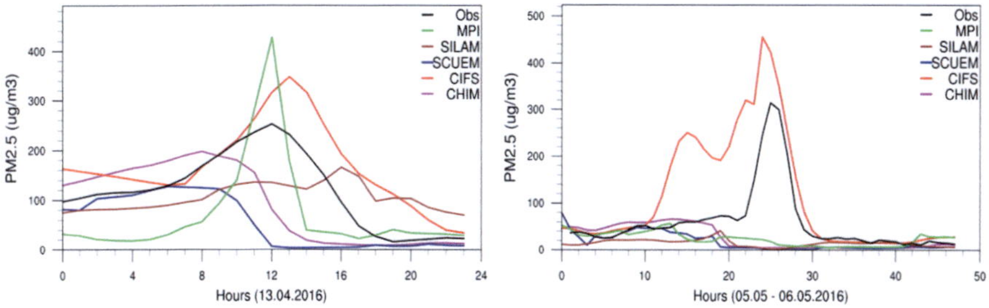

Fig. 18.1 Comparison of $PM_{2.5}$ observations (Obs) with model forecasts for 13.04.2016 (*left*) and for 05–06 Mai 2016 (*right*) in Beijing

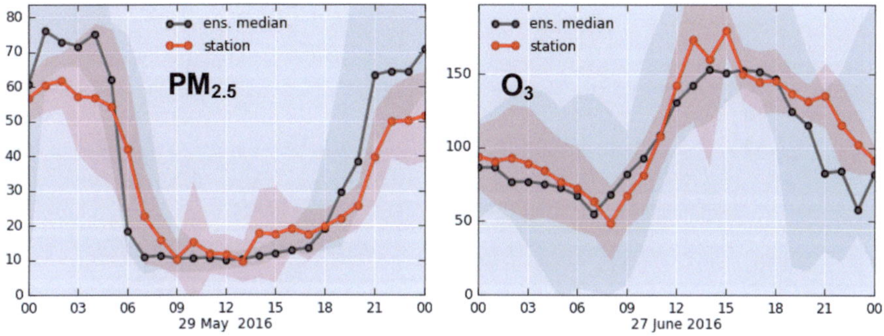

Fig. 18.2 Observed (*red*) and ensemble median forecast (*black*) of $PM_{2.5}$ and O_3 respectively for 29 May and 27 June 2016. *Light red* and *grey* are standard-deviation values

discussed in Sect. 18.4. The model diversity is such that it is not obvious to identify one individual model that outperforms. In fact, any model can perform better in a specific context (e.g. location, forecast time, forecasted pollutant species) but not in other situations.

The application of simple statistical parameters, specifically the calculation of mean or median values derived from a multi-model ensemble forecast, tends to reduce the influence of outliers and hence provides better quality and accuracy forecasts. The ensemble mean/median forecast (EMF) provides also useful indications on the differences among the different individual models. This is illustrated in Fig. 18.2, which shows the EMF for $PM_{2.5}$ and O_3 in Beijing for specific days (here for 29 May and 27 June 2016 respectively), and the standard-deviation calculated from all individual model forecasts. The calculated standard-deviation indicates the differences in individual model forecasts which can be of an order of magnitude while the EMF shows a good agreement with the observations.

Figure 18.3 displays the correlation coefficients calculated between model forecasts and surface observations of O_3 and $PM_{2.5}$ in China in June 2016. The results show that for June 2016, some individual model forecasts (e.g. CHIM and CIFS) have a good O_3 linear relationship with the observations (generally with correlation

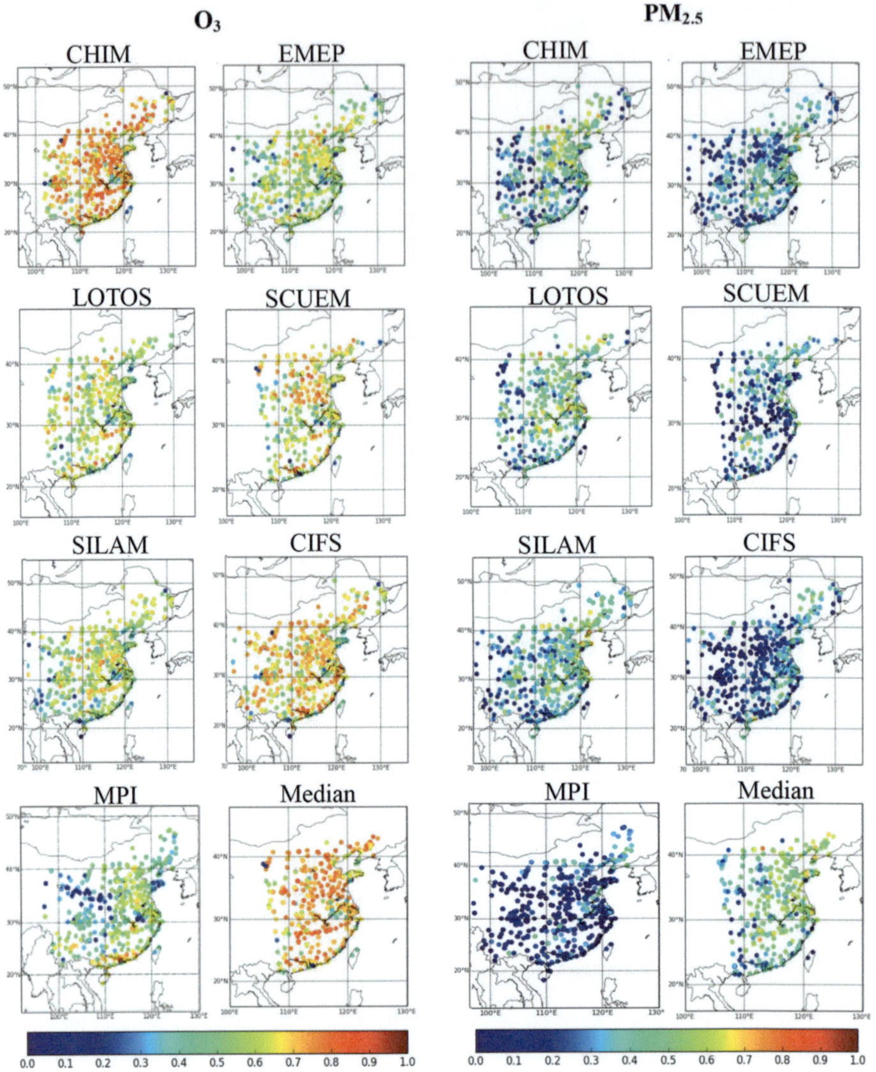

Fig. 18.3 Correlation coefficients between predicted and observed O_3 and $PM_{2.5}$ in June 2016. Calculated correlation coefficients for the ensemble median are also displayed

coefficient $r > 0.7$) while other models (e.g. SILAM, MPI and EMEP) show lower performance (correlation coefficient $r < 0.6$). Similarly, good linear relationships with $PM_{2.5}$ observations were found in some models (e.g. SILAM and LOTOS with $r > 0.6$) while other models (e.g. CIFS, SCUEM and MPI) show lower skills ($r < 0.55$). The calculated EMF (median in this case) indicates generally good performance at most of the stations with correlation coefficients >0.7 and >0.6 respectively for O_3 and $PM_{2.5}$. Similar analyses for other chemical or aerosol

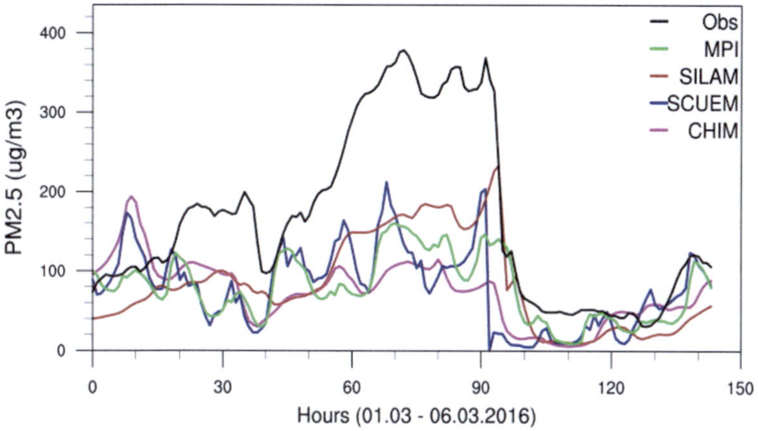

Fig. 18.4 Comparison of modeled and observed $PM_{2.5}$ in Beijing in 01.03–06.03.2016

species or other time periods have also showed significant differences in the skills of individual models. Some models showed better performance for some species (e.g. O_3 in CHIM and CIFS) but not for other species. Generally, when considering the statistical skills for all the chemical and aerosol species predicted by the forecasting systems (O_3, NO_2, $PM_{2.5}$ and PM_{10}), the calculated EMF shows better performance than most of the individual models.

The analysis of the behavior of the multi-model forecasting system highlights some systematic issues that affect all or a part of the individual models during specific situations, and hence of the accuracy of the multi-model system. A first issue is the failure of all models to correctly capture some of the extreme pollution events. This is illustrated by Fig. 18.4, which displays observed and modeled $PM_{2.5}$ concentrations in Beijing on 1–6 March 2016, a time period during which extremely high $PM_{2.5}$ concentrations were observed (>350 μg/m^3). None of the models is able to forecast the 350 μg/m^3 of $PM_{2.5}$ observed on 3–4 March 2016: the modeled concentrations are nearly two times lower than the measured values. This discrepancy between the models and the observations is not clearly understood. It is probably not related only to uncertainties in emissions since the models use different emissions inventories. One possible explanation is that the mechanisms describing the formation and growth of secondary aerosols are not satisfactorily treated in current models. Missing contributions to $PM_{2.5}$ in the models such as for example urban secondary organic aerosols or suspension of dust could be other possible explanations.

Another issue identified in several individual forecasting systems is the poor simulation of the diurnal variation for chemical species such as NO_2 and O_3 in some urban areas and specifically in Beijing. This is illustrated by Fig. 18.5 which highlights the significant differences in NO_2 and O_3 forecasts among the models, particularly during nighttime, when photochemistry is not operating. Most of the models overestimate NO_2 particularly during nighttime leading also to

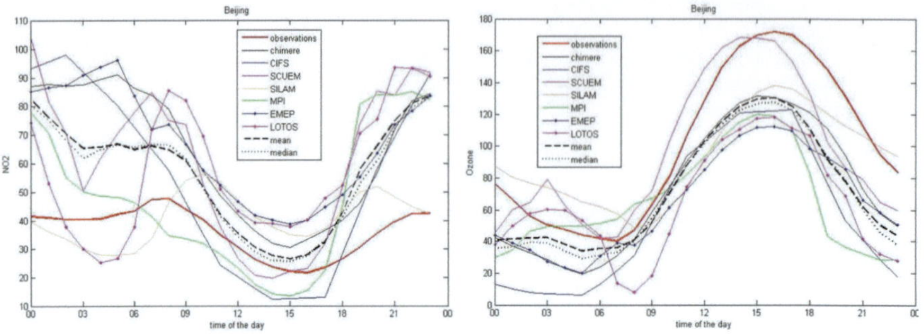

Fig. 18.5 Observed (represented by the *red curve*) and modeled (other *colored curves*) mean diurnal variation of NO_2 and O_3 in Beijing

overestimation of NO_2 in the EMF. Due to titration effects, this discrepancy leads to a general underestimation of O_3. Overestimation of NO_2 emissions or problems with boundary layer height parameterizations are possible explanations.

18.4 Uncertainties in Air Quality Prediction Systems

Air quality forecast systems rely not only on chemical models but also on meteorological models which, despite considerable improvement in recent years, still contain some inaccuracies in the formulation of physical processes. Air quality models, which use meteorological predictions to forecast "chemical weather", and simulate in addition surface processes, chemical transformations and microphysical processes are likely to have greater model errors. In the following sections, we discuss some of the uncertainties leading to such errors.

18.4.1 Uncertainties in Emissions

There are several inventories providing the distribution of emissions of primary pollutants at the global, regional and local scales. Several comparisons of the emission datasets have revealed large differences, not only in the estimated total amounts but also in the spatial and temporal distributions of past, current and future emissions (cf Chap. 6; Granier et al. 2011). Wang et al. (2010) has indicated that the adopted seasonal, day-of-week, diurnal and vertical allocations of anthropogenic emissions can change significantly the concentrations of NO_x, SO_2 and O_3 simulated by mesoscale models. The information on temporal variability of emissions for different sectors (e.g. traffic, industry, etc.) is uncertain and often missing in many current inventories.

Figure 18.6 illustrates the air quality model responses to specific perturbations in emissions through sensitivity studies of model simulations to anthropogenic

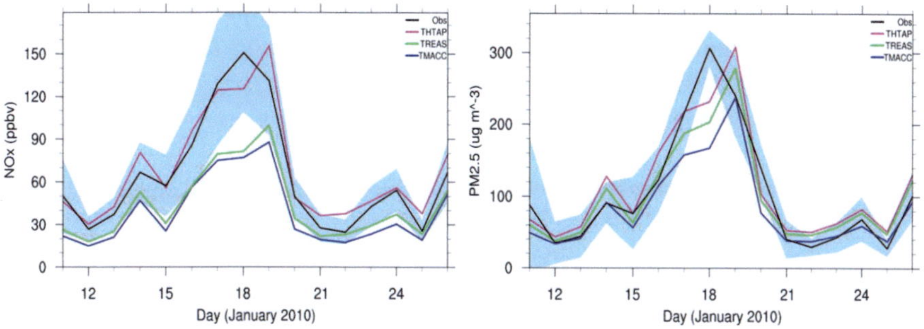

Fig. 18.6 Time-series of observed and simulated daily mean NOx and $PM_{2.5}$ concentrations in Beijing in 12–26 January 2010

emission inventories. The model simulations labelled THTAP, TREAS and TMACC are performed for January 2010 with the WRF-Chem model (MPI configuration) and with HTAPv2 (Janssens-Maenhout et al. 2015), REASv2 (Kurokawa et al. 2013) and MACCity (Granier et al. 2011) anthropogenic emissions data, respectively. The model correctly replicates the variability of NO_x and $PM_{2.5}$ and in particular the rapid increases between 15 and 20 January 2010. The THTAP simulation captures better the NO_x peak values, while TREAS and TMACC simulations show an underestimation of the NO_x concentrations by more than 50 ppbv during the polluted event. THTAP provides concentration values that are closest to the $PM_{2.5}$ observations. The results highlight the significant impacts of differences in emissions on model predictions.

18.4.2 Uncertainties in Model Parameterizations

Modeling present and near future air quality requires the chemistry transport models to accurately represent a large number of atmospheric chemical and physical processes with different degrees of complexity:

- Removal of primary pollutants by chemical reactions, deposition and scavenging
- Dispersion of pollutants from pollution sources into downwind regions
- Formation and removal of secondary pollutants by chemical reactions and microphysical processes as well as the deposition and scavenging of these species
- Regional and intercontinental contribution to local pollution
- Interactions between the boundary layer and the free troposphere
- Transformation of chemical species in general, such as detailed degradation mechanisms of VOCs (e.g. isoprene), and heterogeneous processing in clouds

These processes are not always carefully formulated in air quality models, which adopt simplified assumptions rather than including complex mathematical description. Such differences in model approaches and complexities often explain the differences among models even when same model settings including emissions and initial conditions are adopted.

To illustrate this particular issue, Fig. 18.7 shows monthly mean surface concentrations of NOx for January 2010 simulated by four different models using the same anthropogenic emissions data (HTAPv2 in this case) and the same initial and boundary chemical conditions (MACC). Despite the similarities in the input conditions, the predicted concentrations show quite significant differences among the models: SILAM predicted much higher NOx concentrations compared to other model results; while the EMEP model provides concentrations that are a factor 2 lower. On the other hand, the MPI and SCUEM results show almost similar patterns with some differences in the NOx peak concentrations which can be

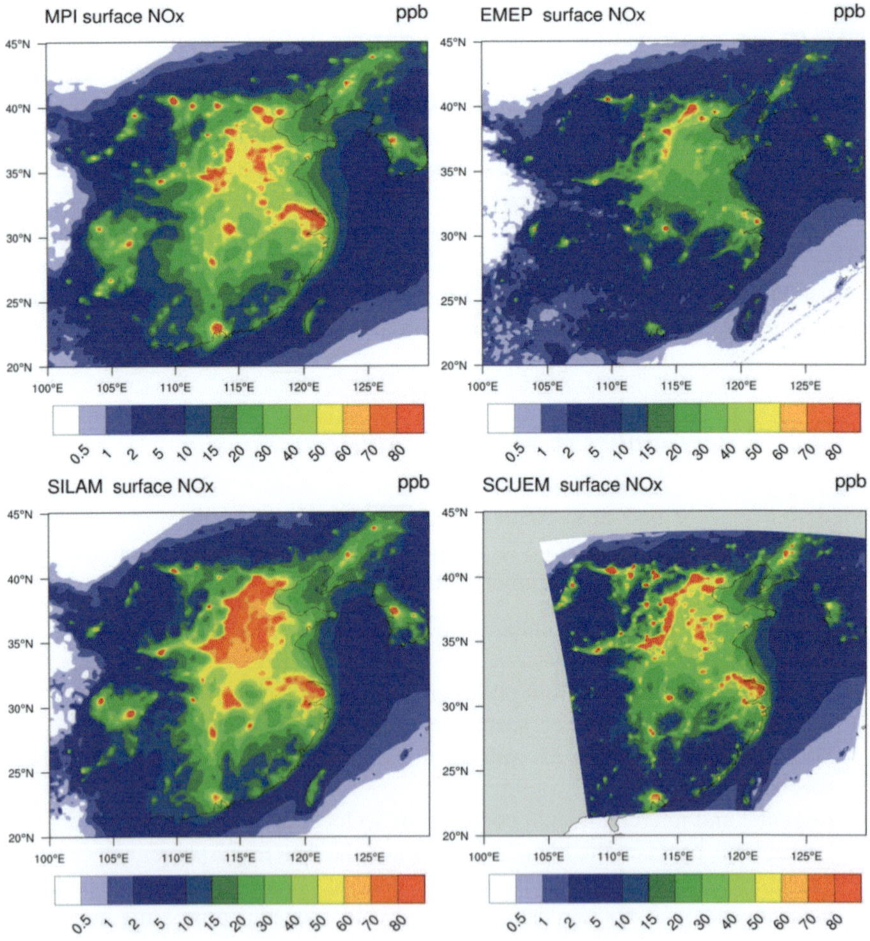

Fig. 18.7 Simulated monthly mean NOx surface concentrations for January 2010 by EMEP, SILAM, MPI and SCUEM models

attributed to the differences in the horizontal resolution used by the two models (20 km and 06 km, respectively).

Adequate representation of meteorological conditions is essential for accurate air quality model predictions. In fact, the atmospheric dispersion and transport of air pollutants are controlled by different parameters such as wind speed and direction, boundary layer depth and turbulence, surface parameters, precipitations, etc. However, the meteorological parameters provided by numerical weather prediction models which are used as input by chemical transport models are also sources of errors that substantially affect the quality of air quality predictions. Many of these parameters are characterized by small spatial and temporal scales that are not captured by the CTMs and are parameterized by formulations based on statistical information. Even when the air quality and numerical weather prediction models are coupled (on-line models), the errors in meteorological parameters constitute a major source of errors in the predictions.

Other sources of uncertainties for air quality model predictions are related to model treatments of processes affecting pollutants lifetime and their formation which depend on the details of the chemical mechanisms which are adopted. Guo et al. (2014), for example, showed the complexity of the aerosol nucleation processes and growth that are responsible for the formation of $PM_{2.5}$ in Beijing. Based on observations, Huang et al. (2014) have analyzed the chemical nature and source of $PM_{2.5}$ at some urban locations during the extremely severe haze pollution in January 2013, and found a high contribution of secondary aerosols to this haze pollution event. Wang et al. (2014) also concluded that secondary aerosol formation is an important mechanism in the development of heavy pollution episodes. Such detailed mechanisms of pollution formation are not considered in many of the current air quality models and if they are considered, their parameterization is simplified using assumptions that increase uncertainties.

18.5 Improving Air Quality Predictions

As discussed in the previous section, the different uncertainties in emissions, physical and chemical processes controlling day-to-day levels of pollutants make the development of accurate short-term air quality forecasts challenging. To address this issue, it is necessary to improve the accuracy of model inputs including emissions and initial and boundary conditions. Physical and chemical parameterizations should also be improved. Despite recent improvements in quantifying the emission sources at higher resolution and in understanding the different chemical and physical processes that affect air pollution, uncertainties remain in many of these processes. Since these processes are usually complex and cover scales that are not always resolved by the models, errors will never be fully eliminated. Different mathematical and statistical approaches have therefore been developed to improve the quality of the model predictions. Among these techniques, data assimilation of atmospheric observations from satellites (Inness et al. 2013; Benedetti et al. 2009;

Massart et al. 2014; Marécal et al. 2015) provide improved initial conditions for some key chemical species (e.g. O_3, CO, NO_2, SO_2, aerosols) which leads to improved predictions. Various approaches used for data assimilation are described in detail in Carmichael et al. (2008), Sandu and Chai (2011) and Bocquet et al. (2015). They include variational (3D-Var and 4D-Var) (Hooghiemstra et al. 2011; Chai et al. 2007) as well as Kalman Filters assimilation methods (Parrington et al. 2008; Kang et al. 2010; Gaubert et al. 2014).

Among the models used in the present version of the PANDA-MarcoPolo multi-model forecasting system, only the CIFS model constrained its initial conditions through data assimilation procedures. The CIFS system, described and evaluated in detail in Inness et al. (2015), uses the ECMWF's 4D-Var algorithm (Courtier et al. 1994) to assimilate observations of chemical species. To illustrate the performance of the CIFS assimilation procedure, Fig. 18.8 shows observed CO total columns for

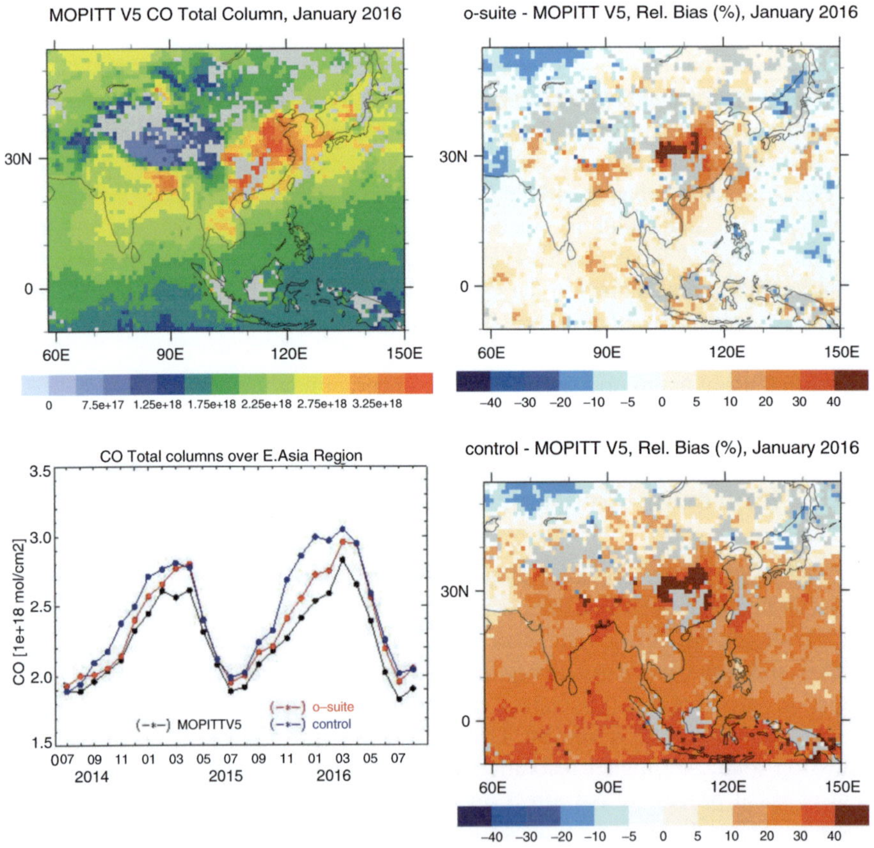

Fig. 18.8 CO total column from MOPITTv5 for January 2016 (*top left*) and relative differences with model prediction with (*top right*) and without (*bottom right*) assimilation of satellite CO observations. Time series (*bottom-left*) of observed (*black*) and modeled CO total columns over East Asia with (*red*) and without (*blue*) assimilation of satellite observations

January 2016 from the MOPITTv5 instrument and bias of model predictions with and without assimilation of satellite CO observations (‘o-suite’ and ‘control’, respectively). In general, the ‘o-suite’ experiment exhibits a small bias which does not exceed 10% over most regions in Asia except in central China where positive biases of more than 30% are found. The ‘control’ simulation exhibits higher positive biases of more than 20% in almost all East and South-East Asia with higher discrepancies in central China (>40%). The time-series of observed and predicted CO total columns in East Asia in Fig. 18.8 indicate also that the assimilation of satellite CO observation improves the quality of model predictions.

Post-processing methods of model predictions can improve the forecasts of surface variables, and various methods of bias correction almost similar to the post-processing methods used in weather forecasts have been developed for air quality predictions. Among these methods, analog-based approaches can provide reliable uncertainty quantification of surface air quality predictions. The analog is defined as a past prediction that matches selected features of the current forecast for a given location and time. The analog-based approach has previously been applied to meteorological forecasts (Delle Monache et al. 2011, 2013) and is being used also in predictions of ground level O_3 and $PM_{2.5}$ (e.g. Djalalova et al. 2015). It can lead to a significant reduction of the systematic and random errors which are common to air quality model forecasts.

18.6 Summary

Operational air quality prediction systems based on CTMs are being increasingly used to alert the population on elevated levels of pollution. Such systems provide information on expected pollution levels and associated air quality and health indices that are communicated to the public through websites, mobile applications and other media tools. This information is also useful for decision-makers in charge of emissions control strategies.

Accurate prediction of present and near future air quality situations is a challenging task for model systems as they need to accurately represent not only several complex chemical and physical processes but also meteorological conditions. Some of the differences in individual model forecasts result from the formulation of such processes, which can range from simple empirical assumptions to more complex mathematical descriptions. As shown in this chapter, even when the models use same initial settings, their predicted air pollution concentrations can exhibit significant differences.

Accurate surface emissions data of primary pollutants constitute essential inputs for reliable air quality model predictions. Several emission inventories are available in East Asia with large differences not only in the estimated total amounts but also in the spatial and temporal distributions of past, current and future emissions. As shown in this chapter, model simulations can differ substantially depending on the emission inventory that is adopted. Addressing the uncertainty on emissions in East Asia is therefore key to improve the skills of operational predictions.

The combination of several air quality prediction systems into an ensemble is a pragmatic approach that can be used to reduce the uncertainties that affect the performance of individual model predictions. Within the EU-funded PANDA and MarcoPolo projects, a multi-model forecasting system has been constructed using seven models and has been operational since summer 2016. This multi-model ensemble approach increases significantly the skill of the air quality forecasts in East Asia, in comparison with individual model predictions. Introducing data-assimilation procedures and bias correction techniques in some of the individual models are expected to further improve the performance of the system.

Acknowledgments This work has received research funding from the European Community's Seventh Framework Programme (FP7) under the PANDA project grant agreement n° 606719. The authors gratefully acknowledge the computing time granted by the John von Neumann Institute for Computing (NIC) and provided on the supercomputer JURECA (JURECA 2016) at Jülich Supercomputing Centre (JSC).

References

Baklanov, A., Korsholm, U., Mahura, A., Petersen, C., & Gross, A. (2008). ENVIRO-HIRLAM: Online coupled modelling of urban meteorology and air pollution. *Advances in Science and Research, 2*, 41–46. doi:10.5194/asr-2-41-2008.

Benedetti, A., Morcrette, J.-J., Boucher, O., Dethof, A., Engelen, R. J., Fisher, M., Flentje, H., Huneeus, N., Jones, L., Kaiser, J. W., Kinne, S., Mangold, A., Razinger, M., Simmons, A. J., & Suttie, M. (2009). Aerosol analysis and forecast in the European Centre for medium-range weather forecasts integrated forecast system: 2. Data assimilation. *Journal of Geophysical Research, 114*(D13), 2156–2202. doi:10.1029/2008JD011115.

Bocquet, M., Elbern, H., Eskes, H., Hirtl, M., Žabkar, R., Carmichael, G. R., Flemming, J., Inness, A., Pagowski, M., Pérez Camaño, J. L., Saide, P. E., San Jose, R., Sofiev, M., Vira, J., Baklanov, A., Carnevale, C., Grell, G., & Seigneur, C. (2015). Data assimilation in atmospheric chemistry models: Current status and future prospects for coupled chemistry meteorology models. *Atmospheric Chemistry and Physics, 15*, 5325–5358. doi:10.5194/acp-15-5325-2015.

Carmichael, G. R., Sandu, A., Chai, T., Daescu, D., Constantinescu, E., & Tang, Y. (2008). Predicting air quality: Improvements through advanced methods to integrate models and measurements. *Journal of Computational Physics, 227*, 3540–3571.

Chai, T., Carmichael, G., Tang, Y., Sandu, A., Hardesty, M., Pilewskie, P., Whitlow, S., Browell, E., Avery, M., Thouret, V. et al. (2007), Four dimensional data assimilation experiments with ICARTT (international consortium for atmospheric transport and transformation) ozone measurements. Journal of Geophysical Research, 112, D12S15.

Cobourn, W. G. (2007). Accuracy and reliability of an automated air quality forecast system for ozone in seven Kentucky metropolitan area. *Atmospheric Environment, 41*, 5863–5875.

Courtier, P., Thépaut, J.-N., & Hollingsworth, A. (1994). A strategy for operational implementation of 4D-Var, using an incremental approach. *Quarterly Journal of the Royal Meteorological Society, 120*, 1367–1388.

Delle Monache, L., Eckel, T., Rife, D., & Nagarajan, B. (2013). Probabilistic weather prediction with an analog ensemble. *Monthly Weather Review, 141*, 3498–3516.

Delle Monache, L., Nipen, T., Liu, Y., Roux, G., & Stull, R. (2011). Kalman filter and analog schemes to post-process numerical weather predictions. *Monthly Weather Review, 139*, 3554–3570.

Djalalova, I., Delle Monache, L., & Wilczak, J. (2015). PM2.5 analog forecast and Kalman filter post-processing for the Community Multiscale Air Quality (CMAQ) model. *Atmospheric Environment, 108*, 76–87.

Flemming, J., Inness, A., Flentje, H., Huijnen, V., Moinat, P., Schultz, M. G., & Stein, O. (2009). Coupling global chemistry transport models to ECMWF's integrated forecast system. *Geoscientific Model Development, 2*, 253–265. doi:10.5194/gmd-2-253-2009.

Flemming, J., Huijnen, V., Arteta, J., Bechtold, P., Beljaars, A., Blechschmidt, A.-M., Diamantakis, M., Engelen, R. J., Gaudel, A., Inness, A., Jones, L., Josse, B., Katragkou, E., Marecal, V., Peuch, V.-H., Richter, A., Schultz, M. G., Stein, O., & Tsikerdekis, A. (2015). Tropospheric chemistry in the integrated forecasting system of ECMWF. *Geoscientific Model Development, 8*, 975–1003. doi:10.5194/gmd-8-975-2015.

Gaubert, B., Coman, A., Foret, G., Meleux, F., Ung, A., Rouil, L., Ionescu, A., Candau, Y., & Beekmann, M. (2014). Regional scale ozone data assimilation using an ensemble Kalman filter and the CHIMERE chemical transport model. *Geoscientific Model Development, 7*, 283–302. doi:10.5194/gmd-7-283-2014.

Granier, C., Bessagnet, B., Bond, T., D'Angiola, A., Denier van der Gon, H., Frost, G. J., Heil, A., Kainuma, M., Kaiser, J., Kinne, S., Klimont, Z., Kloster, S., Lamarque, J.-F., Liousse, C., Matsui, T., Meleux, F., Mieville, A., Ohara, T., Raihi, K., Schultz, M., Smith, S. J., Thomson, A. M., van Aardenne, J., & van der Werf, G. (2011). Evolution of anthropogenic and biomass burning emissions of air pollutants at global and regional scales during the 1980–2010 period. *Climatic Change, 109*, 163–190. doi:10.1007/s10584-011-0154-1.

Grell, G. A., Peckham, S. E., Schmitz, R., McKeen, S. A., Frost, G., Skamarock, W. C., & Eder, B. (2005). Fully coupled online chemistry within the WRF model. *Atmospheric Environment, 39*, 6957–6975.

Guo, S., Hu, M., Zamora, M. L., Peng, J., Shang, D., Zheng, J., Du, Z., Wu, Z., Shao, M., Zeng, L., Molina, M. J., & Zhang, R. (2014). Elucidating severe urban haze formation in China. *Proceedings of the National Academy of Sciences of the United States of America, 111*(49), 17373–17378.

Hooghiemstra, P. B., Krol, M. C., Meirink, J. F., Bergamaschi, P., van der Werf, G. R., Novelli, P. C., Aben, I., & Röckmann, T. (2011). Optimizing global CO emission estimates using a four-dimensional variational data assimilation system and surface network observations. *Atmospheric Chemistry and Physics, 11*, 4705–4723. doi:10.5194/acp-11-4705-2011.

Huang, K., Zhuang, G., Wang, Q., Fu, J. S., Lin, Y., Liu, T., Han, L., & Deng, C. (2014). Extreme haze pollution in Beijing during January 2013: Chemical characteristics, formation mechanism and role of fog processing. *Atmospheric Chemistry and Physics Discussions, 14*, 7517–7556. doi:10.5194/acpd-14-7517-2014.

Inness, A., Baier, F., Benedetti, A., Bouarar, I., Chabrillat, S., Clark, H., Clerbaux, C., Coheur, P., Engelen, R. J., Errera, Q., Flemming, J.,George, M., Granier, C., Hadji-Lazaro, J., Huijnen, V., Hurtmans, D., Jones, L., Kaiser, J. W., Kapsomenakis, J., Lefever, K., Leitão, J., Razinger, M., Richter, A., Schultz, M. G., Simmons, A. J., Suttie, M., Stein, O., Thépaut, J.-N., Thouret, V., Vrekoussis, M., Zerefos, C., & the MACC team. (2013). The MACC reanalysis: An 8 yr data set of atmospheric composition. Atmospheric Chemistry and Physics, 13, 4073–4109. doi:10.5194/acp-13-4073-2013.

Inness, A., Blechschmidt, A.-M., Bouarar, I., Chabrillat, S., Crepulja, M., Engelen, R. J., Eskes, H., Flemming, J., Gaudel, A., Hendrick, F., Huijnen, V., Jones, L., Kapsomenakis, J., Katragkou, E., Keppens, A., Langerock, B., de Mazière, M., Melas, D., Parrington, M., Peuch, V. H., Razinger, M., Richter, A., Schultz, M. G., Suttie, M., Thouret, V., Vrekoussis, M., Wagner, A., & Zerefos, C. (2015). Data assimilation of satellite retrieved ozone, carbon monoxide and nitrogen dioxide with ECMWF's composition-IFS. *Atmospheric Chemistry and Physics, 15*, 5275–5303. doi:10.5194/acp-15-5275-2015.

Janssens-Maenhout, G., Crippa, M., Guizzardi, D., Dentener, F., Muntean, M., Pouliot, G., Keating, T., Zhang, Q., Kurokawa, J., Wankmüller, R., Denier van der Gon, H., Kuenen, J. J. P., Klimont, Z., Frost, G., Darras, S., Koffi, B., & Li, M. (2015). HTAP_v2.2: A mosaic of regional and global emission grid maps for 2008 and 2010 to study hemispheric transport of air pollution. *Atmospheric Chemistry and Physics, 15*, 11411–11432. doi:10.5194/acp-15-11411-2015.

JURECA, Jülich Supercomputing Centre. (2016). JURECA: General-purpose supercomputer at Jülich Supercomputing Centre. *Journal of Large-Scale Research Facilities, 2*, A62. http://dx.doi.org/10.17815/jlsrf-2-121.

Kang, D., Mathur, R., & Trivikrama Rao, S. (2010). Assessment of bias-adjusted PM air quality forecasts over the continental United States during 2007. *Geoscientific Model Development, 3*, 309–320. doi:10.5194/gmd-3-309-2010.

Kurokawa, J., Ohara, T., Morikawa, T., Hanayama, S., Janssens-Maenhout, G., Fukui, T., Kawashima, K., & Akimoto, H. (2013). Emissions of air pollutants and greenhouse gases over Asian regions during 2000–2008: Regional Emission inventory in ASia (REAS) version 2. *Atmospheric Chemistry and Physics, 13*, 11019–11058. doi:10.5194/acp-13-11019-2013.

Marécal, V., Peuch, V.-H., Andersson, C., Andersson, S., Arteta, J., Beekmann, M., Benedictow, A., Bergström, R., Bessagnet, B., Cansado, A., Chéroux, F., Colette, A., Coman, A., Curier, R. L., Denier van der Gon, H. A. C., Drouin, A., Elbern, H., Emili, E., Engelen, R. J., Eskes, H. J., Foret, G., Friese, E., Gauss, M., Giannaros, C., Guth, J., Joly, M., Jaumouillé, E., Josse, B., Kadygrov, N., Kaiser, J. W., Krajsek, K., Kuenen, J., Kumar, U., Liora, N., Lopez, E., Malherbe, L., Martinez, I., Melas, D., Meleux, F., Menut, L., Moinat, P., Morales, T., Parmentier, J., Piacentini, A., Plu, M., Poupkou, A., Queguiner, S., Robertson, L., Rouïl, L., Schaap, M., Segers, A., Sofiev, M., Tarasson, L., Thomas, M., Timmermans, R., Valdebenito, Á., van Velthoven, P., van Versendaal, R., Vira, J., & Ung, A. (2015). A regional air quality forecasting system over Europe: The MACC-II daily ensemble production. *Geoscientific Model Development, 8*, 2777–2813. doi:10.5194/gmd-8-2777-2015.

Massart, S., Agusti-Panareda, A., Aben, I., Butz, A., Chevallier, F., Crevoisier, C., Engelen, R., Frankenberg, C., & Hasekamp, O. (2014). Assimilation of atmospheric methane products into the MACC-II system: From SCIAMACHY to TANSO and IASI. *Atmospheric Chemistry and Physics, 14*, 6139–6158. doi:10.5194/acp-14-6139-2014.

Menut, L., Bessagnet, B., Khvorostyanov, D., Beekmann, M., Blond, N., Colette, A., Coll, I., Curci, G., Foret, G., Hodzic, A., Mailler, S., Meleux, F., Monge, J.-L., Pison, I., Siour, G., Turquety, S., Valari, M., Vautard, R., & Vivanco, M. G. (2013). CHIMERE 2013, a model for regional atmospheric composition modelling. *Geoscientific Model Development, 6*, 981–1028. doi:10.5194/gmd-6-981-2013.

Parrington, M., Jones, D. B. A., Bowman, K. W., Horowitz, L. W., Thompson, A. M., Tarasick, D. W., & Witte, J. C. (2008). Estimating the summertime tropospheric ozone distribution over North America through assimilation of observations from the Tropospheric Emission Spectrometer. *Journal of Geophysical Research, 113*, D18307. doi:10.1029/2007JD009341.

Perez, P., & Reyes, J. (2006). An integrated neural network model for PM forecasting. *Atmospheric Environment, 40*, 2845–2851. doi:10.1016/j.atmosenv.2006.01.010.

Sandu, A., & Chai, T. (2011). Chemical data assimilation – an overview. *Atmosphere, 2*(3), 426–463. doi:10.3390/atmos2030426.

Schaap, M., Timmermans, R. M. A., Sauter, F. J., Roemer, M., Velders, G. J. M., Boersen, G. A. C., Beck, J. P., & Builtjes, P. J. H. (2008). The LOTOS-EUROS model: Description, validation and latest developments. *International Journal of Environment and Pollution, 32*, 270–289.

Simpson, D., Benedictow, A., Berge, H., Bergström, R., Emberson, L. D., Fagerli, H., Flechard, C. R., Hayman, G. D., Gauss, M., Jonson, J. E., Jenkin, M. E., Nyíri, A., Richter, C., Semeena, V. S., Tsyro, S., Tuovinen, J.-P., Valdebenito, Á., & Wind, P. (2012). The EMEP MSC-W chemical transport model – technical description. *Atmospheric Chemistry and Physics, 12*, 7825–7865. doi:10.5194/acp-12-7825-2012.

Sofiev, M., Galperin, M. V., & Genikhovich, E. (2008). Construction and evaluation of Eulerian dynamic core for the air quality and emergency modeling system SILAM. In C. Borrego & A. I. Miranda (Eds.), *NATO science for peace and security serties C: Environmental security, Air pollution modelling and its application* (Vol. XIX, pp. 699–701). Berlin: Springer.

Sofiev, M., Berger, U., Prank, M., Vira, J., Arteta, J., Belmonte, J., Bergmann, K.-C., Chéroux, F., Elbern, H., Friese, E., Galan, C., Gehrig, R., Khvorostyanov, D., Kranenburg, R., Kumar, U., Marécal, V., Meleux, F., Menut, L., Pessi, A.-M., Robertson, L., Ritenberga, O., Rodinkova, V., Saarto, A., Segers, A., Severova, E., Sauliene, I., Siljamo, P., Steensen, B. M., Teinemaa, E., Thibaudon, M., & Peuch, V.-H. (2015). MACC regional multi-model ensemble simulations of birch pollen dispersion in Europe. *Atmospheric Chemistry and Physics, 15*, 8115–8130. doi:10.5194/acp-15-8115-2015.

Vlachogianni, A., Kassomenos, P., Karppinen, A., Karakitsios, S., & Kukkonen, J. (2011). Evaluation of a multiple regression model for the forecasting of the concentrations of NOx and PM_{10} in Athens and Helsinki. *Science of the Total Environment, 409*, 1559–1571.

Wang, X., et al. (2010). WRF-Chem simulation of East Asian air quality: Sensitivity to temporal and vertical emissions distributions. *Atmospheric Environment, 44*, 660–669. doi:10.1016/j.atmosenv.2009.11.011.

Wang, Y., Yao, L., Wang, L., Ji, D., Tang, G., Zhang, J., Sun, Y., Hu, B., & Xin, J. (2014). Mechanism for the formation of the January 2013 heavy haze pollution episode over central and eastern China. *Science China Earth Sciences, 57*, 14. doi:10.1007/s11430-013-4773-4.

Wolff, G. T., & Lioy, P. J. (1978). An empirical model for forecasting maximum daily ozone levels in the northeastern United States. *Journal of the Air Pollution Control Association, 28*, 1034–1038.

Chapter 19
Chemical Weather Forecasting for Eastern China

Ying Xie, Jianming Xu, Guangqiang Zhou, Luyu Chang, Yixuan Gu, and Baode Chen

Abstract Based on the Shanghai Meteorological Service WRF ADAS Real-time Modeling System (SMS-WARMS) and the Community Multi-scale Air Quality (CMAQ) model, a chemical weather forecasting system is being established for Eastern China. Anthropogenic emissions are based on Hemispheric Transport of Air Pollution (HTAP) 2010 dataset, and biogenic emissions are estimated using the Model of Emissions of Gases and Aerosols from Nature (MEGAN). Forecasting results during 7 August to 1 September 2016 are evaluated for Shanghai and eight provincial cities in Eastern China. The results exhibit large differences in model performance for different regions. The overestimation of NO_2 and $PM_{2.5}$ for the cities within or adjacent to the Yangtze River Delta (YRD) region suggests air quality measures taken in preparation of the 2016 G20 Summit were effective in reducing pollutants concentrations. The results also show that large reductions in HTAP 2010 NO_x and SO_2 emissions are required, suggesting that regular updates of emission inventories are necessary to capture the rapid changes in pollutant emissions in Eastern China.

Keywords Chemical weather • Forecasting • Eastern China

Y. Xie (✉) • J. Xu • G. Zhou • L. Chang • Y. Gu
Yangtze River Delta Center for Environmental Meteorology Prediction and Warning, Shanghai 200030, People's Republic of China

Shanghai Key Laboratory of Meteorology and Health, Shanghai Meteorological Service, Shanghai 200030, People's Republic of China
e-mail: yxie33@outlook.com; metxujm@163.com; zhougq21@163.com; changluyu1989@126.com; guyixuan@mail.iap.ac.cn

B. Chen
Shanghai Typhoon Institute of China Meteorological Administration, Shanghai, People's Republic of China
e-mail: baode@mail.typhoon.gov.cn

I. Bouarar et al. (eds.), *Air Pollution in Eastern Asia: An Integrated Perspective*, ISSI Scientific Report Series 16, DOI 10.1007/978-3-319-59489-7_19

19.1 Introduction

With the rapid economic growth, increase in industrial production and urbanization processes in recent years, air pollution in Eastern China has posed strong adverse effects on public health and ecosystems. Real-time air quality forecasting system are useful tools for predicting pollution episodes, so that control measures can be taken when the air quality poses harming effects and the general public can limit their exposure to high levels of pollutants. A number of such forecasting systems have been established in Europe and North America (Marécal et al. 2015; Zhang et al. 2012). In China, large efforts have also been made in recent years to establish real-time forecasting systems (Wang et al. 2009; Zhou et al. 2015).

It is well known that air quality modeled concentrations are highly sensitive to meteorological parameters (e.g. wind, temperature, relative humidity, PBL height, cloud). Errors in meteorological fields have large impacts on predicted concentrations (Menut and Bessagnet 2010). For the current available forecasting systems for Eastern China, however, there are often missing linkages between meteorology and chemical transport modeling. Model evaluations have mainly focused on particulates and ozone concentrations only. Biases in the concentration fields are often corrected solely by adjusting emissions with little investigation on the meteorological predictions. Furthermore, these systems often tend to show degraded performance during highly polluted episodes and weather transitional time periods.

The Shanghai Meteorological Service (SMS) is developing and maintaining a suite of operational meso-scale meteorological models based on WRF and ADAS data assimilation scheme. The systems include: (1) a 9-km WRF ADAS Real-time Modeling System (SMS-WARMS), which produces 72-h forecasts two times per day with ADAS data assimilation; (2) a 3-km WRF ADAS 3DVAR Rapid Refresh system (SMS-WARR), producing forecast up to 12-h with hourly data assimilation cycle and its lateral boundary condition taken from the 9-km SMS-WARMS; and (3) a 9-km SMS-EnWARMS system (ensemble WARMS), which makes meso-scale probabilistic forecast up to 120-h. These modeling systems have been extensively evaluated and are providing strong support for weather predictions in Eastern China. Meanwhile, the SMS is actively developing a 1-km resolution modeling system covering Eastern China with focus on "gray-zone" physics (e.g. grid-resolved vs. subgrid turbulence and convection).

In this work, a chemical weather forecasting system is being established based on the 9-km SMS-WARMS regional meteorological model and CMAQ chemical transport model (as illustrated to Fig. 19.1). The objective is to build an integrated forecasting system with state-of-the-science models, and to upgrade key model processes as well as input datasets, so that overall better prediction ability could be achieved. A key advance here is to use meteorological fields based on the SMS-WARMS forecast, which has shown relatively good ability in predicting meteorological parameters, boundary layer process, and land surface interactions. Ongoing efforts include the improvement of anthropogenic and biogenic emissions, pollutants chemical transformation processes, and chemical boundary conditions.

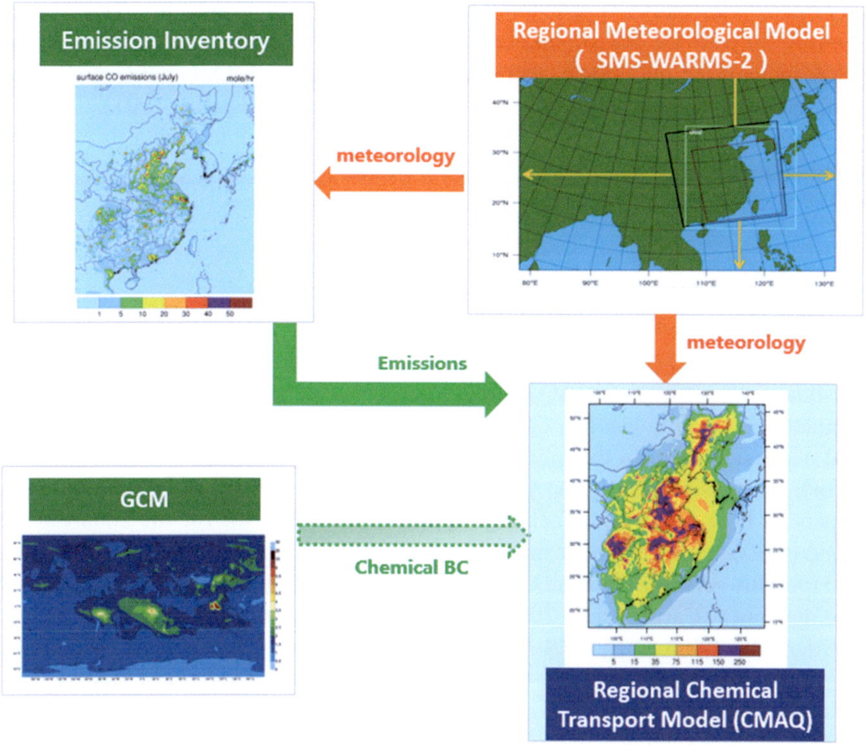

Fig. 19.1 Components of the chemical weather forecasting system (Apply chemical boundary condition provided by GCM is undergoing)

In this work, we present the components of a new chemical weather forecasting system, together with a preliminary evaluation of model performance. In Sect. 19.2, we describe the modeling system, anthropogenic and biogenic emissions. In Sect. 19.3, we evaluate the model performance for nine cities in Eastern China during a three-week period in August 2016. Conclusions and discussion are presented in Sect. 19.4.

19.2 Forecasting System

19.2.1 Meteorological Model

We use the 3-D meteorology fields produced by the 9-km SMS-WARMS system to drive the CMAQ chemical transport model (Fig. 19.1). The National Centers for Environmental Prediction (NCEP) Global Forecast System (GFS) output is used as a background for data assimilation scheme which ingests many local observations (e.g. radar and buoy), and to provide lateral boundary conditions. The physics options adopted for the simulations include the YSU PBL scheme (Hong et al. 2006), Thompson microphysics (Thompson et al. 2008), RRTMG radiation (Iacono et al. 2008), and the Noah Land Surface Model (Chen and Dudhia 2001). Cumulus parameterization is turned off for the runs. The modeling domain consists of 760 by 600 horizontal grids and 51 layers in the vertical.

19.2.2 Chemical Transport Model

The CMAQ model (Byun and Schere 2006) version 5.0.2 is used for photochemical air quality modeling. CMAQ employs an Eulerian grid structure to explicitly simulate emissions, gas-phase, aqueous, and mixed-phase chemistry, advection and dispersion, aerosol thermodynamics and physics, and wet and dry deposition. A more complete description and evaluation of the CMAQ are available in the papers by Foley et al. (2010) and by Appel et al. (2016). We use the 2005 Carbon Bond (CB05) chemical mechanism (Yarwood et al. 2005) for gas-phase chemistry and the fifth-generation modal CMAQ aerosol model (aero5) for aerosol chemistry and dynamics (Carlton et al. 2010).

As a subdomain of the SMS-WARMS run, the CMAQ domain consists of 430 by 370 horizontal grids, covering large areas of China and a portion of Eastern Asia (Fig. 19.1). In the vertical, 26 layers are applied. The model is initialized daily at 12 UTC (20 LST) to produce a 72-h forecast. The model evaluation presented here is based on a 24-h forecast from the daily run.

The chemical boundary conditions are currently based on the default vertical profiles of gaseous species and aerosols in CMAQ that represent clean air conditions. It should be noted however that, in order to capture the long-range transport effects of pollutants, a chemical downscaling approach using key gas-phase and particles concentrations from a global air quality model is preferred. Examples of chemical downscaling for China could be found in air quality predictions made available by the PANDA/MarcoPolo projects, where chemical prediction from the European Centre for Medium-Range Weather Forecasts (ECMWF) global model is used to provide boundary conditions to regional models such as WRF-Chem.

19.2.3 Emissions

The anthropogenic emissions are based on monthly HTAP v2 dataset (http://edgar.jrc.ec.europa.eu/htap_v2) (Janssens-Maenhout et al. 2015) for year 2010. HTAP emissions are developed jointly by US-EPA, the MICS-Asia group, EMEP/TNO, the REAS and the EDGAR group to provide hemispheric emissions. The dataset consists of monthly pollutants maps from different sectors at 0.1° resolution. For China, HTAP emissions are compiled based on the MIX inventory (Li et al. 2015), which is available at http://www.meicmodel.org/dataset-mix. In this work according to operational forecasting results, adjustments are applied to HTAP NO_x (reduced by 55%), SO_2 (reduced by 85%), and CO (increased by 30%) emissions to account for likely changes due to rapid economic growth in the region.

Biogenic emissions are estimated by the MEGAN model version 2.10 (Guenther et al. 2012). Leaf Area Index data version 2.0 and plant functional type (PFT) dataset with the Community Land Model 16 PFT classification are applied. Global emission factor input files are based on version 2011 datasets, available at http://lar.wsu.edu/megan/guides.html. The anthropogenic and biogenic emissions are spatially gridded to the CMAQ domain.

19.3 Forecasting Results

The modeling system produces the time evolution of the concentration for several species (NO_2, SO_2, CO, O_3, $PM_{2.5}$ and PM_{10}). To illustrate the forecasts produced by the model, Fig. 19.2 shows concentrations calculated for the urban area of Jinan based on mean of available stations. Observations of real-time hourly pollutant concentrations measured by the China National Environmental Monitoring Center are also plotted for comparison. Such time series are available for a large number of cities in Eastern China.

19.3.1 Modeled Spatial Pattern of Pollutants

The spatial distribution of the mean concentrations of predicted surface layer NO_2, O_3, and $PM_{2.5}$ over the study period (20 UTC 7 Aug to 20 UTC 1 September 2016) is shown in Fig. 19.3. It can be seen that model predicts elevated NO_2 concentrations over the North China Plain and major city clusters at YRD, Perl River Delta (PRD), and Sichuan Basin. O_3 levels are relatively high for Central China, part of Northern China Plain, and Northwest China. NO_x appears to play titration effects over some regions, leading to lower O_3 at locations with higher NO_2 (e.g. Shanghai, Jiangsu, and Shandong provinces). The spatial pattern of $PM_{2.5}$ concentrations is similar to that of NO_2, with relatively higher concentrations (~60 $\mu g/m^3$) over the North China Plain, Central China, and Sichuan Basin.

19.3.2 Evaluation of Criteria Pollutants

We evaluate the model results for the forecasting period from 20 UTC Aug 7 to 20 UTC 1 September 2016. Observational dataset are based on the real-time hourly pollutant concentrations measured by the China National Environmental

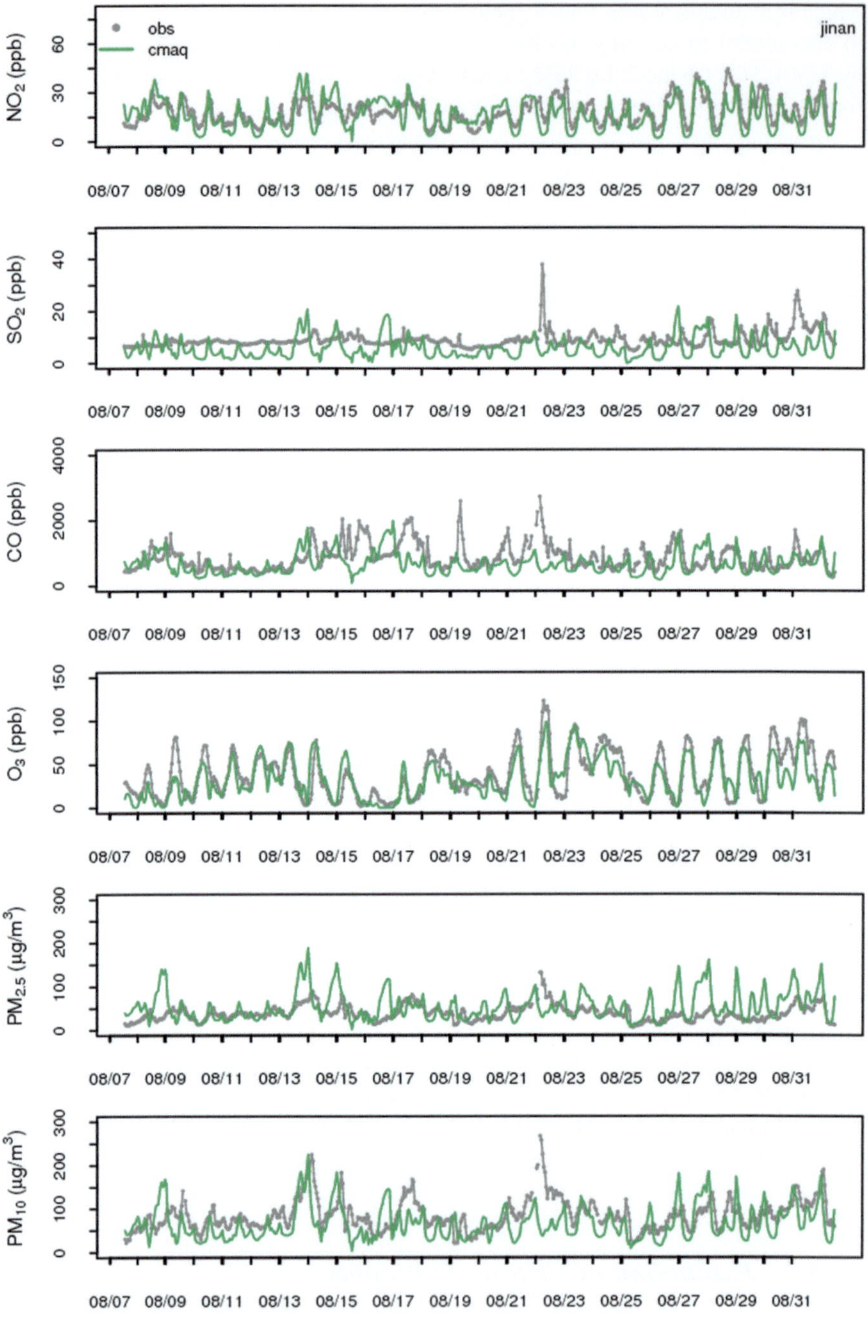

Fig. 19.2 Observation and CMAQ calculated concentrations for Jinan (based on mean of available stations)

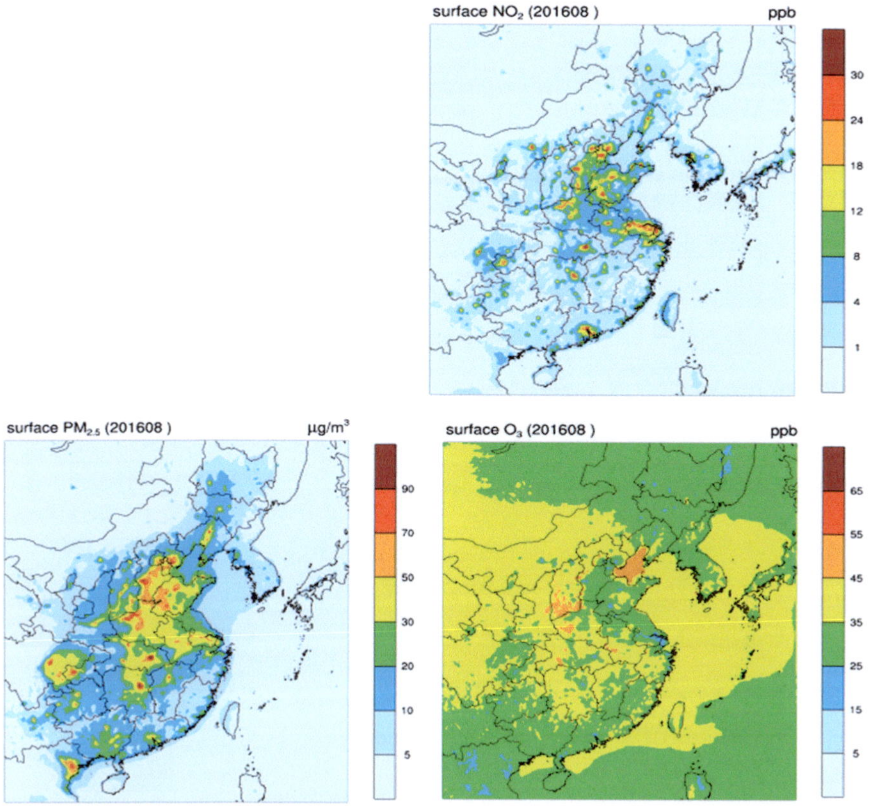

Fig. 19.3 The spatial distribution of the mean concentrations of predicted surface layer NO_2, O_3, and $PM_{2.5}$ (20 UTC 7 Aug to 20 UTC 1 September 2016)

Monitoring Center. Median of CMAQ modeled criteria pollutants concentrations are compared with the median of observed concentrations for Shanghai and eight provincial cities in Eastern China (i.e. Nanjing, Hefei, Nanchang, Hangzhou, Jinan, Zhengzhou, Fuzhou, and Guangzhou) (Fig. 19.4 and Table 19.1). The nine cities are located at different geographical regions, with five cities (Shanghai, Nanjing, Hefei, Nanchang, and Hangzhou) within or adjacent to YRD, Jinan and Zhengzhou of north the YRD, and Fuzhou and Guangzhou south of it.

For the five cities within or close to YRD, the results show that modeled CO and PM_{10} are mostly in good agreement with observation. SO_2 are slightly underestimated for most cities. NO_2 are generally overestimated, with Shanghai and Nanjing showing the largest overestimation (bias of 91% and 67%, respectively). Meanwhile, O_3 concentrations are consistently underestimated at these cities (ranging from −24% to −57%) with the largest bias found in Shanghai and Nanjing, indicating that the bias in O_3 is likely related to the overestimation of NO_x and its titration effects. The region generally experienced low $PM_{2.5}$ concentration

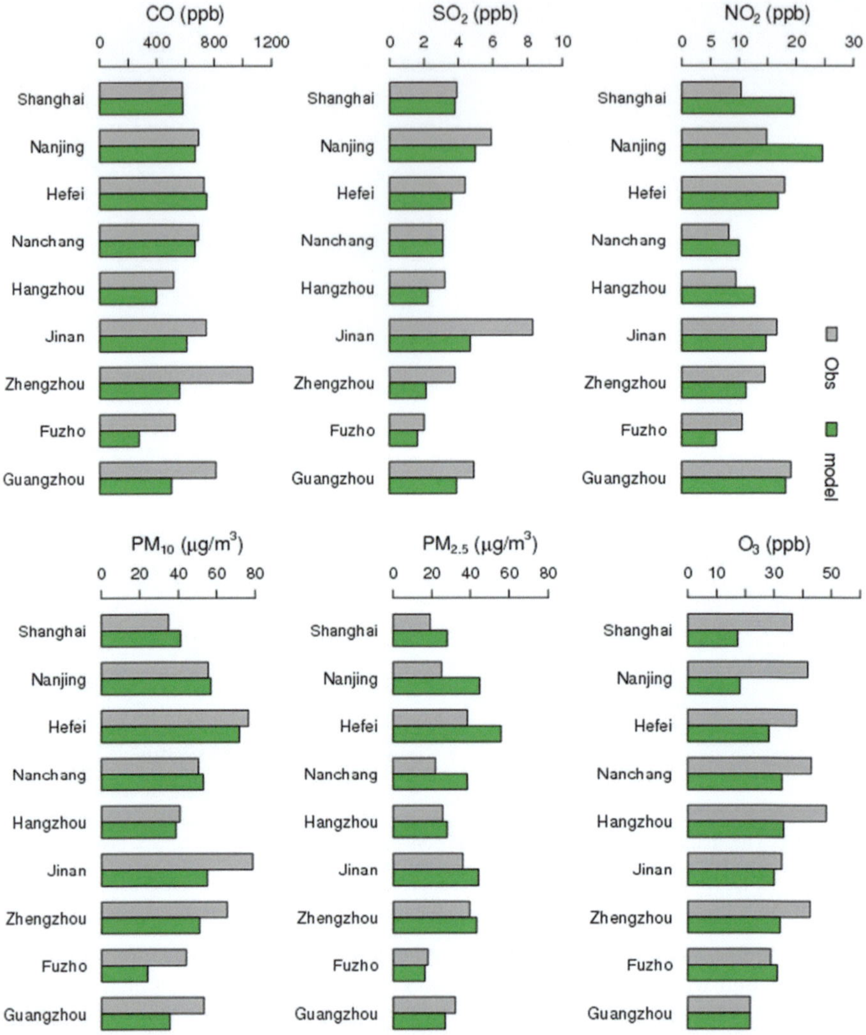

Fig. 19.4 Median of CMAQ predicted pollutants concentrations compared with median of observations for shanghai and eight provincial cities in Eastern China during the study period

during this study period, with observed median concentrations in between 19 μg/m^3 (Shanghai) and 38 μg/m^3 (Hefei). Compared to the observation, $PM_{2.5}$ is also generally overestimated in the region (bias of 9–79%).

It should be pointed out that emission reduction measures were taken in Hangzhou and Zhejiang province as well as the surrounding regions such as Shanghai and Jiangsu province in preparation for the upcoming G20 Summit. Therefore, the overestimation of NO_2, $PM_{2.5}$ and underestimation of O_3 for these cities is most likely due to emission reductions taken at this occasion. Remarkable drops in

Table 19.1 Model bias at selected cities during the study period

Cities	NO_2 (%)	SO_2 (%)	CO (%)	O_3 (%)	$PM_{2.5}$ (%)	PM_{10} (%)
Shanghai	91	−2	0	−53	48	18
Nanjing	67	−16	−3	−57	79	3
Hefei	−6	−18	2	−25	45	−6
Nanchang	23	1	−4	−24	76	5
Hangzhou	35	−33	−23	−31	9	−5
Jinan	−11	−43	−18	−8	22	−30
Zhengzhou	−23	−45	−48	−25	9	−22
Fuzhou	−43	−20	−48	8	−9	−46
Guangzhou	−5	−21	−38	0	−17	−33

Reported as the median of the CMAQ values minus observations, divided by the median of the observations

emissions and pollutant concentrations under air quality measures for major international events have been reported by a number of studies (Hao et al. 2011; Wang et al. 2010; Mijling et al. 2013). For example, a drop of about 40% in NO_x emissions was reported in Beijing during the summer 2008 Olympic Games.

For Jinan and Zhengzhou, which located north to YRD in the North China Plain, the model showed relative good performance for NO_2, O_3 and $PM_{2.5}$, while large underestimation of SO_2 (about −40%), and PM_{10} (about −30%) are found for the two cities. For Fuzhou and Guangzhou, which is located south to YRD, O_3 and $PM_{2.5}$ concentrations are well predicted, whereas NO_2, CO and PM_{10} tend to be underestimated.

19.4 Discussion and Conclusions

Based on the SMS regional meteorological model (SMS-WARMS) and the CMAQ chemical transport model, a new chemical weather forecasting system is being established for Eastern China. The anthropogenic emissions are based on HTAP 2010 dataset. Large reductions in NO_x and SO_2 emissions are applied for the forecast in order to account for likely rapid changes in economic activity. Biogenic emissions are estimated using the MEGAN model with global land cover maps and environmental variables provided by the SMS-WARMS.

Model forecasts of different atmospheric pollutants performed from 7 August to 1 September 2016 have been evaluated. The preliminary results reported here show substantial differences in model performance for different regions. For the five cites located in YRD and its adjacent regions, large overestimation of NO_2, $PM_{2.5}$ and underestimation of O_3 are found, likely due to the air quality measures taken in preparation for the G20 Summit at Hangzhou in September 2016. For the provincial cites located north (Jinan, Zhengzhou) and south (Fuzhou, Guangzhou) of YRD, modeled O_3 and $PM_{2.5}$ show relatively good agreement with observation.

Nevertheless, some gas-phase pollutants (e.g. SO_2 and CO) as well as PM_{10} tend to be underestimated in these cities.

The overestimation of NO_2 and $PM_{2.5}$ in YRD suggests air quality measures taken in preparation of the G20 Summit were fairly effective in reducing pollutants concentrations. Meanwhile, it should be noted that we have applied relatively large reductions in the 2010 HTAP NO_x (reduced by 55%) and SO_2 (reduced by 85%) emissions in our forecasting. The predicted NO_2 is in relatively close agreement with observations or exhibits some positive bias for most cites evaluated here. The resulted SO_2 show some underestimation, but mostly within −20%. The required large reduction suggests that regular updates of emission inventory might be necessary to capture the rapid changes in pollutant emissions in Eastern China. Top-down emission inventories based on satellite observations should provide valuable insights.

We should point out that this preliminary study is based on a relatively short time period of model forecasting results, and that the model evaluation has only been performed for a few provincial cities in Eastern China. Further extensive evaluation (e.g. key metrological parameters, diurnal variations and seasonal variations of pollutant concentrations, performance over urban vs. rural locations) will fully capture the temporal and spatial patterns in model biases and errors. Such evaluation should be performed as a routine procedure in real-time forecasting. Furthermore, the diagnostic evaluations suggested by Dennis et al. (2010) will identify specific model processes or model inputs responsible for model biases or errors.

Acknowledgments The authors are grateful to the financial support from the European Community's Seventh Framework Programme (FP7) under the PANDA project grant agreement n° 606719.

References

Appel, K. W., Napelenok, S. L., Foley, K. M., Pye, H. O. T., Hogrefe, C., Luecken, D. J., Bash, J. O., Roselle, S. J., Pleim, J. E., Foroutan, H., Hutzell, W. T., Pouliot, G. A., Sarwar, G., Fahey, K. M., Gantt, B., Gilliam, R. C., Heath, N. K., Kang, D., Mathur, R., Schwede, D. B., Spero, T. L., Wong, D. C., & Young, J. O. (2017). Description and evaluation of the Community Multiscale Air Quality (CMAQ) modeling system version 5.1. *Geoscientific Model Development, 10*, 1703–1732. doi:10.5194/gmd-10-1703-2017.

Byun, D., & Schere, K. L. (2006). Review of the governing equations, computational algorithms, and other components of the Models-3 Community Multiscale Air Quality (CMAQ) modeling system. *Applied Mechanics Reviews, 59*, 51–77.

Carlton, A. G., Bhave, P. V., Napelenok, S. L., Edney, E. O., Sarwar, G., Pinder, R. W., Pouliot, G. A., & Houyoux, M. (2010). Model representation of secondary organic aerosol in CMAQv4.7. *Environmental Science and Technology, 44*, 8553–8560.

Chen, F., & Dudhia, J. (2001). Coupling an advanced land-surface/hydrology model with the Penn State/NCAR MM5 modeling system. Part I: Model description and implementation. *Monthly Weather Review, 129*, 569–585.

Dennis, R., Fox, T., Fuentes, M., Gilliland, A., Hanna, S., Hogrofe, C., Irwin, J., Trivikrama, R., Scheffe, R., Schere, K., Steyn, D., & Venkatram, A. (2010). A framework for evaluating regional-scale numerical photochemical modeling systems. *Environmental Fluid Mechanics*. doi:10.1007/s10652-009-9163-2.

Foley, K. M., Roselle, S. J., Appel, K. W., Bhave, P. V., Pleim, J. E., Otte, T. L., Mathur, R., Sarwar, G., Young, J. O., Gilliam, R. C., Nolte, C. G., Kelly, J. T., Gilliland, A. B., & Bash, J. O. (2010). Incremental testing of the Community Multiscale Air Quality (CMAQ) modeling system version 4.7. *Geoscientific Model Development, 3*, 205–226. doi:10.5194/gmd-3-205-2010.

Guenther, A. B., Jiang, X., Heald, C. L., Sakulyanontvittaya, T., Duhl, T., Emmons, L. K., & Wang, X. (2012). The Model of Emissions of Gases and Aerosols from Nature version 2.1 (MEGAN2.1): An extended and updated framework for modeling biogenic emissions. *Geoscientific Model Development, 5*, 1471–1492. doi:10.5194/gmd-5-1471-2012.

Hao, N., Valks, P., Loyola, D., Cheng, Y. F., & Zimmer, W. (2011). Space-based measurements of air quality during the World Expo 2010 in Shanghai. *Environmental Research Letters, 6*, 044004. doi:10.1088/1748-9326/6/4/044004.

Hong, S.-Y., Noh, Y., & Dudhia, J. (2006). A new vertical diffusion package with an explicit treatment of entrainment processes. *Monthly Weather Review, 134*, 2318–2341.

Iacono, M. J., Delamere, J. S., Mlawer, E. J., Shephard, M. W., Clough, S. A., & Collins, W. D. (2008). Radiative forcing by long-lived greenhouse gases: Calculations with the AER radiative transfer models. *Journal of Geophysical Research, 113*, D13103. doi:10.1029/2008JD009944.

Janssens-Maenhout, G., Crippa, M., Guizzardi, D., Dentener, F., Muntean, M., Pouliot, G., Keating, T., Zhang, Q., Kurokawa, J., Wankmüller, R., Denier van der Gon, H., Kuenen, J. J. P., Klimont, Z., Frost, G., Darras, S., Koffi, B., & Li, M. (2015). HTAP_v2.2: A mosaic of regional and global emission grid maps for 2008 and 2010 to study hemispheric transport of air pollution. *Atmospheric Chemistry and Physics, 15*, 11411–11432. doi:10.5194/acp-15-11411-2015.

Li, M., Zhang, Q., Kurokawa, J.-I., Woo, J.-H., He, K., Lu, Z., Ohara, T., Song, Y., Streets, D. G., Carmichael, G. R., Cheng, Y., Hong, C., Huo, H., Jiang, X., Kang, S., Liu, F., Su, H., & Zheng, B. (2017). MIX: A mosaic Asian anthropogenic emission inventory under the international collaboration framework of the MICS-Asia and HTAP. *Atmospheric Chemistry and Physics, 17*, 935–963. doi:10.5194/acp-17-935-2017.

Marécal, V., Peuch, V.-H., Andersson, C., Andersson, S., Arteta, J., Beekmann, M., Benedictow, A., Bergström, R., et al. (2015). A regional air quality forecasting system over Europe: The MACC-II daily ensemble production. *Geoscientific Model Development, 8*, 2777–2813. doi:10.5194/gmd-8-2777-2015.

Menut, L., & Bessagnet, B. (2010). Atmospheric composition forecasting in Europe. *Annales de Geophysique, 28*, 61–74. doi:10.5194/angeo-28-61-2010.

Mijling, B., van der A, R. J., & Zhang, Q. (2013). Regional nitrogen oxides emission trends in East Asia observed from space. *Atmospheric Chemistry and Physics, 13*, 12003–12012. doi:10.5194/acp-13-12003-2013.

Thompson, G., Field, P. R., Rasmussen, R. M., & Hall, W. D. (2008). Explicit forecasts of winter precipitation using an improved bulk microphysics scheme. Part II: Implementation of a new snow parameterization. *Monthly Weather Review, 136*, 5095–5115.

Wang, Z., Wu, Q., Gbaguidi, A., Yan, P., Zhang, W., Wang, W., & Tang, X. (2009). Ensemble air quality multi-model forecast system for Beijing (EMS-Beijing): Model description and preliminary application. *Journal of Nanjing University of Information Science & Technology (Natural Science Edition), 1*(1), 19–26. (in Chinese).

Wang, S., Zhao, M., Xing, J., Wu, Y., Zhou, Y., Lei, Y., He, K., Fu, L., & Hao, J. (2010). Quantifying the air pollutants emission reduction during the 2008 olympic games in Beijing. *Environmental Science & Technology, 44*, 2490–2496.

Yarwood, G., Rao, S., Yocke, M., & Whitten, G. (2005). *Updates to the carbon bond chemical mechanism: CB05*. Final report to the US EPA, EPA Report Number: RT-0400675. Available at: www.camx.com

Zhang, Y., Bocquet, M., Mallet, V., Seigneur, C., & Baklanov, A. (2012). Real-time air quality forecasting, part I: History, techniques, and current status. *Atmospheric Environment, 60*, 632–655.

Zhou, G., Geng, F., Xu, J., Xie, Y., Peng, L., Chen, M., & Tie, X. (2015). Numerical ozone forecasting over Shanghai. *China Environmental Science, 35*(6), 1601–1609. (in Chinese).

Chapter 20
Modelling Assessment of Atmospheric Composition and Air Quality in Eastern and Southern Asia

Mikhail Sofiev, Rostislav Kouznetsov, Julius Vira, Joana Soares, Marje Prank, Jukka-Pekka Jalkanen, Lasse Johansson, and Ari Karppinen

Abstract Current chapter outlines the model-based assessment of air pollution in Eastern and Southern Asia. The chemistry transport model SILAM, which covers the main sources of the air pollutants in the region, was applied to evaluate their influence on spatial and temporal characteristics of the regional pollution pattern. We showed that, apart from the anthropogenic sources, air pollution in several parts of Eastern and Southern Asia is dominated by other sources, such as desert dust and vegetation fires. In particular, South-East Asia and Eastern Russia are heavily impacted by the biomass burning smoke, largely from agriculture fires. Fire-induced pollution is also episodically significant in several provinces of China.

Quality and availability of the emission data for the region is often insufficient. It is demonstrated that emission inversion task can be solved for Asia using satellite information and extended four-dimensional variational data assimilation, finally leading to refined emission estimates. In particular, the inverse problem solution suggests that the seasonal cycle of primary aerosol emission is likely to have two peaks rather than one as assumed in the bulk of inventories. This conclusion, however, has to be taken with care since it can be affected by the lacking summer-time aerosols in the model, especially secondary organics and desert dust.

The model evaluation for the region is largely based on the satellite information. Limited datasets for China and India are available over a comparatively short time period, and a few examples of the SILAM evaluation with these datasets are provided.

Keywords Air pollution in Asia • AQ modelling • Aerosol sources in Asia

---

M. Sofiev (✉) • R. Kouznetsov • J. Vira • J. Soares • M. Prank • J.-P. Jalkanen
L. Johansson • A. Karppinen
Finnish Meteorological Institute, P.O. Box 503, FI-00101 Helsinki, Finland
e-mail: Mikhail.Sofiev@fmi.fi; Rostislav.Kouznetsov@fmi.fi; Julius.Vira@fmi.fi; Joana.Soares@fmi.fi; Marje.Prank@fmi.fi; Jukka-Pekka.Jalkanen@fmi.fi; Lasse.Johansson@fmi.fi; Ari.Karppinen@fmi.fi

I. Bouarar et al. (eds.), *Air Pollution in Eastern Asia: An Integrated Perspective*,
ISSI Scientific Report Series 16, DOI 10.1007/978-3-319-59489-7_20

20.1 Introduction

Global pattern of atmospheric pollution has been gradually changing during recent couple of decades. In Europe and Northern America, there was a clear downward trend in the main pollutants, except, in some regions, near-surface ozone concentrations. In Eastern and Southern Asia, the trends are clearly upwards, except for Sulphur dioxide in the last 3–4 years in China. Extremely high near-surface concentrations, as well as mounting evidence of the pollution uplift to the upper troposphere and the stratosphere suggest a hemispheric-to-global distribution scale of Asian plumes. With the recent recognition of air pollution as a cancer risk factor by the WHO, these call for comprehensive understanding of the situation in the region and development of science-based strategies of sustainable development.

Experience of environmental policy development in Europe and the US highlighted the power of combined usage of observations and atmospheric pollution modelling. Such a combination allows for both monitoring of the current situation and its in-depth analysis followed by development of facts-based abatement measures (e.g. Clean Air Policy Package in Europe, http://ec.europa.eu/environment/air/clean_air_policy.htm, Clean Air Act in the US https://www.epa.gov/clean-air-act-overview, and, especially, European Monitoring and Evaluation Programme, http://www.emep.int, a scientifically-based and policy-driven programme under the UN-ECE Convention on Long-range Transboundary Air Pollution, CLRTAP).

20.2 Input Data for Model Applications

Applications of atmospheric composition models require three types of input data:

- Meteorological information for driving the transport simulations
- Emission data characterizing the sources of air pollution
- Observations of the atmospheric concentrations of the tracer species to provide the reference points for the models

Meteorological information is available either from global models, such as the Integrated Forecasting System IFS of European Centre of Medium-Range Weather Forecast ECMWF or Global Forecast System GFS of National Centers for Environmental Prediction NCEP, or from various regional weather prediction systems operated by national meteorological services.

An overview of the available observations, in-situ and space-born, are presented in the Parts III and IV of this book, respectively. It is shown that despite large parts of the region are not covered by the ground networks, a combination of the in-situ information with remote-sensing data can provide a comparatively comprehensive picture.

Information on emission is arguably more uncertain. The multitude of sources contributing to the current pollution level and limited knowledge on some of them adds to the complexity of the problem. For instance, the large differences between the existing inventories of anthropogenic sources are highlighted in the Part II, "Sources of air pollution", of this book. A disagreement as large as a factor of times exists for several key pollutants even at country- and regional-total levels. Apart from the disagreements of different inventories, all of them are outdated: the region develops so fast that 3–4 years-old inventories can already incorporate errors of several tens of percent for country totals. In some regions, the pollution pattern can become completely different if, e.g., a large industrial or a power-generation facility has been built during these years.

20.3 Emission Refinement via Data Assimilation

The above uncertainties in the emission data pose challenges for applications of atmospheric composition models in the region. From one side, the exceptionally high concentrations of many reactive pollutants may question the completeness and accuracy of the physical and chemical transformation schemes developed and evaluated in less polluted regions, such as Europe or the US. From the other side, interpretation and attribution of the model errors have to account for limited accuracy of the input data, and first of all, emission.

One of the ways to improve the timeliness of the emission data and to stay up to date with the rapid region development is to use the observation-based emission refinement methodologies. There is a multitude of such algorithms, (e.g., Elbern et al. 2007; Stohl et al. 2011; Vira and Sofiev 2012), see also Part II of this book, but they all generally follow the same paradigm: the emission fields are adjusted so that the modelled concentrations meet the observations better than the simulations with the original a-priori emission fields. A strong assumption of the approach is that the errors in the model-measurement comparison are implicitly considered to originate from the emission uncertainties. Application of such top-down methodologies is therefore justified only in the regions where the information on emission is not sufficiently good, so that the bulk of uncertainty indeed comes from the emission data rather than the model shortcuts and/or meteorological data. A large disagreement between the inventories shown in Part II of this book suggest that Eastern and Southern Asia may be one of such regions.

The objectives of the current chapter are to quantify the level of air pollution in Asia as it is seen by the models based on available input data, to estimate the contribution of different sources, and to quantify some of the existing uncertainties and model-measurement (dis)agreements. Keeping in mind the above limitations, the analysis of Asian pollution presented below includes:

(i) General assessment of the large-scale trends and features of the pollution patterns
(ii) Analysis of the most uncertain contributors to the regional emissions
(iii) Refinement of the bulk emission estimates using top-down approach and data assimilation
(iv) Episodic and long-term simulations and sensitivity studies aiming at better understanding the features and root causes of the regional air pollution.

20.4 Atmospheric Composition Assessment for 1980–2015

The information presented in the following sections is based on a global hindcast of atmospheric composition, whose settings are described in the Annex. The simulations were made with the chemical transport model SILAM and covered the period of 1980–2015. In addition, several shorter periods were computed zooming towards the Eastern and Southern Asia. SILAM is a global-to-mesoscale chemical transport model (see Annex for details) that covers both the troposphere and the stratosphere and incorporates all necessary chemical and physical transformations.

A challenge for the long-term hindcast was the homogeneity requirement to all input data: their methodologies and features should stay the same throughout the whole period. Several otherwise very good sources of information were excluded from the considerations because they did not cover the considered period and/or contained strong internal inhomogeneity, switches between strongly different datasets, etc.

The emission input (see Part II for details concerning the specific datasets) was compiled from the MACCITY anthropogenic inventory combined with the ACCMIP biomass burning data, MEGAN model (http://lar.wsu.edu/megan, (Guenther et al. 1995, 2006)) calculations of biogenic isoprene and monoterpene emissions, lightning emission of NO_2 (Price et al. 1997) and RETRO aircraft emission.

The meteorological dataset used for the long-term computations was the ERA-Interim (Dee et al. 2011) reanalysis. For the episodes, we used the high-resolution forecasts of the IFS from the ECMWF (http://www.ecmwf.int).

The in-situ observational data were compiled from several sources. The most extensive real-time collection of basic observations is available from http://www.pm25.in. However, the site does not support archiving of the observations and does not provide the station locations. Real-time data are also shown at http://aqicn.org, though without downloading possibilities. An in-situ dataset covering more than a decade with well-documented and openly available observations at the stations located in several countries is the Acid Deposition Network in East Asia, EANET (http://www.eanet.asia), whose first observations date back to late 1990s (EANET 2000).

Due to limited availability of the in-situ observational data, a stress in the emission refinement and model evaluation was put to remote-sensing data, which are available in near-real-time, cover the whole globe and in some cases go back to 1980s (e.g., the series of ATSR-type satellites provide aerosol optical thickness and hot-spot counts (North et al. 1999)). The main sources of information were MODIS aerosol optical depth AOD (Levy and Hsu 2015) and fire FRP (Justice et al. 2002) products; AATSR AOD (Leeuw et al. 2003), and OMI SO_2 (Fioletov et al. 2013) and NO_2 (Boersma et al. 2011) products were used for the model parameterization and evaluation.

20.5 Changes in Large-Scale Pollution Patterns During Last 35 Years

An illustration of the major changes occurred during the last 35 years is shown in Fig. 20.1, where the mean-annual concentrations near surface for O_3 and black carbon (BC) are shown for 1980 and 2015. The simulations were performed with the MACCITY emission inventory. According to these computations, the main changes occurred in the Northern Hemisphere. In the Southern Hemisphere, a certain increase of ozone in Argentina is probably the only visible trend. Differences visible in southern Africa and Amazonia are mainly due to inter-annual variability of the fire emission.

In the Northern Hemisphere, the changes in pollution pattern are fundamental but depend on the species considered. The major sources in Europe and the US in

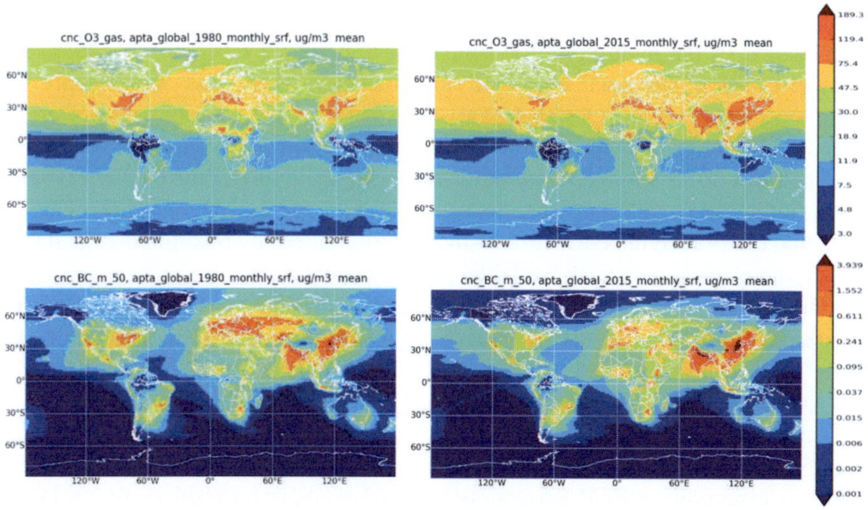

Fig. 20.1 Near-surface mean annual concentration of ozone (*upper row*) and black carbon (*low row*) in 1980 (*left*) and 2015 (*right*), MACCITY emission. Unit [μg m^{-3}]

1980 were strongly reduced or removed by 2000s, which resulted in reduction of the concentrations of most of primary pollutants over these continents. Improvement is also visible over the former Soviet Union countries, which have strongly reduced their emission (for reasons varying from country to country). The impact of these reductions reached a factor of several times for black carbon, whose concentrations respond linearly to changes of the emission strength. For ozone and several other secondary pollutants, the pattern is more complicated. The computations suggest noticeable reduction of ozone only in the US while the European level stayed either almost same or even increased, especially in the north. In Central and Southern Europe, concentrations changed very little.

The only region, where the increase is clear and very significant for all pollutants is Eastern and Southern Asia. For BC, already strong pollution in 1980 in China grew even further by 2015. In the area, the high ozone concentrations (50–60 ppb as annual mean) have almost doubled during this period. Qualitatively, these changes, especially for the primary species, are self-evident consequences of the fast growth of industry and coal-based power production. Quantitative evaluation of these model predictions is not so straightforward: observations in 1980s are very limited even in Europe and the US and essentially absent in Asia.

20.6 Refinement of Emission Information via Top-Down Emission Estimation

Considering the task of refinement of the emission inventories, one has to keep in mind that the efforts towards the comprehensive emission inventories presented in other chapters of this book resulted in a series of inventories with quite accurate localization of the main pollution sources at a moderate resolution of 25–40 km (a size of large industrial or urban agglomerates). Much less accurate are the absolute fluxes and their temporal variations between the seasons, days of week, and diurnal cycle. As mentioned in the introduction, the emission flux in a specific grid cell can differ strongly from the value in the inventory if the dataset misses a (new) industrial installation. Therefore, a task for the top-down emission refinement is to find time-resolving correction to the emission fluxes by assimilating the up-to-date remote-sensing data (if available, also in-situ data).

The technique used for the below refinement is four-dimensional variational assimilation, in more details presented in Vira and Sofiev (2012). The input data were taken from combined MODIS AQUA and MODIS TERRA aerosol optical density retrievals.

An example of the a-priori and a-posteriori emission is shown in Fig. 20.2. The baseline a-priori emission was taken from MACCITY dataset, which, starting from 2010, is essentially a RCP 8.5 emission model prediction (Riahi et al. 2011). As seen from the charts for BC + OC and SO_2, the annual absolute emission is somewhat too low in the RCP but quite close to the a-posteriori evaluation.

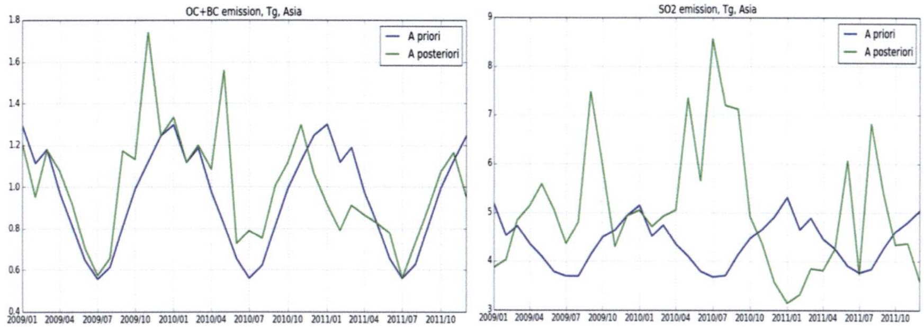

Fig. 20.2 Monthly emission in Eastern and Southern Asia for OC + BC (*left*) and SO_2 (*right*) in 2009–2011, sum over the region Fig. 20.3 unit: Tg mon^{-1}

Seasonality of the a-posteriori emission differs strongly from the a-priori RCP distribution, especially for SO_2: the model suggested a noticeable peak in its emission during summer months, instead of wintertime as the a-priori estimates assume. There are several reasons for this disagreement. Firstly, there seems to be an elevated summer SO_2 release from the power production sector due to the high electricity demand during hot season. This increase in the power generation sector is also reflected in the bottom-up REAS emission inventory (Kurokawa et al. 2013), albeit the seasonality there is more conservative: 10–15%. However, this rise competes with domestic combustion, which peaks in winter due to the residential heating. As a result, the total SO_2 emission of REAS has small seasonality, whereas the inversion suggests it at a level of 25–30% (mean over 3 years). The second reason is artificial: the current simulations underestimate aerosols especially in summer due to missing secondary organic aerosols and mineral dust components (those components were not included in that run for technical reasons). Compensating this deficit, the assimilation procedure could have raised SO_2 emission as the most-efficient way to increase the predicted AOD.

Primary BC + OC emission variation suggested presence of two peaks – in late autumn and early spring – rather than the single mid-winter rise. The pattern repeats itself, almost exactly, in all 3 years but both carbonaceous aerosols and SO_2 were estimated from the same MODIS AOD data, which allows for misattribution of the model-measurement discrepancy. The smaller emission fluxes of carbonaceous aerosols are more sensitive to such artificial effects: a relatively small fraction of misattributed SO_2 would mean substantial changes in BC + OC a-posteriori emission. These characteristics bring our attention to the common limitation of the top-down approach: the model uncertainties propagate into the emission fluxes. Thus, the deficit of particulate matter in summer months caused by a lack of chemical production from organic aerosol precursors was likely offset by the stronger emission correction.

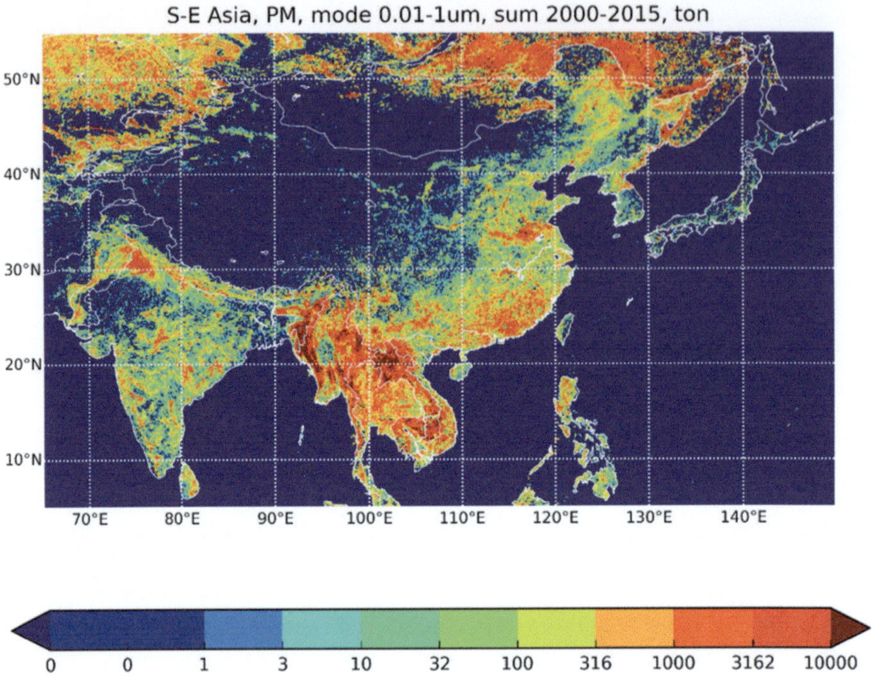

Fig. 20.3 Spatial distribution of the total PM emission from vegetation fires, sum over 2002–2015, IS4FIRES v.2.0. Unit: [ton PM]

20.7 Contribution of Specific Types of Sources

20.7.1 Effect of Fire-Induced PM Emission

Vegetation fires are among the highly varying sources of atmospheric pollution in the region. The 16-years-long IS4FIRES reanalysis based on active-fire records of MODIS instrument, showed that both intensity and location of the fires are strongly different for different regions (Fig. 20.3) and season (Fig. 20.4). The inter-annual variability is about 25%, with particularly strong fires recorded in 2003 in May over the Russian territory to the east of Baikal Lake followed by 2012 and 2015 in March–April south of China, in May in Russia, followed by summertime agriculture fires in China.

From the distribution map in Fig. 20.3, it is well seen that the main fire sources in Eastern and Southern Asia are located outside the industrial areas of China and India. By far the most powerful fires take place over large areas in the north – Russia and Kazakhstan – as well as in the countries south of China: Laos, Burma, etc. The smoke from these fires easily covers the whole area.

Chinese fires are comparatively moderate and largely originate from agriculture burning in the eastern provinces in late summer. But the fire-prone areas in China

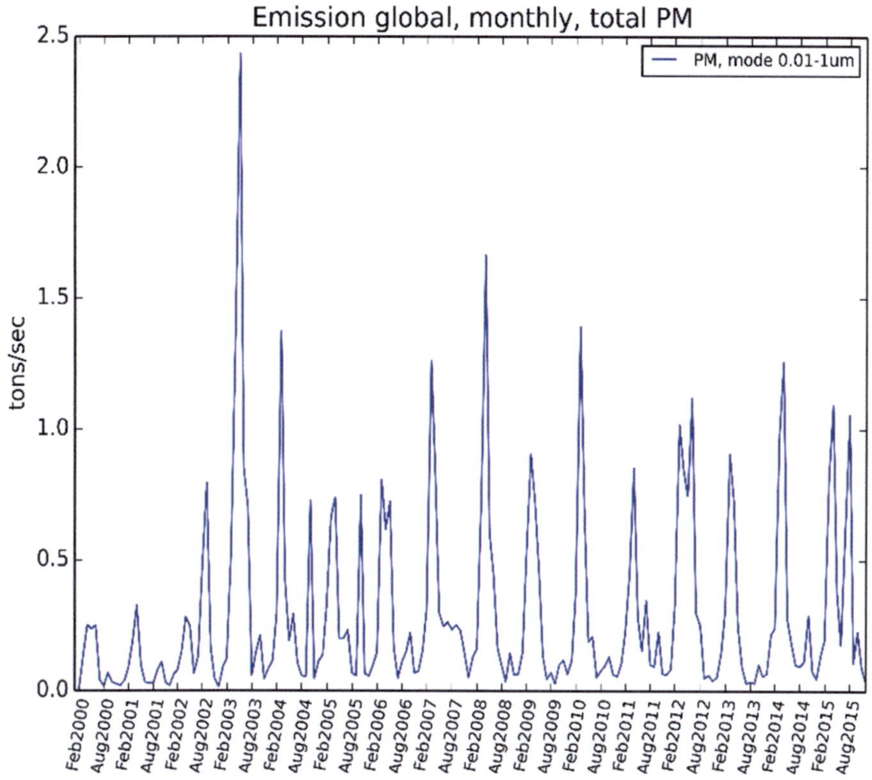

Fig. 20.4 Temporal evolution of the total PM emission from vegetation fires, sum over the domain in Fig. 20.3. Unit: [ton PM mon^{-1}]

are characterized by an intense peak of emission in spring (Fig. 20.4). As an illustration, Fig. 20.5 depicts the August 2007 monthly emissions, where the agriculture fires in Shanghai region release about as much PM as the strong forest fires in the north of the domain. A corresponding contribution to the regional AOD level can exceed 50% as a monthly average.

From this quick overview, one can conclude that the vegetation fires in most cases contribute to the background PM levels over China, dominating the landscape only in the second part of summer (July and August) in the eastern and Eastern and Southern provinces.

20.7.2 Desert Dust

Wind-blown dust is arguably the most-uncertain part of the PM emission. Despite the long history of the dust modelling (first comparatively comprehensive

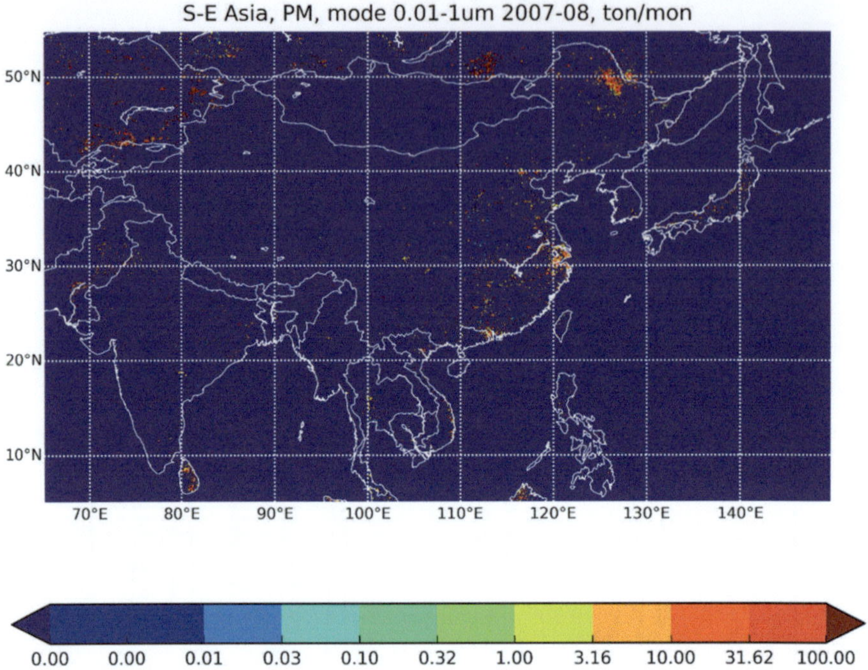

Fig. 20.5 Monthly fire PM emission in August, 2007 according to IS4FIRES v.2.0. Agriculture fires in Shanghai area and forest/grassland burning in the north dominate the emission pattern

approaches were created in the middle of twentieth century in connection with soil erosion – e.g. (Chepil 1945)), presently there is no universally applicable parameterization that would provide similar quality of predictions over all Earth deserts. The basic principles, however, have been settled, so that the models are mainly based on saltation mechanism parameterizations with some variations regarding the functional approximations of observational data and specific numerical constants tuned to improve the agreement with observations (e.g., Marticorena and Bergametti 1995; Zender 2003a, b; Textor et al. 2006). The SILAM emission formulations for wind-blown dust follow similar path, using satellite and Aeronet optical data to identify the numerical constants (see, e.g. intercomparison of operational dust forecasts in http://sds-was.aemet.es/).

Deserts occupy large parts of northern China, yet the bulk of dust emission takes place at comparatively limited areas of Taklimakan and Gobi. In absence of other sources, desert dust completely determines the PM concentrations in those regions. Downwind of Gobi, there are populated and industrialized regions of China, where the PM is emitted by anthropogenic sources, such as industry and traffic. During dust outbreaks, they dominate the coarse aerosol load over the country, whereas fine particles over densely habituated areas are still coming mainly from anthropogenic sources or from gaseous precursors, also largely of anthropogenic origin.

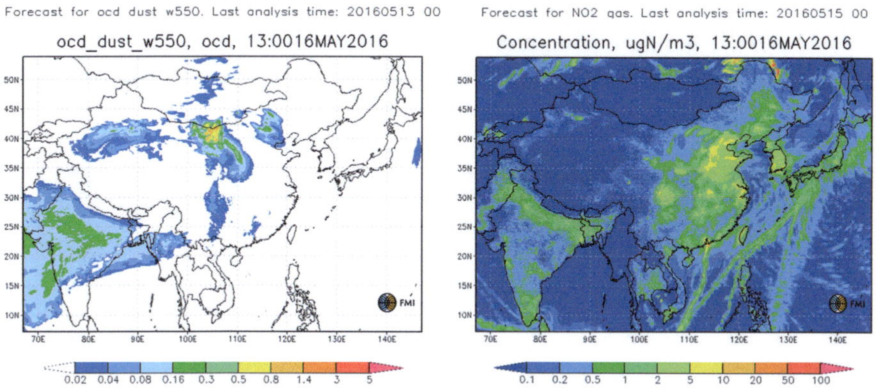

Fig. 20.6 Contribution of different sources to total AOD at 13:00, 16 May 2016. *Left panel*: AOD due to desert dust only. Gobi dust plume reaches towards industrialized parts of China. *Right panel*: Near-surface NO_2 concentration with the effect of Russian forest-fiers visible

An example of episodic dust contribution to AOD over Eastern and Southern Asia is shown in Fig. 20.6. Its left-hand panel depicts a dust plume stretching from Gobi region towards the Central China populated areas. The Taklimakan area is essentially self-polluted, whereas over India the wind-blown dust is predicted to contribute about half of the AOD due to the anthropogenic particles. The strongest but limited in space impact comes from the forest fires in Russia (Fig. 20.6, right-hand panel for NO_2 plumes). South China is essentially self-polluted: impact of all remote sources is predicted to be small for that day.

20.7.3 Marine Sources: Anthropogenic Emission from Ships and Wind-Induced Sea Salt Aerosol

Contribution of ship traffic to the overall pollution pattern is comparatively small in Eastern and Southern Asia, at least according to the emission estimates of MACCITY or PanHam. The reason is the very high emissions from terrestrial sources in Asia, which makes the marine sources all but irrelevant anywhere except for the immediate vicinity of harbors or heavy-marine-traffic lines. This conclusion is illustrated in Fig. 20.6, which shows that the impact of marine sources is restricted to the location of the ship lines. The total AOD over sea is largely dominated by sea salt (not shown), so that the ship lanes are distinguishable from AOD only near harbors and in places with very intense traffic. This is in contrast with Europe, where the model shows that $PM_{2.5}$, NO_2 and SO_2 concentrations induced by ships over marine areas are often higher than over many terrestrial densely populated regions.

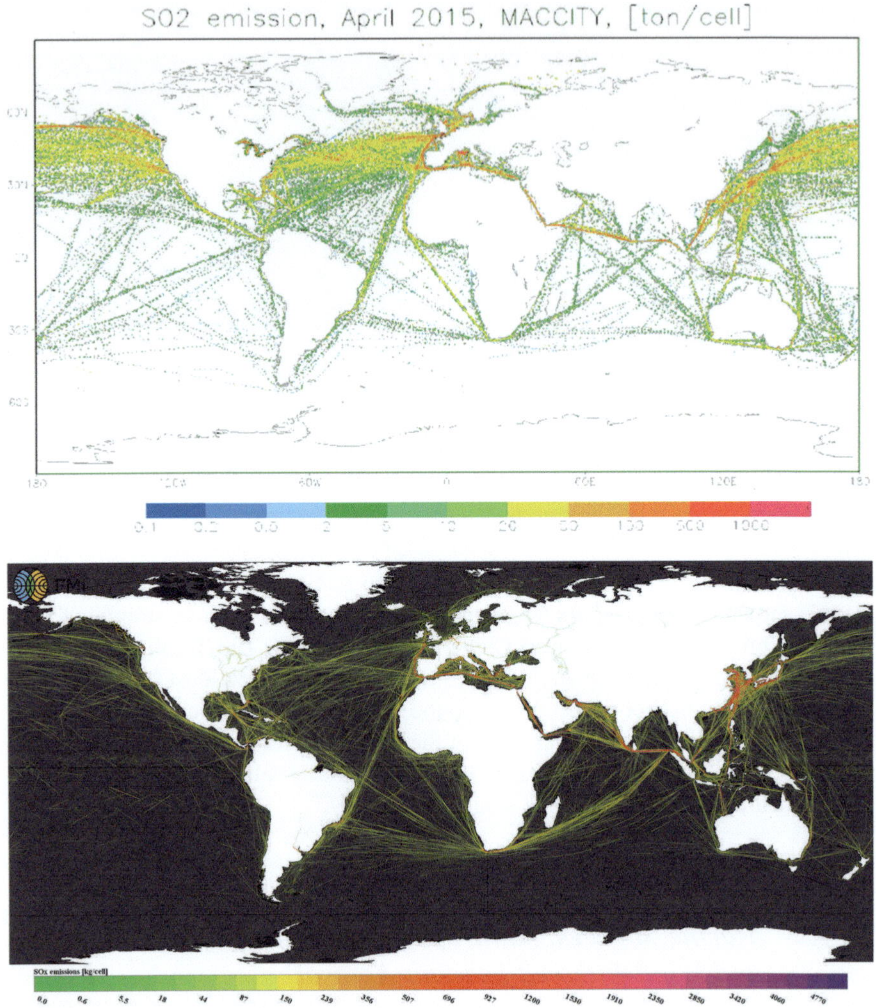

Fig. 20.7 Ship emission of SO_2 in April 2015. *Upper panel*: MACCITY (i.e., RCP 8.5, unit: ton per 0.5 × 0.5° grid cell). *Lower panel*: estimated by the STEAM model based on global AIS data (unit: kg per 10 × 10 km grid cell)

There is however an important uncertainty in the above conclusion: the ship emission inventory in MACCITY is an RCP extrapolation of the economic activity and emission factors over a period of more than a decade. Such estimates are bound to have limited accuracy. For instance, MACCITY suggests noticeably higher emission flux density in Europe than in Asia (Fig. 20.7, upper panel). However, the independent ship emission model STEAM, based on actual ship activity data (Jalkanen et al. 2009) suggests the opposite relation: emission density in coastal seas in Asia is several times higher than that in Europe. If this adjustment is taken into account, the impact of ship traffic on coastal areas can be substantial despite very strong terrestrial emissions. It is, however, worth mentioning that the terrestrial emissions in MACCITY seem to be somewhat under-estimated too (Fig. 20.2).

20.8 How Accurate Are the Modelling Predictions?

In the above sections, we discussed the model predictions and some insight that they can provide into the mechanisms forming the present-day air pollution in Asia. However, these conclusions are only as good as the model and input data used for the simulations. SILAM has been developed in Europe primarily for European applications, which resulted in its extensive evaluation for that region. Global applications are less numerous and primarily address the fire sources (IS4FRIES and the corresponding module in SILAM) and sea salt emission (Soares et al. 2015; Sofiev et al. 2009, 2011).

Variability of the air pollution drivers in Eastern and Southern Asia requires spatially-distributed and species-specific accuracy assessment. The problem is further complicated by a lack of open observations, especially for past time periods. Evaluation of the large-scale phenomena and emission refinement is therefore easier to build using the remote-sensing information – as shown in Sect. 20.5.

Examples of qualitative evaluation against $PM_{2.5}$ observations made at automated AQ stations at the US Embassies in 4 Chinese cities in 2010 (Beijing) and 2013 (Beijing, Chengdu, Guangzhou, Shanghai) are shown in Fig. 20.8. The modelled time series were extracted from the SILAM computations driven by PanHam emission inventory (red lines in the charts). For comparison, the 2010 episode was also computed with MACCITY inventory (black line in the Bejing-2010 chart). This comparison again points out at the under-estimation of the PM-related emission in MACCITY. It also suggests a possibly over-estimated emission in PanHam. Regardless the emission input, temporal correlation is high at Beijing and Shanghai whereas Guangzhou and Chengdu show more disagreement.

20.9 Annex: Models and Input Data

20.9.1 The SILAM Chemistry Transport Model

The computations of this chapter have been performed with the System for Integrated modeLling of Atmospheric coMposition (SILAM, http://silam.fmi.fi, Fig. 20.9, Sofiev et al. 2015), which is a global-to-meso-scale chemical transport model developed at Finnish Meteorological Institute FMI and used in research and operational applications related to air quality and emergency. The transport algorithm is based on the Eulerian advection scheme of M.Galperin (Sofiev et al. 2015),

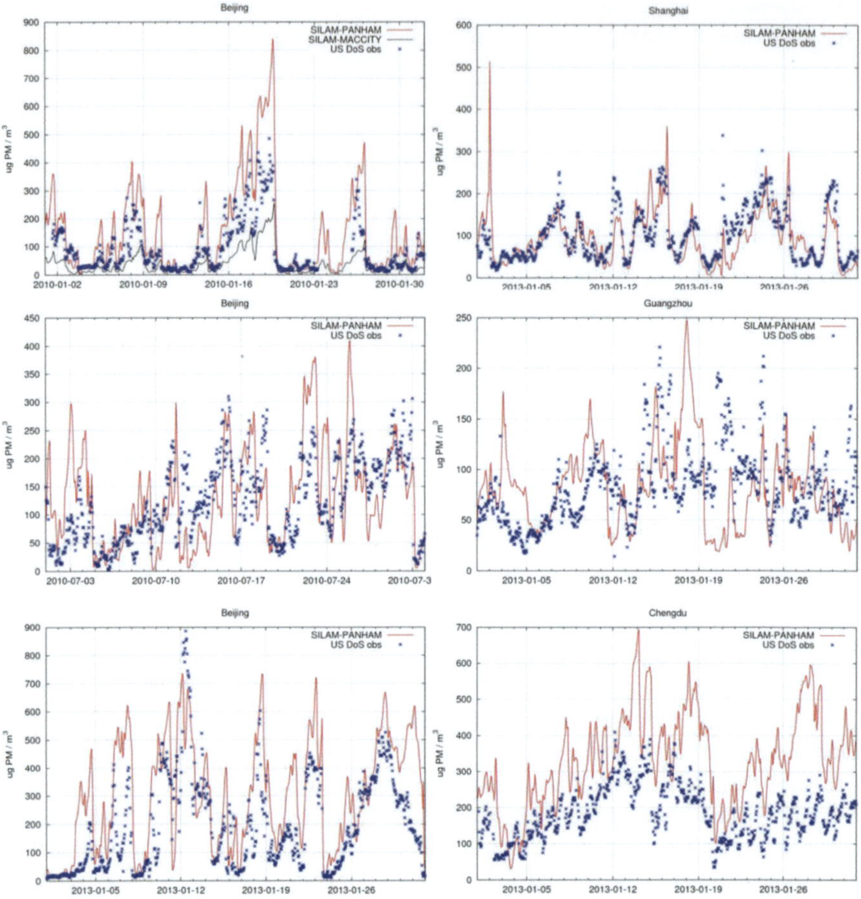

Fig. 20.8 $PM_{2.5}$ time series at 4 US DoS stations in Beijing (January, July 2010 and January 2013, *left column*), Chengdu, Guangzhou, and Shanghai (2013). SILAM time series with PanHam emission are shown as *red lines*. For Beijing 2010, the run with MACCITY is shown (*black line*)

and the adaptive vertical diffusion algorithm of (Sofiev 2002) developed for using thick dispersion layers. The model includes a meteorological pre-processor for diagnosing the basic features of the boundary layer and the free troposphere (diffusivity, similarity scales, and latent and sensible heat fluxes) from the meteorological input fields (Sofiev et al. 2010). For secondary inorganic aerosol formation, the updated chemistry scheme from DMAT model (Sofiev 2000) was used with several extensions, such as the coarse-nitrate formation and refined ammonium nitrate – ammonia – nitric acid equilibrium. The dry deposition scheme for aerosols is described in (Kouznetsov and Sofiev 2012), whereas for gases the simplified approach of the DMAT model is applied.

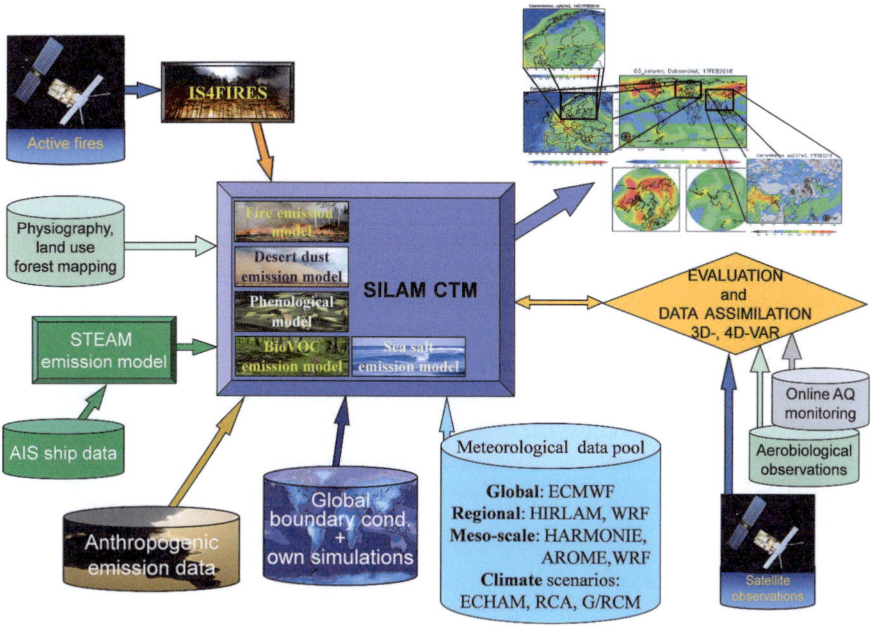

Fig. 20.9 Atmospheric composition and air quality assessment platform of FMI

The SILAM model has been extensively evaluated against air quality observations over Europe and the globe (Huijnen et al. 2010), http://atmosphere.copernicus.eu, (Solazzo et al. 2012a, b). The model has recently been applied to evaluate the dispersion of primary PM2.5 emissions across Europe and in more detail over Finland, and to assess the resulting adverse health impacts (Tainio et al. 2009, 2010; Karvosenoja et al. 2011).

20.9.2 Emission Models and Inventories Used in the Simulations

As shown in Fig. 20.9, SILAM incorporates several emission models, which describe the sources with significant dependencies on meteorological conditions. Thus, the sea-salt emission follows the algorithm of (Sofiev et al. 2011) with improvements described in (Soares et al. 2016).

The new SILAM emission formulations for wind-blown dust follow the path suggested by (Marticorena and Bergametti 1995). Additionally, they involve space-born estimates of the surface roughness (Prigent 2005) and wind gustiness after (Schreur and Geertsema 2008). Application of these new elements required system recalibration that was performed using MODIS and Aeronet AOD data (see, e.g. intercomparison of operational dust forecasts in http://sds-was.aemet.es).

Vegetation fire emissions are produced from active-fire radiative power (FRP) retrieved by the MODIS instrument (Moderate Resolution Imaging Spectroradiometer, http://modis.gsfc.nasa.gov) and converted to emission fluxes following the IS4FIRES v.2 procedure (http://is4fires.fmi.fi, (Sofiev et al. 2009; Soares et al. 2015)). FRP is assumed proportional to the biomass burning rate, which is converted to PM2.5 emission using land-use dependent empirical emission factors (Soares et al. 2015). The Global Land Cover Characterization (GLCC) land-use inventory (Loveland et al. 2000) aggregated to seven major vegetation classes of (Akagi et al. 2011), was used for the fire attribution.

The Ship Traffic Emission Assessment Model (STEAM, (Jalkanen et al. 2009, 2016)) is a global ship emission assessment model based on the bottom-up approach that evaluates emission from each individual ship using its actual position, speed, and other characteristics available from the regional and global AIS systems.

20.9.3 Setup of the Long-Term and Episodic Hindcast

The long-term re-analysis 1980–2015 consisted of several sets of simulations made within the scope of APTA project of Academy of Finland. The results analyzed in the chapter were obtained from global runs including the troposphere and the stratosphere, going up to 10 Pa with 20 uneven hybrid-pressure layers, each combining three corresponding layers of ERA-Interim. The horizontal resolution was taken as half of that of ERA-Interim: 1.44° lon-lat. The internal model time step was 1 h. The output was stored every 3 h for 3-D fields and hourly for surface concentrations and depositions, as well as for the total column load and optical thickness at 380 nm, 550 nm, and 1,020 nm. The chemical transformations were based on CB4 mechanism expanded with the stratospheric chemical scheme of the FinROSE model (Damski et al. 2007).

The episodic regional reanalysis for Eastern and Southern Asia used PanHam emission inventory and IFS archives of operational weather forecasts. It was run with 0.1° lon-lat horizontal resolution and 14 hybrid-pressure layers stacked up to ~335 hPa (~8.5 km height). Lateral and top boundary conditions were taken from the global re-analyses of C-IFS of ECMWF and of the SILAM-APTA global hindcast. The surface fields, column load, and optical thickness at 550 nm were stored at hourly basis. The computations covered China, India, Japan, and surrounding countries: (73E-146E, 17 N–50 N).

References[1]

Akagi, S.K. et al. (2011). Emission factors for open and domestic biomass burning for use in atmospheric models. *Atmospheric Chemistry and Physics, 11*(9), 4039–4072. Available at: http://www.atmos-chem-phys.net/11/4039/2011

Boersma, K. F., et al. (2011). An improved tropospheric NO2 column retrieval algorithm for the ozone monitoring instrument. *Atmospheric Measurement Techniques, 4*(9), 1905–1928.

Chepil, W. S. (1945). Dynamics of wind erosion I, nature of movement of soil by wind. *Soil Science, 60*, 305–320.

Damski, J. et al. (2007). FinROSE: Middle atmospheric chemistry transport model. *Boreal Environment Research, 12*(5), 535–550. Available at: http://cat.inist.fr/?aModele=afficheN&cpsidt=19218894

Dee, D. P., et al. (2011). The ERA-Interim reanalysis: Configuration and performance of the data assimilation system. *Quarterly Journal of the Royal Meteorological Society, 137*(656), 553–597.

EANET (2000). Report on the acid deposition monitoring of EANET during the preparatory phase – its results, major constraints and ways to overcome them – August 2000 Interim Scientific Advisory Group, Japan.

Elbern, H., et al. (2007). Emission rate and chemical state estimation by 4-dimensional variational inversion. *Atmospheric Chemistry and Physics, 7*, 3749–3769.

Fioletov, V. E., et al. (2013). Application of OMI, SCIAMACHY, and GOME-2 satellite SO_2 retrievals for detection of large emission sources. *Journal of Geophysical Research-Atmospheres, 118*(19), 11399–11418.

Guenther, A. et al. (1995). A global model of natural volatile organic compound emissions. *Journal of Geophysical Research, 100*(D5), 8873–8892. Available at: http://www.agu.org/pubs/crossref/1995/94JD02950.shtml

Guenther, A. et al. (2006). Estimates of global terrestrial isoprene emissions using MEGAN (Model of Emissions of Gases and Aerosols from Nature). *Atmospheric Chemistry and Physics, 6*, 3181–3210. Available at: http://www.atmos-chem-phys-discuss.net/6/107/2006

Huijnen, V., et al. (2010). Comparison of OMI NO_2 tropospheric columns with an ensemble of global and European regional air quality models. *Atmospheric Chemistry and Physics, 10*(7), 3273–3296.

Jalkanen, J.-P. et al. (2009). A modelling system for the exhaust emissions of marine traffic and its application in the Baltic Sea area. *Atmospheric Chemistry and Physics, 9*(4), 9209–9223. Available at: http://www.atmos-chem-phys.net/9/9209/2009

Jalkanen, J. P., Johansson, L., & Kukkonen, J. (2016). A comprehensive inventory the ship traffic exhaust emissions in the European sea areas in 2011. *Atmospheric Chemistry and Physics, 16*, 71–84.

Justice, C.O. et al. (2002). The MODIS fire products. *Remote Sensing of Environment, 83*, 244–262. Available at: http://gis-lab.info/docs/justice02_the_modis_fire_products.pdf

Karvosenoja, N., et al. (2011). Integrated modeling assessments of the population exposure in Finland to primary PM2.5 from traffic and domestic wood combustion on the resolutions of 1 and 10 km. *Air Quality, Atmosphere and Health, 4*, 179–188.

Kouznetsov, R., & Sofiev, M. (2012). A methodology for evaluation of vertical dispersion and dry deposition of atmospheric aerosols. *Journal of Geophysical Research, 117*, D01202.

Kurokawa, J., et al. (2013). Emissions of air pollutants and greenhouse gases over Asian regions during 2000–2008: Regional Emission inventory in ASia (REAS) version 2. *Atmospheric Chemistry and Physics, 13*(21), 11019–11058.

[1]All Web sites quoted in the text have been accessed 22.06.2016.

Leeuw, G. D., et al. (2003). Retrieval of aerosol optical depth from satellite measurements using single and dual view algorithms. In O. L.-B. Klaus Schäfer (Ed.), *Remote sensing of clouds and the atmosphere VII* (pp. 275–283). Bellingham: SPIE.

Levy, R. & Hsu, C., et al. (2015). NASA MODIS Adaptive processing system, Available at: http://dx.doi.org/10.5067/MODIS/MOD04_L2.006

Loveland, T., et al. (2000). Development of a global land cover characteristics database and IGBP DISCover from 1 km AVHRR data. *International Journal of Remote Sensing, 21*(607), 1303–1330.

Marticorena, B., & Bergametti, G. (1995). Modelling the atmospheric dust cycle: 1. Design of a soil-derived dust emission scheme. *Journal of Geophysical Research, 100*(D8), 16415–16430.

North, P. R. J., et al. (1999). Retrieval of land surface bidirectional reflectance and aerosol opacity from ATSR-2 multiangle imagery. *IEEE Transactions on Geoscience and Remote Sensing, 37* (1), 526–537.

Price, C. G., Penner, J. E., & Prather, M. J. (1997). NOx from lightning, Part I: Global distribution based on lightning physics. *Journal of Geophysical Research, 102*(D5), 5929–5941.

Prigent, C. (2005). Estimation of the aerodynamic roughness length in arid and semi-arid regions over the globe with the ERS scatterometer. *Journal of Geophysical Research, 110*(D9), D09205. Available at: http://doi.wiley.com/10.1029/2004JD005370

Riahi, K., et al. (2011). RCP 8.5-A scenario of comparatively high greenhouse gas emissions. *Climatic Change, 109*(1), 33–57.

Schreur, B. & Geertsema, G. (2008). Theory for a TKE based parameterization of wind gusts. *HIRLAM Newsletter, 54*, 177. Available at: https://scholar.google.com.tr/scholar?start=20&q=allintitle:+gust+OR+gusts&hl=tr&as_sdt=0,5&as_ylo=2008&as_yhi=2008#2

Soares, J. et al. (2016). Impact of climate change on the production and transport of sea salt aerosol on European seas. *Atmospheric Chemistry and Physics Discussions, 16*, 1–52. Available at: http://www.atmos-chem-phys-discuss.net/acp-2015-1056

Soares, J., Sofiev, M., & Hakkarainen, J. (2015). Uncertainties of wild-land fires emission in AQMEII phase 2 case study. *Atmospheric Environment, 115*, 361–370.

Sofiev, M. (2000). A model for the evaluation of long-term airborne pollution transport at regional and continental scales. *Atmospheric Environment, 34*(15), 2481–2493.

Sofiev, M. (2002). Extended resistance analogy for construction of the vertical diffusion scheme for dispersion models. *Journal of Geophysical Research-Atmospheres, 107*(D12), ACH 10–1.

Sofiev, M., et al. (2009). An operational system for the assimilation of the satellite information on wild-land fires for the needs of air quality modelling and forecasting. *Atmospheric Chemistry and Physics, 9*(18), 6833–6847.

Sofiev, M., et al. (2010). Diagnosing the surface layer parameters for dispersion models within the meteorological-to-dispersion modeling interface. *Journal of Applied Meteorology and Climatology, 49*(2), 221–233.

Sofiev, M. et al. (2011). A regional-to-global model of emission and transport of sea salt particles in the atmosphere. *Journal of Geophysical Research, 116*(D21302), 25. Available at: http://doi.wiley.com/10.1029/2010JD014713

Sofiev, M. et al. (2015). Construction of an Eulerian atmospheric dispersion model based on the advection algorithm of M. Galperin: Dynamic cores v.4 and 5 of SILAM v.5.5. *Geoscientific Model Development, 8*, 3497–3522. Available at: http://www.geosci-model-dev-discuss.net/8/1/2015/

Solazzo, E., Bianconi, R., Vautard, R., et al. (2012a). Model evaluation and ensemble modelling of surface-level ozone in Europe and North America in the context of AQMEII. *Atmospheric Environment, 53*, 60–74.

Solazzo, E., Bianconi, R., Pirovano, G., et al. (2012b). Operational model evaluation for particulate matter in Europe and North America in the context of AQMEII. *Atmospheric Environment, 53*, 75–92.

Stohl, A. et al. (2011), Determination of time- and height-resolved volcanic ash emissions and their use for quantitative ash dispersion modeling: The 2010 Eyjafjallajökull eruption.

Atmospheric Chemistry and Physics, 11(9), 4333–4351. .Available at: http://www.atmos-chem-phys.net/11/4333/2011/. Accessed 10 Aug 2011.

Tainio, M., et al. (2009). Evaluation of the European population intake fractions for European and Finnish anthropogenic primary fine particulate matter emissions. *Atmospheric Environment, 43* (19), 3052–3059.

Tainio, M., et al. (2010). Uncertainty in health risks due to anthropogenic primary fine particulate matter from different source types in Finland. *Atmospheric Environment, 44*(17), 2125–2132.

Textor, C. et al. (2006). Analysis and quantification of the diversities of aerosol life cycles within AeroCom. *Atmospheric Chemistry and Physics, 6*, 1777–1813. Available at: http://www.atmos-chem-phys.net/6/1777/2006

Vira, J. & Sofiev, M. (2012). On variational data assimilation for estimating the model initial conditions and emission fluxes for short-term forecasting of SOx concentrations. *Atmospheric Environment, 46*, 318–328. Available at: http://linkinghub.elsevier.com/retrieve/pii/S1352231011010296

Zender, C.S. (2003a). Mineral dust entrainment and deposition (DEAD) model: Description and 1990s dust climatology. *Journal of Geophysical Research, 108*(D14), 4416. Available at: http://doi.wiley.com/10.1029/2002JD002775

Zender, C.S. (2003b). Spatial heterogeneity in aeolian erodibility: Uniform, topographic, geomorphic, and hydrologic hypotheses. *Journal of Geophysical Research, 108*(D17), 4543. Available at: http://doi.wiley.com/10.1029/2002JD003039

Chapter 21
Chemical and Meteorological Feedbacks in the Formation of Intense Haze Events

Meng Gao, Gregory R. Carmichael, Yuesi Wang, Pablo E. Saide, Zirui Liu, Jinyuan Xin, Yunpeng Shan, and Zifa Wang

Abstract Intense haze events in China provide ideal opportunities to study meteorological and chemical feedbacks due to extremely high aerosol loadings. In this chapter, an online coupled meteorology-chemistry model, WRF-Chem, is applied to simulate impacts of aerosol feedbacks on meteorology and air quality during the January 2010 haze event over the North China Plain (NCP). The results show that the model reasonably reproduces well most meteorological, chemical and optical variables. Aerosols during this haze event can reduce surface downward shortwave radiation by 25.7% and planetary boundary layer height by 14.9%. Due to aerosol feedbacks, $PM_{2.5}$ concentrations in urban Beijing increase by 11.2% at 14:00. The severe haze also enhances cloud droplet number concentrations, which can further affect cloud chemistry. These results indicate that aerosol feedbacks in the NCP, especially in urban regions, are important and should be considered when develop air pollution control and climate mitigation strategies.

Keywords Aerosol feedbacks • Haze • $PM_{2.5}$ • WRF-Chem

M. Gao (✉) • G.R. Carmichael
Center for Global and Regional Environmental Research, The University of Iowa, Iowa City, IA 52242, USA
e-mail: meng-gao@uiowa.edu; gcarmich@engineering.uiowa.edu

Y. Wang • Z. Liu • J. Xin • Z. Wang
State Key Laboratory of Atmospheric Boundary Layer Physics and Atmospheric Chemistry (LAPC), Institute of Atmospheric Physics, Chinese Academy of Sciences, Beijing 100029, China
e-mail: wys@mail.iap.ac.cn; lzr@dq.cern.ac.cn; xjy@mail.iap.ac.cn; zifawang@mail.iap.ac.cn

P.E. Saide
Atmospheric Chemistry observations and Modeling (ACOM) Lab, National Center for Atmospheric Research (NCAR), Boulder, CO, USA
e-mail: saide@ucar.edu

Y. Shan
Division of Atmospheric Sciences, Desert Research Institute, Reno, NV, USA
e-mail: YunPeng.Shan@dri.edu

I. Bouarar et al. (eds.), *Air Pollution in Eastern Asia: An Integrated Perspective*, ISSI Scientific Report Series 16, DOI 10.1007/978-3-319-59489-7_21

21.1 Introduction

It is known that aerosols play a key role in the climate system. Aerosols affect the climate system through the following two ways: First, they scatter and absorb solar and thermal infrared radiation to alter radiative balance, which is referred to as direct effects (Haywood and Boucher 2000; Lohmann and Feichter 2005). Second, aerosols serve as cloud condensation nuclei (CCN) and ice nuclei (IN) to modify the cloud microphysical characteristics and hence the lifetime and radiative properties of clouds, which is called the indirect effect (Haywood and Boucher 2000; Lohmann and Feichter 2005). The indirect effects can be further classified into first indirect and second indirect. An increase in droplet concentration and a decrease in droplet size due to aerosols are named first indirect effect (Twomey 1977; Haywood and Boucher 2000). Cloud albedo might be enhanced due to these changes, so the first indirect effect is also called "cloud albedo" effect (Haywood and Boucher 2000). An increase in liquid water content, cloud lifetime and changes in precipitation efficiency are named second indirect effect, which is also called "cloud lifetime" effect (Haywood and Boucher 2000). Absorbing aerosols may heat the air to reduce cloud droplet, which is referred to as the semi-direct aerosol effect (Hansen et al. 1997; Lohmann and Feichter 2005).

It is of great importance to understand and quantify the complex influences of aerosols on meteorology and air quality. Several online-coupled meteorology and chemical transport models, including WRF-Chem (Grell et al. 2005), WRF-CMAQ (Yu et al. 2014; Wong et al. 2012) and GEM-AQ (Kaminski et al. 2008), have been developed to operationally predict air quality and to study aerosol-meteorology interactions (Baklanov et al. 2014). Zhang et al. (2010) used the WRF-Chem model to simulate the feedbacks over the continental U.S. in January and July 2010, and found that incoming solar radiation was reduced by up to −9% in January and −16% in July due to aerosols. They also found that planetary boundary layer (PBL) heights were reduced by up to −23% in January and −24% in July. Forkel et al. (2012) used the same model to investigate the impacts of aerosols on meteorology over Europe, and found that the inclusion of indirect aerosol effects led to less cloud water content by up to 70% and more rain water content over the North Atlantic. Saide et al. (2015) studied the impact of smoke on tornadoes using the WRF-Chem model, and revealed that the presence of smoke can lead to lower cloud bases and stronger low-level wind shear, which are conducive to severe thunderstorm development. Other studies also indicate that dust aerosols can change monsoon circulation and precipitation over the North America monsoon and the West Africa monsoon (Zhao et al. 2011, 2012). Most of previous aerosol feedbacks studies were conducted on long-time scale, and the studies on short-time scale events with high aerosol loadings, such as haze events, are limited.

A series of severe haze events have been happening in the North China Plain (NCP), and these events attracted great public attention with extremely high $PM_{2.5}$ concentrations. In January 2010, a regional haze event occurred in the NCP from 16 to 19 January. During this haze event, the highest hourly $PM_{2.5}$ concentration reached 445.6 and 318.1 μg/m^3 in Beijing and Tianjin and the areas with low visibility covered most eastern China regions (Zhao et al. 2013). Several haze episodes occurred during the winter of 2012–2013 in east China and the January 2013 month was reported as the haziest month for the past 60 years in Beijing (Gao et al. 2015, 2016a, b; Wang et al. 2014). During that month, the maximum hourly $PM_{2.5}$ concentrations exceeded 1000 μg/m^3 at some urban measurement sites in the NCP (Zheng et al. 2015). The extremely high aerosol loadings during these haze events offer a great opportunity to study the chemical meteorological feedbacks. In this work, the fully online-coupled WRF-Chem model is employed to simulate aerosol feedbacks for the January 2010 haze event.

21.2 Model Configurations and Simulation Design

The WRF-Chem model simulates meteorological and chemical components simultaneously (Grell et al. 2005) and we apply the version 3.5.1 in this study. As shown in Fig. 21.1, three nested domains are used and grid resolutions are 81 $\times$ 81 km, 27 $\times$ 27 km and 9 $\times$ 9 km, respectively from outermost to innermost. The first domain is configured to cover most areas of the East Asia region, including China, Korea, Japan and Mongolia. The innermost nested domain is configured by setting Beijing as the center. Twenty-seven vertical layers up to 100 hPa are used in this study. The Carbon Bond Mechanism version Z (CBMZ) (Zaveri and Peters 1999) is used as the gas-phase chemical mechanism and the 8-bin sectional MOSAIC model with aqueous chemistry (Zaveri et al. 2008) is used as the aerosol module. MOSAIC simulates all the important aerosol species, including sulfate, nitrate, chloride, ammonium, sodium, BC, primary organic mass, liquid water and other inorganic mass (Zaveri et al. 2008). The major chosen physical parameterization options include the Morrison double-moment cloud-microphysics scheme (Morrison et al. 2005), the RRTM long wave radiation scheme (Mlawer et al. 1997), the Goddard short wave radiation scheme (Chou et al. 1998), the Noah land surface model, and the Yonsei University planetary boundary layer parameterization scheme (Hong et al. 2006). In WRF-Chem, the prognostic size and number distribution, composition and aerosol water are used by aerosol optical property module to affect radiation. The prognostic mass, number, composition and size distribution are also used to calculate aerosol activation and cloud droplet number, which are furtherly used by double-moment Morrison scheme to calculate cloud-microphysics.

Anthropogenic emissions are taken from the 2010 MIX inventory, which is a mosaic Asian anthropogenic emission inventory developed to support the MICS-Asia and the HTAP projects (Li et al. 2015). This inventory includes emissions of sulfur dioxide (SO_2), nitrogen oxides (NO_x), Carbon Monoxide (CO), non-methane volatile organic compounds (NMVOC), NH_3, black carbon (BC), organic carbon (OC), $PM_{2.5}$, PM_{10}, and carbon dioxide (CO_2) by five sectors (power generation, industry, residential, transportation and agriculture). The spatial resolution of this inventory is 0.25° × 0.25°. Biogenic emissions are calculated online using the MEGAN model (Guenther et al. 2006). The National Centers for Environmental Prediction (NCEP) Final Analysis (FNL) dataset was used to provide model meteorological initial and boundary conditions. Chemical initial and boundary conditions are obtained from MOZART-4 forecasts (Emmons et al. 2010).

In order to explore the impacts of aerosol feedbacks on meteorology and air quality, three WRF-Chem simulations were conducted. The WF simulation includes full interactions between aerosols and meteorology (i.e., solar radiation is affected by aerosol concentrations and aerosols interact with clouds). The sensitivity simulation (NF) eliminates aerosols' effects on radiation. The difference between WF and NF cases is used to represent the influences of aerosol direct and semi-direct effects. Currently, WRF-Chem is not able to quantify the impacts of indirect effect. For indirect effect comparison, we simulated another case (MET) by turning off aerosol and chemistry modules. In the MET case, cloud droplet number is prescribed as a constant value ($250\#/cm^3$). The period from 11 to 24 January 2010 was chosen as the modeling period, covering the 2010 NCP haze period (from 16 to 19 January 2010). To overcome the impacts of initial conditions, 3 days were simulated and considered as spin-up time.

21.3 Model Evaluations

The model results are evaluated against meteorological, chemical and optical measurements. The meteorological variables, including temperature at 2 m (T_2), relative humidity at 2 m (RH_2) and wind speeds at 10 m (WS_{10}), in four stations (Beijing, Tianjin, Baoding and Chengde) were used. Surface concentrations of $PM_{2.5}$, NO_2, SO_2 in three sites (Beijing, Tianjin and Xianghe), and Aerosol Optical Depth (AOD) in two sites (Beijing city, Beijing forest) are also used in the evaluation against measurements. The locations of these measurement stations are shown in Fig. 21.1. $PM_{2.5}$ and AOD are typical variables to represent severity of haze pollution.

Figure 21.2 shows the temporal variations of simulated and observed 24-h mean T_2 (a–d), RH_2 (e–h) and WS_{10} (i–l) in the Beijing, Tianjin, Baoding and Chengde stations. These observations are obtained from the China Meteorological Data Sharing Service System (CMDSSS) dataset. From normal days to haze days (haze days are from 16 to 19 January), temperature and relative humidity increased and wind speeds decreased. In general, the temporal variations and magnitudes of T_2, RH_2 are captured well by the model. However, WS_{10} values are overestimated in most of the stations throughout the whole period. This is a common problem of the WRF-Chem and WRF model and has been reported in previous modelling studies (Zhang et al. 2010).

Figure 21.3 displays the temporal variations of the simulated and observed $PM_{2.5}$, NO_2 and SO_2 in the Beijing (a–c), Tianjin (d–f) and Xianghe (g–i) stations. Both the model results and the observations show that the buildups of pollution during the haze event are similar in these three sites, occurring over a large

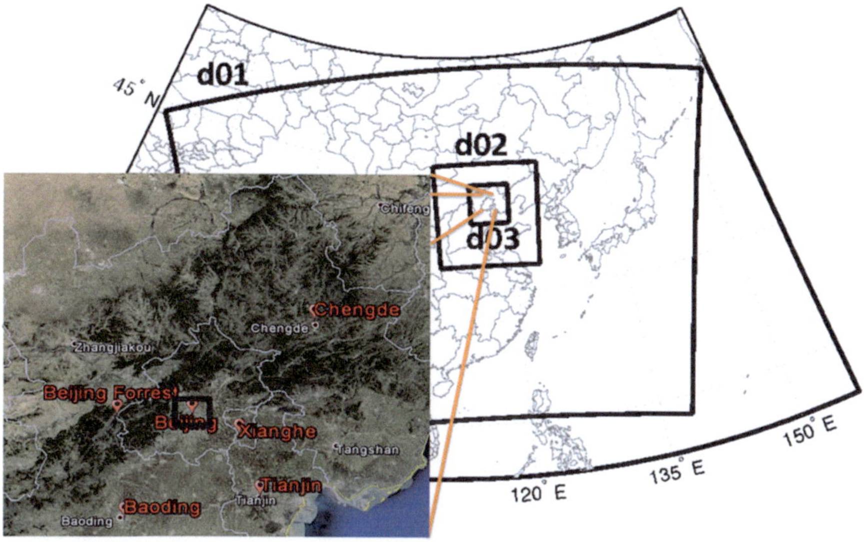

Fig. 21.1 Model domain settings and locations of measurement sites

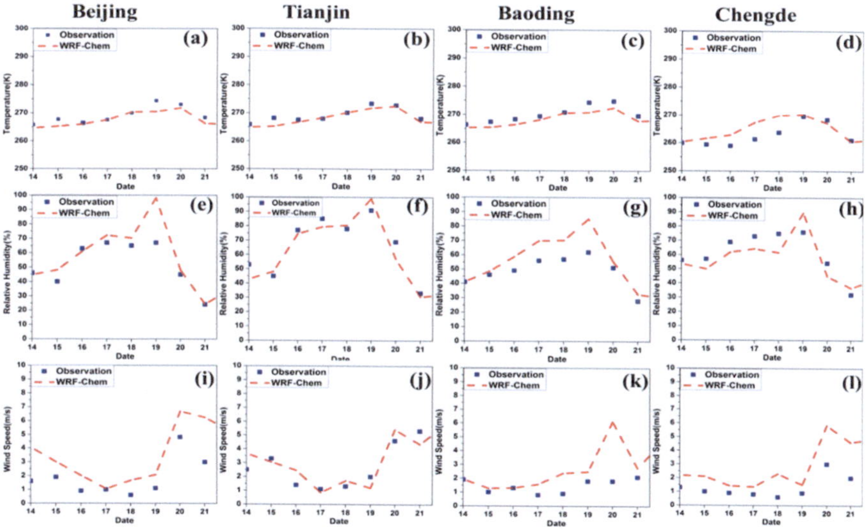

Fig. 21.2 Temporal variations of observed and simulated 24 h averaged temperature (**a–d**), relative humidity (**e–h**) and wind speed (**i–l**) in the Beijing, Tianjin, Baoding, and Chengde stations

geographical region at the same time. The temporal variations of $PM_{2.5}$ and NO_2 are reproduced well by the model. However, SO_2 concentrations are overestimated in the Beijing station, which might be due to the large uncertainties in SO_2 emission inventory. The statistics of the model performance had been calculated and

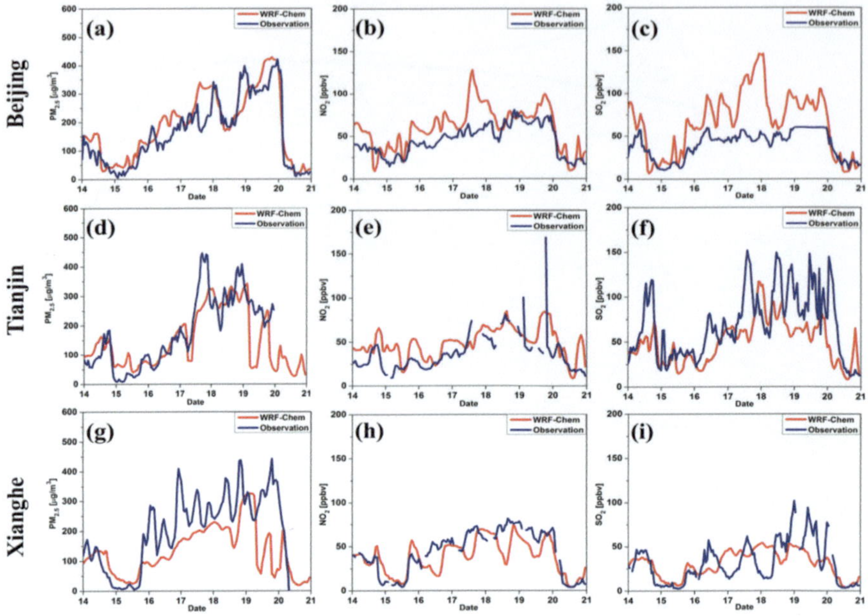

Fig. 21.3 Temporal variations of the simulated and observed $PM_{2.5}$, NO_2 and SO_2 at Beijing (**a**–**c**), Tianjin (**d**–**f**) and Xianghe (**g**–**i**) stations

compared to proposed metrics in Gao et al. (2016c). Although the model performance for $PM_{2.5}$ is satisfactory (Gao et al. 2016c), biases still exist, especially during severe haze days. Reasons for the biases might be errors in meteorological variables, large uncertainties in emissions, effects of horizontal and vertical resolutions, and incomplete treatments of atmospheric chemistry.

Figure 21.4 shows the time series of simulated and observed daily Aerosol Optical Depth (AOD) at 500 nm in the Beijing City and Beijing Forrest stations. The AOD observations are from the Chinese Sun Hazemeter Network (CSHNET). In the WRF-Chem model, aerosol optical properties are calculated using the Mie theory at four specific wavelengths, 300, 400, 600, and 1000 nm. To compare the model AOD to the CSHNET observations, we derive AOD at 500 nm based on the Angstrom exponent relation (Schuster et al. 2006). In severe haze days, AOD could not be retrieved, so the observed AOD data in some days (i.e.g., 19, 20 January) are missing. As shown in Fig. 21.4, the model reasonably reproduces the variations from non-haze days (14–15 January) to haze days (16–18 January) in both the Beijing City and the Beijing Forrest stations. The correlation coefficients between the simulated and observed daily AOD are 0.96 and 0.57 in the Beijing City and the Beijing Forrest stations, respectively. As seen in Gao et al. (2016c), the model had also been compared to CALIPSO satellite data and been proved to capture the vertical distributions of aerosols well.

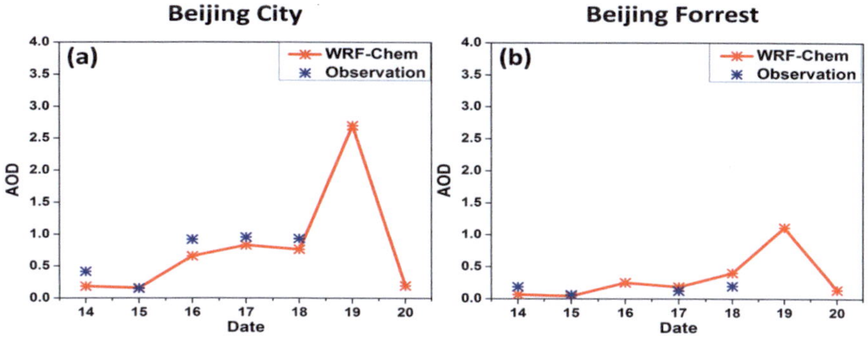

Fig. 21.4 Temporal variations of the simulated and observed daily AOD at 500 nm in the Beijing City (**a**) and Beijing Forrest (**b**) sites

21.4 Results and Discussions

The WRF-Chem model has been shown to be capable of simulating the evolution of the January 2010 haze event. In this section, we explore the impacts of aerosol feedbacks on meteorology and air quality using the outputs from above-mentioned three scenarios.

21.4.1 The Radiative Feedbacks on Meteorology

21.4.1.1 The Impacts on Radiation Budget and PBL Height

During severe haze days, the shortwave radiation reaching the surface may significantly change due to absorbing and scattering of high aerosol loadings. Figure 21.5a, d displays the averaged downwards shortwave radiation (SWDOWN) from 16 to 19 January for the NF case and the differences due to aerosol radiative effects. The domain maximum SWDOWN in the NF case is 141.2 W/m^2, which occurs along the southwestern boundary of the domain. Due to strong aerosol effects during haze, SWDOWN values decrease over the entire domain and the decreasing ratio is about 25.7%. The domain largest decrease is 60.3 W/m^2, which occurs in south Heibei, south Beijing, and west Tianjin regions, above which large aerosols suspend. Due to suspending aerosols, long-wave radiation emitted from the earth can be trapped in the lower atmosphere, and as a result, the outgoing long-wave radiation at the top of the atmosphere (TOA) will be affected. As shown in Fig. 21.5e, outgoing long-wave radiation (OLR) also decreased over the entire domain, but the decreasing ratio is only 0.8%, which is far lower than the decrease of SWDOWN. These results imply that the aerosol feedbacks have less influence on radiation budget at the TOA, but more near the surface. The radiation budget near the surface is an important driver of PBL height evolution.

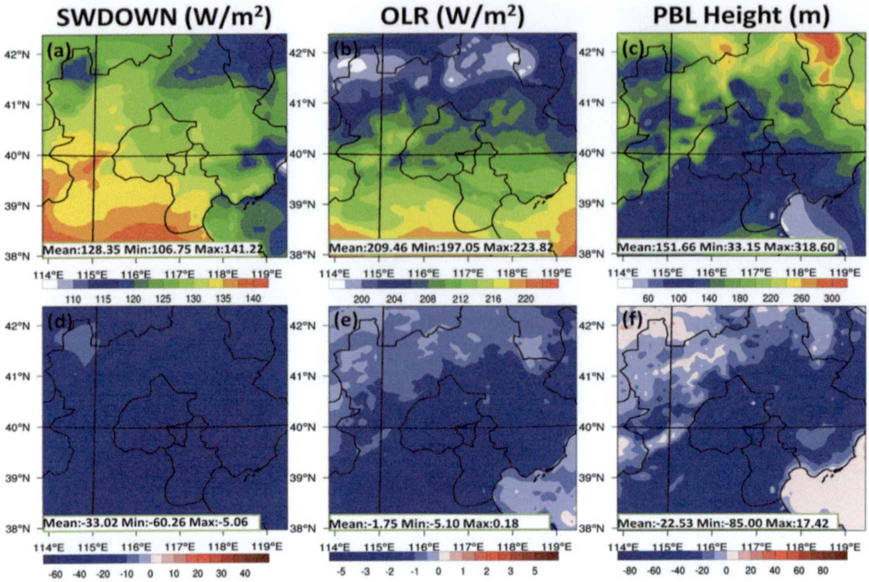

Fig. 21.5 Mean downwards shortwave radiation, outgoing long-wave radiation, and PBL heights during haze days for the NF case (**a–c**), and differences due to aerosol radiative effects (**d–f**)

The responses of PBL heights due to aerosol radiative feedbacks are shown in Fig. 21.5f. The PBL heights decrease over most land surface, but increase along the Bohai Sea rim. Compared to domain averaged PBL heights for the NF case (Fig. 21.5c), the domain mean PBL heights decrease by 14.9% due to aerosol feedbacks. Aerosols affect PBL heights through two ways. First, surface solar radiation is reduced and temperature is reduced at the surface. Second, light-absorbing particles, such as black carbon, brown carbon, heat the atmosphere over the surface. As a result, temperature inversion is enhanced and PBL evolution is suppressed. The decrease of PBL heights over the land is up to 85.0 m. As shown in Fig. 21.5c, averaged PBL heights in south Hebei, Beijing and Tianjin are generally lower than 100 m, which is already unfavorable for diffusions of air pollutants. The decreases in PBL heights additionally aggravate the stagnant weather condition.

The changes of PBL heights shown in Fig. 21.5f are averaged values over both daytime and nighttime, and the changes might be more significant during daytime since PBL heights have strong diurnal trends. Figure 21.6a shows the diurnal variations of PBL heights (m) averaged over the urban Beijing area (defined using the black box shown in Fig. 21.1) and the haze days (16–19 January) in the WF and the NF cases. During nighttime (18:00–23:00 and 0:00–9:00), the averaged PBL heights in both the WF and the NF cases are almost identical and below 50 m. During daytime (10:00–17:00), PBL heights in the WF case are substantially depressed by aerosol feedbacks, and the largest difference is larger than 150 m at 14:00.

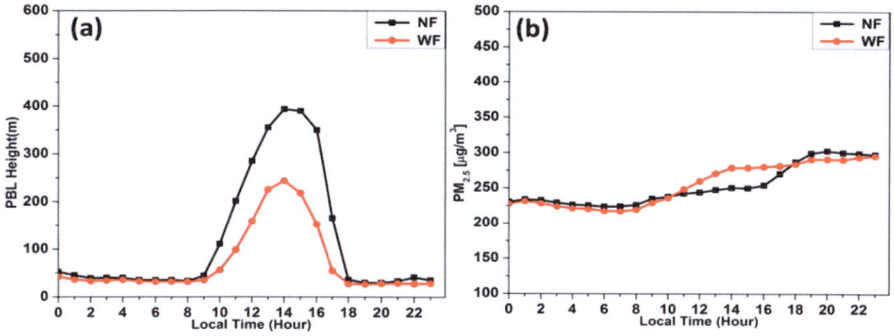

Fig. 21.6 Diurnal variations of PBLH (m) (**a**) and $PM_{2.5}$ concentrations (μg/m^3) averaged over the urban Beijing area and the haze days (16–19 January) in the WF and the NF cases

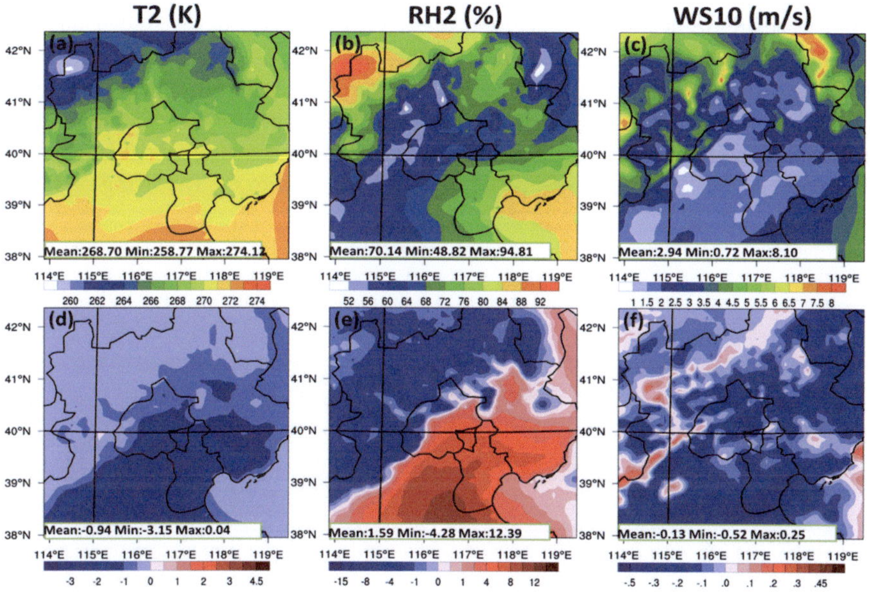

Fig. 21.7 Mean temperature at 2 m, relative humidity at 2 m, and wind speeds at 10 m during haze days for the NF case (**a**–**c**), and differences due to aerosol radiative effects (**d**–**f**)

21.4.1.2 The Impacts on Temperature, Relative Humidity and Wind Speeds

The changes in surface radiation budget can directly lead to changes in near-surface temperature. The spatial changes of temperature at 2 m (T_2) are displayed in Fig. 21.7d, showing similar patterns with the changes of SWDOWN (Fig. 21.5d), with domain mean decrease of 0.9 K due to aerosol radiative effects. The changes of relative humidity at 2 m (RH_2) and wind speeds at 10 m (WS_{10}) are shown in Fig. 21.7e, f. The RH_2 values generally increases over most of the domain. The areas with increasing RH_2 correspond to the areas with large decreases in T_2, suggesting that the increases RH_2 might be related to the decreases in T_2. If the amount of water vapor is constant, decreasing temperature generally increases RH. The domain maximum increase in RH_2 is 12.4%. Due to aerosol radiative effects, WS_{10} is reduced by up to 0.5 m/s, and the domain averaged WS_{10} values decrease by 4.4%, compared to the NF case. The slight decrease of wind speeds may be related to the decrease of temperature since temperature is the driving force of winds.

Figure 21.8a–c shows the temporal variations of vertical profiles of temperature (a), RH (b), and wind speeds (c) differences in Beijing between the WF and NF cases. Due to aerosol radiative effects, temperature near the surface in Beijing decreases by up to 5°, and temperature in the higher air increases by up to 1° (Fig. 21.8a). The changes of RH are opposite to the changes of temperature, with up to 16% increase near the surface and up to 12% decrease in the higher air. The changes of wind speeds also show diurnal variations. As seen Fig. 21.8c, near surface wind speeds are reduced during daytime, but are enhanced during nighttime, because of aerosol radiative feedbacks. The changes in temperature and RH can affect chemical reaction rates, and changes in winds can affect the diffusion and transport of gases and aerosols, which will be discussed in the following section.

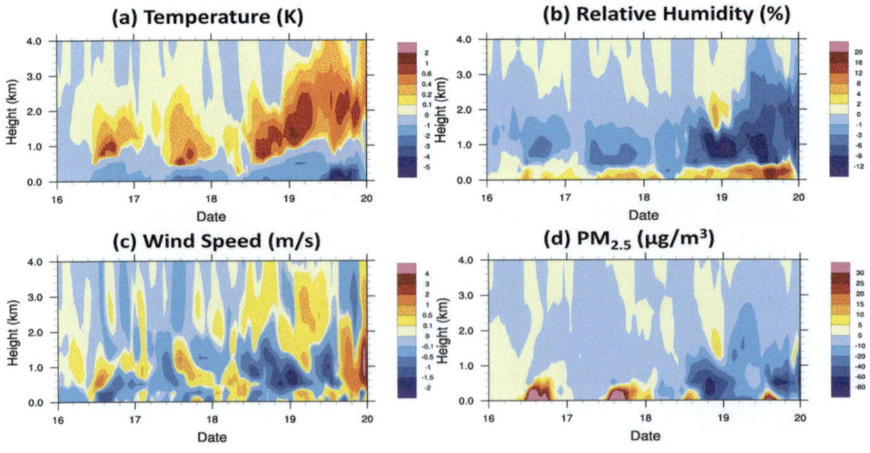

Fig. 21.8 Temporal variations (local time) of vertical profiles of temperature (**a**), RH (**b**), wind speeds (**c**), and $PM_{2.5}$ (**d**) differences (WF-NF)

21.4.2 The Radiative Feedbacks on Air Quality

The above-mentioned changes in meteorology due to aerosol feedbacks can have significant impacts on air quality. Figure 21.9 displays surface SO_2, NO_x, and CO concentrations during haze days for the NF case (a–c), and differences due to aerosol radiative effects (d–f). In the NF case, the spatial distributions of SO_2, NO_x, and CO concentrations are similar except that NO_x concentrations are mainly located in urban areas, since NO_x are mostly from transportation sources. Due to the inclusion of aerosol radiative feedbacks, domain averaged surface NO_x and CO concentrations increase by about 5.9% and 2.9%, respectively. The spatial distributions of their changes are similar, corresponding to the changes in PBL heights shown in Fig. 21.5f. Although surface SO_2 concentrations increase in urban areas, the domain averaged change in surface SO_2 concentrations is negative. This is probably because large SO_2 emissions are not in urban regions, and the increases of PBL heights in suburb areas, as shown in Fig. 21.5f, lead to decreases in surface SO_2 concentrations.

The responses in surface ozone are highly related to changes in radiation and temperature. As shown in Fig. 21.10c, a decrease in surface ozone is found in Beijing, Tianjin and south Hebei areas. Decreases in surface radiation and temperature can lead to lower photochemical activity, and as a result, surface ozone concentrations decrease. The increases of NO_x concentrations in Beijing, Tianjin and south Hebei areas (shown in Fig. 21.9e) may additionally decrease surface ozone concentration since the NCP region is VOC-limited in winter. Conversely,

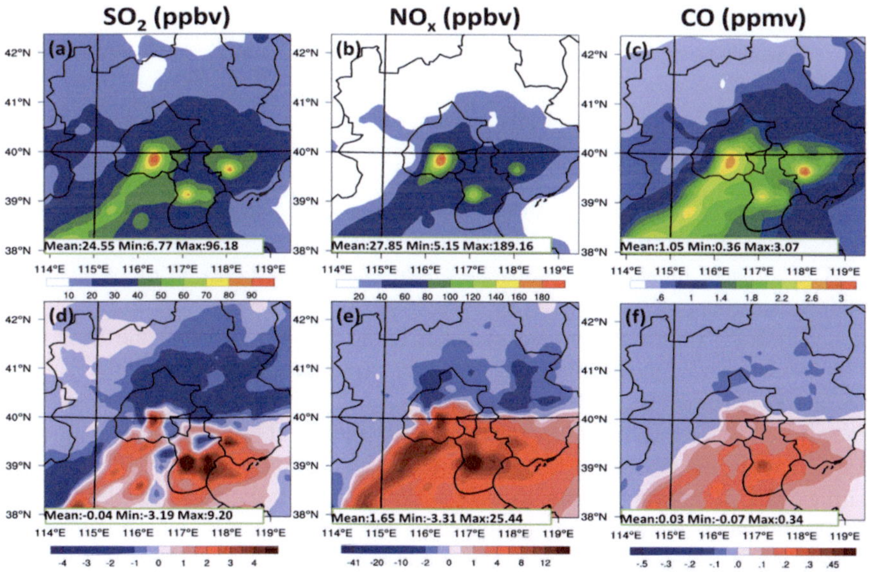

Fig. 21.9 Mean surface SO_2, NO_x, and CO concentrations during haze days for the NF case (**a**–**c**), and differences due to aerosol radiative effects (**d**–**f**)

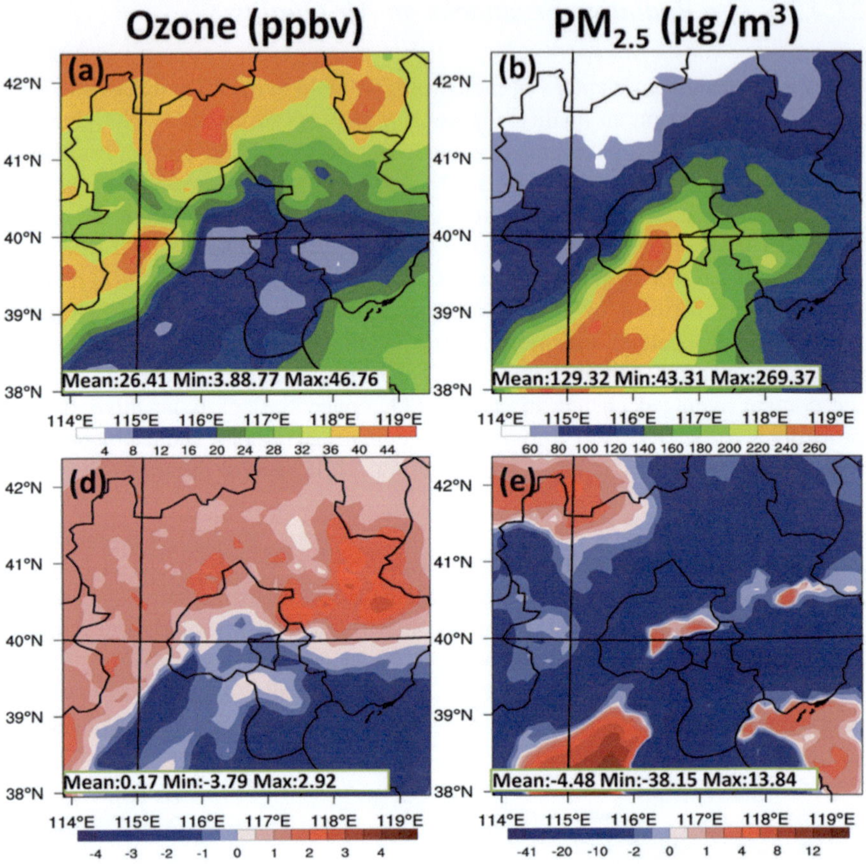

Fig. 21.10 Mean surface ozone and $PM_{2.5}$ concentrations during haze days for the baseline case (**a**, **b**), and differences due to aerosol radiative effects (**c**, **d**)

the decreases of NO_x concentrations in north Hebei (shown in Fig. 21.9e) can lead to increases in surface ozone concentrations, which have been found in Fig. 21.10d.

The impacts of aerosol feedbacks on surface $PM_{2.5}$ concentrations in urban Beijing are illustrated in Fig. 21.6b, which shows the diurnal variations of $PM_{2.5}$ concentrations (μg/m^3) averaged over the urban Beijing area and the haze days (16–19 January) for the WF and the NF cases. During daytime, the aerosol feedbacks generally increase surface $PM_{2.5}$ concentrations. The averaged increases in surface $PM_{2.5}$ concentrations can reach 28.0 μg/m^3 (+11.2%) at 14:00. Although PBL heights do not show significant increases during nighttime (Fig. 21.6a), surface $PM_{2.5}$ concentrations in urban Beijing decrease during nighttime. At 20:00, the decrease reaches 11.9 μg/m^3 (−3.9%) (Fig. 21.6b). The decreases of surface $PM_{2.5}$ concentrations in urban Beijing may be caused by increases in near surface horizontal wind speeds during nighttime induced by aerosol feedbacks, as shown in Fig. 21.8c.

The spatial distributions of changes in surface $PM_{2.5}$ concentrations averaged over haze days (both daytime and nighttime) due to aerosol feedbacks are shown in Fig. 21.10d. Overall, aerosol feedbacks increase surface $PM_{2.5}$ concentrations in urban regions (i.e., urban Beijing, urban Tianjin, and Shijiazhuang city), but the increases are less than 5%. The overall insignificant increases are mainly due to the different response directions in day and night. Aerosol feedbacks decrease surface $PM_{2.5}$ concentrations in rural areas because the changes in wind fields dominate the total impacts on meteorology in those regions.

21.4.3 Aerosol Indirect Effects

To explore the impacts of aerosols during haze days on cloud formation, we compared the cloud droplet number concentrations (CDNC) and cloud water concentrations from the MET and WF simulations. In the MET simulation, CDNC is prescribed as a constant value (250#/cm^3, default setting of the Morrison double-moment microphysics scheme), but CDNC values are predicted based on aerosol concentrations in the WF simulation. As shown in Fig. 21.11a, near surface CDNC values over areas of land around the Bohai Sea are generally over 400#/cm^3 on January 19 08:00, when severe haze occurred. At that moment, domain maximum predicted CDNC value is above 1400#/cm^3, which is about five times higher than the default setting in the MET simulation. The corresponding cloud water concentrations are shown in Fig. 21.11b, c. In the MET simulation, although CDNC is identical everywhere, cloud water only forms over the sea and over the land around the sea because water vapor is sufficient over there. However, in the WF simulation, cloud water concentrations are very low over the sea since aerosol concentrations over there are not sufficient to form CDNC (as shown in Fig. 21.11a), although water vapor over the sea is sufficient. Due to the high aerosol loadings during haze, more near surface clouds form in the WF simulation, compared to the MET simulation. The enhanced clouds may produce more aerosols

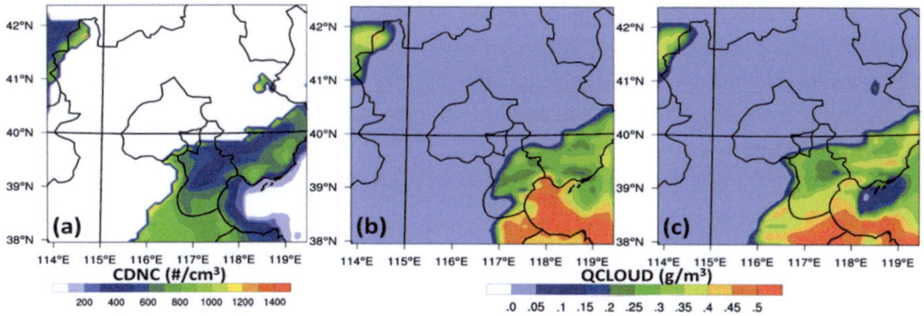

Fig. 21.11 Near surface cloud droplet number concentration (*CDNC*) from the WF simulation (**a**) and near surface cloud water concentration from the MET simulation (**b**) and the WF simulation (**c**) at January 19 08:00 AM

from cloud chemistry. The aerosol-cloud interactions are not very significant during winter since it is drier than summer, so it is better to explore more using a summer case in the future.

21.5 Conclusions

Severe haze events in China, especially in the NCP, have attracted great attention recently, but the roles of aerosol feedbacks in the formation of them have not been well understood. In this study, the fully online coupled WRF-Chem model is applied to investigate aerosol feedbacks on meteorology and air quality, focusing on a haze episode from 16 to 19 January 2010. Three simulations were conducted (WF, NF and MET) to evaluate the impacts of aerosol radiative feedbacks on meteorology and air quality, and the impacts of aerosols on cloud formation.

Due to aerosol radiative feedbacks, SWDOWN is decreased by 25.7% on domain average and OLR is only decreased by 0.8%, suggesting that the changes in near surface radiation budget are more important. The domain averaged PBL heights are reduced by 14.9%. In urban Beijing, the largest difference is larger than 150 m at 14:00. On the other hand, aerosol radiative effects increase T_2 by 0.9 K on domain average, and the domain maximum increase in RH_2 is 12.4%. WS_{10} is reduced by up to 0.5 m/s, and the domain averaged WS_{10} values decrease by 4.4% due to significant aerosol radiative feedbacks. The changes in near surface wind speeds show diurnal variations, with decreasing trend during daytime, but increasing trend during nighttime in urban Beijing.

The suppressed PBL heights lead to increase of near surface NO_x (+5.9%) and CO (+2.9%). In urban Beijing, the averaged increase in surface $PM_{2.5}$ concentrations can reach 28.0 $\mu g/m^3$ (+11.2%) at 14:00. During nighttime, surface $PM_{2.5}$ concentrations in urban Beijing decrease due to enhanced near surface horizontal wind speeds induced by aerosol radiative feedbacks. It has also been shown that high aerosol loadings during haze can promote cloud formation, which might favors aerosol formations from cloud chemistry.

All above discussions are based on the January 2010 winter haze case. Winter is usually drier than summer because less water evaporates due to lower temperature. As a result, the aerosol-cloud interactions are not very significant. It would be more interesting to study more about aerosol indirect effects based on a summer haze case in the future.

References

Baklanov, A., Schlünzen, K., Suppan, P., Baldasano, J., Brunner, D., Aksoyoglu, S., Carmichael, G., Douros, J., Flemming, J., Forkel, R., Galmarini, S., Gauss, M., Grell, G., Hirtl, M., Joffre, S., Jorba, O., Kaas, E., Kaasik, M., Kallos, G., Kong, X., Korsholm, U., Kurganskiy, A.,

Kushta, J., Lohmann, U., Mahura, A., Manders-Groot, A., Maurizi, A., Moussiopoulos, N., Rao, S., Savage, N., Seigneur, C., Sokhi, R., Solazzo, E., Solomos, S., Sørensen, B., Tsegas, G., Vignati, E., Vogel, B., & Zhang, Y. (2014). Online coupled regional meteorology chemistry models in Europe: Current status and prospects. *Atmospheric Chemistry and Physics, 14*, 317–398.

Chou, M., Suarez, M., Ho, C., Yan, M., & Lee, K. (1998). Parameterizations for cloud overlapping and shortwave single-scattering properties for use in general circulation and cloud ensemble models. *Journal of Climate, 11*, 202–214.

Emmons, L., Walters, S., Hess, P., Lamarque, J., Pfister, G., Fillmore, D., Granier, C., Guenther, A., Kinnison, D., Laepple, T., Orlando, J., Tie, X., Tyndall, G., Wiedinmyer, C., Baughcum, S., & Kloster, S. (2010). Description and evaluation of the model for ozone and related chemical tracers, version 4 (MOZART-4). *Geoscientific Model Development, 3*, 43–67.

Forkel, R., Werhahn, J., Hansen, A., McKeen, S., Peckham, S., Grell, G., & Suppan, P. (2012). Effect of aerosol-radiation feedback on regional air quality – a case study with WRF/Chem. *Atmospheric Environment, 53*, 202–211.

Gao, M., Guttikunda, S., Carmichael, G., Wang, Y., Liu, Z., Stanier, C., Saide, P., & Yu, M. (2015). Health impacts and economic losses assessment of the 2013 severe haze event in Beijing area. *Science of the Total Environment, 511*, 553–561.

Gao, M., Carmichael, G., Saide, P., Lu, Z., Yu, M., Streets, D., & Wang, Z. (2016a). Response of winter fine particulate matter concentrations to emission and meteorology changes in North China. *Atmospheric Chemistry and Physics, 16*, 11837–11851.

Gao, M., Carmichael, G., Wang, Y., Ji, D., Liu, Z., Wang, Z. (2016b). Improving simulations of sulfate aerosols during winter haze over Northern China: The impacts of heterogeneous oxidation by NO2. *Frontiers of Environmental Science & Engineering, 10*.

Gao, M., Carmichael, G., Wang, Y., Saide, P., Yu, M., Xin, J., Liu, Z., & Wang, Z. (2016c). Modeling study of the 2010 regional haze event in the North China Plain. *Atmospheric Chemistry and Physics, 16*, 1673–1691.

Grell, G., Peckham, S., Schmitz, R., McKeen, S., Frost, G., Skamarock, W., & Eder, B. (2005). Fully coupled "online" chemistry within the WRF model. *Atmospheric Environment, 39*, 6957–6975.

Guenther, A., Karl, T., Harley, P., Wiedinmyer, C., Palmer, P., & Geron, C. (2006). Estimates of global terrestrial isoprene emissions using MEGAN (Model of Emissions of Gases and Aerosols from Nature). *Atmospheric Chemistry and Physics, 6*, 3181–3210.

Hansen, J., Sato, M., & Ruedy, R. (1997). Radiative forcing and climate response. *Journal of Geophysical Research-Atmospheres, 102*, 6831–6864.

Haywood, J., & Boucher, O. (2000). Estimates of the direct and indirect radiative forcing due to tropospheric aerosols: A review. *Reviews of Geophysics, 38*, 513–543.

Hong, S., Noh, Y., & Dudhia, J. (2006). A new vertical diffusion package with an explicit treatment of entrainment processes. *Monthly Weather Review, 134*, 2318–2341.

Kaminski, J., Neary, L., Struzewska, J., McConnell, J., Lupu, A., Jarosz, J., Toyota, K., Gong, S., Côté, J., Liu, X., Chance, K., & Richter, A. (2008). GEM-AQ, an on-line global multiscale chemical weather modelling system: Model description and evaluation of gas phase chemistry processes. *Atmospheric Chemistry and Physics, 8*, 3255–3281.

Li, M., Zhang, Q., Kurokawa, J., Woo, J., He, K., Lu, Z., Ohara, T., Song, Y., Streets, D., Carmichael, G., Cheng, Y., Hong, C., Huo, H., Jiang, X., Kang, S., Liu, F., Su, H., & Zheng, B. (2015). MIX: A mosaic Asian anthropogenic emission inventory for the MICS-Asia and the HTAP projects. *Atmospheric Chemistry and Physics Discussions, 15*, 34813–34869.

Lohmann, U., & Feichter, J. (2005). Global indirect aerosol effects: A review. *Atmospheric Chemistry and Physics, 5*, 715–737.

Mlawer, E., Taubman, S., Brown, P., Iacono, M., & Clough, S. (1997). Radiative transfer for inhomogeneous atmospheres: RRTM, a validated correlated-k model for the longwave. *Journal of Geophysical Research-Atmospheres, 102*, 16663–16682.

Morrison, H., Curry, J., Shupe, M., & Zuidema, P. (2005). A new double-moment microphysics parameterization for application in cloud and climate models. Part II: Single-column modeling of arctic clouds. *Journal of the Atmospheric Sciences, 62*, 1678–1693.

Saide, P., Spak, S., Pierce, R., Otkin, J., Schaack, T., Heidinger, A., da Silva, A., Kacenelenbogen, M., Redemann, J., & Carmichael, G. (2015). Central American biomass burning smoke can increase tornado severity in the U.S. *Geophysical Research Letters, 42*, 956–965.

Schuster, G., Dubovik, O., & Holben, B. (2006). Angstrom exponent and bimodal aerosol size distributions. *Journal of Geophysical Research, 111*, D07207.

Twomey, S. (1977). The influence of pollution on the shortwave albedo of clouds. *Journal of the Atmospheric Sciences, 34*, 1149–1152.

Wang, L., Wei, Z., Yang, J., Zhang, Y., Zhang, F., Su, J., Meng, C., & Zhang, Q. (2014). The 2013 severe haze over southern Hebei, China: Model evaluation, source apportionment, and policy implications. *Atmospheric Chemistry and Physics, 14*, 3151–3173.

Wong, D., Pleim, J., Mathur, R., Binkowski, F., Otte, T., Gilliam, R., Pouliot, G., Xiu, A., Young, J., & Kang, D. (2012). WRF-CMAQ two-way coupled system with aerosol feedback: Software development and preliminary results. *Geoscientific Model Development, 5*, 299–312.

Yu, S., Mathur, R., Pleim, J., Wong, D., Gilliam, R., Alapaty, K., Zhao, C., & Liu, X. (2014). Aerosol indirect effect on the grid-scale clouds in the two-way coupled WRF–CMAQ: Model description, development, evaluation and regional analysis. *Atmospheric Chemistry and Physics, 14*, 11247–11285.

Zaveri, R., & Peters, L. (1999). A new lumped structure photochemical mechanism for large-scale applications. *Journal of Geophysical Research-Atmospheres, 104*, 30387–30415.

Zaveri, R., Easter, R., Fast, J., & Peters, L. (2008). Model for simulating aerosol interactions and chemistry (MOSAIC). *Journal of Geophysical Research, 113*, 1–29.

Zhang, Y., Wen, X., & Jang, C. (2010). Simulating chemistry–aerosol–cloud–radiation–climate feedbacks over the continental U.S. using the online-coupled Weather Research Forecasting Model with chemistry (WRF/Chem). *Atmospheric Environment, 44*, 3568–3582.

Zhao, C., Liu, X., Ruby Leung, L., & Hagos, S. (2011). Radiative impact of mineral dust on monsoon precipitation variability over West Africa. *Atmospheric Chemistry and Physics, 11*, 1879–1893.

Zhao, C., Liu, X., & Leung, L. (2012). Impact of the Desert dust on the summer monsoon system over Southwestern North America. *Atmospheric Chemistry and Physics, 12*, 3717–3731.

Zhao, X., Zhao, P., Xu, J., Meng, W., Pu, W., Dong, F., He, D., & Shi, Q. (2013). Analysis of a winter regional haze event and its formation mechanism in the North China Plain. *Atmospheric Chemistry and Physics, 13*, 5685–5696.

Zheng, B., Zhang, Q., Zhang, Y., He, K., Wang, K., Zheng, G., Duan, F., Ma, Y., & Kimoto, T. (2015). Heterogeneous chemistry: A mechanism missing in current models to explain secondary inorganic aerosol formation during the January 2013 haze episode in North China. *Atmospheric Chemistry and Physics, 15*, 2031–2049.

Chapter 22
Impact of Urbanization on Regional Climate and Air Quality in China

Xuemei Wang, Zhiyong Wu, Qi Zhang, Jason Cohen, and Jiongming Pang

Abstract China has experienced rapid urbanization and economic development during the past 30 years. Changes in land use and land cover (LULC) alter the exchange of energy, momentum, moisture, and trace gases within the vegetation-soil-atmosphere continuum, which in turn affects local and regional circulation and climate, and consequently the dispersion of pollutants and air quality. In this chapter, the integrated WRF/Chem-urban modelling system is described. The urban canopy schemes are tested under different weather conditions and the optimized scheme is obtained in China. The trend of urbanization in China is clarified including land use change, GDP, energy consumption and emissions variations. We present the study on the connections among land cover change, regional climate and air quality. The Pearl River Delta (PRD) and Yangtze River Delta (YRD) Economic Belts were chosen as an example to quantitatively investigate the regional climate and air quality change due to urbanization.

Keywords Urbanization • China • Regional climate • Air quality • WRF/Chem-urban modelling

22.1 Introduction

In 2007, half of the world's population lived in cities and urban population is projected to grow to 60% in 2030 (Greenhalgh et al. 2007). Although Asia has experienced rapid urbanization and land cover change during last 30 years, the current trends of population increase and urban expansion are expected to continue.

X. Wang (✉)
Institute for Environment and Climate Research, Jinan University, Guangzhou, China
e-mail: eciwxm@jnu.edu.cn

Z. Wu • Q. Zhang • J. Pang
School of Environmental Science and Engineering, Sun Yat-sen University, Guangzhou, China
e-mail: wuzhiyong_163@163.com; zhangq358@mail2.sysu.edu.cn; pangjm@foxmail.com

J. Cohen
School of Atmospheric Sciences, Sun Yat-sen University, Guangzhou, China
e-mail: jasonbc@alum.mit.edu

I. Bouarar et al. (eds.), *Air Pollution in Eastern Asia: An Integrated Perspective*,
ISSI Scientific Report Series 16, DOI 10.1007/978-3-319-59489-7_22

Changes in land use and land cover (LULC) alter the exchange of energy, momentum, moisture, and trace gases within the vegetation-soil-atmosphere continuum, which in turn affects local and regional circulation and climate, and consequently the dispersion of pollutants and air quality. The combined effect of global climate change and rapid urban growth, in tandem with economic and industrial development, will induce or exacerbate a number of the urban environmental problems such as extreme weather, sea-level rise, poor air quality with adverse impacts on public health, changes in atmospheric dispersion, and water resources (Fig. 22.1).

Changes of extreme temperatures in megacities are driven by both large-scale or regional warming, and the urban heat island (UHI) effect (Penner et al. 2001). In many cities, the most significant increases in temperature are more greatly affected by the UHI than the global warming signal. Based on global reanalysis data from 1950 to 1999, Kalnay and Cai (2003) suggested that half of the observed decrease in the diurnal temperature range in the Continental United States (0.27 °C per century of surface warming) was caused by urban and other land use changes. Lo et al. (2006), using an atmospheric model and a three-dimensional particle trajectory model, pointed out that urbanization in the PRD can modify regional land-sea-breeze circulations, potentially enhance pollutant trapping, and therefore contribute to the overall poor air quality in the region. Wang et al. (2007) indicated that the effects of urbanization can increase surface ozone concentration by 4–15 ppb in the PRD. It is therefore imperative to understand and project effects of weather and regional climate on cities.

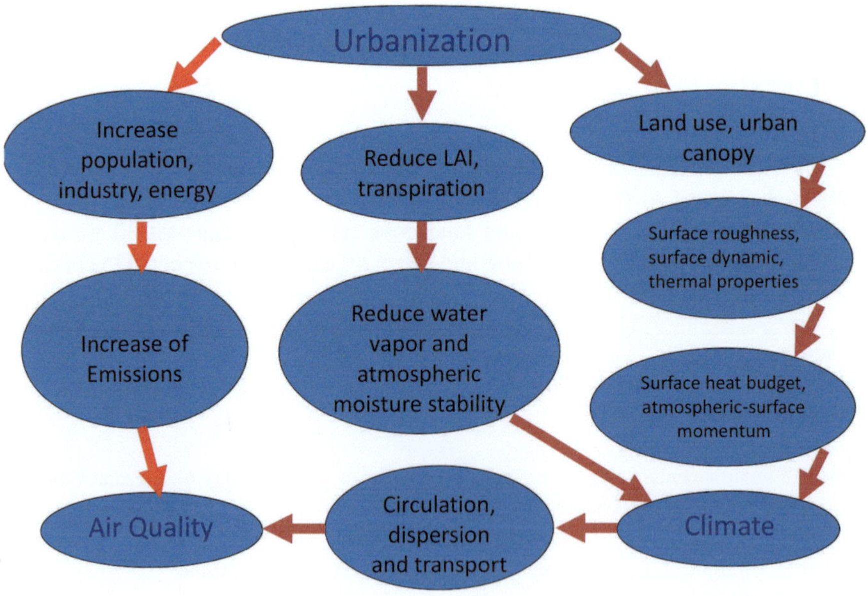

Fig. 22.1 Pathway from urbanization to climate and air quality

Investigating urban-scale physical and chemical processes is the key first step required in order to predict future risks and impacts of various mitigation strategies to improve knowledge of air pollution and influence air quality. One valuable tool for this purpose is a multi-scale atmospheric modeling system (Chen et al. 2011a), which is able to predict/simulate meteorological conditions from regional to building scales, and can be coupled to human-response models. At the same time, some new observing technologies have been explored to obtain atmospheric variables in greater spatial detail both horizontally and vertically to more directly link land cover change with air quality and climate change. This approach also provides an opportunity to integrate models and observations to address urban canopy effects, urban anthropogenic heat, urban precipitation and hydrology, air quality, and interaction between air pollution and climate (radiation and precipitation, temperature, humidity etc.).

22.2 The Integrated WRF/Chem-Urban Modelling System

22.2.1 Modeling System Overview

The Weather Research and Forecasting (WRF) model is a mesoscale non-hydrostatic meteorological model that includes several options for physical parameterizations of the Planetary Boundary Layer (PBL), cloud processes and land surface (www.wrf-model.org; Skamarock and Klemp 2008). The chemistry version is a coupling of WRF with various different chemistry models "online", in which meteorological and chemical components of the model are predicted simultaneously (Grell et al. 2005; http://ruc.fsl.noaa.gov/wrf/WG11/). It has been successfully applied for regional air quality studies (Fast et al. 2006; Archernicholls et al. 2016).

One of the major advantages of WRF/Chem is that different physics and chemistry options can be used. For example, WRF has a wide array of treatments/schemes for solving convective closure, planetary boundary layer (PBL), shortwave radiation, and longwave radiation, including: the Yonsei University (YSU) and a 2.5 level planetary boundary layer (PBL) schemes (Hong and Pan 1996), NCEP (National Center for Environmental Prediction)-5 class microphysics and Kain-Fritsch convective parameterizations (Kain and Fritsch 1993), and the Dudhia (Dudhia 1989) and the Rapid Radiative Transfer Method (RRTM) radiation schemes (Mlawer et al. 1997). Among its options for land surface models (LSMs) there are many including the Community Land Model (CLM) and the community Noah Land-Surface Model (Noah LSM, e.g., Chen et al. 1996; Chen and Dudhia 2001), which have been used in weather prediction models. There are many aerosol and gas phase chemistry schemes, including the Carbon-Bond Mechanism version Z (CBM-Z, Zaveri and Peters 1999), the Model for Simulating Aerosol Interactions and Chemistry (MOSAIC, Zaveri et al. 2008). CBM-Z is a new lumped structure photochemical mechanism extending the original CBM-IV framework to larger spatial and longer timescales, including 52 chemical species and 132 reactions, including isoprene. MOSAIC treats all the major aerosol species important at urban, regional and global scales using a mass conservation approach (Zaveri et al. 2008), where the aerosol size distribution is represented by division into discrete size bins. However each bin is assumed to be internally mixed so that all particles within a bin have the same chemical composition, impacting issues such as water, sulfate, and nitrate uptake on fast scales, or associated changes in particle size due to these reactions. Some of the other models, such as Cohen et al. (2011) try a new approach, whereby the use many fewer chemical species, but attempt to account for the different mixing states for each lognormally distributed size particle, such as which portion is externally mixed, which portion is internally mixed, and which portion is core/shell mixed. This has been shown to produce much better results in urban areas, although it only works for certain critical chemical species and not others. Another interesting approach is the Volatility Basis-set (VBS) method developed by Donahue et al. (2006) for secondary organic aerosols (SOA) simulation, which includes reactions from both heterogeneous pathways and across multiple phases.

To represent the thermal and dynamic effects of urban areas, there are several urban canopy models (UCMs; Kusaka and Kimura 2004; Chen et al. 2011b) coupled with Noah in the WRF/Chem. Such coupling is through the urban percentage (or urban fraction) parameter that represents the proportion of impervious surfaces in the WRF sub-grid scale. For a given WRF grid cell, the Noah model calculates surface fluxes and temperature for vegetated urban areas (trees, parks, etc.) and the UCM provides the fluxes for urban surfaces. Grid-integrated latent heat flux, upward long wave radiation flux, albedo, and emissivity are estimated in the same way. Surface skin temperature is calculated as the averaged value of the artificial and natural surface temperature values. The final values are formed by taking a subsequent area weightage.

22.2.2 Bulk Urban Parameterization

A bulk urban parameterization in Noah can roughly approximate the zero-order effects of urban surfaces on the urban atmosphere by using a reasonable set of parameter values (Wang et al. 2009a). This includes (1) roughness length of 0.8 m to represent turbulence generated by anthropogenic rough elements and drag due to buildings; (2) surface albedo of 0.15 to represent shortwave radiation trapped in urban canyons; (3) volumetric heat capacity of 3.0 J m^{-3} K^{-1} for urban surfaces (walls, roofs, and roads), assumed to be similar in nature to concrete or asphalt; (4) soil thermal conductivity of 3.24 W m^{-1} K^{-1} to represent the large heat storage in urban buildings and roads; and (5) decreased evaporation due to the reduced green-vegetation fraction over urban areas. This approach has been successfully employed in real-time weather forecasts (Liu et al. 2006) and to study the impact of urbanization on land-sea breeze circulations (Lo et al. 2006).

22.2.3 Single-Layer Urban Canopy Model

To address the next order of complexity, a single-layer UCM (SLUCM) was developed by Kusaka et al. (2001) and Kusaka and Kimura (2004). The approach assumes infinitely-long street canyons parameterized to represent urban geometry, but recognizes the vertical nature of urban surfaces. In a street canyon, for example, shadowing, reflections, and trapping of radiation are considered, and an exponential wind profile is prescribed with respect to distance from the surface. This new approach leads to additional prognostic variables: surface skin temperature at the roof, wall, and road (from the surface energy budget) and temperature profile within roof, wall, and road layers (from the thermal conduction equation). Surface sensible heat fluxes from each facet are calculated using Monin-Obukhov similarity theory and Jurges formula (Fig. 22.2), with the total sensible heat flux from roof, wall, roads, and the urban canyon passed to the WRF-Noah model. The total momentum flux is similarly computed and passed, with SLUCM calculating canyon drag coefficient and friction velocity using a similarity stability function for momentum. The total friction velocity is aggregated from urban and non-urban surfaces and passed to the appropriate WRF boundary layer scheme. Anthropogenic heat and its diurnal variation are considered by adding them to the urban canopy layer's sensible heat flux.

22.2.4 Multi-layer Urban Canopy (BEP) and Indoor-Outdoor Exchange (BEM) Models

Unlike the SLUCM, which is only embedded within the first model layer, the multi-layer UCM developed by Martilli et al. (2002), called BEP (Building Effect Parameterization), represents the most sophisticated urban modeling in WRF, allowing a direct interaction with the PBL (Fig. 22.2). BEP recognizes the three-dimensional nature of urban surfaces and the fact that buildings vertically distribute sources and sinks of heat, moisture, and momentum through the whole urban canopy layer, substantially impacting the thermodynamic structure of the urban roughness sub-layer and hence the lower part of the urban boundary layer. It takes into account effects of vertical (walls) and horizontal (streets and roofs) surfaces on momentum (drag force), turbulent kinetic energy (TKE), and potential temperature (Fig. 22.2). The radiation at walls and roads considers shadowing, reflections, and trapping of shortwave and long-wave radiation in street canyons. The Noah-BEP model has been coupled with two turbulence schemes: Bougeault and Lacarrere

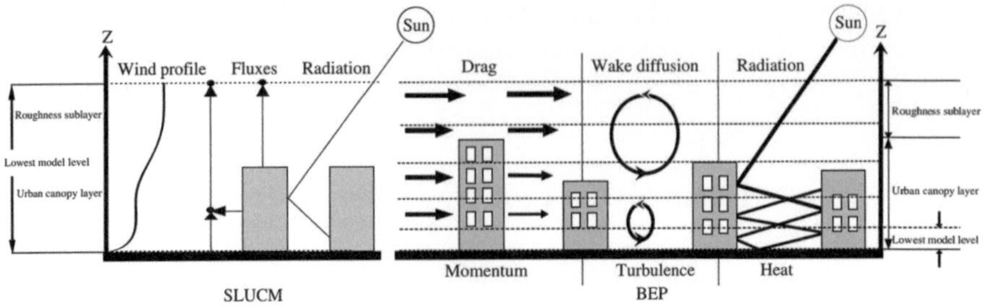

Fig. 22.2 A schematic of the SLUCM and BEP models (From Chen et al. 2011a)

(1989) and Mellor-Yamada-Janjic (Janjic 1994) in WRF by introducing a source term in the TKE equation within the urban canopy and modifying turbulent length scales to account for the presence of buildings.

To take full advantage of BEP, it is necessary to have sufficiently high vertical resolution close to the ground, so that there is more than one model level within the urban canopy. Consequently, this approach is more appropriate for research (when computational demands are not a constraint) than for real-time weather forecasts.

In the standard version of BEP, the internal temperature of the buildings is kept constant. To improve estimation of exchanges of energy between the interior of buildings and the outdoor atmosphere, which can be an important component of the urban energy budget, a simple Building Energy Model (BEM, Salamanca and Martilli 2010) has been developed and linked to BEP. BEM accounts for the diffusion of heat through walls, roofs, and floors, radiation exchanged through windows, longwave radiation exchanged between indoor surfaces, generation of heat due to occupants and equipment, and air conditioning, ventilation, and heating. Buildings of several floors can be considered, with the evolution of indoor air temperature and moisture estimated for each floor, thereby allowing the impact of energy consumption due to air conditioning to be estimated. The coupled BEP +BEM has been tested offline using the basel urban boundary layer experiment (BUBBLE, Rotach et al. 2005) data.

Executing the coupled WRF/Chem-urban modeling system raises three challenges, including initialization of the detailed spatial distribution of UCM state variables (such as temperature profiles within wall, roofs, and roads), specification of a potentially vast number of parameters related to building characteristics (thermal properties, emissivity, albedo, anthropogenic heating, etc.), and anthropogenic heating (AH) inventory. High-resolution routine observations of wall/roof/road temperature are rarely available to initialize the WRF/Chem-urban model, which usually covers a large domain (e.g., ~10^6 km^2) and may include urban areas with a typical size of ~10^2 km^2. Nevertheless, to a large extent, this initialization problem is analogous to that of initializing soil moisture and temperature in a coupled atmospheric-land surface model. One approach is to use observed rainfall, satellite-derived surface solar insolation, and meteorological analyses to drive an uncoupled (off-line) integration of an LSM, so that the evolution of the modeled soil state can be constrained by observed forcing conditions. The North-American Land Data Assimilation System (NLDAS, Mitchell et al. 2004) and the NCAR High-Resolution Land Data Assimilation System (HRLDAS, Chen and Song 2007) are two examples that employ this method.

Using UCMs in WRF requires users to specify at least 20 urban canopy parameters (UCPs). A combination of remote-sensing and in-situ data can be used for this purpose thanks to recent progress in developing UCP data sets (Burian et al. 2004; Feddema et al. 2006; Taha 2008; Ching et al. 2009). While the availability of these data is growing, data sets are currently limited to a few geographical locations, such as the US, since many of the measurements are proxies and relationships need to be formed to connect the measurements to the respective land surface types. Global high-resolution data sets comprising the full suite of UCPs simply do not exist. In anticipation of increased database coverage, three methods are employed to specify UCPs in WRF/Chem-urban, first) urban land-use maps and urban-parameter tables, second) gridded high-resolution UCP data sets, and third) a mixture of the above.

Anthropogenic heating, from buildings, industry/manufacturing, and vehicles, is very important for understanding the total sensible heat budget. It can be estimated either from bottom-up inventory approaches or through direct modeling. In the former approach (e.g., Sailor and Lu 2004), aggregated consumption data are typically gathered for an entire city or utility service territory, often at monthly or annual resolution, and then are mapped onto suitable spatial and temporal profiles. Waste heat emissions from industrial sectors can be obtained at the state or regional level (from sources such as the Federal Energy Regulatory Commission, Elsenstat and Wentworth 2006), but it is difficult to assess the characteristics of these facilities that would enable estimation of higher frequency (i.e. diurnal) anthropogenic flux profiles. Again, there are very limited gridded or categorized AH datasets.

22.3 Urbanization in China

China has been experiencing very rapid urbanization for the last 30 years. The urbanization rate of China was about 19% in 1980 and around 50.0% in 2010, and is still increasing. Predictions have placed China's urbanization rate from 58 to 63% in 2020 to 80% by 2050. Megacities play an important role in China's socioeconomic development, with three massive and interconnected urban areas currently in China, including the Yangtze River Delta (YRD), Pearl River Delta (PRD), and Beijing-Tianjin-Hebei. The three cities occupy less than 3% of the national land area, yet accommodate about 10% of the total population, and contribute more than 40% to the national GDP, in turn leading to more frequent air pollution episodes. There are seven additional sublevel megacities presently in the process of development: the Shandong peninsula (Qingdao, Yantai, Weifang, Zibo, Jinan), central-south Liaoning, central Henan Plain, Wuhan, Chang-Zhu-Tan, Chengdu-Chongqing, and the central Shaanxi Plain. With China's huge population and rapid population growth, there will be continued rapid urbanization in the future, with more people and industries aggregating into existing and new cities. New megacities may shape in the Harbin-Daqing-Qiqihar area, west bank of Taiwan Strait zone, surrounding Poyang Lake, central Yunnan, Guangxi North Bay, Lanzhou-Xining-Germu, and the northern slope region of Tianshan Mountains (Dong et al. 2012). Land use changes show rapid urbanization in the three clusters (Fig. 22.3).

However, urbanization is unbalanced between different regions in China, with the urbanization rate of west China far less than the national average level. In addition, the spatial differentiation in city size in China is obvious. East China has more than 56% of the total number of megacities, over 57% of big cities, and about 50% of medium-sized cities, while Western China has a large proportion of small cities (Dong et al. 2012).

At present, GDP, energy consumption, carbon dioxide, sulfur dioxide and volatile organic compounds emissions in China's urban regions account, respectively, for 85%, 85%, 90%, 98% and 50% of national total amount (Dong et al. 2012). The contradiction between China's urban development and low-carbon transformation has become increasingly prominent, with many challenges in energy conservation and environmental improvement entangled with the ongoing urbanization process. Figure 22.4 shows the energy consumption in China for different sectors. Over the last 30 years (1980–2010), an annual growth of 5% year^{-1} of energy use supported an annual increase of 8% year^{-1} of GDP, while over the next 20 years (2010–2030) a 50% increase in energy is expected to support a doubling of the GDP.

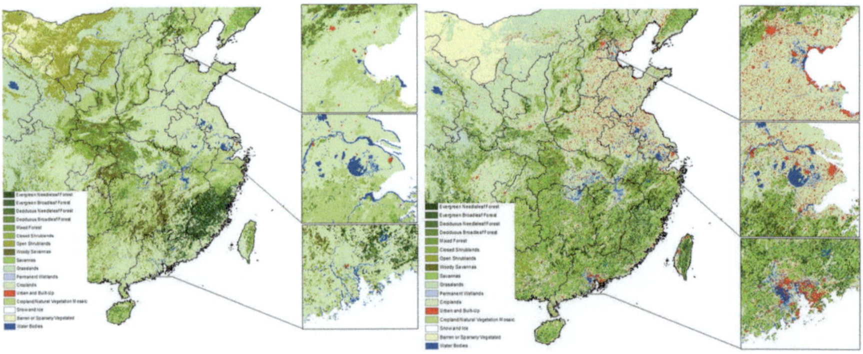

Fig. 22.3 Land use change in three clusters of cities **a** From the International Geosphere-Biosphere Programme (IGBP) data in 1992, and **b** From Environmental and Ecological Science Data Center for West China (WestDC) data in 2001 (From Ran et al. 2010)

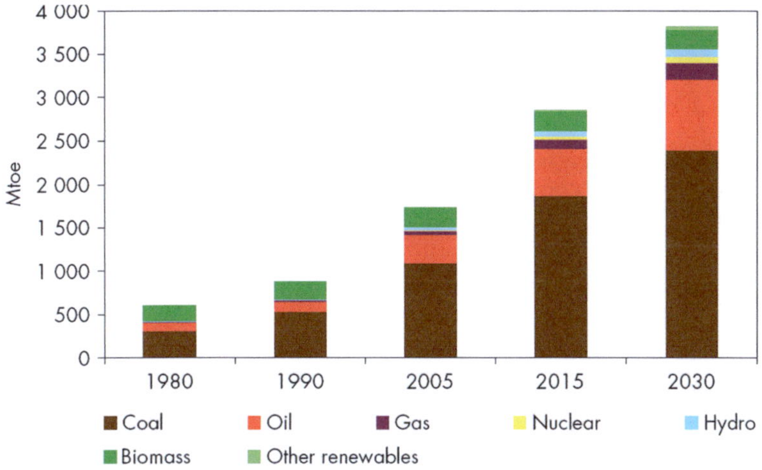

Fig. 22.4 Energy consumption in China for different sectors (Adapted from Dong et al. 2012)

22.4 Impact of Urbanization on Regional Climate

One important goal of developing the integrated WRF/urban modeling system is to use it to understand the effects of urban expansion, and predict or assess the impacts of urbanization on UHI, local and regional climate, circulation change, precipitation, and water resources. Generally, urbanization increase surface temperature while decreasing wind speed and humidity. However, the result is complicated in terms of precipitation. For instance, Miao et al. (2009a) showed that intense urbanization contributed to enhancing a heavy rainfall event in Beijing on 1 August 2006, while Zhang et al. (2009) demonstrated that augmenting urban green vegetation in Beijing could reduce extreme rainfall events in summer. Therefore, it is possible that intense urban development might have exacerbated the intensity of specific strong thunderstorms. Furthermore, rapid urban expansion of the YRD in eastern China and the PRD in southern China have been shown to change the local climate, increasing surface temperature and boundary-layer mixing (Wang et al. 2009a), consistent with the fact that these are the most two advanced economic regions in China, and therefore have experienced the most development and urbanization in the past two decades. The current urban fractions are roughly 70% in the PRD and 65% in YRD, with most of the change being from converted farmlands, a very strong alteration which has been demonstrated to modify local and regional meteorological conditions (Wang et al. 2014; Zhang et al. 2010). Figure 22.5 shows the diurnal variation of 2 m temperature in the YRD, with the difference revealing that the influence of urbanization is greater in summer (1.9 $\pm$ 0.55 °C) than in winter (0.45 $\pm$ 0.43 °C). Consistently, in PRD the differences of 2-m temperature are greater (1.4 $\pm$ 0.3 °C) in summer than the winter (0.9 $\pm$ 0.4 °C).

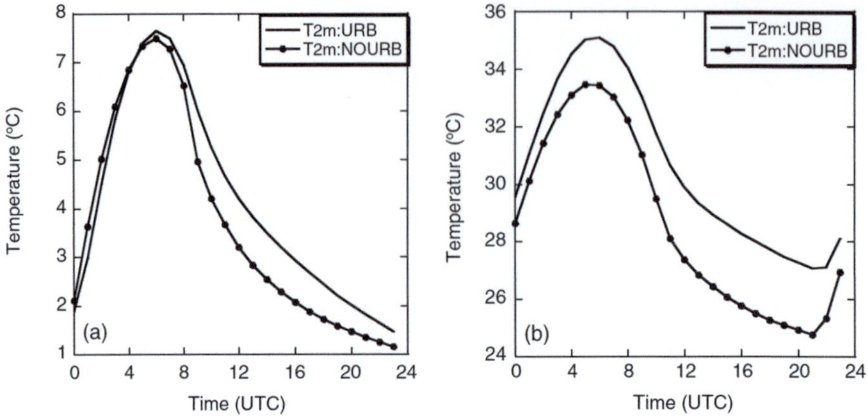

Fig. 22.5 Diurnal variation of 2 m temperature over urban areas in the YRD region for (**a**) January, and (**b**) July (From Zhang et al. 2010)

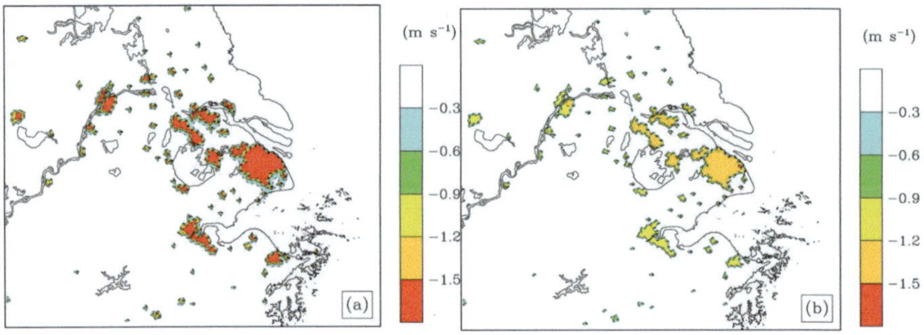

Fig. 22.6 Influence of urbanization in YRD (the urban case minus the NOURB) on 10-m wind speed in: (**a**) January, and (**b**) July (From Zhang et al. 2010)

Urbanization also increases the surface roughness over the urban area, with the friction and drag of buildings decreasing near surface wind speed in the urban area. Figure 22.6 shows the wind speed decreases in both winter and summer with the wind speed reducing by up to 1.5 m s^{-1} in the YRD and 1.2–1.5 m s^{-1} in the PRD. This is a significant fraction of the annual average simulated surface wind speed of about 3–4 m s^{-1} in the non-urbanized case: the urbanization leads to a 50% decrease in wind speed loss over the urban area in the YRD and a 37% decrease in the wind speed over the PRD.

Given that mixing and transport within the boundary layer are important mechanisms influencing air pollutants, the impact of the large increase in surface roughness due to urbanization is expected to be significant. The computed reductions in wind speed in turn change the venting of air pollution. Studies in the PRD also found urbanization increases the height of the PBL during both the daytime and

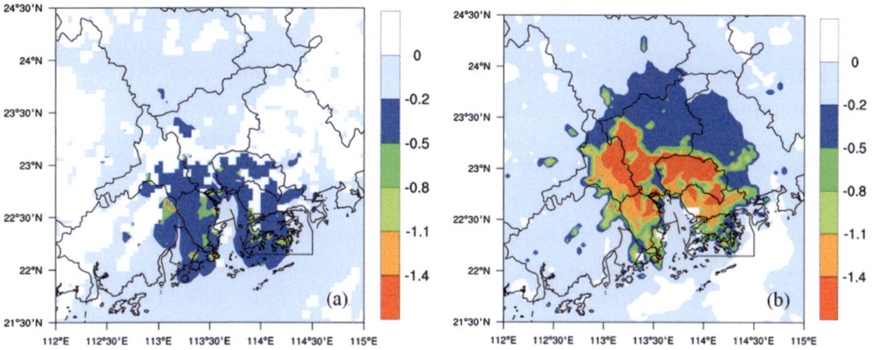

Fig. 22.7 Influence of urbanization in the PRD (Urban case minus the No Urban case) on 2-m water vapor mixing ratio: (**a**) DJF and (**b**) JJA (From Wang et al. 2014)

nighttime, with the urbanized-area average PBL height increasing by 125 ± 75 m in the daytime and by 100 ± 50 m at night. This will have a further effect on the concentration of air pollutants near the surface. Finally, it has been found that these influences are different during different times of the year, with the influence of urbanization on the PBL height strongest in summer (~200 m in the daytime and ~150 m at night) and the weakest in winter (~50 m throughout both the day and night).

Urbanized land types consist of roads and other impervious concrete surface, leading to less water being available for evaporation when compared against natural land surfaces prior to urbanization. Figure 22.7 shows the change of water vapor mixing ratio in the PRD, with the impact of urbanization on water vapor stronger in summer (decrease of 1.5 g kg^{-1}) than the winter (decrease of 0.4 g kg^{-1}). The magnitudes are similarly found over the YRD, with the summertime loss on order of 1.5 g kg^{-1} while the wintertime loss is found to be 0.1 g kg^{-1}.

Precipitation is an important component of urban climatology, and it has been demonstrated, due to the reasons already mentioned, that urbanization has a significant influence on the mesoscale circulations affecting regional convection and precipitation (Oke 1982; Bornstein and Lin 2010; Dixon and Mote 2003; Zhang et al. 2009). Increased precipitation has been observed downwind of the urban area due to an increase in the urban heat island effect (Lin et al. 2008). Furthermore, there has been another recent pathway also found, with a connection between urbanization and enhanced surface convergence (Rozoff et al. 2003).

Generally, the thermal and dynamic effects caused by urbanization play a significant role on regional precipitation formation, with the impact occurring via several different mechanisms. First urban land creates a heat island, inducing thermal circulation, thereby triggering dynamical convection. Second, changes in trapped urban pollutants may act as local cloud nuclei. Third, urbanization reduces local evaporation. Finally, fourth anthropogenic heat release may enhance local convection. However, it is difficult to differentiate these different but related types of influences, since the impacts on not just the total rainfall but also the intensity of the rainfall is quite challenging to predict. The results from work performed over the PRD obtained from both urban and no urban simulations show that the annual total precipitation increases by approximately 30%, mainly due to the increase in heavy and extreme rain events. On the other hand, a similar experiment performed over the YRD, shows an annual increase in precipitation of about 15% over the major cities. In both cases, the built-up surface induces higher air temperature, vertical wind speed and PBL height. Meanwhile, increased surface roughness also enhances mechanical turbulence, which leads to a stronger low-level convergence and increased convection. Together, these processes more effectively lift water vapor to upper layers, and ultimately lead to more intense precipitation.

22.5 Impact of Urbanization on Regional Air Quality

Presently, there are two main methods used to investigate the impact of urbanization on air pollutants over China: Data Statistics (DS), and Modeling Simulation (MS). DS mainly focus on analyzing the correlation between air pollutant and urbanization indicators (land cover, urban size in terms of built-up area, population, secondary industry fraction, power usage, industry exhaust, GDP, difference vegetation index) (Han et al. 2014, 2015a, b; Huang et al. 2013; Gu et al. 2012; Zhang et al. 2011; Lin et al. 2010) or meteorology (Cui and Shi 2012) from in-situ observation and satellite data. Meanwhile, MS mainly focus on investigating how air pollutants and meteorology change after specific changes are made to land use data (Liao et al. 2015; Li et al. 2015; Yu et al. 2012; Wang et al. 2007, 2009a, b), urban canopy schemes (Liao et al. 2014) and emission changes (Penrod et al. 2014) with numerical models. There have been a few new attempts to use results from DS as a means of driving MS, but these are still in their very early stages of development, although they may ultimately bear fruit.

Due to the rapid urbanization, many cities in China suffer from air pollution, with two-thirds of China's cities not attaining the ambient air quality standards applicable to urban residential areas (Grade II). Combination of the satellite data and in-situ observation data (Huang et al. 2013), from 1996 to 2011 have shown that the urban built-up areas in Beijing, Shanghai, and Guangzhou increased by 197%, 148%, and 273%, respectively, while population increased by 87%, 65%, and 25%, respectively. Meanwhile, the tropospheric column of NO_2, a short lived urban air pollutant, over the surrounding areas of that three megacities increased by 82%, 292%, and 307%, respectively. The anthropogenic emissions related to urbanization are a dominant factor in the long-term changes in the yearly and seasonal mean tropospheric columns of NO_2 in Beijing and Shanghai. In the case of Guangzhou, it is more complex, since NO_x emissions stemming from the urbanization of Guangzhou are not the only source, with there being many other major urban areas across the PRD region, and the increased NO_x emissions resulting from urbanization may have been mitigated by control measures instituted across the PRD. On the other hand, observed tropospheric O_3 levels were found to be relatively insensitive to changes in urbanization.

Atmospheric environmental quality in China has been improving due to a variety of programs implemented by the Chinese government in recent decades, such as the reduction in the total number of emission sources from major industries, reduction in emissions of primary fine particles, reduction in SO_2 from power plants, and updated vehicle exhaust standards. Figure 22.8 shows the annual AOD, NO_2, and SO_2 changes between 2005 and 2013 in the PRD (Wang et al. 2016). However, the

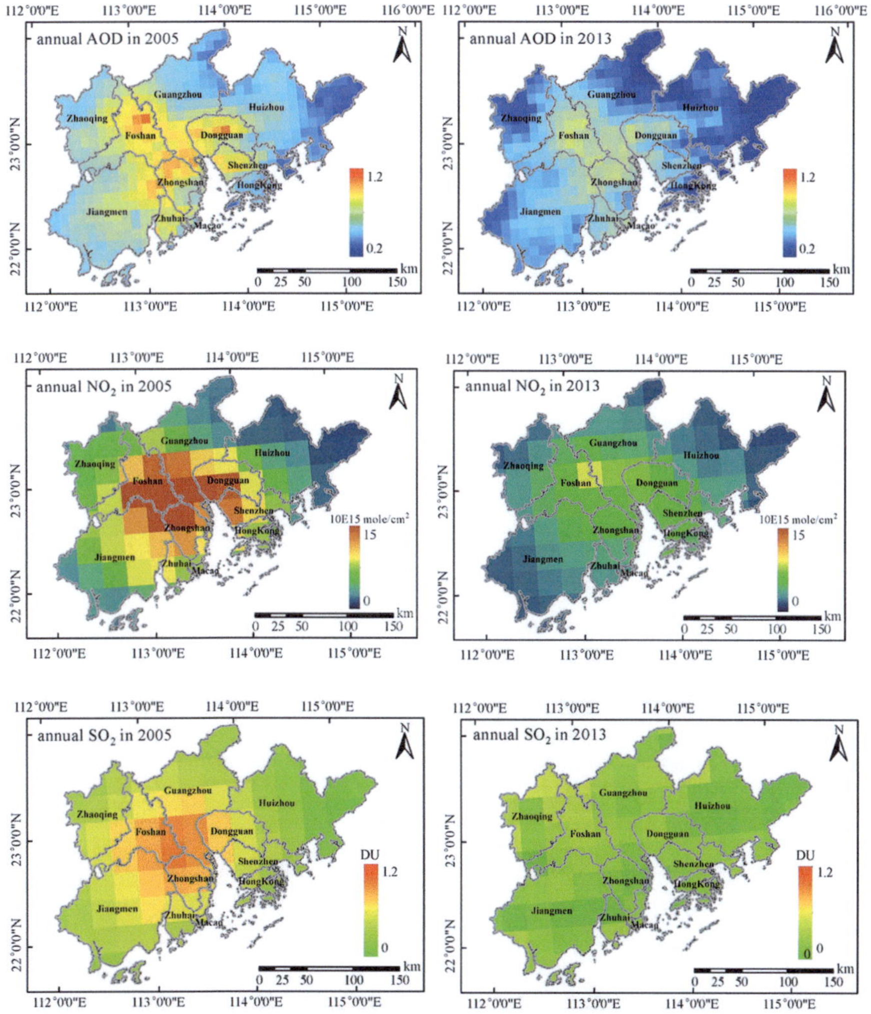

Fig. 22.8 Annual averaged distributions of the satellite retrieved AOD, NO_2, SO_2 in PRD in 2005 (*left panels*) and in 2013 (*right panels*) (From Wang et al. 2016)

present-day energy structure will continue to affect the quality of the atmosphere for a long time. Increased energy efficiency, optimization of existing energy infrastructure, and a generation of a sustainable consumption and production patterns will provide opportunities to resolve regional and the global environmental problems (Fig. 22.9).

The coupled WRF/Chem-Urban model has been applied to major metropolitan regions (e.g., Beijing, Guangzhou/Hong Kong, Houston, New York City, Salt Lake

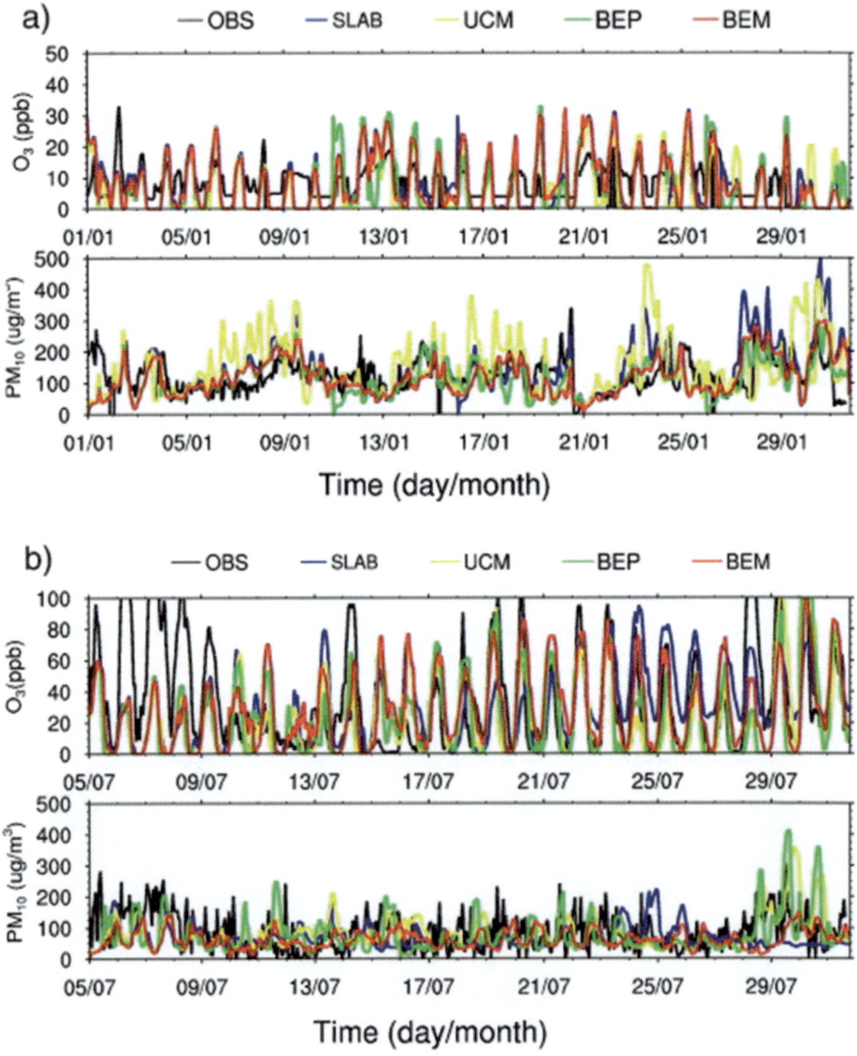

Fig. 22.9 Comparisons between chemical outputs from all experiments and the observations: (**a**) for January, (**b**) for July (From Liao et al. 2014)

City, Taipei, and Tokyo. Miao et al. 2009a, b; Wang et al. 2009b; Gutiérrez et al. 2011; Jiang et al. 2008; Nehrkorn et al. 2011; Lin et al. 2008; Kusaka et al. 2012), and its performance has been evaluated against surface observations. For instance, four regimes of urban canopy schemes with updated USGS land-use data, based on 2004 MODIS observations, have been examined in YRD region: (1) SLAB scheme that does not consider urban canopy parameters (the control experiment); (2) a single-layer urban model with a fixed diurnal profile for anthropogenic heat (UCM); (3) multilayer urban canopy model (BEP-Building effect parameterization); and (4) multilayer urban models with a building energy model including anthropogenic heat due to air conditioning (BEP + BEM). The results (see Figs. 22.10 and 22.11) show the relative differences of the schemes are significant in terms of both PM_{10} and ozone concentration. The BEP + BEM scheme predicts PM_{10} well during January, while the best estimate of PM_{10} is obtained with UCM scheme during July. Figure 22.9 shows the time series of the chemical predictions from all experiments compared with observations. Statistical skills are summarized in Table 22.1.

Under the same emission inventory conditions, the best simulated ozone concentrations for January and July result from BEP + BEM and SLAB schemes. And all these two schemes under-estimated ozone concentrations for the 2 months. However, compared with the SLAB scheme, the UCM scheme calculates higher

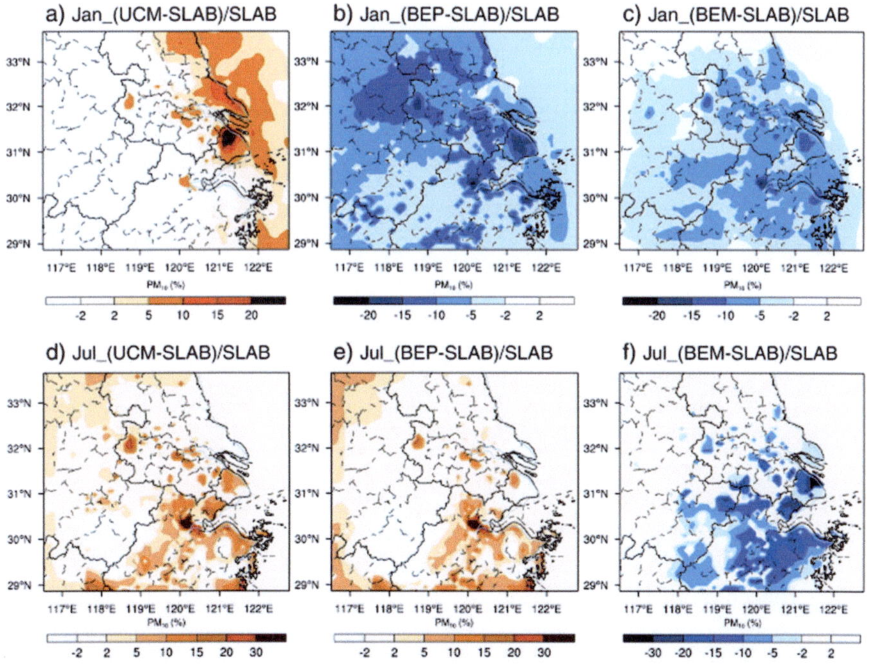

Fig. 22.10 The relative differences of PM_{10} concentration compared with the SLAB scheme: (**a–c**) during January and (**d–f**) during July (From Liao et al. 2014)

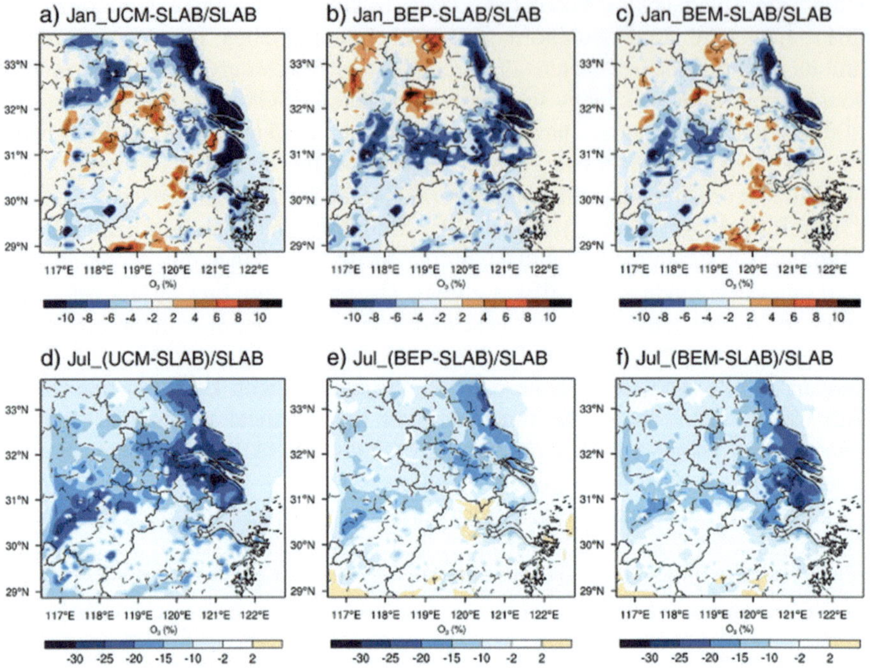

Fig. 22.11 The relative differences of O_3 concentration compared with the SLAB scheme: (**a–c**) in January; and (**d–f**) in July (From Liao et al. 2014)

Table 22.1 PM_{10} and ozone predictions from different urban canopy schemes compared with observations

	JAN				JULY			
Var	EXP	MB	RMSE	Corr[a]	EXP	MB	RMSE	Corr[a]
PM_{10}(μg/m^3)	SLAB	12.1	81.9	0.32		−15.7	71.6	0.21
	UCM	23.7	96.2	0.19	UCM	−0.9	75.4	0.33
	BEP	−11.6	61.1	0.39	BEP	−7.5	75.9	0.24
	BEM	−4.5	66.0	0.33	BEM	−21.0	65.1	0.22
O_3(ppb)	SLAB	−0.3	6.9	0.58	SLAB	−3.2	33.0	0.36
	UCM	−2.3	7.1	0.52	UCM	−11.6	31.4	0.44
	BEP	−1.2	7.5	0.55	BEP	−11.0	29.8	0.52
	BEM	−0.6	7.2	0.56	BEM	−5.4	30.8	0.46

Adapted from Liao et al. (2014)
[a]Statistically significant at 95% confident level

PM_{10} concentrations in both January and July, with the difference being 22.3% (24.4 μg m^{-3}) in January, and 31.4% (17.4 μg m^{-3}) in July, respectively. A difference as large as 32.7% (18.3 μg m^{-3}) of PM_{10} is found over Hangzhou during July. On the other hand, 18.6% (22.1 μg m^{-3}) and 16.7% (24.6 μg m^{-3}) decreases in PM_{10} are found in BEP and BEP + BEM schemes during January. Compared with control experiment during January, 6.5% (2.6 ppb) to 10.4% (4.2 ppb) increases of ozone are computed over urban areas when the urban canopy experiments are done in the winter. On the other hand, all three canopy schemes predict lower ozone concentrations by as large as 30.2% (11.2 ppb) with UCM scheme, 16.5% (6.2 ppb) with the BEP scheme during July, and 7.3% (or 2.9 ppb) in BEP + BEM scheme. Lower PBLH and wind speed reduced the transport of ozone and PM_{10} while higher temperature enhanced ozone generation. The PBLHs from UCM are about 100 m lower than the SLAB scheme both in January and July, and the wind speed is also lower, causing PM_{10} tend to accumulated and is difficult to diffuse over urban areas. On the contrary, both BEP and BEP + BEM scheme have higher PBLHs during July, resulting in the better and conductive diffusion conditions of pollutants. However, the major factor affecting the ozone generation and transport seems more complicated. It was possibly that horizontal transport plays a major role, and other factors (i.e. cloud position and formation, ozone precursors, the canopy parameters involves in analysis of behaviors of the PBL and cloud schemes) also affect ozone concentrations. But the influence of canopy schemes on ozone formation and transport still contains uncertainty (Liao et al. 2014).

It is apparently that canopy schemes significantly change the urban thermal and dynamic characteristics. Furthermore, it proves that realistic urban morphology and parameters used in the urban canopy schemes (i.e. building height, albedo, heat capacity and roughness length) largely influence the surface energy balance and heat absorption. But the calculation of some parameters in SLAB and UCM schemes is not directly dependent on urban morphology, contrary to BEP and BEP + BEM schemes (Salamanca et al. 2011). The SLAB scheme is suitable for real-time weather forecast while the multiple urban canopy scheme is necessary when attempting to quantify the urbanization impacts on regional climate (Liao et al. 2014). In this study, BEP and BEP + BEM seem more reasonable. Overall, these results show that there is no ideal scheme and that further work likely needs to be done to improve our ability to predict the impacts of urbanization on PM_{10} and Ozone.

Land use and land cover change can directly influence the meteorological conditions and in turn impact the transport and diffusion of pollutants. For instance, the Pearl River Delta (PRD) and Yangtze River Delta (YRD) regions of China, have experienced a rapid, if not the most rapid in the world, increase in economic development and urbanization of anywhere in the world over the past two decades. These city clusters, centered around mega cities such as Hong Kong, Guangzhou, Nanjing, and Shanghai (Fig. 22.12), have induced an increase in 2-m temperature and PBL height, and a decrease in 10-m wind speed. Changes in meteorological conditions result in an increase of surface ozone concentrations by 4.7–8.5% during the nighttime and about 2.9–4.2% during the daytime (Fig. 22.13). However, despite the fact that both the PRD and the YRD have experienced similar degrees of urbanization in the last decade, and that both are located in coastal zones, urbanization has had different effects on the surface ozone for the PRD and the YRD, presumably due to differences in their urbanization characteristics, topography, and emission source strength and distribution (e.g., Wang et al. 2009a). In addition to ozone, the changes in meteorological conditions due to land use change

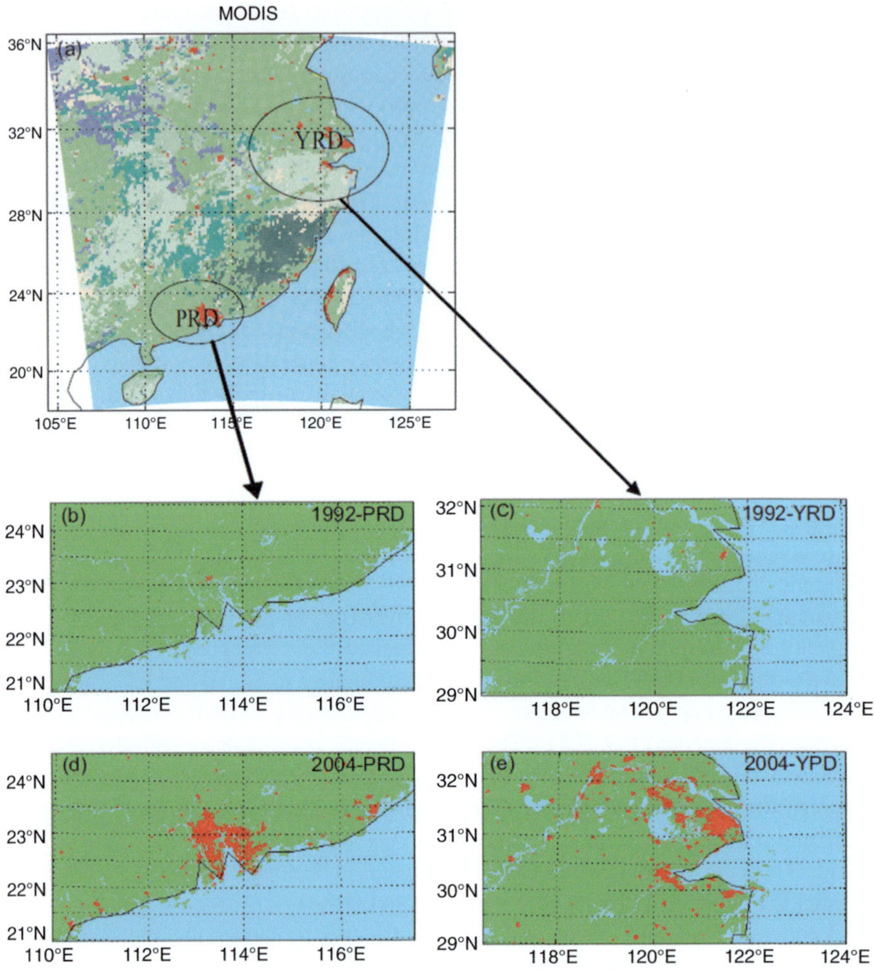

Fig. 22.12 Urban land-use change in the PRD and YRD regions, China, marked in *red* from pre-urbanization (1992–1993) and current (2004): (**a**) WRF-Chem domain with 12-km grid spacing; (**b**) 1992–1993 USGS data for PRD, (**c**) 2004 MODIS data for YRD, (**d**) 1992–1993 USGS data for PRD, and (**e**) 2004 MODIS data for YRD (From Wang et al. 2009a)

have also resulted in detectable concentration changes of NO_x, VOCs, SOAs and NO_3 radicals in the PRD (Wang et al. 2009b). Urbanization decreases surface NO_x and VOC concentrations by a maximum of 4 ppbv and 1.5 ppbv, respectively. Surface NO_3 radical concentrations over major cities increase by about 4–12 pptv. On the other hand, areas with increasing O_3 and NO_3 radical concentrations generally coincide with the areas of temperature increase and wind speed reduction also found from the urbanization. Urbanization has a stronger influence on SOA formation from Aitken mode (−3% to 9%) and stronger influence on aromatics precursors to produce SOA (14% increase), with similar results also found in Jing-Jin-Ji (JJJ) (Yu et al. 2012).

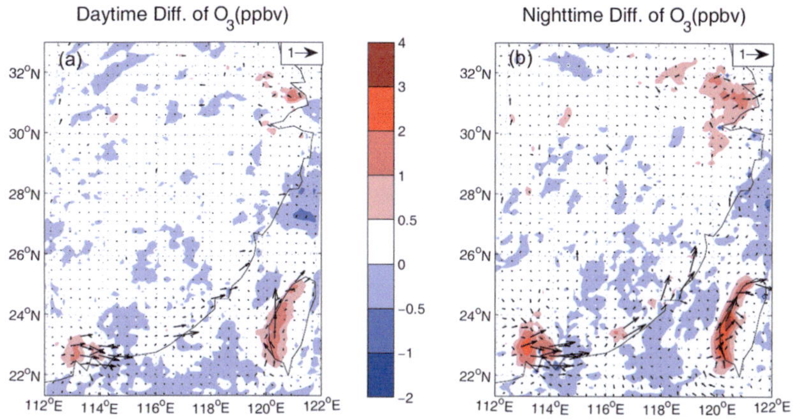

Fig. 22.13 Difference of surface ozone (in ppbv) and relative 10-m wind vectors: (**a**) daytime, (**b**) nighttime (From Wang et al. 2009a)

The WRF-Chem model coupled with UCMs is equally useful to project, for instance, air quality change in cities under future climate change scenarios. For example, the impact of future urbanization on surface ozone in Houston under the future IPCC A1B scenario for 2051–2053 (Jiang et al. 2008) shows generally a 2 °C increase in surface air temperature due to the combined change in climate and urbanization. In this example, the projected 62% increase of urban areas exerted more influence than attributable to climate change alone. The combined effect of the two factors on O_3 concentrations can be up to 6.2 ppbv. Sensitivity experiments performed by Jiang et al. (2008) revealed that future changes in anthropogenic emissions produces a similar order change in O_3 as those induced by changes in climate and urbanization. This reconfirms that a better understanding of the impacts of urbanization are important if we are to improve our understanding of these issues.

In addition, changes in emissions are also expected with changes in urbanization. Some studies that the anthropogenic emissions related to urbanization are a dominant factor in the long-term changes (Huang et al. 2013). Such results reflect that, although the impacts of land use change on air quality are non-negligible, the associated changes in emissions exert a more significant influence on air quality than land use change, although this is not clear from the results summarized here. This provides clear implications for policy makers to control urban air pollution over the developing region in China, especially with respect to spatial urban planning and reasonable emissions controls (Chen et al. 2014).

Acknowledgements This work was supported by European Union FP7 project PANDA (3206429) under the grant agreement n° 606719, National Science Fund for Distinguished Young Scholars (41425020), China Special Fund for Meteorological Research in the Public Interest (GYHY201406031), National Ministry of science and technology project (2014BAC21B02), and the Jiangsu Collaborative Innovation Center for Climate Change.

References

Archernicholls, S., Carter, E. M., Kumar, R., Xiao, Q., Yang, L., Frostad, J., et al. (2016). The regional impacts of cooking and heating emissions on ambient air quality and disease burden in China. *Environmental Science Technology, 50*(17), 9416–9623. doi:10.1021/acs.est.6b02533.

Bornstein, R., & Lin, Q. (2010). Urban heat islands and summertime convective thunderstorms in Atlanta: Three case studies. *Atmospheric Environment, 34*(3), 507–516. doi:10.1016/S1352-2310(99)00374-X.

Bougeault, P., & Lacarrère, P. (1989). Parameterization of orography-induced turbulence in a mesobeta-scale model. *Monthly Weather Review, 117*(8), 1872–1890. doi:10.1175/1520-0493 (1989)117<1872:POOITI>2.0.CO;2.

Burian, S. J., Stetson, S. W., Han, W., Ching, J. et al. (2004). *High-resolution dataset of urban canopy parameters for Houston, Texas* (pp. 23–26). Preprint proceedings, Fifth Symposium on the Urban Environment, AMS.

Chen, F., & Dudhia, J. (2001). Coupling an advanced land-surface/hydrology model with the Penn state/NCAR MM5 modeling system. Part I: Model implementation and sensitivity. *Monthly Weather Review, 129*(4), 569–585. doi:10.1175/1520-0493(2001)129<0569:CAALSH>2.0.CO;2.

Chen, J., & Song, Z. (2007). Description and evaluation of the characteristics of the NCAR high-resolution land data assimilation system. *Journal of Applied Meteorology and Climatology, 46* (6), 694–713. doi:10.1104/pp.82.4.904.

Chen, F., Mitchell, K., Schaake, J., Xue, Y., Pan, H., Koren, V., et al. (1996). Modeling of land surface evaporation by four schemes and comparison with fife observations. *Journal of Geophysical Research – Atmospheres, 101*(D3), 7251–7268. doi:10.1029/95JD02165.

Chen, F., Kusaka, H., Bornstein, R., Ching, J., Grimmond, C. S. B., Grossman-Clarke, S., et al. (2011a). The integrated WRF/urban modelling system: Development, evaluation, and applications to urban environmental problems. *International Journal of Climatology, 31*(2), 273–288. doi:10.1002/joc.2158.

Chen, F., Miao, S., Tewari, M., Bao, J., & Kusaka, H. (2011b). A numerical study of interactions between surface forcing and sea-breeze circulations and their effects on stagnant winds in the greater Houston area. *Journal of Geophysical Research-Atmospheres, 116*(D12), 3–5. doi:10.1029/2010JD015533.

Chen, B., Yang, S., Xu, X., & Zhang, W. (2014). The impacts of urbanization on air quality over the Pearl River Delta in winter: Roles of urban land use and emission distribution. *Theoretical and Applied Climatology, 117*(1), 29–39. doi:10.1007/s00704-013-0982-1.

Ching, J., Brown, M., Burian, S., Chen, F., Cionco, R., Hanna, A., et al. (2009). National urban database and access portal tool. *Bulletin American Meteorology Society, 90*(8), 1157–1168. doi:10.1175/2009BAMS2675.1.

Cohen, J. B., Prinn, R. G., & Wang, C. (2011). The impact of detailed urban-scale processing on the composition, distribution, and radiative forcing of anthropogenic aerosols. *Geophysical Research Letters, 38*(10), 264–265. doi:10.1029/2011GL047417.

Cui, L., & Shi, J. (2012). Urbanization and its environmental effects in Shanghai, China. *Urban Climate, 2*, 1–15. doi:10.1016/j.uclim.2012.10.008.

Dixon, P. G., & Mote, T. L. (2003). Patterns and causes of Atlanta's urban heat island-initiated precipitation. *Journal of Applied Meteorology, 42*(9), 1273–1284. doi:10.1175/1520-0450 (2003)042<1273:PACOAU>2.0.CO;2.

Donahue, N. M., Robinson, A. L., Stanier, C. O., & Pandis, S. N. (2006). Coupled partitioning, dilution, and chemical aging of semivolatile organics. *Environmental Science & Technology, 40*(8), 2635–2643. doi:10.1021/es052297c.

Dong, S., Tao, S., Yang, W., et al. (2012). Impact of climate change on urban agglomerations in coastal region of China. *Chinese Journal Population Resources Environmental, 10*(2), 78–83. doi:10.1080/10042857.2012.10685079.

Dudhia, J. (1989). Numerical study of convection observed during the winter monsoon experiment using a mesoscale two-dimensional model. *Journal of the Atmospheric Sciences, 46*(46), 3077–3107. doi:10.1175/1520-0469(1989)046<3077:NSOCOD>2.0.CO;2.

Elsenstat, L., & Wentworth, M. (2006). FERC's station power policies. *Electric Light & Power, 84*(5), 30.

Fast, J. D., Gustafson, W. I., Easter, R. C., et al. (2006). Evolution of ozone, particulates, and aerosol direct radiative forcing in the vicinity of Houston using a fully coupled meteorology-chemistry-aerosol model. *Journal of Geophysical Research – Atmospheres, 111*, 5173–5182. doi:10.1029/2005JD006721.

Feddema, J., Oleson, K., Bonan, G. (2006). *Developing a global database for the CLM urban model.* Sixth Symposium on the Urban Environment, 86th AMS Annual Meeting, Atlanta, GA. February 1.

Greenhalgh, S., Montgomery, M., Segal, S. J., & Todaro, M. P. (2007). State of world population 2007: Unleashing the potential of urban growth. *Population and Development Review, 33*(3), 639–640.

Grell, G. A., Peckham, S. E., Schmitz, R., Mckeen, S. A., Frost, G., Skamarock, W. C., et al. (2005). Fully coupled "online" chemistry within the WRF model. *Atmospheric Environment, 39*(37), 6957–6975. doi:10.1016/j.atmosenv.2005.04.027.

Gu, B., Dong, X., Peng, C., Luo, W., Chang, J., & Ge, Y. (2012). The long-term impact of urbanization on nitrogen patterns and dynamics in Shanghai, China. *Environmental Pollution, 171*(4), 30–37. doi:10.1016/j.envpol.2012.07.015.

Gutiérrez, E., González, J. E., Bornstein, R., Arend, M., Martilli, A. (2011). *Numerical simulations of a summer 2010 heat wave in New York City using WRF's building energy parameterization.* Abstracts of International Workshop on Urban Weather and Climate: Observation and Modeling.

Han, L., Zhou, W., Li, W., & Li, L. (2014). Impact of urbanization level on urban air quality: A case of fine particles ($PM_{2.5}$) in Chinese cities. *Environmental Pollution, 194*(7), 163–170. doi:10.1016/j.envpol.2014.07.022.

Han, L., Zhou, W., & Li, W. (2015a). City as a major source area of fine particulate ($PM_{2.5}$) in China. *Environmental Pollution, 206*, 183–187. doi:10.1016/j.envpol.2015.06.038.

Han, L., Zhou, W., Pickett, S. T. A., Li, W., & Li, L. (2015b). An optimum city size? The scaling relationship for urban population and fine particulate ($PM_{2.5}$) concentration. *Environmental Pollution, 282*(1), 96–101. doi:10.1016/j.envpol.2015.08.039.

Hong, S., & Pan, H. (1996). Nonlocal boundary layer vertical diffusion in a medium-range forecast model. *Monthly Weather Review, 124*(124), 2322. doi:10.1175/1520-0493(1996)124<2322: NBLVDI>2.0.CO;2.

Huang, J., Zhou, C., Lee, X., Bao, Y., Zhao, X., Fung, J., et al. (2013). The effects of rapid urbanization on the levels in tropospheric nitrogen dioxide and ozone over east China. *Atmospheric Environment, 77*(3), 558–567. doi:10.1016/j.atmosenv.2013.05.030.

Janjic, Z. I. (1994). The step-mountain eta coordinate model: Further development of the convection, viscous sublayer, and turbulent closure schemes. *Monthly Weather Review, 122*, 927–945. doi:10.1175/1520-0493(1994)122<0927:TSMECM>2.0.CO;2.

Jiang, X., Wiedinmyer, C., Chen, F., Yang, Z., & Lo, C. (2008). Predicted impacts of climate and land use change on surface ozone in the Houston, Texas, area. *Journal of Geophysical Research – Atmospheres, 113*(D20), 115–117. doi:10.1029/2008JD009820.

Kain, J. S., & Fritsch, J. M. (1993). Convective parameterization for mesoscale models: The Kain-Fritsch scheme. *Meteorological Monographs, 24*, 165–170.

Kalnay, E., & Cai, M. (2003). Impact of urbanization and land-use change on climate. *Nature, 423* (6939), 528–531.

Kusaka, H., & Kimura, F. (2004). Coupling a single-layer urban canopy model with a simple atmospheric model: Impact on urban heat island simulation for an idealized case. *Journal of the Meteorological Society of Japan, 82*(1), 67–80. doi:10.2151/jmsj.82.67.

Kusaka, H., Kondo, H., Kikegawa, Y., & Kimura, F. (2001). A simple single-layer urban canopy model for atmospheric models: Comparison with multi-layer and slab models. *Boundary-Layer Meteorology, 101*(3), 329–358. doi:10.1023/A:1019207923078.

Kusaka, H., Chen, F., Tewari, M., Dudhia, J., Gill, D. O., et al. (2012). Numerical simulation of urban heat island effect by the WRF model with 4-km grid increment: An inter-comparison study between the urban canopy model and slab model (special issue on downscaling studies for the impact assessments of climate change). *Journal of the Meteorological Society of Japan, 90B*(1), 33–45. doi:10.2151/jmsj.2012-B03.

Li, Y., Li, Y., Westlund, H., & Liu, Y. (2015). Urban–rural transformation in relation to cultivated land conversion in china: Implications for optimizing land use and balanced regional development. *Land Use Policy, 47*(47), 218–224. doi:10.1016/j.landusepol.2015.04.011.

Liao, J., Wang, T., Wang, X., Xie, M., Jiang, Z., Huang, X., et al. (2014). Impacts of different urban canopy schemes in WRF/Chem on regional climate and air quality in Yangtze River Delta, China. *Atmospheric Research, 145–146*, 226–243. doi:10.1016/j.atmosres.2014.04.005.

Liao, J., Wang, T., Jiang, Z., Zhuang, B., Xie, M., Yin, C., et al. (2015). WRF/Chem modeling of the impacts of urban expansion on regional climate and air pollutants in Yangtze River Delta, China. *Atmospheric Environment, 106*, 204–214. doi:10.1016/j.atmosenv.2015.01.059.

Lin, C., Chen, W., Liu, S., Liou, Y., Liu, G., & Lin, T. (2008). Numerical study of the impact of urbanization on the precipitation over Taiwan. *Atmospheric Environment, 42*(13), 2934–2947. doi:10.1016/j.atmosenv.2007.12.054.

Lin, J., Nielsen, C. P., Zhao, Y., Lei, Y., Liu, Y., & Mcelroy, M. B. (2010). Recent changes in particulate air pollution over China observed from space and the ground: Effectiveness of emission control. *Environmental Science & Technology, 44*(20), 7771–7776. doi:10.1021/es101094t.

Liu, Y., Chen, F., Warner, T., & Basara, J. (2006). Verification of a mesoscale data-assimilation and forecasting system for the Oklahoma City area during the joint urban 2003 field project. *Journal of Applied Meteorology and Climatology, 45*(45), 912–929. doi:10.1175/JAM2383.1.

Lo, J. C. F., Lau, A. K. H., Fung, J. C. H., & Fei, C. (2006). Investigation of enhanced cross-city transport and trapping of air pollutants by coastal and urban land-sea breeze circulations. *Journal of Geophysical Research – Atmospheres, 111*(D14), 3085–3098. doi:10.1029/2005JD006837.

Martilli, A., Clappier, A., & Rotach, M. W. (2002). An urban surface exchange parameterization for mesoscale models. *Boundary-Layer Meteorology, 104*(2), 261–304. doi:10.1023/A:1016099921195.

Miao, S., Chen, F., Li, Q., Fan, S. (2009a). *Impacts of urbanization on a summer heavy rainfall in Beijing*. The seventh International Conference on Urban Climate: Proceeding, 708, 29 June – 3 July 2009, Yokohama, Japan, B12-1.

Miao, S., Chen, F., Lemone, M. A., Tewari, M., Li, Q., & Wang, Y. (2009b). An observational and modeling study of characteristics of urban heat island and boundary layer structures in Beijing. *Journal of Applied Meteorology and Climatology, 48*(3), 484–501. doi:10.1175/2008JAMC1909.1.

Mitchell, K. E., Lohmann, D., Houser, P. R., et al. (2004). The multi-institution North American Land Data Assimilation System (NLDAS): Utilizing multiple GCIP products and partners in a continental distribute hydrological modeling system. *Journal of Geophysical Research, 109* (D7), 585–587. doi:10.1029/2003JD003823.

Mlawer, E. J., Taubman, S. J., Brown, P. D., Iacono, M. J., & Clough, S. A. (1997). Radiative transfer for inhomogeneous atmospheres: RRTM, a validated correlated-k model for the longwave. *Journal of Geophysical Research – Atmospheres, 102*(14), 16663–16682. doi:10.1029/97JD00237.

Nehrkorn, T., Leidner, M., Ellis, M., Maher, A., Eluszkiewicz, J. (2011). Modeling the urban circulation in the Salt Lake City area using the WRF urban canopy parameterization. *Meeting of the American Meteorology Society*.

Oke, T. R. (1982). The energetic basis of the urban heat island. *Quarterly Journal Royal Meteorology Society, 108*(455), 1–24. doi:10.1002/qj.49710845502.

Penner, J. E., Andreae, M., Annegarn, H., Barrie, L., Feichter, J., Hegg, D., et al. (2001). Observed climate variability and change. In: *Climate change 2001: The scientific basis*. Intergovernmental Panel on Climate Change. Available via DIALOG. http://ipcc.ch/ipccreports/tar/wg1/pdf/TAR-02.PDF. Accessed 2001.

Penrod, A., Zhang, Y., Wang, K., Wu, S., & Leung, L. R. (2014). Impacts of future climate and emission changes on U.S. air quality. *Atmospheric Environment, 89*(2), 533–547. doi:10.1002/qj.49710845502.

Ran, Y., Li, X., & Lu, L. (2010). Evaluation of four remote sensing based land cover products over China. *International Journal of Remote Sensing, 31*(2), 391–401. doi:10.1080/01431160902893451.

Rotach, M. W., Vogt, R., Bernhofer, C., Batchvarova, E., Christen, A., Clappier, A., et al. (2005). Bubble – an urban boundary layer meteorology project. *Theoretical and Applied Climatology, 81*(3–4), 231–261. doi:10.1007/s00704-004-0117-9.

Rozoff, C. M., Cotton, W. R., & Adegoke, J. O. (2003). Simulation of St. Louis, MO landuse impacts on thunderstorms. *Journal of Applied Meteorology, 42*(6), 716–738. doi:10.1175/1520-0450(2003)0422.0.CO;2.

Sailor, D. J., & Lu, L. (2004). A top–down methodology for developing diurnal and seasonal anthropogenic heating profiles for urban areas. *Atmospheric Environment, 38*(17), 2737–2748. doi:10.1016/j.atmosenv.2004.01.034.

Salamanca, F., & Martilli, A. (2010). A new building energy model coupled with an urban canopy parameterization for urban climate simulations—part II. Validation with one dimension off-line simulations. *Theoretical and Applied Climatology, 99*(3–4), 345–356. doi:10.1007/s00704-009-0143-8.

Salamanca, F., Martilli, A., Tewari, M., et al. (2011). A study of the urban boundary layer using different urban parameterizations and high-resolution urban canopy parameters with WRF. *Journal of Applied Meteorology and Climatology, 50*(5), 1107–1128. doi:10.1175/2010JAMC2538.1.

Skamarock, W. C., & Klemp, J. B. (2008). A time-split nonhydrostatic atmospheric model for weather research and forecasting applications. *Journal of Computational Physics, 227*(7), 3465–3485. doi:10.1016/j.jcp.2007.01.037.

Taha, H. (2008). Episodic performance and sensitivity of the urbanized MM5 (UMM5) to perturbations in surface properties in Houston Texas. *Boundary-Layer Meteorology, 127*(2), 193–218. doi:10.1007/s10546-007-9258-6.

Wang, X., Lin, W., Yang, L., Deng, R., & Lin, H. (2007). A numerical study of influences of urban land-use change on ozone distribution over the pearl river delta region, China. *Tellus Series B, 59*(3), 633–641. doi:10.1111/j.1600-0889.2007.00271.x.

Wang, X., Chen, F., Wu, Z., Zhang, M., Tewari, M., Guenther, A., et al. (2009a). Impacts of weather conditions modified by urban expansion on surface ozone: Comparison between the pearl river delta and Yangtze River Delta regions. *Advances in Atmospheric Sciences, 26*(5), 962–972. doi:10.1007/s00376-009-8001-2.

Wang, X., Wu, Z., & Liang, G. (2009b). WRF/Chem modeling of impacts of weather conditions modified by urban expansion on secondary organic aerosol formation over Pearl river delta. *Particuology, 7*(5), 384–391. doi:10.1016/j.partic.2009.04.007.

Wang, X., Liao, J., Zhang, J., Shen, C., Chen, W., Xia, B., et al. (2014). A numeric study of regional climate change induced by urban expansion in the Pearl River Delta, China. *Journal of Applied Meteorology and Climatology, 53*(2), 346–362. doi:10.1175/JAMC-D-13-054.1.

Wang, Z., Shao, M., Chen, L., Tao, M., Zhong, L., Chen, D., et al. (2016). Space view of the decadal variation for typical air pollutants in the Pearl river delta (PRD) region in China. *Frontiers of Environmental Science & Engineering, 10*(5), 1–14. doi:10.1007/s11783-016-0853-y.

Yu, M., Carmichael, G. R., Zhu, T., & Cheng, Y. (2012). Sensitivity of predicted pollutant levels to urbanization in China. *Atmospheric Environment, 60*(60), 544–554. doi:10.1016/j.atmosenv.2012.06.075.

Zaveri, R. A., & Peters, L. K. (1999). A new lumped structure photochemical mechanism for large-scale applications. *Journal of Geophysical Research – Atmospheres, 104*(D23), 30387–30415. doi:10.1029/1999JD900876.

Zaveri, R. A., Easter, R. C., Fast, J. D., & Peters, L. K. (2008). Model for simulating aerosol interactions and chemistry (MOSAIC). *Journal of Geophysical Research – Atmospheres, 113* (D13), 1395–1400. doi:10.1029/2007JD008782.

Zhang, C., Chen, F., Miao, S., Li, Q., Xia, X., & Xuan, C. (2009). Impacts of urban expansion and future green planting on summer precipitation in the Beijing metropolitan area. *Journal of Geophysical Research – Atmospheres, 114*(D2), 356–360. doi:10.1029/2008JD010328.

Zhang, N., Gao, Z., Wang, X., & Chen, Y. (2010). Modeling the impact of urbanization on the local and regional climate in Yangtze River Delta, China. *Theoretical and Applied Climatology, 102*(3), 331–342. doi:10.1007/s00704-010-0263-1.

Zhang, Y., Zhang, X., & Liu, S. (2011). Correlation analysis on normalized difference vegetation index (NDVI) of different vegetations and climatic factors in southwest China. *Chinese Journal of Applied Ecology, 22*(2), 323–330.

Part VI
Impacts of Air Pollution

Chapter 23
Surface $PM_{2.5}$, Satellite Distribution of Atmospheric Optical Depth and Related Effects on Crop Production in China

Xuexi Tie, Xin Long, Wenting Dai, and Guy P. Brasseur

Abstract The surface concentrations of $PM_{2.5}$ measured by the Chinese National Environmental Monitoring Center (CNEMC) and the aerosol optical depth (AOD) observed by the Moderate-Resolution Imaging Spectroradiometer (MODIS) aboard the Aqua satellite are used to study the spatial and temporal variations of aerosol pollution in eastern China. A solar radiation transfer model (Tropospheric Ultraviolet-Visible - TUV) developed at National Center of Atmospheric Research (NCAR) is applied to study the reduction of solar radiation by aerosol pollution, and the effect of this air pollution on crop production in the major farmlands of eastern China. The study shows that the correlation between the surface $PM_{2.5}$ concentrations and the AOD values is complicated. In addition to $PM_{2.5}$ concentrations, the vertical integration of the aerosol particle concentration, the hygroscopic growth of the particles, and the aerosol optical properties affect the AOD values. As a result, AOD is a better parameter to characterize the effect of aerosol particles on solar radiation in the atmosphere. Because regions with high values of AOD are collocated in many cases with major crop production regions, aerosol pollution has an important impact on the crop production in eastern China. Heavy aerosol load strongly reduces the sunlight reaching the surface in the major farmlands of eastern China, and, as a consequence, is responsible for a substantial decrease in crop

X. Tie (✉)
Key Laboratory of Aerosol Chemistry & Physics, SKLLQG, Institute of Earth Environment, Chinese Academy of Sciences, Xi'an 710061, China

Center for Excellence in Urban Atmospheric Environment, Institute of Urban Environment, Chinese Academy of Sciences, Xiamen 361021, China
e-mail: tiexxacd@163.com

X. Long • W. Dai
Key Laboratory of Aerosol Chemistry & Physics, SKLLQG, Institute of Earth Environment, Chinese Academy of Sciences, Xi'an 710061, China
e-mail: longxin@ieecas.cn; daiwt@ieecas.cn

G.P. Brasseur
National Center for Atmospheric Research, Boulder, CO, USA

Max Planck Institute for Meteorology, Bundesstr. 53, Hamburg 20146, Germany
e-mail: guy.brasseur@mpimet.mpg.de

I. Bouarar et al. (eds.), *Air Pollution in Eastern Asia: An Integrated Perspective*, ISSI Scientific Report Series 16, DOI 10.1007/978-3-319-59489-7_23

production. However, the uncertainty in the estimated reduction of crop yield is large. One of the important uncertainties is associated with the calculation of the partitioning between direct and diffusive sunlight. The study suggests that, under the high diffusive case, the estimated reduction in crop yield is significantly lower than under the low diffusive case. For example, the rice reduction ranges from ~2% in the low diffusive case to ~1% in the high diffusive case. The reduction in the wheat yield ranges from ~6% in the low diffusive case, to ~4% in the high diffusive case. Future studies will have to be performed to reduce the uncertainty in these estimates.

Keywords Crop production • Satellite AOD • $PM_{2.5}$ • Eastern China

23.1 Introduction

In recent years, China has been experiencing severe air pollution problems, especially the formation of high concentrations of aerosol particles. In Beijing, the capital city of China, heavy haze events occurred rather frequently in recent years. During such events, the concentration of $PM_{2.5}$ (particle matter with a particle's radius equal to or less than 2.5 μm) rapidly increases, with a maximum of typically 600 μg/m^3 (Quan et al. 2014). This severe environmental problem has large impacts, for example, on people's health and on climate (Charlson et al. 1987; Ramanathan and Vogelmann 1997; Tegen et al. 2000; Tie et al. 2005).

Many studies have addressed the impact of heavy aerosol pollution on human health (Tie et al. 2009; Cao et al. 2012; Kan et al. 2007), specifically in eastern China, where severe pollution episodes are observed. However, only a few studies have addressed the effects of air pollution on ecosystems, especially on crop production in China.

In recent years, the Chinese National Environmental Monitoring Center (CNEMC) has released surface measurements of $PM_{2.5}$ concentrations. Most monitoring sites are located in eastern China and provide information on the regional distribution of the aerosol load. These measurements show some interesting features: (1) unlike the situation in the developed countries where air pollution is confined in and around large cities, the aerosol pollution in China is spread over large areas outside urban centers; (2) the location of highly polluted regions in China often coincides with regions of intense crop production. These high aerosol concentrations cause a significant reduction in the intensity of the solar radiation reaching the surface, and therefore affect the rate of photosynthesis by plants in the farmlands.

In this Chapter, we first present the spatial distribution of $PM_{2.5}$ concentrations measured by CNEMC and its relationship with the aerosol optical depth (AOD) measured by the Moderate Resolution Imaging Spectroradiometer (MODIS) aboard the Aqua satellite. The differences between the $PM_{2.5}$ and AOD spatial distributions are discussed. Second, we estimate the reduction of solar radiation in major farmland areas due to the heavy aerosol load. The calculation is based on the solar radiation transfer model (Tropospheric Ultraviolet-Visible - TUV) developed at National Center of Atmospheric Research (NCAR) (Madronich and Weller

1990). Third, we derive the effects of the reduction of solar radiation on crop production in the major farmlands of China. Finally, we assess the uncertainty in our estimated crop reduction.

23.2 Methods; Radiative Transfer Model (TUV)

The radiative transfer model employed here to calculate the effect of aerosol particles on the surface solar radiation flux is based on the Tropospheric Ultraviolet-Visible Model (TUV) developed by Madronich and Weller (1990). TUV is a state-of-the-art radiation transfer model, and is widely used by the scientific community. The code and the description of the model are available from http://www.acom.ucar.edu/TUV. The model calculates spectral irradiance, spectral actinic flux for the wavelength range between 121 and 750 nm. The calculation of actinic flux depends on local optical conditions, i.e., the vertical distribution of aerosol particles. The actinic flux varies according to atmospheric absorption as well as molecular (Rayleigh) and particle (Mie) scattering. Atmospheric absorbers include oxygen (Schumann-Runge and Hertzberg structures) and ozone (Hartley, Huggins, and Chappuis structures). A host of other gaseous absorbers (sulfur and nitrogen dioxides) can attenuate the actinic flux under polluted conditions. The effect of Rayleigh scattering on actinic flux is relatively well known and is routinely modeled. However, the calculation of Mie scattering due to the presence of aerosols is more complex since it depends on the chemical composition and size distribution of the particles.

23.3 Results and Discussion

23.3.1 Spatial Distributions of $PM_{2.5}$ Concentrations

In order to understand the effect of aerosol pollution on the solar radiation at the surface, the distributions of surface measurements of $PM_{2.5}$ are analyzed. Since 2013, the Chinese National Environmental Monitoring Center releases nationwide surface measurements of $PM_{2.5}$ concentrations. The monitoring sites are indicated in Fig. 23.1 (squares). The color contour lines represent the interpolation of the measurements. The results indicate that aerosol pollution occurs at a large spatial scale, and therefore extents not only in large cities, but also in large farmland areas. For example, in the main wheat and corn production region of the northern Chinese Plain (NCP), the 3-year (from April 2013 to March 2016) mean concentrations of $PM_{2.5}$ range from 85 to 100 $\mu g/m^3$, which is seven to eight times higher than the US standard of 15 $\mu g/m^3$. These extremely high aerosol concentrations, in addition to their important impact on human's health, especially in large densely populated cities (Tie et al. 2009; Cao et al. 2012; Kan et al. 2007), have substantial direct impacts on

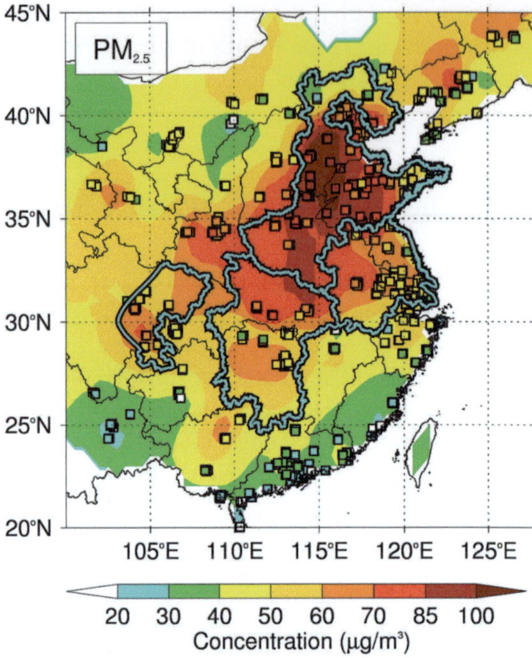

Fig. 23.1 Surface measurements of mean $PM_{2.5}$ concentrations (Apr. 2013 to Mar. 2016) by the Chinese National Environmental Monitoring Center (CNEMC). The *squares* represent the location of the monitoring stations. The *color contour lines* show the interpolation of the $PM_{2.5}$ concentrations

crop productivity in large areas of the Chinese farmlands (i.e., the NCP region). However, the impact of the aerosol pollution on crop production has not yet received enough attention, and requires additional investigations.

23.3.2 *Spatial Distributions of AOD*

The calculation of the effect of aerosol particles on solar radiation at the surface requires that the vertical distribution of aerosols and aerosol properties, including their absorbing, scattering, and hygroscopic properties, be accurately known. Satellite measurements of aerosol optical depth (AOD) provide more useful information than only surface concentrations of $PM_{2.5}$ since AOD is a function of the vertical column of aerosol particles, the hygroscopic growth of aerosol particles, and the aerosol optical properties. Surface $PM_{2.5}$ concentrations only provide mass concentrations. AOD is therefore a more suitable aerosol parameter for radiative transfer calculations than the surface $PM_{2.5}$ concentration. Figure 23.2 shows the 3-year (April 2013 to March 2016) averaged distribution of AOD in eastern China.

An examination of Figs. 23.1 and 23.2 shows that there are similarities and discrepancies between the $PM_{2.5}$ and AOD distributions. In the NCP region (indicated as Region 1), both $PM_{2.5}$ and AOD values are consistently high. However, in central eastern China (indicated as Region 3), and the Si Chuan Basin (indicated as Region 4), the AOD concentrations are high, while the $PM_{2.5}$ concentrations are

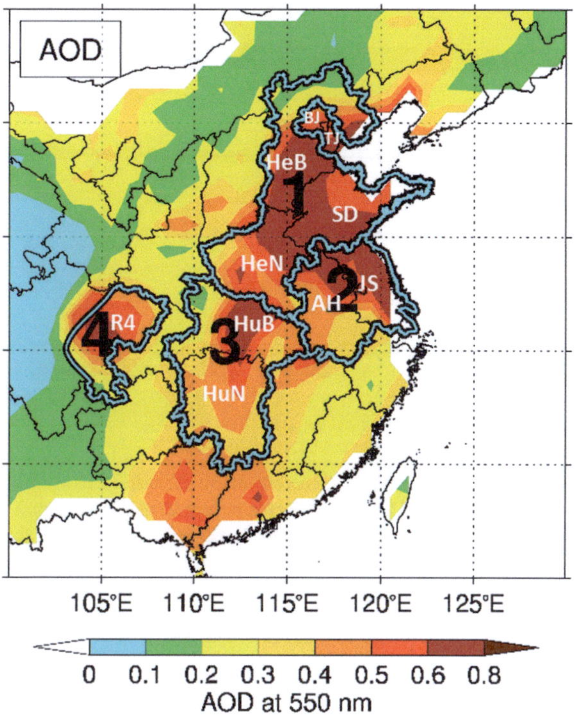

Fig. 23.2 Surface measurements of mean (Apr. 2013 to Mar. 2016) aerosol optical depth (AOD) measured by the Moderate Resolution Imaging Spectroradiometer (MODIS) aboard the Aqua satellite. Four different regions are defined (Region 1 as the North China Plain region (NCP), Region 2 as the Yangtze River Delta region (YRD), Region 3 as the region of central eastern China (CEC), and Region 4 as the Si Chuan Basin (SCB)). The white labels represent the name of provinces (BJ and TJ are two megacities), and R4 is for SCB

relatively low. This result suggests that the correlation between surface $PM_{2.5}$ concentrations and AOD values is complicated. It can be expressed by

$$AOD\tilde{}PM_{2.5} + \delta(Y1, Y2, Y3)$$

where Y1, Y2, and Y3 represent the vertically integrated aerosol amount, the hygroscopic growth of the particles, and their optical property, respectively. For example, in Regions 3 and 4, the relative humidity is generally high, resulting in high hygroscopic growth of aerosol particles and high AOD values.

23.3.3 Relationship Between AOD and Crop Production Areas

As discussed in the above section, the effect of aerosol particles on solar radiation at the surface is best derived from the distribution of AOD rather than from the measured surface concentration of particles. A comparison of Figs. 23.1 and 23.2 suggests that high values of AOD are in many cases collocated with the major crop production regions of China. Figure 23.3 shows the correlation between the crop yield (rice and wheat) and the AOD values in eastern China. The red squares with labels, which generally have high AOD values, correspond to the provinces within

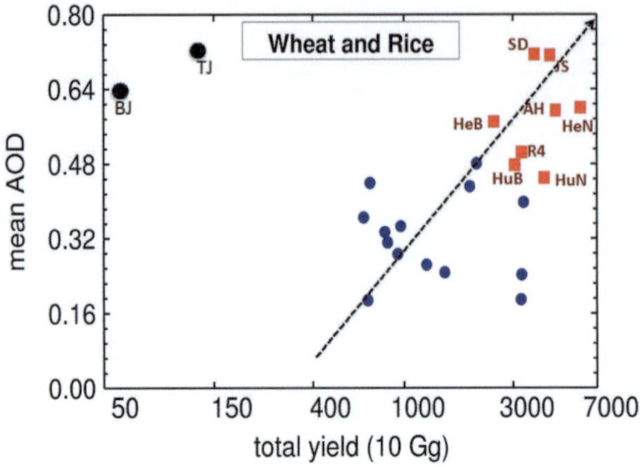

Fig. 23.3 Relation between the crop yield (rice and wheat) and the AOD values in eastern China. The *red squares* with labels correspond to the provinces within the major 4 crop production areas identified in Fig. 23.2. The *blue circles* represent the other provinces outside the 4 regions, which are generally characterized low AOD values. The 2 *black circles* show data for two megacities (BJ-Beijing and TJ-Tianjin) inside the NCP region

the major four crop production regions highlighted in Fig. 23.2. The blue circles with generally low AOD values, represent the other provinces outside the four regions. The two black circles with high AOD and low crop yields refer to the two megacities (BJ-Beijing and TJ-Tianjin) inside the NCP region. Figure 23.3 shows that except in the two megacities, the high AOD values are correlated with the highest crop yields. For example, in SD and JS provinces, the mean AOD values reach about 0.75 with high an annual crop production of about 40–50 million tons.

The strong correlation between the aerosol pollution and crop production suggests that the aerosol particles may have important impacts on the crop production in eastern China, and need to be carefully assessed.

23.3.4 Estimation of Aerosol Impact on the Crop Production

One of the important effects of aerosol pollution is to absorb and scatter solar light in the atmosphere, which reduces the sunlight that reaches the Earth's surface, and affects ecosystems. As shown in Fig. 23.3, because the high crop farmlands are located in heavily polluted regions, the reduction of sunlight can decrease crop production substantially in eastern China (Tie et al. 2016). The quantitative estimate of the reduction in rice and wheat production follows the following steps:

1. The distribution of the reduction of surface sunlight $\Delta S(x, y, t)$ is calculated for different months, using the TUV model and the AOD distributions, where $\Delta S(x, y, t)$ represent the percent reduction of sunlight for different locations (x, y) and different months (t).

2. The calculated value of $\Delta S(x, y, t)$ is used to estimate the rice and wheat production using the correlation factor F from the following relation

$$\Delta C(x, y, t) = F\Delta S(x, y, t) \times C(x, y, t),$$

where $C(x, y, t)$ and $\Delta C(x, y, t)$ represent the crop production and the reduction in crop production for the different locations and the different months, and where F denotes a conversion factor provided by Chameides et al. (1999).

23.3.5 Uncertainties Related to the Estimation

Before we represent our estimate of the crop reduction due to aerosol pollution, some important uncertainties must be stated and estimated. The photosynthesis of crop plants is mainly dependent on three factors: water, nutrients, and sunlight (Nobel 1983). The value of the converting factor F is established by assuming that agricultural conditions are often manipulated to maximize crop yields through irrigation and fertilization. As a result, the surface sunlight has a larger effect than the two other factors for the determination of the crop production. However, this condition may vary as a function of location, which introduces an uncertainty in the estimate.

Another uncertainty is related to the calculation of the diffusive solar light. After the reduction of sunlight by aerosol particles, the reduced sunlight reaching the surface includes two contributions: direct sunlight and diffusive sunlight. According to the study of Mercado et al. (2009), the diffusive light tends to increase the rate of photosynthesis in comparison with the effect of direct light. The impact of direct and diffusive sunlight on photosynthesis is illustrated in Fig. 23.4.

The estimate of direct and diffusive light is complex because it depends on the physical and chemical properties of the aerosols (such as the size, shape, chemical composition, and optical property of aerosol particles). This uncertainty affects the calculated ratio of direct and diffusive light. In the present study, we used a wide range of measured direct and diffuse solar radiation to characterize low and high diffusive cases and to estimate the uncertainty of the calculated crop reduction. We denote by $R = DI/DU$ the ratio between direct and diffuse solar radiation. On the basis of the measurements made in different regions of the world, the value of R ranges from 2.5 to 7.0 (Liepert and Tegen 2002). We adopt therefore $R = 2.5$ and 7.0 to represent high and low diffuse solar radiation, respectively. The crop production for these two cases is estimated according to Mercado et al. (2009).

Figure 23.5 shows the estimated reduction in crop and wheat yields for the low and high diffusive cases, respectively. In the high diffusive case, the estimated reduction of crop is significantly lower than in the low diffusive case. For example, the reduction in the rice production ranges from 35 to 50 (10^4 tons/year) in Region 2 under the low diffusive case (corresponding to ~2%), while the reduction ranges only from 20 to 35 (10^4 tons/year) in the high diffusive case (corresponding to ~1%). The reduction in the production of wheat is larger than 150 (10^4 tons/year) in

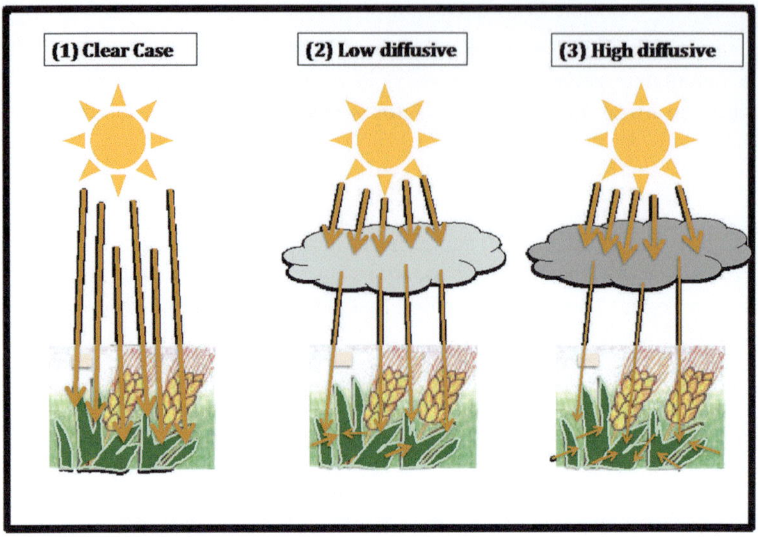

Fig. 23.4 Schematic illustration of the effects solar light reduction on the photosynthesis of crop plants. The *left panel* shows the sunlight reaching the plant's leaves without considering aerosol effects. The *middle* and *right panels* show the sunlight reaching the plant's leaves with aerosol effects for the low and high diffusive cases, respectively

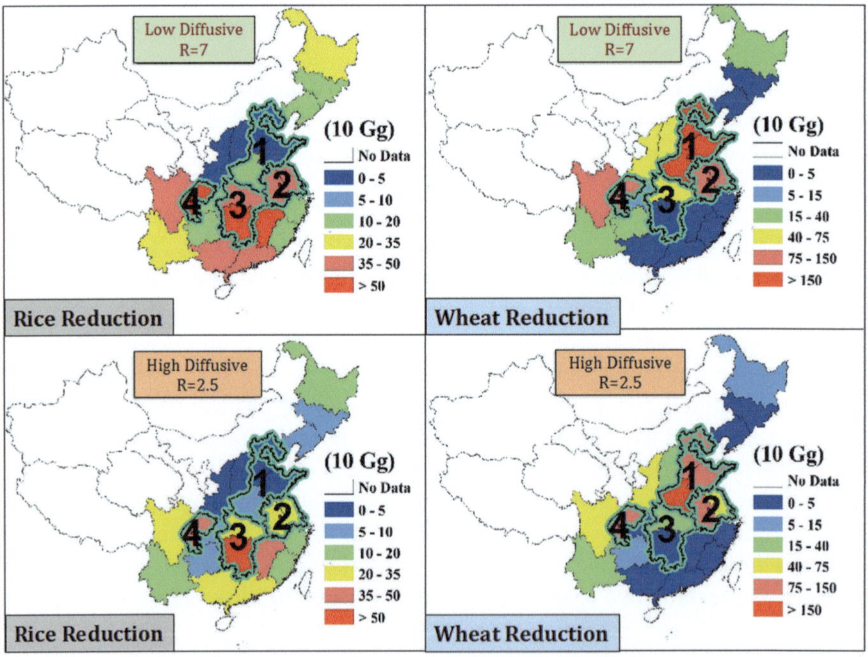

Fig. 23.5 Estimated decrease in rice and wheat production (10^4 tons/year) in different regions of eastern China for low and high diffusive sunlight. The *left panels* represent the reduction in rice production, while the *right panels* represent the reduction in wheat production. The *upper* and *lower panels* refer to the low and high diffusive cases, respectively

Region 1 for the low diffusive case (corresponding to ~6%), and is of the order of 75–150 (10^4 tons/year) in the high diffusive case (corresponding to ~4%). Our study suggests therefore that the effect of aerosol pollution on the crop production is substantial in eastern China, but, at the same time, that the uncertainty on this estimate is large. Future studies are therefore necessary.

23.4 Summary

In recent years, China has been experiencing severe air pollution problems, especially the presence of high concentrations of aerosol particles in the atmosphere. Studies on the impact of heavy aerosol pollution events have highlighted primarily the adverse effects of aerosols on human health. The effect of the air pollution on the crop production has received less attention.

We have used measurements of surface $PM_{2.5}$ by the Chinese National Environmental Monitoring Center (CNEMC) and the observations of aerosol optical depth (AOD) from the Moderate Resolution Imaging Spectroradiometer (MODIS) aboard the Aqua satellite to study the spatial and temporal distributions of aerosol pollution in eastern China. A solar radiation transfer model (Tropospheric Ultraviolet-Visible - TUV) developed at National Center of Atmospheric Research (NCAR) has been applied to assess the effects of the reduction in solar radiation on crop production in the major farmland areas of China.

The findings of this study can be summarized as follows: (1) The correlation between surface $PM_{2.5}$ concentrations and AOD values is complex. In addition to surface aerosol mass concentrations, one has to consider the vertical column of aerosol particles, the hygroscopic growth of aerosol particles, and the aerosol optical properties. As a result, AOD is a more appropriate parameter for calculating the effect of aerosol particles on the penetration of solar radiation in the atmosphere. (2) High values of AOD (pollution situations) coexist with regions of intense crop production in China. As a result, aerosol pollution has an important impact on crop production in eastern China. (3) Heavy aerosol pollution strongly reduces the sunlight intensity in the major farmlands of eastern China, and reduces crop production in these regions. However, there are large quantitative uncertainties in the estimated reductions. One of the factors that contribute to these uncertainties is the poorly quantified partitioning between direct and diffusive sunlight. According to previous studies, diffusive light tends to increase photosynthesis in comparison with direct light. The estimate of the relative intensity of direct and diffusive light is complex, because it depends on the physical and chemical properties of aerosols (such as the size, shape, chemical composition and optical properties of aerosol particles).

In this study, low and high diffusive cases are defined according to the measurements in different regions of the world. For the high diffusive case, the estimated reduction of crop is significantly lower than the reduction determined for the low diffusive case. For example, in the intense rice production region of China, the reduction in the rice yield ranges from ~2% in the low diffusive case to

~1% in the high diffusive case. For wheat production, the corresponding reduction ranges from ~6% to ~4%. Additional studies should further reduce the uncertainties on the calculated impacts of air pollution on crop productivity.

Acknowledgments This work was supported by the National Natural Science Foundation of China (NSFC) under Grant Nos. 41275168, 41430424, and 41503117. The National Center for Atmospheric Research is sponsored by the National Science Foundation.

References

Cao, J. J., Xu, H. M., Xu, Q., Chen, B. H., & Kan, H. D. (2012). Fine particulate matter constituents and cardiopulmonary mortality in a heavily polluted Chinese City. *Environmental Health Perspectives, 120*, 373–378.

Chameides, W. L., Yu, H., Liu, S. C., Bergin, M., Zhou, X., Mearns, L., Wang, G., Kiang, C. S., Saylar, R. D., Lou, C., Huang, Y., Steiner, A., & Giorgi, F. (1999). Case study of the effects of atmospheric aerosols and regional haze on agriculture: An opportunity to enhance crop yields in China through emission controls? *Proceedings of the National Academy of Sciences, 96*, 13626–13633.

Charlson, R. J., Lovelock, J. E., Andreae, M. O., & Warren, S. G. (1987). Oceanic phytoplankton, atmospheric sulfur, cloud albedo and climate. *Nature, 326*, 655–661.

Kan, H., London, S. J., Chen, G., Zhang, Y., Song, G., Zhao, N., et al. (2007). Differentiating the effects of fine and coarse particles on daily mortality in Shanghai. *China Environment International, 33*(3), 376–384.

Liepert, B. & Tegen, I. (2002). Multidecadal solar radiation trends in the United States and Germany and direct tropospheric aerosol forcing. *Journal of Geophysical Research 107*, doi:10.1029/2001JD000760.

Madronich, S., & Weller, G. (1990). Numerical integration errors in calculated tropospheric photodissociation rate coefficients. *Journal of Atmospheric Chemistry, 10*, 289–300.

Mercado, L. M., Bellouin, N., Sitch, S., Boucher, O., Huntingford, C., Wild, M., & Cox, P. (2009). Impact of changes in diffuse radiation on the global land carbon sink. *Nature, 458*, 1014–1018.

Nobel, P. S. (1983). *Biophysical plant physiology and ecology*. San Francisco: WH Freeman and company.

Ramanathan, V., & Vogelmann, A. M. (1997). Greenhouse effect, atmospheric solar absorption, and the Earth's radiation budget: From the Arrhenius–Lanely era to the 1990s. *Ambio, 26*(1), 38–46.

Quan, J. N., Tie, X., Zhang, Q., Liu, Q., Li, X., Yang, G., & Zhao, D. (2014). Characteristics of heavy aerosol pollution during the 2012–2013 winter in Beijing, China. *Atmospheric Environment, 88*, 83–89. doi:10.1016/j.atmosenv.2014.01.058.

Tegen, I., Koch, D., Lacis, A. A., & Sato, M. (2000). Trends in tropospheric aerosol loads and corresponding impact on direct radiative forcing between 1950 and 1990: A model study. *Journal of Geophysical Research, 105*, 26971–26990.

Tie, X., Madronich, S., Walters, S., Edwards, D.P., Ginoux, P., Mahowald, N., Zhang, R.Y., Lou, C., Brasseur, G. (2005). Assessment of the global impact of aerosols on tropospheric oxidants. *Journal of Geophysical Research*, 110 (D03204), 10.1029/2004JD00535.

Tie, X., Wu, D., & Brasseur, G. (2009). Lung cancer mortality and exposure to atmospheric aerosol particles in Guangzhou, China. *Atmospheric Environment, 43*, 2375–2377.

Tie, X., Huang, R. J., Dai, W. T., Cao, J. J., Long, X., Su, X. L., Zhao, S. Y., Wang, Q. Y., & Li, G. H. (2016). Effect of heavy haze and aerosol pollution on rice and wheat productions in China. *Scientific Reports, 6*, 29612. doi:10.1038/srep29612.

Chapter 24
Research Perspectives on Air Pollution and Human Health in Asia

Benjamin Guinot and Isabella Annesi-Maesano

Abstract The scientific challenges abound to better assess the relationship between air pollutants, their biological effects on the human health and the responses of the exposed populations. In Asia, atmospheric chemists have to consider much higher levels of air pollution but also an original aerosol chemical composition which results from multiple sources, and specific climate and meteorological patterns. As for health researchers, the ways of exposure and the biological responses differ from what have been observed in other regions of the world, and should lead to new air pollution dose-response functions particular to Asia. Such challenges create opportunities to bring together chemists, biologists and epidemiologists in order to go beyond the management of current interdisciplinary projects, by creating integrated research units where original cross methodologies and joint technical approaches could be efficiently developed.

Keywords Air pollution • Asia • Aerosols • Exposure • Response

24.1 Introduction

Atmospheric pollution drives major health risks on humans. In Asia, where air pollution can reach 5–10 times higher levels than in Europe or North America, attempts to estimate mortality, morbidity and costs related to pollution have led to alarming figures. Generally, the pollution-health models used to produce such estimations were based on figures obtained in other regions of the world and may thus not be well relevant in Asia. Comparative epidemiological studies between

B. Guinot (✉)
Laboratoire d'Aérologie, Université de Toulouse, CNRS, UPS, 14 avenue Edouard-Belin, F31400 Toulouse, France
e-mail: benjamin.guinot@gmail.com

I. Annesi-Maesano
UPMC Université Paris 06, INSERM, Institut Pierre Louis d'Epidémiologie et de Santé Publique (IPLESP UMRS 1136), Epidemiology of Allergic and Respiratory diseases department (EPAR), Medical School Saint-Antoine, Sorbonne Universités, F75012 Paris, France
e-mail: isabella.annesi-maesano@inserm.fr

I. Bouarar et al. (eds.), *Air Pollution in Eastern Asia: An Integrated Perspective*,
ISSI Scientific Report Series 16, DOI 10.1007/978-3-319-59489-7_24

European and Chinese pupils have shown discrepancies in their response to air pollution exposure: higher prevalence of respiratory symptoms but lower prevalence of asthma, since asthma is a multifactorial disease, including environmental and genetic factors (Zhao et al. 2008a, b). Incidentally, recent epidemiological results obtained in China have suggested a rupture of linearity in the exposure-response function at high particulate matter (PM) concentrations (Lai et al. 2013).

Particulate matter (PM) or aerosols, are the key characteristic of air pollution in Asia. They have various sizes and a complex chemical composition which result from their multiple sources, including traffic, industries, coal-fired thermal plants, as well as biomass burning from agricultural activities or for domestic usages. They also have biogenic origins, in particular desert dust. Indoor concentrations of particles are found at a high level too, since outdoor ambient air can react with possible indoor chemicals (like formaldehyde) and biological aerosols. The complex atmospheric processes, on one hand, and the qualification of the various pollution effects on health, on the other hand, still remain poorly understood. Thus, our knowledge still faces key limitations to attempt a quantification of such effects in Asia, today and in the future.

This chapter points out the specificities of air pollution in Asia. Aerosols and gases of anthropogenic and biogenic origins can be transported and transformed at local, regional or even intercontinental scales, which results in a complicated and probably unique mix of air pollutants. Billions of people breathe it in, according to various exposure modes, which are combined with a variety of individual responses. From these observations, research directions for the future are finally suggested in order to evaluate adverse health effects of air pollution and establish proper exposure-dose functions for Eastern and Southern Asia.

24.2 Asian Characteristics

24.2.1 Air Pollutants

Eastern and Southern Asia gathers the most populated regions in the world, thus a major source of anthropogenic materials in the Northern Hemisphere. But natural features significantly determine air pollution processes and transport in this part of the globe: topography, geology, vegetation and climates.

The Himalayas act as a wall against which air pollutants from Northern India accumulate. The North China Plain is like a reservoir, which is literally filled in by anthropogenic pollutants under anticyclonic conditions. Long rivers and large deltas lead to swampy regions where secondary aerosol formation is favoured by humidity, pools of precursors and photochemistry.

Deserts are widely considered to be the major sources for Asian dust according to present-day observations and rain-dust records in historical documents (Merrill et al. 1994, 1989; Prospero 1981; Zhang 1984). The most important sources for Asian soil dust have been identified in Northern China (Badain Jaran and Tengger deserts, mainly), Western China (Taklimakan), Southern Mongolia (Gobi) and the Loess Plateau further south mainly over the Shanxi and Shaanxi provinces (Zhang et al. 1997). High winds inject hundreds of teragrams of mineral matter into the atmosphere each year (Zhang et al. 2003, 1996a, b). Some experts have pointed out the significant contribution to dust emissions from degraded grasslands and open farmlands (Tegen and Fung 1995). However, the model simulations indicate that meteorology and climate have had a greater influence than desertification on the Asian dust emissions and associated Asian dust storm occurrences (Zhang et al. 2003).

There has been increasing interest in connections among Asian dust, marine biogeochemical cycles, and climate, as reported by the Intergovernmental Panel on Climate Change (IPCC). Interestingly, observations in urban environments have considered the impregnation of the atmosphere by mineral dust and suggested original interactions with anthropogenic pollutants in both the coarse and fine aerosol size fractions (Roger et al. 2009). Fine dust particles would act as a core component to form an aerosol mix coated by neutralized light organic and inorganic acids, besides other fine aerosol modes made from combustion aerosols only (Fig. 24.1). The resulting aerosol model would hence lead to six modes. It is a far more complicated representation than the widely used two- or three-mode (fine and coarse, or two fine and one coarse modes) aerosol models, but it may possibly better account for the multiplicity of aerosol sources in Asia, both anthropogenic and biogenic, like dust.

Vegetation in Asia emits large amounts of biogenic volatile organic carbon (BVOCs), which have been shown responsible for the so-called "blue hazes" documented for the first time by Went (1960) in forest areas. Steiner et al. (2002), in their attempt to compare present-time BVOC emissions in East Asia with a pre-disturbed land-cover scenario, led to the conclusion that deforestation drastically decreased BVOC. But a recent study for China by Fu and Liao (2014) suggested that between the late 1980s and the mid-2000s, while urban areas have expanded, areas covered by croplands overall decreased and those covered by

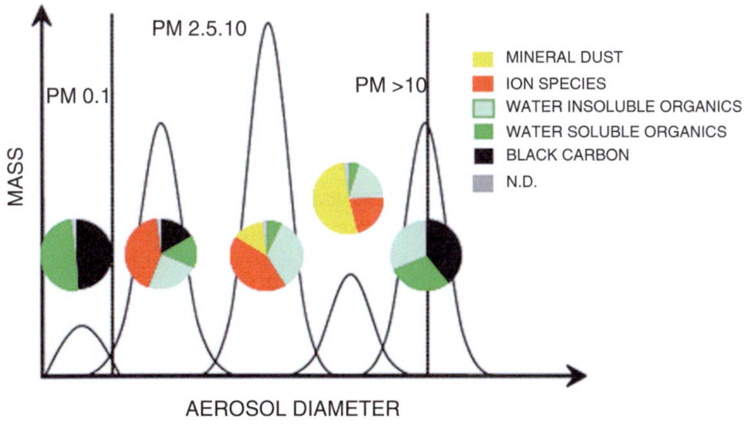

Fig. 24.1 The size-distributed aerosol chemical composition model suggested for Beijing by Roger et al. (2009)

forests increased again. Those authors showed that the resulting increase in BVOCs led to a change in the production of secondary organic aerosol (SOA) by −20% to 30%, and of surface ozone by −10% to 20%.

But vegetation is not only a possible contributor to secondary air pollutants, it also acts a source of bioaerosols. A bioaerosol is defined as a collection of airborne biological particles: viruses, bacteria, the spores of fungi, algae, pollens and other microorganisms. All are emitted naturally and can severely affect human respiratory health. Noticeably, some Chinese authors have reported an increase in the prevalence of allergic asthma and rhinitis over the last two decades (Wang et al. 2003; Li et al. 2009; Zhang and Zhang 2014). Agricultural activities like crop farming and animal production are sources of bacteria and fungi, mainly, with evidenced effects on respiratory health (Lacey and Dutkiewicz 1994; Lee et al. 2006; Millner 2009).

Climatology is the last but not the least of the natural phenomena that determine air pollution pattern in Asia. The northern regions are under the influence of continental climate conditions, leading to cold winters and heavy smog episodes due to huge coal consumption rates driven by heating purposes, and hot and humid summers favouring the photochemical haze.

The South Asian summer monsoon is one of the most important features of the boreal summer atmosphere in the tropics. It is characterized by a persistent large-scale anticyclonic structure in the upper troposphere centred over the Indian subcontinent. Surface air that has been rapidly transported upwards through deep convection becomes trapped within the strong anticyclonic circulation (Baker et al. 2012). The summer monsoon provides an important pathway of tropospheric pollutants into the lower stratosphere (Ploeger et al. 2015).Monsoon regimes regulate the atmospheric conditions encountered by the southern regions of Asia. Both ozone (O_3) and black carbon (BC) showed well-defined seasonal cycles with maxima during pre-monsoon and minima during the summer monsoon. The increase of BC occurs in the pre-monsoon period, when the O_3 diurnal variability is strongly related to the transport of polluted air-mass rich on BC (Marinoni et al. 2013).

The trans-Eurasian transport of air masses is another long-range transport of pollutants in Asia which leads to significant ozone levels in the sparsely populated regions of Western China (Li et al. 2014). Dust storms can severely affect air quality, but simulations showed they can act as a sink for trace gases, and can consequently significantly perturb their spatial and vertical distributions at the regional scale (Kumar et al. 2014).

Large-scale biomass burning events occur in Southeast Asia during the winter monsoon. They predominantly originate from agriculture fires caused by corporations as well as small-scale farmers who use the slash-and-burn method to clear vegetation. The fires often turn out of control and spread into protected forested areas. The burnt land also becomes drier, which makes it more likely to catch fire the next time there are slash-and-burn clearings. The haze engendered by fires usually measures hundreds of kilometres across. It spreads to Malaysia, Singapore, Thailand and the Philippines, causing a significant deterioration in air quality. Periods of haze, can be prolonged thanks to dry weather in Indonesia caused by the El-Niño climate phenomenon. Health studies show that such massive and intense inputs of carbonaceous aerosols in the atmosphere are dramatic on human health in this region (Bowman and Johnston 2005; Moore et al. 2006; Jayachandran 2009).

24.2.2 *Exposure*

The 2016 World Health Organization (WHO)'s Urban Ambient Air Pollution database that included 175 towns in Eastern and Southern Asia provides an overview about their air quality. Only 2% of the PM_{10} or $PM_{2.5}$ (particulate matter which aerodynamic diameters are less than 10 µm and 2.5 µm, respectively) levels met the WHO Air Quality Guidelines (ACG) annual mean values of 20 $\mu g/m^3$ for PM_{10} and 10 $\mu g/m^3$ for $PM_{2.5}$. Cities produce about 80% of Asia's gross domestic product (GDP) through their activities, most of which engender air pollution. But the socioeconomic cost of the associated air pollution is huge. It has been estimated that air pollution alone can negatively affect GDP by about 2–4% (for China, see Matus et al. 2012). About 30% of the Asian urban population lives in slums – over 200 million people in 2016. Another 1.1 billion people in Asia will become urban in the 20 coming years. Consequently, Asian cities will hence be also the main contributors to air pollution costs over the next 20 years if no actions are taken to reduce the emissions. Exposure levels to air pollutants vary depending on the socioeconomic conditions. They often determine the level of crowding as well as the proximity of the households to road axis or industrial settings, for instance, that have been related to adverse health effects. But socioeconomic conditions also drive the environmental awareness, the access to health care or the practice of Asian medicines, and finally the lifestyle, such as tobacco smoking, nutrition, and physical exercise.

Besides the urban challenges, the exposure issues of the rural populations cannot be neglected. In 2015, the World Bank estimated that 67% of Southern Asia and 43% of Eastern Asia's populations were rural. In China, their income is estimated to be 3–4 times less than urban dwellers', which leads to an increasing social inequity (World Bank 2013).

24.2.3 *Responses*

It is well recognized that air pollution is a major environmental risk to human health and quality of life through the development of several adverse health effects due to inhalation, dermal absorption and ingestion (Fig. 24.2). Current levels of major air pollutants have been associated with an increased risk of both short and long-term morbidity and mortality. However, scarce data exist for Eastern and Southern Asia where air pollution levels are the highest in the world.

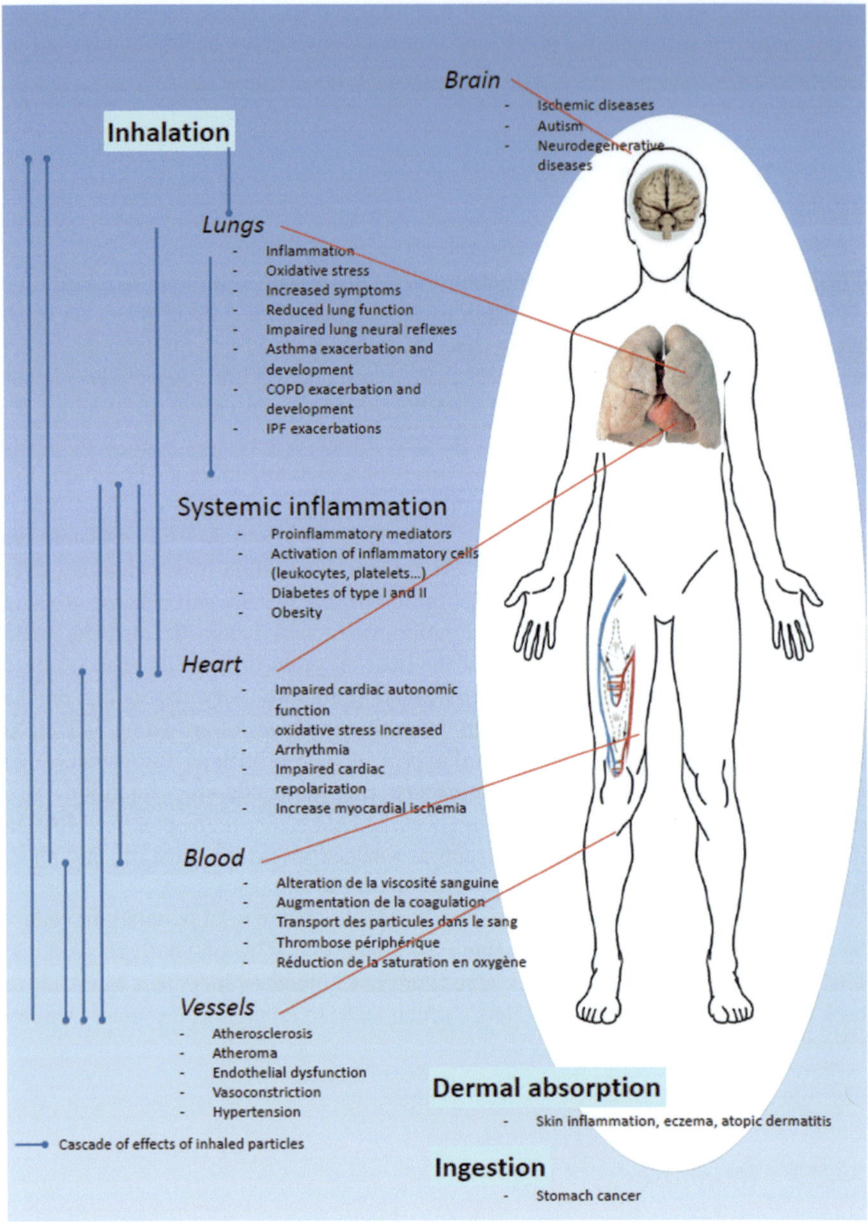

Fig. 24.2 Health effects of particulate matter

Recently, the WHO estimated outdoor air pollution from cities and rural areas causes 3.7 million premature deaths worldwide. Some 88% of those premature deaths occurred in low- and middle-income countries, and the greatest numbers arise from the regions of Western Pacific and Eastern and Southern Asia, with 1463 thousands and 862 thousands deaths in children <5 years-old and people aged >30 years-old, respectively (Lelieveld et al. 2015). Those deaths are due to urban and rural air pollution.

Fire haze has been blamed in Indonesia for deaths and respiratory illnesses in around 500,000 people according to official sources. The latest burden estimates reflect the very significant role played by air pollution in the occurrence of cardio-vascular illness and premature deaths. Projections according to a business-as-usual scenario indicate that, in 2050, figures will almost triple with 2470 thousands deaths due to air pollution in Eastern and Southern Asia in these age groups (Lelieveld et al. 2015).

In China, premature mortalities attributed to $PM_{2.5}$ nationwide were 1.37 million in total, and 0.69, 0.38, 0.13, and 0.17 million for stroke, ischemic heart disease, lung cancer, and chronic obstructive pulmonary disease, respectively (Liu et al. 2016). High population density areas exhibited the highest health risks attributed to air pollution. Still in China, the severity of health threat from air pollution has been estimated comparable to that of tobacco smoking in 74 leading cities in 2013, where, if the WHO Air Quality Guideline (AQG) targets for $PM_{2.5}$ could be achieved, the $PM_{2.5}$-related all-cause mortality would be reduced by 95%, respectively, compared with that of 2013 (Fang et al. 2016).

Unexpectedly, a meta-analysis of several studies conducted in China found an inverse linear relationship between the risk of mortality and the annual PM_{10} and NO_2 concentration. This inverse relationship indicates greater risk at lower than at higher particulate concentrations and could be related to a saturation mechanism that may reduce the exposure-response gradients (Lai et al. 2013). Mechanisms underlying the response to air pollutant aggression, including inflammation and oxidative stress may become saturated at even relatively low doses of cigarette smoke.

Besides outdoor air pollution, indoor smoke is a serious health risk for some three billion people who cook and heat their homes with biomass fuels and coal, mostly in Africa and Asia, as reported by the WHO. As a result, in China for instance, when rural air pollution is taken into consideration in the calculations, the burden on health led to similar contribution from ambient air than from indoor pollution (World Bank 2007). Not because of volatile compounds emitted indoors from paintings or new furniture, like modern households in new buildings rightly pay attention to, but from combustion pollutants emitted from very basic heating and/or cooking stoves using biofuels in rural areas (Fig. 24.3).

In adults, these are deaths from non-communicable diseases including stroke, ischaemic heart disease, chronic obstructive pulmonary disease (COPD) and lung cancer. Women and children are at higher risk; more than 50% of premature deaths due to pneumonia among children under 5 are also caused by the particulate matter inhaled from household air pollution (Kodgule and Salvi 2012).

Although health effects of air pollution are well described, the underlying biological mechanisms remain poorly understood. Among them, DNA methylations and other epigenetic variations have emerged as promising biomarkers for air pollution-related diseases, because they depend directly on exposure to

Fig. 24.3 Hmong people in a rural household, Laos (Courtesy of J.-F. Rozis)

environmental stressors including air pollutants and may be involved in disease risk through their impact on gene activities. Whether Asia, due to its elevated air pollution, constitutes a performant model for investigating the role of epigenetics in the development of air pollution-related diseases has to be investigated.

But already, some results in the region have been obtained. In an Indonesian study, fine particulate matter ($PM_{2.5}$) and polycyclic aromatic hydrocarbon (PAH) have been shown to be at the origin of methylations within candidate genes of interest in cardiovascular and respiratory diseases and cancer (Pavagadhi et al. 2013). Among workers of the largest steelworks, oil refinery, and petrochemical complex in Thailand, PAH exposure was associated with hypomethylation in genes involved in DNA repair, cell cycle arrest, as well as cell death and inflammation that have negative consequences for health (Peluso et al. 2012). In a longitudinal study conducted in Shanghai, among COPD patients, organic carbon, element carbon, NO_3^- and NH_4^+ attached to $PM_{2.5}$ were shown responsible for the decreased DNA methylation of its encoding gene (NOS2A), and for an elevated concentration of fractional exhaled nitric oxide (FeNO), an established biomarker of airway inflammation (Chen et al. 2015). In the Beijing Truck Driver Air Pollution Study conducted shortly before the 2008 Beijing Olympic Games, the methylation of one DNA sequence (NBL2) was positively associated with concentrations of silicon and calcium found in the $PM_{2.5}$ inhaled by drivers (Hou et al. 2014). Finally, in a very recent intervention study, exposure to purified air from $PM_{2.5}$ induced rapid increases in DNA methylation and, consequently, partly mediated PM effects on cardiovascular biomarkers (Chen et al. 2016).

24.3 Research Directions

For decades, atmospheric chemists, toxicologists and epidemiologists have addressed air pollution issues from their expertise viewpoint, using the knowledge acquired by others to further address their own scientific questions (Fig. 24.4). A lot remains to be achieved in this way, but new collaboration frames may well open the way to more advanced scientific challenges.

24.3.1 Atmospheric Chemistry

The quality of the emission inventory of primary pollutants like BC, organic carbon and trace metals, determines the relevance of the modeling results of exposure and health risks. While field experiments conducted over the last 15 years have provided emission factors (EF) for tropical biomass burning, EF related to biofuels and fossil fuel burning are still few and their relevance questioned. In particular, the limited amount of data available for Asian countries cannot properly account for the variety of fuels and combustion processes. Rural emissions, especially indoors, represent a difficult but an essential part of the task. As epidemiological observations have pointed out the predominant impact of chronic rather than acute events, it seems important in Asia, where the background pollution is maintained at high levels by the multiplicity of sources, to enhance source apportionment means and thus better identify and monitor the variability of the most harmful background sources.

Air pollutants and pollens should lead to joint monitoring strategies. Indeed, there is a growing body of evidence that components of air pollution interact with inhalant allergens carried by pollen grains and may enhance the risk of both atopic sensitization and exacerbation of symptoms in sensitized subjects, since urban air pollution affects both airborne allergenic pollen and the airways of exposed

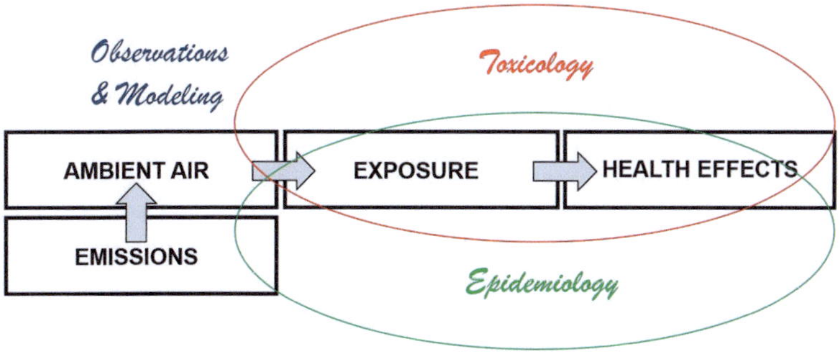

Fig. 24.4 Scheme of the interactions between atmospheric sciences, toxicology and epidemiology in the field of air pollution

subjects. Pollen grains or plant-derived paucimicronic cytoplasmic components carry allergens that can produce allergic symptoms. They may also interact with air pollution (particulate matter, ozone, nitrogen oxides) in producing these effects by facilitating the bioavailability of airborne pollen allergens. Studies have demonstrated that urbanization and high levels of vehicle emissions are correlated with the increasing frequency of pollen-induced respiratory allergy (Cecchi et al. 2013), and people who live in urban areas tend to be more affected by pollen-induced respiratory allergy than those of rural areas (Ishizaki et al. 1987). Recent Chinese findings show that short-term exposure of oak pollen to high concentrations of SO_2 or NO_2 significantly increases their fragility and disruption, leading to subsequent release of pollen cytoplasmic granules into the atmosphere (Ouyang et al. 2016). These results suggest that heightened air pollution during the pollen season may possibly increase the incidence of allergic airway disease in sensitized individuals. Furthermore, airway mucosal damage and impaired mucociliary clearance induced by air pollution may facilitate the access of inhaled allergens to the cells of the immune system.

Finally, vegetation reacts with air pollution and environmental conditions, a phenomenon that may influence the plant allergenicity. Several factors influence this interaction, including type of air pollutants, plant species, nutrient balance, climatic factors, and degree of airway sensitization and hyperresponsiveness of exposed subjects. However, few data derive from investigations in Eastern and Southern Asia. In the megacity of Kolkata, asthma-related hospital admissions shows two seasonal peaks that can be correlated with outdoor grass/weed pollen and respirable PM concentrations (Ghosh et al. 2012). In contrast, so far, ambient fungal spore were not found to be significantly associated to air pollution levels.

As for modeling, besides the vast works to be undertaken in the atmospheric compartment, among which an enhanced consideration of the gas phase of the semi volatile compounds, there is a need for an original adaptation of the aerosol deposition models in the respiratory tract. The International Commission for Radiological Protection (ICRP) has developed such a model that is regularly updated (Valentin 2002; Guha et al. 2014), and widely used to estimate doses of pollutants deposited in the different respiratory tract compartments. However, the ICRP reference values have been built upon populations which morphological and physiological characteristics generally differ from those of Asian populations (Fig. 24.5). It requires therefore significant adjustments to be relevant for Asia.

At last, the necessary development of the monitoring capacity in Asia could open the way to original opportunities. The traditional networks of certified monitoring stations, usually operated by governmental authorities, cannot respond to both of the needs: (i) exhaustive observations, (ii) at good time and space resolutions. Supersites using real time gas and aerosol speciation analyzers provide a large database from which source apportionment calculations can lead to useful results. But their cost and the spatial representativeness of their outputs are limiting for epidemiology. Innovation is expected to take benefit from the recent advances in electronics, sensors and microfabrication, to succeed in commercializing robust and relevant microsensors at a much lower cost. The current monitoring stations would

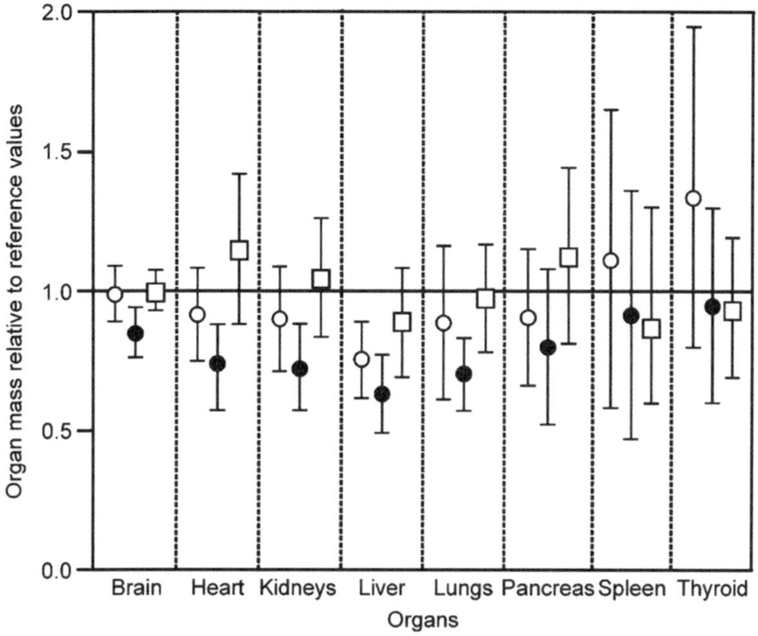

Fig. 24.5 Normalised comparison of organ masses for adult males from three Asian countries with the International Commission on Radiological Protection reference values (ICRP values = 1.0) (From Valentin 2002). ○, China; ●, India; and □, Japan

remain necessary to provide certified data, but a secondary network of numerous microsensors, more largely spread in the environment, would complement the observations, providing a higher space coverage, as well as the monitoring of new pollution indicators. Among those indicators, aerosol number concentration, black carbon, or a selection of volatile organic carbon compounds would represent a significant progress in the source regions of Asia.

24.3.2 Toxicology

Air pollution inhalation can lead to inflammation mechanisms that originate from oxidizing chemical species, which are able to disturb the cell redox balance, and thus generate an oxidative stress to the organism. The oxidative potential is an intrinsic indicator of the PM capacity to oxidize a biological media, as well as an acellular proxy of the health impact. It correlates well with toxicological markers but studies combining aerosol chemistry and epidemiological observations are rare, to be able to validate the relevance of the oxidative potential at a population scale and have not been conducted in Asia.

Among trace metals, Zn and Ni have been investigated worldwide, but coal combustion among other air pollution sources in Asia may emit other metals, like Pb or Hg for instance, with important impacts on human health. Another scientific interest may lie on the association of such metals for their effects on cells. Admittedly, no metal is targeted to act as a key disruptor of the cell biology, but it results from a lack of data about of their effective toxicity, and calls research works in that direction.

Metabolomics is a recent approach in the toxicological and epidemiological fields. It may offer the possibility to monitor markers of air pollution among the metabolites of given organs. In the case of air pollution, of particular interest is the analysis of exhaled breath condensate, a non-invasive technique. However, much remains to be done in this new approach: the choice of the organs, of the metabolites, and the technical implementation for monitoring.

24.3.3 Epidemiology

There is as an urgent need for longer term investigations in Eastern and Southern Asia using robust measures of air pollution exposure with different population groups that include a wider range of indoor and outdoor air pollutants and health outcomes, including microsensors and early health indicators. Microsensors will allow assessing individual exposure in real time to air pollution indoors and outdoors, thus including exposure to biomass and volatile organic compounds. The extent to which the problem of air pollution and human health will impact the countries of Asia will depend largely on the information available to inform policy and programs, which are still lacking, in view of a suitable social and economic development.

Air pollution effects may differ by subpopulation according to host's susceptibility and vulnerability; however, a few studies only have investigated this issue in Asia. Susceptibility is the diversity in the manner individuals will respond to air pollution. It depends on genetic background, age, sex, health status. Vulnerability is the diminished capacity of an individual or group to anticipate and cope with exposure to air pollution because of gender, socioeconomic status (SES) or proximity to air pollution sources. Recent findings provided evidence that some populations differ in the responsiveness to air pollution in Seoul, Korea. Specifically, associations for total or cardiovascular mortality were higher for males, those 65–74 years, and those with indicators of low SES, such as no education or manual occupation, for some air pollutants (Son et al. 2012). Prevention has to target these subpopulations.

Finally, the exposomic approach taking the totality of exposure of an individual during his/her lifespan has to be considered to identify environmental factors interacting with air pollutants in order to identify the proper role of each pollutant.

24.3.4 Interdisciplinary Challenges

BC, sulfates, organic carbon and trace metals are of primary interest for their role on toxicological processes, and they are found at particularly high levels in Asia. Therefore, the record of their variations over time in a more systematic manner would allow more cross approaches with toxicological and epidemiological studies. More generally, the creation of database providing the aerosol chemical composition data obtained from published observation campaigns would offer the opportunity to complete data series and to achieve meta-analyses.

Epidemiologists are in continuous search of new markers of exposure. Air pollution exposure has to be assessed at the individual level in order to better identify the underlying mechanisms. Markers of exposure and of effects should be relevant from a toxicological viewpoint, have efficient metrological specifications (repeatability, accuracy), be low-cost but robust in order to extend their use to the largest possible number of participants, and finally, they should be user-friendly to facilitate their acceptation by the surveyed populations.

24.4 Conclusions

In Asia, air pollution and the population physiological and genetic profiles present specificities that require an enhanced knowledge in order to prioritize and optimize the pollution control tasks. Interdisciplinary projects are encouraged but their implementation often happens to be difficult, and their outputs disappointing. More than anywhere else in the world, integrated research groups should be created in Asia to increase efficiency by a better mutual understanding and a co-development of methodologies between atmospheric chemists, biologists and epidemiologists.

In the field of cellular responsiveness to air pollutants, if biologists have made recent advances in cell culture and their exposure to air pollutants, they however face limitation in their capacity to generate and characterize aerosols, secondary pollutants in particular, before exposure to cells. Another common challenge to toxicologists and atmospheric chemists is the corona effect on PM, that is, the coating of proteins adsorbed at the PM surface when aerosols enter the human organism. The corona effect greatly determines the PM interaction with the living matter. Multiphasic chambers offer opportunities to control the generation of primary and secondary aerosols and observe toxicological processes related to aerosol aging and secondary organic aerosol formation in particular.

In a close future, the innovative monitoring of markers of exposure and/or of its effects at the individual scale, or at least of its immediate environment, may open the way to more systematic cohort surveys in time. Asia is often presented like an open chemical reactor to study the complex issues related to particles and other air pollutants, the effective exposure of people to these pollutants (personal exposure), and their observed health impacts. The high academic quality of the Asian experts is another feature that supports the creation of integrated research units, with proper analytical, instrumental, modeling and development capacities.

References

Baker, A.K., Schuck, T.J., Brenninkmeijer, C.A., Rauthe-Schöch, A., Slemr, F., Velthoven, P.F., Lelieveld, J. (2012). Estimating the contribution of monsoon-related biogenic production to methane emissions from South Asia using CARIBIC observations. *Geophysical Research Letters*, 39(10).

Bowman, D. M., & Johnston, F. H. (2005). Wildfire smoke, fire management, and human health. *EcoHealth, 2*(1), 76–80.

Cecchi, L., D'Amato, G., Annesi-Maesano, I. (2013). Climate, urban air pollution, and respiratory allergy.

Chen, R., et al. (2015). Fine particulate matter constituents, nitric oxide synthase DNA methylation and exhaled nitric oxide. *Environmental Science & Technology, 49*(19), 11859–11865.

Chen, R., et al. (2016). DNA hypomethylation and its mediation in the effects of fine particulate air pollution on cardiovascular biomarkers: A randomized crossover trial. *Environment International, 94*, 614–619.

Fang, D., Wang, Q. G., Li, H., Yu, Y., Lu, Y., Qian, X. (2016). Mortality effects assessment of ambient PM 2.5 pollution in the 74 leading cities of China. Science of the Total Environment 1;569–570.

Fu, Y., & Liao, H. (2014). Impacts of land use and land cover changes on biogenic emissions of volatile organic compounds in China from the late 1980s to the mid-2000s: Implications for tropospheric ozone and secondary organic aerosol. *Tellus B, 66.*

Ghosh, D., Chakraborty, P., Gupta, J., Biswas, A., Roy, I., Das, S., & Gupta-Bhattacharya, S. (2012). Associations between pollen counts, pollutants, and asthma-related hospital admissions in a high-density indian metropolis. *Journal of Asthma, 49*(8), 792–799.

Guha, S., Hariharan, P., & Myers, M. R. (2014). Enhancement of ICRP's lung deposition model for pathogenic bioaerosols. *Aerosol Science and Technology, 48*(12), 1226–1235.

Hou, L., et al. (2014). Altered methylation in tandem repeat element and elemental component levels in inhalable air particles. *Environmental and Molecular Mutagenesis, 55*(3), 256–265.

Ishizaki, T., Koizumi, K., Ikemori, R., Ishiyama, Y., & Kushibiki, E. (1987). Studies of prevalence of Japanese cedar pollinosis among the residents in a densely cultivated area. *Annals of Allergy, 58*(4), 265.

Jayachandran, S. (2009). Air quality and early-life mortality evidence from Indonesia's wildfires. *Journal of Human Resources, 44*(4), 916–954.

Kodgule, R., & Salvi, S. (2012). Exposure to biomass smoke as a cause for airway disease in women and children. *Current Opinion in Allergy and Clinical Immunology, 12*(1), 82–90.

Kumar, R., Barth, M. C., Madronich, S., Naja, M., Carmichael, G. R., Pfister, G. G., & Sarangi, T. (2014). Effects of dust aerosols on tropospheric chemistry during a typical pre-monsoon season dust storm in northern India. *Atmospheric Chemistry and Physics, 14*(13), 6813–6834.

Lacey, J., & Dutkiewicz, J. (1994). Bioaerosols and occupational lung disease. *Journal of Aerosol Science, 25*(8), 1371–1404.

Lai, H. K., Tsang, H., & Wong, C. M. (2013). Meta-analysis of adverse health effects due to air pollution in Chinese populations. *BMC Public Health, 13*(1), 1.

Lee, S. A., Adhikari, A., Grinshpun, S. A., McKay, R., Shukla, R., & Reponen, T. (2006). Personal exposure to airborne dust and microorganisms in agricultural environments. *Journal of Occupational and Environmental Hygiene, 3*(3), 118–130.

Lelieveld, J., Evans, J. S., Fnais, M., Giannadaki, D., & Pozzer, A. (2015). The contribution of outdoor air pollution sources to premature mortality on a global scale. *Nature, 525*(7569), 367–371.

Li, J., Sun, B., Huang, Y., Lin, X., Zhao, D., Tan, G., & Zhong, N. (2009). A multicentre study assessing the prevalence of sensitizations in patients with asthma and/or rhinitis in China. *Allergy, 64*(7), 1083–1092.

Li, X., Liu, J., Mauzerall, D. L., Emmons, L. K., Walters, S., Horowitz, L. W., & Tao, S. (2014). Effects of trans-Eurasian transport of air pollutants on surface ozone concentrations over Western China. *Journal of Geophysical Research: Atmospheres, 119*(21).

Liu, J., Han, Y., Tang, X., Zhu, J., & Zhu, T. (2016). Estimating adult mortality attributable to PM 2.5 exposure in China with assimilated PM 2.5 concentrations based on a ground monitoring network. *Science of the Total Environment, 568*, 1253–1262.

Marinoni, A., Cristofanelli, P., Laj, P., Duchi, R., Putero, D., Calzolari, F., ..., Bonasoni, P. (2013). High black carbon and ozone concentrations during pollution transport in the Himalayas: Five years of continuous observations at NCO-P global GAW station. *Journal of Environmental Sciences, 25*(8), 1618–1625.

Matus, K., Nam, K. M., Selin, N. E., Lamsal, L. N., Reilly, J. M., & Paltsev, S. (2012). Health damages from air pollution in China. *Global Environmental Change, 22*(1), 55–66.

Merrill, J. T., Uematsu, M., & Bleck, R. (1989). Meteorological analysis of long range transport of mineral aerosols over the North Pacific. *Journal of Geophysical Research: Atmospheres, 94* (D6), 8584–8598.

Merrill, J., Arnold, E., Leinen, M., & Weaver, C. (1994). Mineralogy of aeolian dust reaching the North Pacific Ocean: 2. Relationship of mineral assemblages to atmospheric transport patterns. *Journal of Geophysical Research: Atmospheres, 99*(D10), 21025–21032.

Millner, P. D. (2009). Bioaerosols associated with animal production operations. *Bioresource Technology, 100*(22), 5379–5385.

Moore, D., Copes, R., Fisk, R., Joy, R., Chan, K., & Brauer, M. (2006). Population health effects of air quality changes due to forest fires in British Columbia in 2003: Estimates from physician-visit billing data. *Canadian Journal of Public Health/Revue Canadienne de Sante'e Publique, 97*, 105–108.

Ouyang, Y., Xu, Z., Fan, E., Li, Y., & Zhang, L. (2016). Effect of nitrogen dioxide and sulfur dioxide on viability and morphology of oak pollen. *International Forum of Allergy & Rhinology, 6*(1), 95–100.

Pavagadhi, S., Betha, R., Venkatesan, S., Balasubramanian, R., & Hande, M. P. (2013). Physico-chemical and toxicological characteristics of urban aerosols during a recent Indonesian biomass burning episode. *Environmental Science and Pollution Research, 20*(4), 2569–2578.

Peluso, M., et al. (2012). DNA methylation differences in exposed workers and nearby residents of the Ma Ta Phut industrial estate, Rayong, Thailand. *International Journal of Epidemiology.*, dys129, *41*, 1753–1760.

Ploeger, F., et al. (2015). A potential vorticity-based determination of the transport barrier in the Asian summer monsoon anticyclone. *Atmospheric Chemistry and Physics, 15*(22), 13145–13159.

Prospero, J. M. (1981). 21. Eolian transport to the World Ocean. In C. Emiliani (Ed.), *The oceanic lithosphere The Sea, 7*, 801–874.

Roger, J. C., Guinot, B., Cachier, H., Mallet, M., Dubovik, O., & Yu, T. (2009). Aerosol complexity in megacities: From size-resolved chemical composition to optical properties of the Beijing atmospheric particles. *Geophysical Research Letters, 36*(18).

Son, J. Y., Lee, J. T., Kim, H., Yi, O., & Bell, M. L. (2012). Susceptibility to air pollution effects on mortality in Seoul, Korea: A case-crossover analysis of individual-level effect modifiers. *Journal of Exposure Science and Environmental Epidemiology, 22*(3), 227–234.

Steiner, A., Luo, C., Huang, Y., & Chameides, W. L. (2002). Past and present-day biogenic volatile organic compound emissions in East Asia. *Atmospheric Environment, 36*(31), 4895–4905.

Tegen, I., & Fung, I. (1995). Contribution to the atmospheric mineral aerosol load from land surface modification. *Journal of Geophysical Research: Atmospheres, 100*(D9), 18707–18726.
Valentin, J. (2002). Basic anatomical and physiological data for use in radiological protection: Reference values: ICRP Publication 89. *Annals of the ICRP, 32*(3), 1–277.
Wang, M., et al. (2003). A common IL-13 Arg130Gln single nucleotide polymorphism among Chinese atopy patients with allergic rhinitis. *Human Genetics, 113*(5), 387–390.
Went, F. (1960). Blue haze in the atmosphere. *Nature, 187*, 641–643.
World Bank. (2007). *Cost of pollution in China: Economic estimates of physical damages*: Available at http://siteresources.worldbank.org/INTEAPREGTOPENVIRONMENT/Resources/China_Cost_of_Pollution.pdf
World Bank. (2013). Poverty reduction and equity department: www.worldbank.org/poverty 2, 2: August 2013.
Zhang, D. E. (1984). Synoptic-climatic studies of dust fall in China since historic times. *Science in China Series B-Chemistry, Biological, Agricultural, Medical & Earth Sciences, 27*(8), 825–836.
Zhang, X., Shen, Z., Zhang, G., Chen, T., & Liu, H. (1996a). Remote mineral aerosols in Westerlies and their contributions to the Chinese loess. *Science in China Series D-Earth Sciences, 39*(2), 134–143.
Zhang, X., Zhang, G., Zhu, G., Zhang, D. E., An, Z. S., Chen, T., & Huang, X. P. (1996b). Elemental tracers for Chinese source dust. *Science in China Series D-Earth Sciences, 39*, 512–521.
Zhang, X. Y., Arimoto, R., & An, Z. S. (1997). Dust emission from Chinese desert sources linked to variations in atmospheric circulation. *Journal of Geophysical Research: D. Atmospheres, 102*, 28–041.
Zhang, X.Y., Gong, S.L., Zhao, T.L., Arimoto, R., Wang, Y.Q., Zhou, Z.J. (2003). Sources of Asian dust and role of climate change versus desertification in Asian dust emission. *Geophysical Research Letters, 30*(24).
Zhang, Y., & Zhang, L. (2014). Prevalence of allergic rhinitis in China. *Allergy, Asthma & Immunology Research, 6*(2), 105–113.
Zhao, Z., Sebastian, A., Larsson, L., Wang, Z., Zhang, Z., & Norbäck, D. (2008a). Asthmatic symptoms among pupils in relation to microbial dust exposure in schools in Taiyuan, China. *Pediatric Allergy and Immunology, 19*(5), 455–465.
Zhao, Z., Zhang, Z., Wang, Z., Ferm, M., Liang, Y., & Norbäck, D. (2008b). Asthmatic symptoms among pupils in relation to winter indoor and outdoor air pollution in schools in Taiyuan, China. *Environmental Health Perspectives*, 90–97.
Zhao, Z., Cao, J., Shen, Z., Xu, B., Zhu, C., Chen, L. W. A., & Ho, K. (2013). Aerosol particles at a high-altitude site on the Southeast Tibetan Plateau, China: Implications for pollution transport from South Asia. *Journal of Geophysical Research: Atmospheres, 118*(19).

MIX
Papier aus verantwortungsvollen Quellen
Paper from responsible sources
FSC® C105338

If you have any concerns about our products,
you can contact us on
ProductSafety@springernature.com

In case Publisher is established outside the EU,
the EU authorized representative is:
Springer Nature Customer Service Center GmbH
Europaplatz 3, 69115 Heidelberg, Germany

Printed by Libri Plureos GmbH
in Hamburg, Germany